T0199916

CODE OF FEDERAL REGULATIONS

Title 7

Agriculture

Parts 300 to 399

Revised as of January 1, 2022

Containing a codification of documents
of general applicability and future effect

As of January 1, 2022

Published by the Office of the Federal Register
National Archives and Records Administration
as a Special Edition of the Federal Register

Code of Federal Regulations

Table of Contents

Cite this Code: CFR

To cite the regulations in this volume use title, part and section number. Thus, 7 CFR 300.1 refers to title 7, part 300, section 1.

Explanation

The Code of Federal Regulations is a codification of the general and permanent rules published in the Federal Register by the Executive departments and agencies of the Federal Government. The Code is divided into 50 titles which represent broad areas subject to Federal regulation. Each title is divided into chapters which usually bear the name of the issuing agency. Each chapter is further subdivided into parts covering specific regulatory areas.

Each volume of the Code is revised at least once each calendar year and issued on a quarterly basis approximately as follows:

Title 1 through Title 16...as of January 1
Title 17 through Title 27 ....as of April 1
Title 28 through Title 41 ....as of July 1
Title 42 through Title 50 ....as of October 1

The appropriate revision date is printed on the cover of each volume.

LEGAL STATUS

The contents of the Federal Register are required to be judicially noticed (44 U.S.C. 1507). The Code of Federal Regulations is prima facie evidence of the text of the original documents (44 U.S.C. 1510).

HOW TO USE THE CODE OF FEDERAL REGULATIONS

The Code of Federal Regulations is kept up to date by the individual issues of the Federal Register. These two publications must be used together to determine the latest version of any given rule.

To determine whether a Code volume has been amended since its revision date (in this case, January 1, 2022), consult the "List of CFR Sections Affected (LSA)," which is issued monthly, and the "Cumulative List of Parts Affected," which appears in the Reader Aids section of the daily Federal Register. These two lists will identify the Federal Register page number of the latest amendment of any given rule.

EFFECTIVE AND EXPIRATION DATES

Each volume of the Code contains amendments published in the Federal Register since the last revision of that volume of the Code. Source citations for the regulations are referred to by volume number and page number of the Federal Register and date of publication. Publication dates and effective dates are usually not the same and care must be exercised by the user in determining the actual effective date. In instances where the effective date is beyond the cutoff date for the Code a note has been inserted to reflect the future effective date. In those instances where a regulation published in the Federal Register states a date certain for expiration, an appropriate note will be inserted following the text.

OMB CONTROL NUMBERS

The Paperwork Reduction Act of 1980 (Pub. L. 96–511) requires Federal agencies to display an OMB control number with their information collection request.

Many agencies have begun publishing numerous OMB control numbers as amendments to existing regulations in the CFR. These OMB numbers are placed as close as possible to the applicable recordkeeping or reporting requirements.

PAST PROVISIONS OF THE CODE

Provisions of the Code that are no longer in force and effect as of the revision date stated on the cover of each volume are not carried. Code users may find the text of provisions in effect on any given date in the past by using the appropriate List of CFR Sections Affected (LSA). For the convenience of the reader, a "List of CFR Sections Affected" is published at the end of each CFR volume. For changes to the Code prior to the LSA listings at the end of the volume, consult previous annual editions of the LSA. For changes to the Code prior to 2001, consult the List of CFR Sections Affected compilations, published for 1949-1963, 1964-1972, 1973-1985, and 1986-2000.

"[RESERVED]" TERMINOLOGY

The term "[Reserved]" is used as a place holder within the Code of Federal Regulations. An agency may add regulatory information at a "[Reserved]" location at any time. Occasionally "[Reserved]" is used editorially to indicate that a portion of the CFR was left vacant and not dropped in error.

INCORPORATION BY REFERENCE

What is incorporation by reference? Incorporation by reference was established by statute and allows Federal agencies to meet the requirement to publish regulations in the Federal Register by referring to materials already published elsewhere. For an incorporation to be valid, the Director of the Federal Register must approve it. The legal effect of incorporation by reference is that the material is treated as if it were published in full in the Federal Register (5 U.S.C. 552(a)). This material, like any other properly issued regulation, has the force of law.

What is a proper incorporation by reference? The Director of the Federal Register will approve an incorporation by reference only when the requirements of 1 CFR part 51 are met. Some of the elements on which approval is based are:

(a) The incorporation will substantially reduce the volume of material published in the Federal Register.

(b) The matter incorporated is in fact available to the extent necessary to afford fairness and uniformity in the administrative process.

(c) The incorporating document is drafted and submitted for publication in accordance with 1 CFR part 51.

What if the material incorporated by reference cannot be found? If you have any problem locating or obtaining a copy of material listed as an approved incorporation by reference, please contact the agency that issued the regulation containing that incorporation. If, after contacting the agency, you find the material is not available, please notify the Director of the Federal Register, National Archives and Records Administration, 8601 Adelphi Road, College Park, MD 20740-6001, or call 202-741-6010.

CFR INDEXES AND TABULAR GUIDES

A subject index to the Code of Federal Regulations is contained in a separate volume, revised annually as of January 1, entitled CFR INDEX AND FINDING AIDS. This volume contains the Parallel Table of Authorities and Rules. A list of CFR titles, chapters, subchapters, and parts and an alphabetical list of agencies publishing in the CFR are also included in this volume.

An index to the text of "Title 3—The President" is carried within that volume.

The Federal Register Index is issued monthly in cumulative form. This index is based on a consolidation of the "Contents" entries in the daily Federal Register.

A List of CFR Sections Affected (LSA) is published monthly, keyed to the revision dates of the 50 CFR titles.

REPUBLICATION OF MATERIAL

There are no restrictions on the republication of material appearing in the Code of Federal Regulations.

INQUIRIES

For a legal interpretation or explanation of any regulation in this volume, contact the issuing agency. The issuing agency's name appears at the top of odd-numbered pages.

For inquiries concerning CFR reference assistance, call 202-741-6000 or write to the Director, Office of the Federal Register, National Archives and Records Administration, 8601 Adelphi Road, College Park, MD 20740-6001 or e-mail *fedreg.info@nara.gov*.

THIS TITLE

Title 7—AGRICULTURE is composed of fifteen volumes. The parts in these volumes are arranged in the following order: Parts 1–26, 27–52, 53–209, 210–299, 300–399, 400–699, 700–899, 900–999, 1000–1199, 1200–1599, 1600–1759, 1760–1939, 1940–1949, 1950–1999, and part 2000 to end. The contents of these volumes represent all current regulations codified under this title of the CFR as of January 1, 2022.

The Food and Nutrition Service current regulations in the volume containing parts 210–299, include the Child Nutrition Programs and the Food Stamp Program. The regulations of the Federal Crop Insurance Corporation are found in the volume containing parts 400–699.

All marketing agreements and orders for fruits, vegetables and nuts appear in the one volume containing parts 900–999. All marketing agreements and orders for milk appear in the volume containing parts 1000–1199.

For this volume, Michele Bugenhagen was Chief Editor. The Code of Federal Regulations publication program is under the direction of John Hyrum Martinez, assisted by Stephen J. Frattini.

Title 7—Agriculture

(This book contains parts 300 to 399)

SUBTITLE B—REGULATIONS OF THE DEPARTMENT OF AGRICULTURE (CONTINUED)

1

Subtitle B—Regulations of the Department of Agriculture (Continued)

CHAPTER III—ANIMAL AND PLANT HEALTH INSPECTION SERVICE, DEPARTMENT OF AGRICULTURE

PART 300—INCORPORATION BY REFERENCE

Subpart—Materials Incorporated by Reference

AUTHORITY: 7 U.S.C. 7701–7772 and 7781–7786; 7 CFR 2.22, 2.80, and 371.3.

SOURCE: 67 FR 8463, Feb. 25, 2002, unless otherwise noted.

EDITORIAL NOTE: Nomenclature changes to part 300 appear at 69 FR 18803, Apr. 9, 2004.

§ 300.1 [Reserved]

§ 300.2 Dry Kiln Operator's Manual.

(a) The Dry Kiln Operator's Manual, which was published in August 1991 as Agriculture Handbook No. 188 by the United States Department of Agriculture, Forest Service, has been approved for incorporation by reference in 7 CFR chapter III by the Director of the Office of the Federal Register in accordance with 5 U.S.C. 552(a) and 1 CFR part 51.

(b) The kiln drying schedules specified in the Dry Kiln Operator's Manual provide a method by which certain articles regulated by "Subpart I—Logs, Lumber, and Other Wood Articles" (7 CFR 319.40–1 through 319.40 11) may be imported into the United States.

(c) *Availability.* Copies of the Dry Kiln Operator's Manual:

(1) Are available for inspection at the National Archives and Records Administration (NARA). For information on the availability of this material at NARA, call 202–741–6030, or go to: *http://www.archives.gov/federal_register/code_of_federal_regulations/ibr_locations.html;* or

(2) Are for sale as ISBN 0–16–035819–1 by the U.S. Government Printing Office, Superintendent of Documents, Mail Stop: SSOP, Washington, DC 20402–9328.

[67 FR 8463, Feb. 25, 2002 as amended at 84 FR 2427, Feb. 7, 2019]

§ 300.3 Reference Manual A.

(a) The Reference Manual for Administration, Procedures, and Policies of the National Seed Health System, which was published on February 25, 2000, by the National Seed Health System (NSHS), has been approved for incorporation by reference in 7 CFR chapter III by the Director of the Office of the Federal Register in accordance with 5 U.S.C. 552(a) and 1 CFR part 51.

(b) *Availability.* Copies of Reference Manual A:

(1) Are available for inspection at the APHIS Library, U.S. Department of Agriculture, 4700 River Road, Riverdale, MD or at the National Archives and Records Administration (NARA). For information on the availability of this material at NARA, call 202–741–6030, or go to: *http://www.archives.gov/federal_register/code_of_federal_regulations/ibr_locations.html;* or

(2) May be obtained by writing to Phytosanitary Issues Management, Operational Support, PPQ, APHIS, 4700 River Road Unit 140, Riverdale, MD 20737–1236; or

(3) May be viewed on the APHIS Web site at *http://www.aphis.usda.gov/ppq/pim/accreditation.*

§ 300.4 Reference Manual B.

(a) The Reference Manual for Seed Health Testing and Phytosanitary Field Inspection Methods, which was published on February 27, 2001, by the National Seed Health System (NSHS), has been approved for incorporation by reference in 7 CFR chapter III by the Director of the Office of the Federal Register in accordance with 5 U.S.C. 552(a) and 1 CFR part 51.

(b) *Availability.* Copies of Reference Manual B:

(1) Are available for inspection at the APHIS Library, U.S. Department of Agriculture, 4700 River Road, Riverdale, MD or at the National Archives and Records Administration (NARA). For information on the availability of this material at NARA, call 202–741–6030, or go to: *http://www.archives.gov/federal_register/code_of_federal_regulations/ibr_locations.html;* or

(2) May be obtained by writing to Phytosanitary Issues Management,

Operational Support, PPQ, APHIS, 4700 River Road Unit 140, Riverdale, MD 20737–1236; or

(3) May be viewed on the APHIS Web site at *http://www.aphis.usda.gov/ppq/ pim/accreditation.*

§ 300.5 International Standards for Phytosanitary Measures.

(a) The International Standards for Phytosanitary Measures Publication No. 4, "Requirements for the Establishment of Pest Free Areas," which was published February 1996 by the International Plant Protection Convention of the United Nations' Food and Agriculture Organization has been approved for incorporation by reference in 7 CFR chapter III by the Director of the Office of the Federal Register in accordance with 5 U.S.C. 552(a) and 1 CFR part 51.

(b) *Availability.* Copies of International Standards for Phytosanitary Measures Publication No. 4:

(1) Are available for inspection at the National Archives and Records Administration (NARA). For information on the availability of this material at NARA, call 202–741–6030, or go to: *http:// www.archives.gov/federal_register/ code_of_federal_regulations/ ibr_locations.html*; or

(2) May be obtained by writing to Phytosanitary Issues Management, Operational Support, PPQ, APHIS, 4700 River Road Unit 140, Riverdale, MD 20737–1236; or

(3) May be viewed on the APHIS Web site at *http://www.aphis.usda.gov/ppq/ pim/standards/.*

[68 FR 37915, June 25, 2003]

PART 301—DOMESTIC QUARANTINE NOTICES

Subpart A—Preemption and Special Need Requests

Subpart B—Imported Plants and Plant Parts

Subpart C—Fruit Flies

Subpart D—Black Stem Rust

Subpart E—Gypsy Moth

AUTHORITY: 7 U.S.C. 7701–7772 and 7781–7786; 7 CFR 2.22, 2.80, and 371.3.

Section 301.75–15 issued under Sec. 204, Title II, Public Law 106–113, 113 Stat. 1501A–293; sections 301.75–15 and 301.75–16 issued under Sec. 203, Title II, Public Law 106–224, 114 Stat. 400 (7 U.S.C. 1421 note).

11

Subpart A—Preemption and Special Need Requests

SOURCE: 73 FR 63064, Oct. 23, 2008, unless otherwise noted. Redesignated at 84 FR 2427, Feb. 7, 2019

§ 301.1 Purpose and scope.

(a) Under section 436 of the Plant Protection Act (7 U.S.C. 7756), a State or political subdivision of a State may not impose prohibitions or restrictions upon the movement in interstate commerce of articles, means of conveyance, plants, plant products, biological control organisms, plant pests, or noxious weeds if the Secretary has issued a regulation or order to prevent the dissemination of the biological control organism, plant pest, or noxious weed within the United States. The only exceptions to this are:

(1) If the prohibitions or restrictions issued by the State or political subdivision of a State are consistent with and do not exceed the regulations or orders issued by the Secretary, or

(2) If the State or political subdivision of a State demonstrates to the Secretary and the Secretary finds that there is a special need for additional prohibitions or restrictions based on sound scientific data or a thorough risk assessment.

(b) The regulations in this subpart provide for the submission and consideration of special need requests when a State or a political subdivision of a State seeks to impose prohibitions or restrictions on the movement in interstate commerce of articles, means of conveyance, plants, plant products, biological control organisms, plant pests, or noxious weeds that are in addition to the prohibitions or restrictions imposed by this part or by a Federal Order.

§ 301.1–1 Definitions.

Administrator. The Administrator, Animal and Plant Health Inspection Service (APHIS), or any person authorized to act for the Administrator.

Animal and Plant Health Inspection Service (APHIS). The Animal and Plant Health Inspection Service of the United States Department of Agriculture.

Biological control organism. Any enemy, antagonist, or competitor used to control a plant pest or noxious weed.

Interstate commerce. Trade, traffic, or other commerce

(1) From one State into or through any other State or

(2) Within the District of Columbia, Guam, the Virgin Islands of the United States, or any other territory or possession of the United States.

Move (moved, movement). Shipped, offered to a common carrier for shipment, received for transportation or transported by a common carrier, or carried, transported, moved or allowed to be moved.

Noxious weed. Any plant or plant product that can directly or indirectly injure or cause damage to crops (including nursery stock or plant products), livestock, poultry, or other interests of agriculture, irrigation, navigation, the natural resources of the United States, the public health or the environment.

Plant pest. Any living stage of any insects, mites, nematodes, slugs, snails, protozoa, or other invertebrate animals, bacteria, fungi, other parasitic plants or reproductive parts thereof, viruses, or any organisms similar to or allied with any of the foregoing, or any infectious substances which can directly or indirectly injure or cause disease or damage in any plants or parts thereof or any processed, manufactured, or other products of plants.

State. The District of Columbia, Puerto Rico, the Northern Mariana Islands, or any State, territory, or possession of the United States.

§ 301.1–2 Criteria for special need requests.

(a) A special need request, as described in § 301.1, may be generated by a State or a political subdivision of a State. If the request is generated by a political subdivision of a State, the request must be submitted to APHIS through the State. States may also collaborate with other States to submit multi-State special need requests. However, if submitted, the multi-State special need request must include information in sufficient detail to allow APHIS to analyze the impacts on each

State on an individual basis. All special need requests must be signed by the executive official or officials or by a plant protection official or officials of the State(s) making the request and must contain the following:

(1) Data drawn from a scientifically sound detection survey, showing that the biological control organism, noxious weed, or plant pest of concern does not exist in the State or political subdivision or, if already present in the State or political subdivision, the distribution of the biological control organism, noxious weed, or plant pest of concern;

(2) If the biological control organism, noxious weed, or plant pest is not present in the State or political subdivision, a risk analysis or other scientific data showing that the biological control organism, noxious weed, or plant pest could enter the State or political subdivision and become established;

(3) Specific information showing that, if introduced into or allowed to spread within the State or political subdivision, the biological control organism, noxious weed, or plant pest would harm or injure the environment or agricultural resources in the State or political subdivision. The request should contain detailed information, including quantitative estimates, if available, about what harm or injury would result from the introduction or dissemination of the biological control organism, noxious weed, or plant pest in the State or political subdivision;

(4) Specific information showing that the State or political subdivision has characteristics that make it particularly vulnerable to the biological control organism, noxious weed, or plant pest, such as unique plants, diversity of flora, historical concerns, or any other special basis for the request for additional restrictions or prohibitions; and

(5) Information detailing the proposed additional prohibitions or restrictions and scientific data demonstrating that the proposed additional prohibitions or restrictions are necessary and adequate, and that there is no less drastic action that is feasible and that would be adequate, to prevent the introduction or spread of the biological control organism, noxious weed,

or plant pest in the State or political subdivision.

(b) All special need requests must be submitted to the Deputy Administrator for Plant Protection and Quarantine, APHIS, USDA, Jamie L. Whitten Federal Building, 14th Street and Independence Avenue, SW., Room 301–E, Washington, DC 20250.

§ 301.1–3 **Action on special need requests.**

(a) Upon receipt of a complete special need request submitted in accordance with §301.1–2, APHIS will publish a notice in the FEDERAL REGISTER to inform the public of the special need request and to make the request and its supporting information available for review and comment for at least 60 days.

(b) Following the close of the comment period, APHIS will publish another notice announcing the Administrator's decision to either grant or deny the special need request. The Administrator's determination will be based upon the evaluation of the information submitted by the State or political subdivision of a State in support of its request and would take into account any comments received.

(1) If the Administrator grants the special need request, the State or political subdivision of a State will be authorized to impose only the specific prohibitions or restrictions identified in the request and approved by APHIS. APHIS will coordinate with the State, or with the State on behalf of the political subdivision of the State, to ensure that the additional prohibitions or restrictions are in accord with the special need exception granted by the Administrator.

(2) If the Administrator denies the special need request, the State or political subdivision of a State will be notified in writing of the reason for the denial and may submit any additional information the State or political subdivision of a State may have in order to request a reconsideration.

(c) If granted, a special need exception will be applicable for 2 years, at the end of which the State or political subdivision of a State must submit a request for renewal of the exception. A special need renewal request must address the same criteria as the initial request submitted under § 301.1–2 and must show that a special need still exists that warrants the continuation of the special need exception. The renewal must be submitted no sooner than 6 months and no later than 3 months prior to the end of the 2-year applicability period for the initial exception. Once a special need renewal request has been received, APHIS will follow the same notice and comment process outlined in paragraphs (a) and (b) of this section. If, by the end of the 2-year applicability period, the State or political subdivision of a State does not submit a special need renewal request, the State's or political subdivision's special need exception will lapse and the State or political subdivision of a State will have to reapply for the special need exception.

(d) If the Administrator determines that there is a need for the withdrawal of a special need exception before the renewal date of the special need exception, the reasons for the withdrawal would be communicated to the State or to the political subdivision of the State and APHIS will publish a notice in the FEDERAL REGISTER to inform the public of the withdrawal of the special need exception and to make the information supporting the withdrawal available for review and comment for at least 60 days. Reasons for withdrawal of approval of a special need exception may include, but are not limited to, the availability of new scientific data or changes in APHIS regulations. Following the close of the comment period, APHIS will publish another notice announcing the Administrator's decision to either withdraw or uphold the special need exception. The Administrator's determination will be based upon the evaluation of the information submitted in support of the withdrawal and would take into account any comments received.

(Approved by the Office of Management and Budget under control number 0579–0291)

Subpart B—Imported Plants and Plant Parts

SOURCE: 62 FR 61212, Nov. 17, 1997, unless otherwise noted. Redesignated at 84 FR 2428, Feb. 7, 2019.

§ 301.10   Definitions.

Move (moved, movement). Shipped, offered to a common carrier for shipment, received for transportation or transported by a common carrier, or carried, transported, moved, or allowed to be moved.

State. Any State, territory, district, or possession of the United States.

§ 301.11   Notice of quarantine; prohibition on the interstate movement of certain imported plants and plant parts.

(a) In accordance with part 319 of this chapter, some plants and plant parts may only be imported into the United States subject to certain destination restrictions. That is, under part 319, some plants and plant parts may be imported into some States or areas of the United States but are prohibited from being imported into, entered into, or distributed within other States or areas, as an additional safeguard against the introduction and establishment of foreign plant pests and diseases.

(b) Under this quarantine notice, whenever any imported plant or plant part is subject to destination restrictions under part 319:

(1) The State(s) or area(s) into which the plant or plant part is allowed to be imported is quarantined with respect to that plant or plant part; and

(2) No person shall move any plant or plant part from any such quarantined State or area into or through any State or area not quarantined with respect to that plant or plant part.

Subpart C—Fruit Flies

SOURCE: 73 FR 32432, June 9, 2008, unless otherwise noted. Redesignated at 84 FR 2428, Feb. 7, 2019.

§ 301.32 Restrictions on interstate movement of regulated articles.

(a) No person may move interstate from any quarantined area any regulated article except in accordance with this subpart.[1]

(b) Section 414 of the Plant Protection Act (7 U.S.C. 7714) provides that the Secretary of Agriculture may, under certain conditions, hold, seize, quarantine, treat, apply other remedial measures to, destroy, or otherwise dispose of any plant, plant pest, plant product, article, or means of conveyance that is moving, or has moved into or through the United States or interstate if the Secretary has reason to believe the article is a plant pest or is infested with a plant pest at the time of movement.

§ 301.32–1 Definitions.

Administrator. The Administrator, Animal and Plant Health Inspection Service, or any person authorized to act for the Administrator.

Animal and Plant Health Inspection Service. The Animal and Plant Health Inspection Service (APHIS) of the United States Department of Agriculture.

Certificate. A document in which an inspector or person operating under a compliance agreement affirms that a specified regulated article is free of fruit flies and may be moved interstate to any destination.

Commercially produced. Fruits and vegetables that an inspector identifies as having been produced for sale and distribution in mass markets. Such identification will be based on a variety of indicators, including, but not limited to: Quantity of produce, monocultural practices, pest management programs, good sanitation practices including destruction of culls, type of packaging, identification of grower or packinghouse on the packaging, and documents consigning the shipment to a wholesaler or retailer.

Compliance agreement. A written agreement between APHIS and a person engaged in growing, handling, or moving regulated articles, wherein the person agrees to comply with this subpart.

Core area. The area within a circle surrounding each site where fruit flies have been detected using a ½-mile radius with the detection site as a center point.

Day degrees. A unit of measurement used to measure the amount of heat required to further the development of fruit flies through their life cycle. Day-degree life cycle requirements are calculated through a modeling process specific for each species of fruit fly.

Departmental permit. A document issued by the Administrator in which he or she affirms that interstate movement of the regulated article identified on the document is for scientific or experimental purposes and that the regulated article is eligible for interstate movement in accordance with § 301.32–4(c).

Dripline. The line around the canopy of a plant.

Fruit fly (fruit flies). The melon fruit fly, Mexican fruit fly, Mediterranean fruit fly, Oriental fruit fly, peach fruit fly, sapote fruit fly, or West Indian fruit fly, or other species of insects found in the family Tephritidae, collectively.

Infestation. The presence of fruit flies or the existence of circumstances that makes it reasonable to believe that fruit flies are present.

Inspector. Any employee of APHIS or other person authorized by the Administrator to enforce this subpart.

Interstate. From any State into or through any other State.

Limited permit. A document in which an inspector or person operating under a compliance agreement affirms that the regulated article identified on the document is eligible for interstate movement in accordance with § 301.32–5(b) only to a specified destination and only in accordance with specified conditions.

Mediterranean fruit fly. The insect known as Mediterranean fruit fly, *Ceratitis capitata* (Wiedemann), in any stage of development.

Melon fruit fly. The insect known as the melon fruit fly, *Bactrocera cucurbitae* (Coquillett), in any stage of development.

---

[1] Permit and other requirements for the interstate movement of any of the fruit flies regulated under this subpart are contained in part 330 of this chapter.

Mexican fruit fly. The insect known as Mexican fruit fly, *Anastrepha ludens* (Loew), in any stage of development.

Move (moved, movement). Shipped, offered to a common carrier for shipment, received for transportation or transported by a common carrier, or carried, transported, moved, or allowed to be moved.

Oriental fruit fly. The insect known as Oriental fruit fly, *Bactrocera dorsalis* (Hendel), in any stage of development.

Peach fruit fly. The insect known as peach fruit fly, *Anastrepha zonata* (Saunders), in any stage of development.

Person. Any individual, partnership, corporation, association, joint venture, or other legal entity.

Plant Protection and Quarantine. The organizational unit within the Animal and Plant Health Inspection Service that has been delegated responsibility for enforcing provisions of the Plant Protection Act and related legislation, quarantines, and regulations.

Quarantined area. Any State, or any portion of a State, designated as a quarantined area in accordance with § 301.32–3.

Regulated article. Any article listed in § 301.32–2 or otherwise designated as a regulated article in accordance with § 301.32–2(d).

Sapote fruit fly. The insect known as the sapote fruit fly, *Anastrepha serpentina*, in any stage of development.

State. Any of the several States of the United States, the Commonwealth of the Northern Mariana Islands, the Commonwealth of Puerto Rico, the District of Columbia, Guam, the Virgin Islands of the United States, or any other territory or possession of the United States.

West Indian fruit fly. The insect known as the West Indian fruit fly, *Anastrepha obliqua* (Macquart), in any stage of development.

§ 301.32–2　Regulated articles.

(a) In the following table, the berry, fruit, nut, or vegetable listed in each row in the left column is a regulated article for each of the fruit fly species listed in that row in the right column, unless the article is canned, dried, or frozen below −17.8 °C (0 °F):

Botanical name	Common name(s)	Fruit fly
Abelmoschus esculentus = Hibiscus esculentus.	Okra	Melon, Peach.
Acca sellowiana = Feijoa sellowiana	Pineapple guava	Mediterranean, Oriental, Peach.
Actinidia chinensis	Kiwi	Mediterranean.
Aegle marmelos	Indian bael	Peach.
Anacardium occidentale	Cashew	Oriental.
Annona cherimola	Cherimoya	Mexican, Oriental, Peach.
Annona glabra	Pond-apple	Sapote.
Annona muricata	Soursop	Melon, Oriental, Peach.
Annona reticulata	Custard apple, Annona	Melon, Mexican, Oriental, Peach.
Annona squamosa	Custard apple	Peach.
Artocarpus altilis	Breadfruit	Oriental.
Artocarpus heterophyllus	Jackfruit	Oriental.
Averrhoa carambola	Carambola, Country gooseberry	Oriental, West Indian.
Benincasa hispida	Melon, Chinese	Melon.
Brassica juncea	Mustard, leaf	Melon.
Brassica oleracea var. botrytis	Cauliflower	Melon.
Brosimum alicastrum	Ramón	West Indian.
Byrsonima crassifolia	Nance	Sapote.
Calophyllum inophyllum	Alexandrian-laurel, Laurel	Oriental.
Cananga odorata	Ylang-Ylang	Oriental.
Capsicum annum	Pepper, chili	Mediterranean, Melon, Oriental.
Capsicum frutescens	Pepper, tabasco	Mediterranean, Melon.
Capsicum frutescens abbreviatum	Oriental bush red pepper	Oriental.
Capsicum frutescens var. grossum	Pepper, sweet	Oriental.
Carica papaya	Papaya	Mediterranean, Melon, Oriental, Peach.
Carissa grandiflora	Natal plum	Oriental.
Carissa macrocarpa	Natal plum	Mediterranean.
Casimiroa edulis	Sapote, white	Mediterranean.
Casimiroa greggii = Sargentia greggii	Sargentia, yellow chapote	Mexican.
Casimiroa spp	Sapote	Mexican.
Cereus coerulescens	Cactus	Oriental.
Chrysophyllum cainito	Star apple	Oriental, Sapote.
Chrysophyllum oliviforme	Calmitillo	Oriental.
Citrofortunella japonica	Orange, calamondin	Peach.

Botanical name	Common name(s)	Fruit fly
Citrullus colocynthis	Colocynth	Melon.
Citrullus lanatus = Citrullus vulgaris	Watermelon	Melon, Peach.
Citrullus spp	Melon	Melon.
Citrus aurantiifolia	Lime	Mediterranean, Mexican,[1] Oriental, Peach.
Citrus aurantium	Orange, sour	Mediterranean, Mexican, Oriental, Peach.
Citrus jambhiri	Lemon, Rough	Mediterranean.
Citrus latifolia	Lime, Persian	Oriental.
Citrus limon	Lemon	Mediterranean,[2] Mexican,[3] Oriental, Peach.
Citrus limon × reticulata	Lemon, Meyer	Mediterranean.
Citrus madurensis = xCitrofortunella mitis	Orange, Panama	Sapote.
Citrus maxima = Citrus grandis	Pummelo or Shaddock	Mediterranean, Mexican, Oriental, Peach.
Citrus medica	Citrus citron	Mediterranean, Mexican, Peach.
Citrus paradisi	Grapefruit	Mediterranean, Melon, Mexican, Oriental, Peach.
Citrus reticulata	Mandarin orange, tangerine	Mediterranean, Mexican, Oriental, Peach.
Citrus reticulata var. Unshu	Orange, Unshu	Mediterranean, Oriental.
Citrus reticulata × C. sinensis = Citrus nobilis.	Orange, king	Mediterranean, Melon, Oriental, Peach.
Citrus reticulata × Fortunella	Orange, calamondin	Mediterranean, Mexican, Oriental.
Citrus sinensis	Orange, sweet	Mediterranean, Melon, Mexican, Oriental, Peach.
Citrus spp	Citrus	Sapote.
Clausena lansium	Wampi	Oriental.
Coccinia spp	Gourds	Melon, Peach.
Coccoloba uvifera	Seagrape	Oriental.
Coffea arabica	Coffee, Arabian	Oriental.
Cresentia spp	Gourds	Melon, Peach.
Cucumis melo and Cucumis melo var. Cantalupensis.	Cantaloupe	Melon, Peach.
Cucumis melo var. conomon	Melon, oriental pickling	Melon.
Cucumis pubescens and Cucumis trigonus.	Cucurbit	Melon.
Cucumis sativus	Cucumber	Melon, Oriental, Peach.
Cucumis utilissimus	Melon, long	Peach.
Cucurbita maxima	Squash	Melon.
Cucurbita moschata	Pumpkin, Canada	Melon.
Cucurbita pepo	Pumpkin	Melon.
Cydonia oblonga	Quince	Mexican, Mediterranean, Oriental, Peach, Sapote.
Cyphomandra betaceae	Tomato, tree	Melon.
Diospyros digyna	Black sapote	Sapote.
Diospyros discolor	Velvet apple	Oriental.
Diospyros khaki	Japanese persimmon	Mediterranean, Oriental.
Diospyros spp	Sapote	Sapote, West Indian.
Dovyalis hebecarpa	Kitembilla	Oriental, Sapote, West Indian.
Dracena draco	Dragon tree	Oriental.
Elaeocarpus angustifolius	Blue marbletree; New Guinea quandong	Peach.
Elaeocarpus grandiflorus	Lily of the valley tree	Peach.
Elaeocarpus madopetalus	Ma-kok-nam	Peach.
Eriobotrya japonica	Loquat	Mediterranean, Oriental, Peach, West Indian.
Eugenia brasiliensis = E. dombeyi	Brazil-cherry, grumichama	Mediterranean, Oriental, Peach.
Eugenia malaccensis	Malay apple	Oriental.
Eugenia uniflora	Surinam cherry	Mediterranean, Oriental, Peach.
Euphoria longan	Longan	Oriental.
Ficus benghalensis	Fig, Banyan	Peach.
Ficus carica	Fig	Mediterranean, Melon, Oriental, Peach.
Ficus macrophylla	Fig, Moreton Bay	Peach.
Ficus retusa	Fig, glossy leaf	Peach.
Ficus rubiginosa	Fig, Port Jackson	Peach.
Ficus spp	Fig	Peach.
Fortunella japonica	Chinese Orange, Kumquat	Mediterranean, Oriental, Peach.
Garcinia celebica	Gourka	Oriental.
Garcinia mangostana	Mangosteen	Oriental.
Grewia asiatica	Phalsa	Peach.
Jubaea chilensis = Jubaea spectabilis	Syrup palm	Oriental.
Juglans hindsii	Walnut	Oriental.
Juglans regia	Walnut, English	Oriental.
Juglans spp	Walnut with husk	Mediterranean.

17

Botanical name	Common name(s)	Fruit fly
Lablab purpureus subsp. *purpureus* = *Dolichos lablab.*	Bean, hyacinth	Melon.
Lagenaria spp	Gourds	Melon, Peach.
Luffa acutangula	Gourd, ribbed or ridged, luffa	Peach.
Luffa aegyptiaca	Gourd, smooth luffa, sponge	Peach.
Luffa spp	Gourds	Melon, Peach.
Luffa vulgaris	Gourd	Peach.
Lychee chinensis	Lychee nut	Oriental
Lycopersicon esculentum	Tomato	Mediterranean, [4] Melon, Oriental, Peach.
Madhuca indica = *Bassia latifolia*	Mahua, mowra-buttertree	Peach.
Malpighia glabra	Cherry, Barbados	Oriental, West Indian.
Malpighia punicifolia	West Indian cherry	Oriental.
Malus sylvestris	Apple	Mediterranean, Melon, Mexican, Oriental, Sapote, Peach.
Mammea americana	Mammy apple	Mexican, Oriental, Peach, Sapote.
Mangifera foetida	Mango, Bachang	Peach.
Mangifera indica	Mango	All.
Mangifera odorata	Kuine	Peach.
Manilkara hexandra	Sapodilla, balata	Peach.
Manilkara jaimiqui subsp. *emarginata*	Sapodilla, wild	Peach.
Manilkara zapota	Sapodilla, chiku	Oriental, Peach, Sapote, West Indian.
Mimusops elengi	Spanish cherry	Mediterranean, Oriental.
Momordica balsamina	Balsam apple, hawthorn	Peach.
Momordica charantia	Balsam pear, bitter melon	Peach.
Momordica cochinchinensis	Balsam apple, gac	Peach.
Momordica spp	Gourds	Melon, Peach.
Morus nigra	Mulberry	Oriental.
Murraya exotica	Mock orange	Mediterranean, Oriental.
Musa × *paradisiaca* = *Musa paradisiaca* subsp. *sapientum.*	Banana	Oriental.
Musa acuminata = *Musa nana*	Banana, dwarf	Oriental.
Ochrosia elliptica	Orange, bourbon	Peach.
Olea europea	Olive	Mediterranean.
Opuntia ficus-indica = *Opuntia megacantha.*	Prickly pear	Oriental.
Opuntia spp	Opuntia cactus	Mediterranean.
Passiflora edulis	Passionflower, passionfruit, yellow lilikoi	Melon, Oriental, West Indian.
Passiflora laurifolia	Lemon, water	Melon.
Passiflora ligularis	Granadilla, sweet	Oriental.
Passiflora quadrangularis	Granadilla, giant	West Indian.
Passiflora tripartita var. *mollissima*	Passionflower, softleaf	Oriental.
Persea americana	Avocado	Mediterranean, Melon, Mexican, Oriental, Peach, Sapote.
Phaseolus lunatus = *Phaseolus limensis*	Bean, lima	Melon.
Phaseolus vulgaris	Bean, mung	Melon.
Phoenix dactylifera	Date palm	Mediterranean, Melon, Oriental, Peach.
Planchonia careya = *Careya arborea*	Patana oak, kumbhi	Peach.
Pouteria caimito	Abiu	Sapote.
Pouteria campechiana	Eggfruit tree	Oriental, Sapote.
Pouteria obovata	Lucmo	Sapote.
Pouteria viridis	Sapote, green	Sapote.
Prunus americana	Plum, American	Mediterranean, Mexican, Oriental, Peach.
Prunus armeniaca	Apricot	Mediterranean, Mexican, Oriental, Peach.
Prunus avium	Sweet cherry	Mediterranean, Peach.
Prunus cerasus	Sour cherry	Mediterranean, Peach.
Prunus domestica	Plum, European	Mediterranean, Mexican, Oriental, Peach.
Prunus dulcis = *P. amygdalus*	Almond with husk	Mediterranean, Peach [5].
Prunus ilicifolia	Cherry, Catalina	Oriental, Peach.
Prunus lusitanica	Cherry, Portuguese	Oriental, Peach.
Prunus persica	Peach	All.
Prunus persica var. *nectarine*	Nectarine	Mediterranean, Mexican, Oriental, Peach.
Prunus salicina	Japanese plum	Mediterranean, Mexican, Peach, West Indian.
Prunus salicina × *Prunus cerasifera*	Methley plum	Peach.
Psidium cattleianum	Strawberry guava, Cattley guava	Mediterranean, Melon, Oriental.
Psidium cattleianum var. *cattleianum* f. *lucidum.*	Yellow strawberry guava	Peach.
Psidium cattleianum var. *littorale*	Red strawberry guava	Oriental, West Indian, Peach.
Psidium guajava	Guava	All.
Punica granatum	Pomegranate	Mediterranean, Mexican, Oriental, Peach.

Botanical name	Common name(s)	Fruit fly
Pyrus communis	Pear	All.
Pyrus pashia	Kaeuth	Peach.
Pyrus pyrifolia	Pear, sand	Peach.
Rhodomyrtus tomentosa	Myrtle, downy rose	Oriental.
Sandoricum koetjape	Santol	Oriental.
Santalum album	Sandalwood, white	Oriental.
Santalum paniculatum	Sandalwood	Oriental.
Sapotaceae	Sapota, Sapodilla	Mexican.
Sechium edule	Chayote	Melon.
Sesbania grandiflora	Scarlet wisteria tree	Melon.
Sicyes sp	Cucumber, bur	Melon.
Solanum aculeatissimum	Nightshade	Peach.
Solanum mauritianum = S. auriculatum	Tobacco, wild	Peach.
Solanum melongena	Eggplant	Mediterranean, [6] Melon, Peach.
Solanum muricatum	Pepino	Oriental, Peach.
Solanum pseudocapsicum	Jerusalem cherry	Oriental, Peach.
Solanum seaforthianum	Nightshade, Brazilian	Peach.
Solanum verbascifolium	Nightshade, Mullein	Peach.
Spondias dulcis = Spondias cytherea	Otaheite apple, Jew plum	Oriental, West Indian.
Spondias mombin	Hog-plum	Sapote, West Indian.
Spondias purpurea	Red mombin	Sapote, West Indian.
Spondias spp	Spanish plum, purple mombin or Ciruela	Mexican.
Spondias tuberose	Imbu	Oriental.
Syzygium aquem	Water apple, watery roseapple	Peach.
Syzygium cumini	Java plum, jambolana	Peach.
Syzygium jambos = Eugenia jambos	Rose apple	Mediterranean, Mexican, Oriental, Peach, West Indian.
Syzygium malaccense = Eugenia malaccensis.	Mountain apple, Malay apple	Mediterranean, Peach, West Indian.
Syzygium samarangense	Java apple	Peach.
Terminalia bellirica	Myrobalan, belleric	Peach.
Terminalia catappa	Tropical almond	Oriental, Peach.
Terminalia chebula	Myrobalan, black or chebulic	Mediterranean, Oriental, Peach.
Thevetia peruviana	Yellow oleander	Mediterranean, Oriental.
Trichosanthis spp	Gourds	Melon, Peach.
Vaccinium spp	Blueberry	Mediterranean.
Vigna unguiculata	Cowpea	Melon.
Vitis spp	Grapes	Mediterranean, Oriental.
Vitis trifolia	Grape	Melon.
Wikstroemia phillyreifolia	Akia	Oriental.
Ziziphus mauritiana	Chinese date, jujube	Peach.

[1] Sour limes are not regulated articles for Mexican fruit fly.
[2] Only yellow lemons are regulated articles for Mediterranean fruit fly.
[3] Eureka, Lisbon, and Villa Franca cultivars (smooth-skinned sour lemon) are not regulated articles for Mexican fruit fly.
[4] Only pink and red ripe tomatoes are regulated articles for Mediterranean fruit fly.
[5] Harvested almonds with dried husks are not regulated articles for peach fruit fly.
[6] Commercially produced eggplants are not regulated articles for Mediterranean fruit fly.

(b) Plants of the following species in the family Curcurbitaceae are regulated articles for the melon fruit fly only:

Cantaloupe (Cucumis melo)
Chayote (Sechium edule)
Colocynth (Citrullus colocynthis)
Cucumber (Cucumis sativus)
Cucumber, bur (Sicyes spp.)
Cucurbit (Cucumis pubescens and C. trigonus)
Cucurbit, wild (Cucumis trigonus)
Gherkin, West India (Cucumis angaria)
Gourds (Coccinia, Cresentia, Lagenaria, Luffa, Momordica, and Trichosanthis spp.)
Gourd, angled luffa (Luffa acutangula)
Gourd, balsam apple (Momordica balsaminia)
Gourd, ivy (Coccinia grandis)
Gourd, kakari (Momordica dioica)
Gourd, serpent cucumber (Trichosanthis anguina)
Gourd, snake (Trichosanthis cucumeroides)
Gourd, sponge (Luffa aegyptiaca)
Gourd, white flowered (Lagenaria siceraria)
Melon, Chinese (Benincasa hispida)
Melon, long (Cucumis utilissimus)
Pumpkin (Cucurbita pepo)
Pumpkin, Canada (Cucurbita moschata)
Squash (Cucurbita maxima)
Watermelon (Citrullus lanatus = Citrullus vulgaris)

(c) Soil within the dripline of the plants listed in paragraph (b) of this section or plants that are producing or have produced any article listed in paragraph (a) of this section.

(d) Any other product, article, or means of conveyance not listed in paragraphs (a), (b), or (c) of this section that an inspector determines presents a risk of spreading fruit flies, when the inspector notifies the person in possession of the product, article, or means of conveyance that it is subject to the restrictions of this subpart.

[73 FR 34232, June 9, 2008, as amended at 75 FR 12962, Mar. 18, 2010]

§ 301.32–3 Quarantined areas.

(a) *Designation of quarantined areas.* In accordance with the criteria listed in paragraph (c) of this section, the Administrator will designate as a quarantined area each State, or each portion of a State, in which a fruit fly population subject to the regulations in this subpart has been found by an inspector, or in which the Administrator has reason to believe that a fruit fly population is present, or that the Administrator considers necessary to quarantine because of its inseparability for quarantine enforcement purposes from localities in which a fruit fly population has been found. The Administrator will publish the description of the quarantined area on the Plant Protection and Quarantine Web site, *http://www.aphis.usda.gov/ plant_health/plant_pest_info/ fruit_flies/index.shtml.* The description of the quarantined area will include the date the description was last updated and a description of the changes that have been made to the quarantined area. The description of the quarantined area may also be obtained by request from any local office of PPQ; local offices are listed in telephone directories. After a change is made to the quarantined area, we will publish a notice in the FEDERAL REGISTER informing the public that the change has occurred and describing the change to the quarantined area.

(b) *Designation of an area less than an entire State as a quarantined area.* Less than an entire State will be designated as a quarantined area only if the Administrator determines that:

(1) The State has adopted and is enforcing restrictions on the intrastate movement of the regulated articles that are equivalent to those imposed by this subpart on the interstate movement of regulated articles; and

(2) The designation of less than the entire State as a quarantined area will prevent the interstate spread of the fruit fly.

(c) *Criteria for designation of a State, or a portion of a State, as a quarantined area.* A State, or a portion of a State, will be designated as a quarantined area when a fruit fly population has been found in that area by an inspector, when the Administrator has reason to believe that the fruit fly is present in that area, or when the Administrator considers it necessary to quarantine that area because of its inseparability for quarantine enforcement purposes from localities in which the fruit fly has been found.

(d) *Removal of a State, or a portion of a State, from quarantine.* A State, or a portion of a State, will be removed from quarantine when the Administrator determines that sufficient time has passed without finding additional flies or other evidence of infestation in the area to conclude that the fruit fly no longer exists in that area.

§ 301.32–4 Conditions governing the interstate movement of regulated articles from quarantined areas.

Any regulated article may be moved interstate from a quarantined area[2] only if moved under the following conditions:

(a) With a certificate or limited permit issued and attached in accordance with §§ 301.32–5 and 301.32–8;

(b) Without a certificate or limited permit if:

(1) The regulated article originated outside the quarantined area and is either moved in an enclosed vehicle or is completely enclosed by a covering adequate to prevent access by fruit flies (such as canvas, plastic, or other closely woven cloth) while moving through the quarantined area; and

(2) The point of origin of the regulated article is indicated on the waybill, and the enclosed vehicle or the enclosure that contains the regulated article is not opened, unpacked, or unloaded in the quarantined area; and

(3) The regulated article is moved through the quarantined area without stopping except for refueling or for

---

[2] Requirements under all other applicable Federal domestic plant quarantines and regulations must also be met.

traffic conditions, such as traffic lights or stop signs.

(c) Without a certificate or limited permit if the regulated article is moved:

(1) By the United States Department of Agriculture for experimental or scientific purposes;

(2) Pursuant to a permit issued by the Administrator for the regulated article;

(3) Under conditions specified on the permit and found by the Administrator to be adequate to prevent the spread of fruit flies; and

(4) With a tag or label bearing the number of the permit issued for the regulated article attached to the outside of the container of the regulated article or attached to the regulated article itself if not in a container.

(d) Hass avocados that are grown or packed in an area quarantined for Mediterranean, Mexican, or sapote fruit fly and that are moving interstate from such an area are subject to the following additional requirements:

(1) *Orchard sanitation and safeguarding requirements.* (i) Hass avocado fruit that has fallen from the trees may not be included in field boxes of fruit to be packed for shipping.

(ii) Harvested Hass avocados must be placed in field boxes or containers of field boxes that are marked to show the location of the orchard. The avocados must be moved from the orchard to the packinghouse within 3 hours of harvest or they must be protected from fruit fly infestation until moved.

(iii) Hass avocados must be protected from fruit fly infestations during their movement from the orchard to the packinghouse and must be accompanied by a field record indicating the location of the orchard where the avocados originated.

(2) *Packinghouse requirements for Hass avocados packed within a quarantined area.* (i) All openings to the outside of the packinghouse must be covered by screening with openings of not more than 1.6 mm or by some other barrier that prevents insects from entering the packinghouse.

(ii) The packinghouse must have double doors at the entrance to the facility and at the interior entrance to the area where the avocados are packed.

(iii) If the Hass avocados were grown in an orchard within the quarantined area, the identity of the avocados must be maintained from field boxes or containers to the shipping boxes in the packinghouse so that the avocados can be traced back to the orchard in which they were grown. The avocados must be packed in boxes or crates that are clearly marked with the identity of the grower and the packinghouse.

(iv) Any boxes of Hass avocados packed in the quarantined area must be placed in a refrigerated truck or refrigerated container and remain in that truck or container while in transit through the quarantined area. Prior to leaving the packinghouse, the truck or container must be secured with a seal that will be broken when the truck or container is opened. Once sealed, the refrigerated truck or refrigerated container must remain unopened until it is outside the quarantined area.

(v) Any avocados that have not been packed or loaded into a refrigerated truck or refrigerated container by the end of the workday must be kept inside the screened packinghouse.

(3) *Packinghouse requirements for Hass avocados packed outside a quarantined area but grown within a quarantined area.* Hass avocados grown in an orchard within a quarantined area but packed in a packinghouse outside the quarantined area must meet the requirements of paragraph (d)(2)(iii) of this section.

(Approved by the Office of Management and Budget under control numbers 0579–0000 and 0579–0336)

[73 FR 32432, June 9, 2008, as amended at 74 FR 31159, June 30, 2009; 75 FR 12962, Mar. 18, 2010; 76 FR 43807, July 22, 2011]

§301.32–5 Issuance and cancellation of certificates and limited permits.

(a) A certificate may be issued by an inspector[3] for the interstate movement of a regulated article if the inspector determines that:

(1)(i) The regulated article has been treated under the direction of an inspector in accordance with §301.32–10; or

---

[3] Services of an inspector may be requested by contacting local PPQ offices, which are listed in telephone directories.

(ii) Based on inspection of the premises of origin, the premises are free from fruit flies; or

(iii) Based on inspection of the regulated article, the regulated article is free of fruit flies; or

(iv) The regulated articles are Hass variety avocados that have been harvested, safeguarded, and packed in accordance with the conditions in § 301.32-4(d); and

(2) The regulated article will be moved through the quarantined area in an enclosed vehicle or will be completely enclosed by a covering adequate to prevent access by fruit flies; and

(3) The regulated article is to be moved in compliance with any additional emergency conditions the Administrator may impose under section 414 of the Plant Protection Act (7 U.S.C. 7714) to prevent the spread of fruit flies; and

(4) The regulated article is eligible for unrestricted movement under all other Federal domestic plant quarantines and regulations applicable to the regulated article.

(b) An inspector[4] will issue a limited permit for the interstate movement of a regulated article if the inspector determines that:

(1) The regulated article is to be moved interstate to a specified destination for specified handling, processing, or utilization (the destination and other conditions to be listed in the limited permit), and this interstate movement will not result in the spread of fruit flies because life stages of the fruit flies will be destroyed by the specified handling, processing, or utilization;

(2) The regulated article is to be moved in compliance with any additional emergency conditions the Administrator may impose under section 414 of the Plant Protection Act (7 U.S.C. 7714) to prevent the spread of fruit flies; and

(3) The regulated article is eligible for interstate movement under all other Federal domestic plant quarantines and regulations applicable to the regulated article.

---

[4] See footnote 3.

(c) Certificates and limited permits for the interstate movement of regulated articles may be issued by an inspector or person operating under a compliance agreement. A person operating under a compliance agreement may issue a certificate for the interstate movement of a regulated article if an inspector has determined that the regulated article is eligible for a certificate in accordance with paragraph (a) of this section. A person operating under a compliance agreement may issue a limited permit for interstate movement of a regulated article when an inspector has determined that the regulated article is eligible for a limited permit in accordance with paragraph (b) of this section.

(d) Any certificate or limited permit that has been issued may be withdrawn, either orally or in writing, by an inspector if he or she determines that the holder of the certificate or limited permit has not complied with all conditions in this subpart for the use of the certificate or limited permit. If the withdrawal is oral, the withdrawal and the reasons for the withdrawal will be confirmed in writing as promptly as circumstances allow. Any person whose certificate or limited permit has been withdrawn may appeal the decision in writing to the Administrator within 10 days after receiving the written notification of the withdrawal. The appeal must state all of the facts and reasons upon which the person relies to show that the certificate or limited permit was wrongfully withdrawn. As promptly as circumstances allow, the Administrator will grant or deny the appeal, in writing, stating the reasons for the decision. A hearing will be held to resolve any conflict as to any material fact. Rules of practice concerning a hearing will be adopted by the Administrator.

(Approved by the Office of Management and Budget under control numbers 0579-0088 and 0579-0336)

[73 FR 32432, June 9, 2008, as amended at 74 FR 31160, June 30, 2009]

§ 301.32-6 Compliance agreements and cancellation.

(a) Any person engaged in growing, handling, or moving regulated articles may enter into a compliance agreement when an inspector determines that the person is aware of this subpart, agrees to comply with its provisions, and agrees to comply with all the provisions contained in the compliance agreement.[5]

(b) Any compliance agreement may be canceled, either orally or in writing, by an inspector whenever the inspector finds that the person who has entered into the compliance agreement has failed to comply with any of the conditions of this subpart or with any of the provisions of the compliance agreement. If the cancellation is oral, the cancellation and the reasons for the cancellation will be confirmed in writing as promptly as circumstances allow. Any person whose compliance agreement has been canceled may appeal the decision, in writing, within 10 days after receiving written notification of the cancellation. The appeal must state all of the facts and reasons upon which the person relies to show that the compliance agreement was wrongfully canceled. As promptly as circumstances allow, the Administrator will grant or deny the appeal, in writing, stating the reasons for the decision. A hearing will be held to resolve any conflict as to any material fact. Rules of practice concerning a hearing will be adopted by the Administrator.

§ 301.32–7 Assembly and inspection of regulated articles.

(a) Any person, other than a person authorized to issue certificates or limited permits under § 301.32–5(c), who desires to move a regulated article interstate accompanied by a certificate or limited permit must notify an inspector[6] as far in advance of the desired interstate movement as possible, but no less than 48 hours before the desired interstate movement.

(b) The regulated article must be assembled at the place and in the manner

---

[5] Compliance agreement forms are available without charge from the Animal and Plant Health Inspection Service, Plant Protection and Quarantine, Emergency and Domestic Programs, 4700 River Road Unit 134, Riverdale, MD 20737–1236, and from local PPQ offices, which are listed in telephone directories.

[6] See footnote 3 to § 301.32–5(a).

the inspector designates as necessary to comply with this subpart.

§ 301.32–8 Attachment and disposition of certificates and limited permits.

(a) A certificate or limited permit required for the interstate movement of a regulated article must, at all times during the interstate movement, be:

(1) Attached to the outside of the container containing the regulated article; or

(2) Attached to the regulated article itself if not in a container; or

(3) Attached to the consignee's copy of the accompanying waybill: Provided, however, that if the certificate or limited permit is attached to the consignee's copy of the waybill, the regulated article must be sufficiently described on the certificate or limited permit and on the waybill to identify the regulated article.

(b) The certificate or limited permit for the interstate movement of a regulated article must be furnished by the carrier to the consignee listed on the certificate or limited permit upon arrival at the location provided on the certificate or limited permit.

(Approved by the Office of Management and Budget under control number 0579–0088)

§ 301.32–9 Costs and charges.

The services of the inspector during normal business hours (8 a.m. to 4:30 p.m., Monday through Friday, except holidays) will be furnished without cost. The user will be responsible for all costs and charges arising from inspection and other services provided outside normal business hours.

§ 301.32–10 Treatments.

Regulated articles may be treated in accordance with part 305 of this chapter to neutralize fruit flies. The following treatments also may be used for the regulated articles indicated:

(a) *Soil within the dripline of plants that are producing or have produced regulated articles listed § 301.32(a) or (b).* The following soil treatments may be used: Apply diazinon at the rate of 5 pounds active ingredient per acre to the soil within the dripline with sufficient water to wet the soil to at least a depth of 0.5 inch. Both immersion and pour-on treatment procedures are also acceptable.

(b) *Premises.* Fields, groves, or areas that are located within a quarantined area but outside the infested core area and that produce regulated articles may receive regular treatments with either malathion or spinosad bait spray as an alternative to treating fruits and vegetables as provided in part 305 of this chapter. These treatments must take place at 6- to 10-day intervals, starting a sufficient time before harvest (but not less than 30 days before harvest) to allow for development of fruit fly egg and larvae. Determination of the time period must be based on the day degrees model for the specific fruit fly. Once treatment has begun, it must continue through the harvest period. The malathion bait spray treatment must be applied by aircraft or ground equipment at a rate of 2.4 oz of technical grade malathion and 9.6 oz of protein hydrolysate per acre. The spinosad bait spray treatment must be applied by aircraft or ground equipment at a rate of 0.01 oz of a USDA-approved spinosad formulation and 48 oz of protein hydrolysate per acre. For ground applications, the mixture may be diluted with water to improve coverage.

[73 FR 32432, June 9, 2008, as amended at 75 FR 4240, Jan. 26, 2010]

Subpart D—Black Stem Rust

SOURCE: 54 FR 32791, Aug. 10, 1989, unless otherwise noted. Redesignated at 84 FR 2428, Feb. 7, 2019.

§ 301.38 Notice of quarantine; restrictions on interstate movement of regulated articles.

The conterminous 48 States and the District of Columbia are quarantined in order to prevent the spread of black stem rust. No person shall move interstate any regulated article except in accordance with this subpart.[1]

[54 FR 32791, Aug. 10, 1989, as amended at 66 FR 21050, Apr. 27, 2001]

___

[1] Any properly identified employee of the Animal and Plant Health Inspection Service is authorized to stop and inspect persons and means of conveyance, and to seize, quarantine, treat, apply other remedial measures to destroy, or otherwise dispose of regulated

§ 301.38-1 Definitions.

In this subpart the following definitions apply:

Administrator. The Administrator, Animal and Plant Health Inspection Service (APHIS), or any person authorized to act for the Administrator.

Animal and Plant Health Inspection Service (APHIS). The Animal and Plant Health Inspection Service of the United States Department of Agriculure.

Black stem rust. The disease commonly known as the black stem rust of grains (*Puccinia graminis*).

Certificate. A document in which an inspector, or a person operating under a compliance agreement, affirms that a specified regulated article has met the criteria in § 301.38-5(b) of this subpart and may be moved interstate to any destination.

Clonally propagated. Reproduced asexually through cuttings, tissue culture, suckers, or crown division. For the purposes of this subpart, a *Berberis* plant will be considered clonally propagated only if its parent stock is, or was derived from, a seed-propagated black stem rust-resistant plant of more than 2 years' growth.

Compliance agreement. A written agreement between a State that is a protected area or that encompasses a protected area and a person who moves regulated articles interstate, or in a non-protected area between APHIS and such person, in which that person agrees to comply with this subpart.

Departmental permit. A document issued by the Administrator in which he or she affirms that interstate movement of the regulated article identified on the document is for scientific or experimental purposes, and that the regulated article is eligible for interstate movement under the conditions specified on the Departmental permit and found by the Administrator to be adequate to prevent the introduction of rust-susceptible varieties of the genera *Berberis*, *Mahoberberis*, and *Mahonia* into protected areas.

___

articles as provided in sections 414 and 421 of the Plant Protection Act (7 U.S.C. 7714 and 7731).

Inspector. Any APHIS employee or other person authorized by the Administrator in accordance with law to enforce this subpart.

Interstate. From any State into or through any other State.

Limited permit. A document issued by an inspector to allow the interstate movement into or through a protected area of regulated articles not eligible for certification under this subpart to a specified destination outside the protected area.

Moved (movement, move). Shipped, offered to a common carrier for shipment, received for transportation or transported by a common carrier, or carried, transported, moved, or allowed to be moved. "Movement" and "move" shall be construed in accordance with this definition.

Person. Any association, company, corporation, firm, individual, joint stock company, partnership, society, or any other legal entity.

Protected area. Those States or counties designated in § 301.38–3(d) of this subpart.

Rust-resistant plants. All plants of the genera *Berberis, Mahoberberis,* and *Mahonia,* and their progeny, that have proven resistant to black stem rust during testing by the United States Department of Agriculture,[2] and that are listed as rust-resistant under § 301.38–2 (a)(1) and (a)(2).

Rust-susceptible plants. All plants of the genera *Berberis, Mahoberberis,* and *Mahonia* not listed as rust-resistant under § 301.38–2 (a)(1) and (a)(2).

---

[2] Testing is performed by the Agricultural Research Service of USDA as follows: In a greenhouse, the suspect plant, or test subject, is placed under a screen with a control plant, *i.e.,* a known rust-susceptible variety of *Berberis, Mahoberberis,* or *Mahonia.* Infected wheat stems, a primary host of black stem rust, are placed on top of the screen. The plants are moistened and maintained in 100% humidity, causing the spores to swell and fall on the plants lying under the screen. The plants are then observed for 7 days at 20–80% relative humidity. This test procedure is repeated 12 times. If in all 12 tests, the rust-susceptible plant shows signs of infection after 7 days and the test plants do not, USDA will declare the test plant variety rust-resistant. The tests must be performed on new growth, just as the leaves are unfolding.

Regulated article. Any article listed in § 301.38–2 (a)(1) through (a)(3) of this subpart or otherwise designated as a regulated article in accordance with § 301.38–2(a)(4) of this subpart.

Seedling. Any plant of the genera *Berberis, Mahoberberis,* and *Mahonia* grown from seed and having less than 2 years' growth.

State. The District of Columbia, Puerto Rico, the Northern Mariana Islands, or any State, territory or possession of the United States.

Two years' growth. The growth of a plant during all growing seasons of 2 successive calendar years.

[54 FR 32791, Aug. 10, 1989; 54 FR 38494, Sept. 18, 1989; 67 FR 8178, Feb. 22, 2002; 71 FR 5778, Feb. 3, 2006]

§ 301.38–2 Regulated articles.

(a) The following are regulated articles:[3]

(1) All plants, seeds, fruits, and other plant parts capable of propagation from the following rust-resistant *Berberis* species and varieties.

B. aggregata × *B. wilsoniae* 'Pirate King'

B. 'Amstelveen'

B. aridocalida

B. beaniana

B. buxifolia

B. buxifolia nana

B. calliantha

B. candidula

B. candidula 'Amstelveen'

B. candidula × *B. verruculosa* 'Amstelveen'

B. cavullieri

B. chenaulti

B. chanaulti 'Apricot Queen'

B. circumserrata

B. concinna

B. coxii

B. darwini

B. dasystachya

B. dubia

B. feddeana

B. formosana

B. franchetiana

B. gagnepainii

B. gagnepaini 'Chenault'

B. gilgiana

---

[3] Permit and other requirements for the insterstate movement of black stem rust organisms are contained in part 330 of this chapter.

B. gladwynensis
B. gladwynensis 'William Penn'
B. gyalaica
B. heterophylla
B. horvathi
B. hybrido-gagnepaini
B. insignis
B. integerrima 'Wallichs Purple'
B. julianae
B. julianae 'Nana'
B. julianae 'Spring Glory'
B. koreana
B. koreana × *B. thunbergii* hybrid *Bailsel*
B. koreana × *B. thunbergii* hybrid *Tara*
B. lempergiana
B. lepidifolia
B. linearifolia
B. linearifolia var. 'Orange King'
B. lologensis
B. lologensis 'Mystery Fire'
B. manipurana
B. media 'Dual Jewel'
B. media 'Park Jewel'
B. media 'Red Jewel'
B. mentorensis
B. pallens
B. poirettii 'BJG 073', 'MTA'
B. potanini
B. Renton
B. replicata
B. sanguinea
B. sargentiana
B. sikkimensis
B. soulieana 'Claret Cascade'
B. stenophylla
B. stenophylla diversifolia
B. stenophylla gracilis
B. stenophylla irwini
B. stenophylla nana compacta
B. taliensis
B. telomaica artisepala
B. thunbergii
B. thunbergii 'Ada'
B. thunbergii 'Admiration'
B. thunbergii 'Amera'
B. thunbergii 'Antares'
B. thunbergii argenteo marginata
B. thunbergii 'Arlene'
B. thunbergii atropurpurea
B. thunbergii atropurpurea erecta
B. thunbergii atropurpurea erecta Marshalli
B. thunbergii atropurpurea 'Golden Ring'
B. thunbergii atropurpurea 'Intermedia'
B. thunbergii atropurpurea 'Knight Burgundy'

B. thunbergii atropurpurea 'Moretti Select'
B. thunbergii atropurpurea nana
B. thunbergii atropurpurea 'Redbird'
B. thunbergii atropurpurea 'Rose Glow'
B. thunbergii atropurpurea x *B. x media* H2011–085–006
B. thunbergii aurea
B. thunbergii 'Aurea Nana'
B. thunbergii 'Bagatelle'
B. thunbergii 'BailAnna' Moscato
B. thunbergii 'BailElla' Lambrusco
B. thunbergii 'BailErin' Limoncello
B. thunbergii 'Bailgreen' (Jade Carousel ™)
B. thunbergii 'BailJulia' Toscana
B. thunbergii 'Bailone'
B. thunbergii 'Bailone' (Ruby Carousel ®)
B. thunbergii 'Bailtwo'
B. thunbergii 'Bailtwo' (Burgundy Carousel ®)
B. thunbergii 'Benita'
B. thunbergii 'Bonanza Gold'
B. thunbergii 'Breval 8'
B. thunbergii 'Celeste'
B. thunbergii 'Chloe'
B. thunbergii 'Concorde'
B. thunbergii 'Crimson Ruby'
B. thunbergii 'Crimson Pygmy'
B. thunbergii 'Criruzam' Crimson Ruby ™
B. thunbergii 'Daybreak'
B. thunbergii 'Diabolicum'
B. thunbergii 'Della'
B. thunbergii 'Dwarf Jewell'
B. thunbergii 'Edda'
B. thunbergii erecta
B. thunbergii 8–8–13
B. thunbergii 'Fay'
B. thunbergii 'Fireball'
B. thunbergii 'Gail'
B. thunbergii 'globe'
B. thunbergii 'golden'
B. thunbergii 'Golden Carpet'
B. thunbergii 'Golden Devine'
B. thunbergii 'Golden Pygmy'
B. thunbergii 'Golden Rocket'
B. thunbergii 'Golden Ruby'
B. thunbergii 'Golden Torch'
B. thunbergii 'Green Carpet'
B. thunbergii 'Grhozam' (Green Hornet ™)
B. thunbergii H2007–001–031
B. thunbergii 'Harlequin'
B. thunbergii 'Helen'
B. thunbergii 'Helmond Pillar'
B. thunbergii 'Joyce'

B. *thunbergii* 'Kasia'
B. *thunbergii* 'Kobold'
B. *thunbergii* 'Koren'
B. *thunbergii* 'Lime Glow'
B. *thunbergii* 'Lotty'
B. *thunbergii* 'Lustre Green'
B. *thunbergii* 'Maria'
B. *thunbergii* 'Martha'
B. *thunbergii maximowiczi*
B. *thunbergii* 'Midruzam' Midnight Ruby™
B. *thunbergii* 'Mimi'
B. *thunbergii minor*
B. *thunbergii* 'Monlers'
B. *thunbergii* 'Monomb'
B. *thunbergii* 'Monry'
B. *thunbergii* 'NCBT1'
B. *thunbergii* 'O'Byrne'
B. *thunbergii* 'Orange Rocket'
B. *thunbergii* Orange Torch
B. *thunbergii* 'Painter's Palette'
B. *thunbergii* 'Phoebe'
B. *thunbergii* 'Pink Queen'
B. *thunbergii pluriflora*
B. *thunbergii* 'Pow Wow'
B. *thunbergii* 'Pyruzam' (Pygmy Ruby™)
B. *thunbergii* 'Red Carpet'
B. *thunbergii* 'Red Rocket'
B. *thunbergii* Red Torch
B. *thunbergii* 'Rosy Rocket'
B. *thunbergii* 'Royal Burgundy'
B. *thunbergii* 'Royal Cloak'
B. *thunbergii* 'Ruth'
B. *thunbergii* 'Sparkle'
B. *thunbergii* 'Sparkler'
B. *thunbergii* Striking Gold
B. *thunbergii* 'Talago'
B. *thunbergii* 'Thornless'
B. *thunbergii* 'Tiny Gold'
B. *thunbergii* '24kagozam' (24 Karat Gold™)
B. *thunbergii* UCONNBT039
B. *thunbergii* UCONNBT048
B. *thunbergii* UCONNBT113
B. *thunbergii* UCONNBTCP4N
B. *thunbergii* UCONNtrispecific
B. *thunbergii* 'Upright Jewell'
B. *thunbergii* 'Velglozam' (Velvet Glow™)
B. *thunbergii variegata*
B. *thunbergii xanthocarpa*
B. *thunbergii x B. sieboldii* H2010-079-012
B. *thunbergii x calliantha* ''NCBX3'
B. *thunbergii x media* 'NCBX1'
B. *thunbergii x media* 'NCBX2'
B. *thunbergii* × 'Bailsel' (Golden Carousel®)

B. *thunbergii* × 'Tara' (Emerald Carousel®)
B. *triacanthophora*
B. *triculosa*
B. *verruculosa*
B. *verruculosa x gagnepainii x vulgaris* Trispecific#2
B. *virgatorum*
B. *workingensis*
B. *xanthoxylon*
B. *x media x thunbergii atropurpurea* H2011-165-002
B. × *carminea* 'Pirate King'
B. × *frikartii* 'Amstelveen'
(2) All plants, seedlings, seeds, fruits, and other plant parts capable of propagation from the following rust-resistant *Mahoberberis* and *Mahonia* species and varieties, except *Mahonia* cuttings for decorative purposes:
(i) Genus *Mahoberberis:*
M. *aqui-candidula*
M. *aquifolium* 'Smaragd'
M. *aqui-sargentiae*
M. *miethkeana*
M. × 'Magic'
(ii) Genus *Mahonia:*
M. *amplectens*
M. *aquifolium*
M. *aquifolium atropurpurea*
M. *aquifolium compacta*
M. *aquifolium compacta* 'John Muir'
M. *aquifolium* 'Donewell'
M. *aquifolium* 'Kings Ransom'
M. *aquifolium* 'Orangee Flame'
M. *aquifolium* 'Undulata'
M. *aquifolium* 'Winter Sun'
M. 'Arthur Menzies'
M. *bealei*
M. 'Bokasio' Sioux
M. 'Bokrafoot' Blackfoot
M. *dictyota*
M. *eurybracteata* Soft Caress (PPAF)
M. *fortunei*
M. 'Golden Abundance'
M. *japonica*
M. *japonica* × M. *lomariifolia* 'Charity'
M. *lomarifolia*
M. *media* Marvel (PPAF)
M. *nervosa*
M. *pinnata*
M. *pinnata* 'Ken Hartman'
M. *piperiana*
M. *pumila*
M. *repens*
M. × *media* 'Charity'
M. × *media* 'Lionel Fortescue'
M. × *media* 'Winter Sun'
(3) All plants, seeds, fruits, and other plant parts capable of propagation

from rust-susceptible species and varieties of the genera *Berberis*, *Mahoberberis*, and *Mahonia*, except *Mahonia* cuttings for decorative purposes.

(4) Any other product or article not listed in paragraphs (a)(1) through (a)(3) of this section that an inspector determines presents a risk of spread of black stem rust. The inspector must notify the person in possession of the product or article that it is subject to the provisions of this subpart.

(b) A person may request that an additional rust-resistant variety be added to paragraph (a)(1) or (a)(2) of this section. The person requesting that a rust-resistant variety be added to paragraph (a)(1) or (a)(2) of this section must provide APHIS with a description of the variety, including a written description and color pictures that can be used by an inspector to clearly identify the variety and distinguish it from other varities.

(Approved by the Office of Management and Budget under control number 0579–0186)

[67 FR 8179, Feb. 22, 2002, as amended at 71 FR 5778, Feb. 3, 2006; 72 FR 32167, June 12, 2007; 72 FR 72233, Dec. 20, 2007; 75 FR 29193, May 25, 2010; 75 FR 54462, Sept. 8, 2010; 78 FR 27856, May 13, 2013; 81 FR 3702, Jan. 22, 2016; 82 FR 41826, Sept. 5, 2017]

§ 301.38–3 **Protected areas.**

(a) The Administrator may designate as a protected area in paragraph (d) of this section any State that has eradicated rust-susceptible plants of the genera *Berberis*, *Mahoberberis*, and *Mahonia* under the cooperative Federal-State eradication program. In addition, the State must employ personnel with responsibility for the issuance and withdrawal of certificates in accordance with § 301.38–5, and maintain and enforce an inspection program under which every plant nursery within the State is inspected at least once each year to ensure that they are free of rust-susceptible plants. During the requisite nursery inspections, all nursery stock shall be examined to determine that it consists only of rust-resistant varieties of the genera *Berberis*, *Mahoberberis*, and *Mahonia*, and that the plants are true to type. Plants that do not meet this criteria must be destroyed.

(b) The Administrator may designate as a protected area any county within a State, rather than the entire State, if areas within the State have eradicated rust-susceptible plants of the genera *Berberis*, *Mahoberberis*, and *Mahonia* under the cooperative Federal-State program, and;

(1) The State employs personnel with responsibility for the issuance and withdrawal of certificates in accordance with § 301.38–5;

(2) The State is enforcing restrictions on the intrastate movement of the regulated articles that are equivalent to those imposed by this subpart on the interstate movement of regulated articles, as determined by the Administrator; and

(3) The State maintains and enforces an inspection program under which every plant nursery within the county is inspected at least once each year to ensure that plant nurseries within that area are free of rust-susceptible plants of the genera *Berberis*, *Mahoberberis*, and *Mahonia*. During the requisite nursery inspections, all nursery stock shall be examined to determine that it consists only of rust-resistant varieties of the genera *Berberis*, *Mahoberberis*, and *Mahonia*, and that the plants are true to type. Plants that do not meet this criteria must be destroyed.

(c) All seed used to propagate plants of the genera *Berberis*, *Mahoberberis*, and *Mahonia* in protected areas, and all seed used to propagate plants of genera *Berberis*, *Mahoberberis*, and *Mahonia* that are certified as rust-resistant for interstate movement into protected areas, must be produced at properties where a State inspector has verified that no wild or domesticated rust-susceptible plants are growing at or within one-half mile of the property.[4]

----

[4] Persons performing the inspections must be able to recognize rust-susceptible varieties of *Berberis*, *Mahoberberis*, and *Mahonia*. Inspectors must work side by side, 10 to 20 feet apart, and walk outward away from the property a distance of one-half mile measured from the edge of the property, and observe all plants growing in the half-mile band. The distance between the inspectors may vary within this range, depending upon the visibility of the plant growth. In areas

(d) The following are designated as protected areas:

(1) The States of Illinois, Indiana, Iowa, Kansas, Michigan, Minnesota, Missouri, Montana, Nebraska, North Dakota, Ohio, Pennsylvania, South Dakota, West Virginia, Wisconsin, and Wyoming.

(2) The following counties in the State of Washington: Adams, Asotin, Benton, Chelan, Columbia, Douglas, Ferry, Franklin, Garfield, Grant, Kittitas, Klickitat, Lincoln, Okanogan, Pend Oreille, Spokane, Stevens, Walla Walla, Whitman, Yakima.

(e) Each State that is a protected area or that encompasses a protected area must submit annually to the Administrator a written statement, signed by an inspector, assuring APHIS that all nursery inspections have been performed in accordance with this section. The statement must be submitted by January 1st of each year, and must include a list of the nurseries inspected and found free of rust-susceptible plants.

(f) The Administrator may remove a protected area from the list of designated protected areas in paragraph (d) of this section if he or she determines that it no longer meets the criteria of paragraph (a) or (b)(1) through (3) of this section. A hearing will be held to resolve any conflict as to any material fact. Rules of practice for the hearing shall be adopted by the Administrator.

[54 FR 32791, Aug. 10, 1989, as amended at 55 FR 29558, July 20, 1990; 57 FR 3118, Jan. 28, 1992; 71 FR 5778, Feb. 3, 2006]

_____

with low brush and flat terrain, the inspectors may be the maximum distance of 20 feet apart if they can observe all plants growing within 10 feet of them. In areas of high plant growth or hilly terrain, the inspectors must be closer together due to limited or obstructed visibility. Inspectors must observe all plants growing between themselves and the mid-point of the distance between themselves and the next inspector. This process must be repeated so that the entire band, measured from the border of the property to the circumference of an imaginary circle having the property as its mid-point, is visually inspected in this manner.

§ 301.38–4 Interstate movement of regulated articles.

(a) *Non-protected areas.* Interstate movement of regulated articles into or through any State or area that is not designated as a protected area under § 301.38–3(d) is allowed without restriction under this subpart.

(b) *Protected areas—*(1) *Prohibited movement.* The following regulated articles are prohibited from moving interstate into or through any protected area:

(i) All rust-susceptible *Berberis, Mahoberberis,* and *Mahonia* plants, seeds, fruits, and other plant parts capable of propagation, except Mahonia cuttings for decorative purposes.

(ii) All seed-propagated plants of the *Berberis* species and varieties designated as rust-resistant in § 301.38–2(a)(1) of this subpart that are of less than 2 years' growth, and any seeds, fruits, and other plant parts capable of propagation from such plants.

(2) *Restricted movement.* The following regulated articles may be moved interstate into or through a protected area with a certificate issued and attached in accordance with §§ 301.38–5 and 301.38–7 of this subpart:

(i) Seed-propagated plants of at least 2 years' growth, clonally propagated plants of any age, seeds, fruits, and other plant parts capable of propagation of the *Berberis* species and varieties designated as rust-resistant in § 301.38–2(a)(1) of this subpart;

(ii) Plants, seeds, fruits, and other plant parts capable of propagation of the *Mahoberberis* and *Mahonia* species and varieties designated as rust-resistant in § 301.38–2(a)(2) of this subpart.

(c) An inspector may issue a limited permit to allow a regulated article not eligible for certification under § 301.38–4(b)(2) to move interstate into or through a protected area to a specified destination that is stated in the permit and is outside the protected area, if the requirements of all other applicable Federal domestic plant quarantines are met. A regulated article moved interstate under a limited permit must be placed in a closed sealed container that prevents unauthorized removal of the regulated article, and that remains sealed until the regulated article reaches the final destination stated in the permit. At the final destination, the sealed container must be opened only in the presence of an inspector or with the authorization of an inspector obtained expressly for that shipment.

(d) The United States Department of Agriculture may move any regulated article interstate into or through a protected area in accordance with the conditions determined necessary to prevent the introduction or spread of black stem rust in protected areas, as specified in a Departmental permit issued for this purpose.

[54 FR 32791, Aug. 10, 1989, as amended at 67 FR 8180, Feb. 22, 2002; 71 FR 5778, Feb. 3, 2006]

§ 301.38–5 Assembly and inspection of regulated articles: issuance and cancellation of certificates.

(a) Any person, other than a person authorized to issue certificates under paragraph (c) of this section, who desires to move interstate a regulated article that must be accompanied by a certificate under § 301.38–4(b), shall, as far in advance of the desired interstate movement as possible (and no less than 48 hours before the desired interstate movement), request an inspector[5] to issue a certificate. To expedite the issuance of a certificate, an inspector may direct that the regulated articles be assembled in a manner that facilitates inspection.

(b) An inspector may issue a certificate for the interstate movement of a regulated article if he or she:

(1) Determines, upon examination, that the regulated article may be moved interstate in accordance with this subpart; and

(2) Determines that the regulated article may be moved interstate in accordance with all other Federal domestic plant quarantines and regulations applicable to the regulated article.

---

[5] Services of an inspector may be requested by contacting a local APHIS office (listed in telephone directories under Animal and Plant Health Inspection Service (APHIS), Plant Protection and Quarantine). The addresses and telephone numbers of local offices may also be obtained by writing to the Animal and Plant Health Inspection Service, Plant Protection and Quarantine, Domestic and Emergency Operations, 4700 River Road Unit 134, Riverdale, Maryland 20737–1236.

(c) Certificates for interstate movement of regulated articles may be issued by an inspector to a person operating under a compliance agreement for use with subsequent shipments of regulated articles to facilitate their movement. A person operating under a compliance agreement must make the determinations set forth in paragraph (b) of this section before shipping any regulated articles.

(d) Any certificate that has been issued may be withdrawn by an inspector, orally or in writing, if he or she determines that the holder of the certificate has not complied with the conditions of this subpart for the use of the certificate. If the withdrawal is oral, the inspector will confirm the withdrawal and the reasons for the withdrawal, in writing, within 20 days of oral notification of the withdrawal. Any person whose certificate has been withdrawn may appeal the decision, in writing within 10 days after receiving written notification of the withdrawal. The appeal must state all of the facts and reasons upon which the person relies to show that the certificate was wrongfully withdrawn. A hearing will be held to resolve any conflict as to any material fact. An appeal shall be granted or denied, in writing, as promptly as circumstances allow, and the reasons for the decision shall be stated. In a non-protected area, appeal shall be made to the Administrator. The Administrator shall adopt rules of practice for the hearing. The certificate will remain withdrawn pending decision of the appeal.

[54 FR 32791, Aug. 10, 1989, as amended at 59 FR 67608, Dec. 30, 1994; 67 FR 8180, Feb. 22, 2002]

§ 301.38–6 Compliance agreements and cancellation.

(a) Any State may enter into a written compliance agreement with any person who grows or handles regulated articles in a protected area, or moves interstate regulated articles from a protected area, under which that person agrees to comply with this subpart, to provide inspectors with information concerning the source of any regulated articles acquired each year, and to prevent the unauthorized use of certificates issued for future use under the compliance agreement.[6]

(b) A compliance agreement may be cancelled by an inspector, orally or in writing, whenever he or she determines that the person who has entered into the compliance agreement has failed to comply with the agreement or this subpart. If the cancellation is oral, the cancellation and the reasons for the cancellation will be confirmed, in writing, within 20 days of oral notification of the cancellation. Any person whose compliance agreement has been cancelled may appeal the decision, in writing, within 10 days after receiving written notification of the cancellation. The appeal must state all of the facts and reasons upon which the person relies to show that the compliance agreement was wrongfully cancelled. A hearing will be held to resolve any conflict as to any material fact. An appeal shall be granted or denied, in writing, as promptly as circumstances allow, and the reasons for the decision shall be stated. In a non-protected area, appeal shall be made to the Administrator. The Administrator shall adopt rules of practice for the hearing. The compliance agreement will remain cancelled pending decision of the appeal.

[54 FR 32791, Aug. 10, 1989; 54 FR 38494, Sept. 18, 1989, as amended at 57 FR 3118, Jan. 28, 1992; 59 FR 67608, Dec. 30, 1994]

§ 301.38-7 **Attachment and disposition of certificates.**

(a) The certificate required for the interstate movement of a regulated article must, at all times during the interstate movement, be attached to the outside of the container containing the regulated article except as follows:

(1) The certificate may be attached to the regulated article itself if it is not in container; or

---

[6]In non-protected areas, compliance agreements may be arranged by contacting a local office of the Animal and Plant Health Inspection Service (APHIS), Plant Protection and Quarantine, or by writing to the Animal and Plant Health Inspection Service, Plant Protection and Quarantine, Domestic and Emergency Operations, 4700 River Road Unit 134, Riverdale, Maryland 20737–1236.

(2) The certificate may be attached to the accompanying waybill or other shipping document if the regulated article is identified and described on the certificate or waybill.

(b) The carrier must furnish the certificate to the consignee at the destination of the regulated article.

§ 301.38-8 **Costs and charges.**

The services of an inspector[4] during normal business hours, Monday through Friday, 8 a.m. to 4:30 p.m., will be furnished without cost to persons requiring the services. The United States Department of Agriculture will not be responsible for any other costs or charges.

[54 FR 32791, Aug. 10, 1989; 54 FR 38494, Sept. 18, 1989]

Subpart E—Gypsy Moth

SOURCE: 58 FR 39423, July 23, 1993, unless otherwise noted. Redesignated at 84 FR 2428, Feb. 7, 2019.

§ 301.45 **Notice of quarantine; restriction on interstate movement of specified regulated articles.**

(a) *Notice of quarantine.* Pursuant to the provisions of , sections 411, 412, 414, 431, and 434 of the Plant Protection Act (7 U.S.C. 7711, 7712, 7714, 7751, and 7754), the Secretary of Agriculture hereby quarantines the States of Connecticut, Delaware, District of Columbia, Illinois, Indiana, Maine, Maryland, Massachusetts, Michigan, New Hampshire, New Jersey, New York, North Carolina, Ohio, Pennsylvania, Rhode Island, Vermont, Virginia, West Virginia, and Wisconsin in order to prevent the spread of the gypsy moth, *Lymantria dispar* (Linnaeus), a dangerous insect injurious to forests and shade trees and not theretofore widely prevalent or distributed within or throughout the United States; and establishes regulations governing the interstate movement from generally infested areas of the quarantined States of regulated articles and outdoor household articles defined in § 301.45–1.

(b) *Restrictions on the interstate movement of regulated articles and outdoor household articles.* No common carrier or other person may move interstate from any generally infested area any regulated article or outdoor household article except in accordance with the conditions prescribed in this subpart.

[58 FR 39423, July 23, 1993, as amended at 62 FR 29287, May 30, 1997; 63 FR 38280, July 16, 1998; 66 FR 21050, Apr. 27, 2001; 66 FR 37114, July 17, 2001]

§ 301.45–1 Definitions.

Terms used in the singular form in this subpart shall be construed as the plural, and vice versa, as the case may demand. The following terms, when used in this subpart, shall be construed, respectively, to mean:

Administrator. The Administrator, Animal and Plant Health Inspection Service, or any person authorized to act for the Administrator.

Animal and Plant Health Inspection Service. The Animal and Plant Health Inspection Service of the U.S. Department of Agriculture (APHIS).

Associated equipment. Articles associated and moved with mobile homes and recreational vehicles, such as, but not limited to, awnings, tents, outdoor furniture, trailer blocks, and trailer skirts.

Bark. The tough outer covering of the woody stems of trees, shrubs, and other woody plants as distinguished from the cambium and inner wood.

Bark products. Products containing pieces of bark including bark chips, bark nuggets, bark mulch, and bark compost.

Certificate. A Plant Protection and Quarantine-approved form, stamp, or document issued and signed by an inspector, or by a qualified certified applicator or by any other person operating in accordance with a compliance agreement, affirming that a specified regulated article is eligible for interstate movement in accordance with this subpart.

Compliance agreement. A written agreement between APHIS and a person engaged in growing, handling, or moving regulated articles, in which the person agrees to comply with the provisions of this subpart.

Effectively diminishing. An eradication program is considered to be effectively diminishing the gypsy moth population of an area if the results of two successive annual Federal or State delimiting trapping surveys of the area conducted in accordance with Section II, "Survey Procedures—Gypsy Moth," of the Gypsy Moth Treatment Manual show that the average number of gypsy moths caught per trap in the second delimiting survey (when comparable geographical areas and trapping densities are used) is: (1) Less than 10, and (2) less than the average number of gypsy moths caught per trap in the first survey.

Eradication program. A program that uses pesticide application, biological controls, or other methods with the goal of eliminating gypsy moth from a particular area.

General infestation. (1) The detection of gypsy moth egg masses through visual inspection by an inspector during a 10-minute walk through the area; however, it does not include the presence of gypsy moth egg masses which are found as a result of hitchhiking on transitory means of conveyance; or

(2) The detection of gypsy moth through multiple catches of adult gypsy moths at multiple trapping locations in the area over a period of 2 or more consecutive years, if the Administrator determines, after consulting with the State plant regulatory official, that gypsy moth is established in the area.

Generally infested area. Any State, or portion thereof, listed as a generally infested area in § 301.45–3 or temporarily designated as a generally infested area in accordance with § 301.45–2(c).

Gypsy moth. The live insect known as the gypsy moth, *Lymantria dispar* (Linnaeus), in any life stage (egg, larva, pupa, adult).

Inspector. Any employee of APHIS, a State government, or any other person, authorized by the Administrator in accordance with law to enforce the provisions of the quarantine and regulations in this subpart. A person operating under a compliance agreement is not an inspector.

Interstate. From any State into or through any other State.

Limited permit. A document in which an inspector or a person operating under a compliance agreement affirms that the regulated article identified on the document is eligible for interstate movement in accordance with § 301.45–5 only to the specified destination and only in accordance with the specified conditions.

Mobile home. Any vehicle, other than a recreational vehicle, designed to serve, when parked, as a dwelling or place of business.

Move (movement, moved). Shipped, offered for shipment to a common carrier, received for transportation or transported by a common carrier, or carried, transported, moved, or allowed to be moved by any means. "Movement" and "moved" shall be construed in accordance with this definition.

OHA document. The self-inspection checklist portion of USDA–APHIS Program Aid Number 2065, "Don't Move Gypsy Moth," completed and signed by the owner of an outdoor household article (OHA) affirming that the owner has inspected the OHA for life stages of gypsy moth in accordance with the procedures in the program aid.

Outdoor household articles. Articles associated with a household that have been kept outside the home such as awnings, barbecue grills, bicycles, boats, dog houses, firewood, garden tools, hauling trailers, outdoor furniture and toys, recreational vehicles and associated equipment, and tents.

Person. Any individual, partnership, corporation, company, society, association, or other organized group.

Qualified certified applicator. Any individual (1) certified pursuant to the Federal Insecticide, Fungicide, and Rodenticide Act (FIFRA) (7 U.S.C. 136i) as a certified commercial applicator in a category allowing use of the restricted use pesticides Spray N Kill (EPA Registration No. 8730–30), Ficam W (EPA Registration No. 45639–1), and acephate (Orthene®); (2) who has attended and completed a workshop approved by the Administrator on the identification and treatment of gypsy moth life stages on outdoor household articles and mobile homes; and (3) who has entered into a compliance agreement in accordance with §301.45–6 of this part for the purpose of inspecting, treating, and issuing certificates for

the movement of outdoor household articles and mobile homes.[1]

Recreational vehicles. Highway vehicles, including pickup truck campers, one-piece motor homes, and travel trailers, designed to serve as temporary places of dwelling.

Regulated articles. (1) Trees without roots (e.g., Christmas trees), trees with roots, and shrubs with roots and persistent woody stems, unless they are greenhouse grown throughout the year.

(2) Logs, pulpwood, and bark and bark products.

(3) Mobile homes and associated equipment.

(4) Any other products, articles, or means of conveyance, of any character whatsoever, when it is determined by an inspector that any life stage of gypsy moth is in proximity to such articles and the articles present a high risk of artificial spread of gypsy moth infestation and the person in possession thereof has been so notified.

State. Any State, Territory, or District of the United States including Puerto Rico.

Treatment manual. The provisions currently contained in the Gypsy Moth Program Manual.[2]

Under the direction of. Monitoring treatments to assure compliance with the requirements in this subpart.

Under the direct supervision of a qualified certified applicator. An inspection or treatment is considered to be applied under the direct supervision of a qualified certified applicator if the inspection or treatment is performed by a person acting under the instructions of a qualified certified applicator who is available if and when needed, even though such qualified certified applicator is not physically present at the time and place the inspection or treatment occurred.

[58 FR 39423, July 23, 1993, as amended at 59 FR 67608, Dec. 30, 1994; 67 FR 8464, Feb. 25, 2002; 70 FR 33268, June 7, 2005; 71 FR 40878, July 19, 2006; 72 FR 70764, Dec. 13, 2007; 78 FR 24666, Apr. 26, 2013]

---

[1] Names of qualified certified applicators may be obtained from State departments of agriculture.

[2] The Gypsy Moth Program Manual may be viewed on the Internet at *http:// www.aphis.usda.gov/import_export/plants/ manuals/domestic/downloads/gypsy_moth.pdf.*

§ 301.45–2 Authorization to designate and terminate designation of generally infested areas.

(a) *Generally infested areas.* The Administrator shall list as generally infested areas in § 301.45–3 each State or each portion thereof in which a gypsy moth general infestation has been found by an inspector, or each portion of a State which the Administrator deems necessary to regulate because of its proximity to infestation or its inseparability for quarantine enforcement purposes from infested localities; Except that, an area shall not be listed as a generally infested area if the Administrator has determined that:

(1) The area is subject to a gypsy moth eradication program conducted by the Federal government or a State government in accordance with the Eradication, Suppression, and Slow the Spread alternative of the Final Environmental Impact Statement (FEIS) on Gypsy Moth Suppression and Eradication Projects that was filed with the United States Environmental Protection Agency on January 16, 1996; and,

(2) State or Federal delimiting trapping surveys conducted in accordance with Section II, "Survey Procedures—Gypsy Moth" of the Gypsy Moth Treatment Manual show that the average number of gypsy moths caught per trap is less than 10 and that the trapping surveys show that the eradication program is effectively diminishing the gypsy moth population of the area.

(b) Less than an entire State will be designated as a generally infested area only if the Administrator has determined that:

(1) The State has adopted and is enforcing a quarantine or regulation which imposes restrictions on the intrastate movement of the regulated articles which are substantially the same as those which are imposed with respect to the interstate movement of such articles under this subpart; and,

(2) The designation of less than the entire State as a generally infested area will be adequate to prevent the artificial interstate spread of infestations of the gypsy moth.

(c) *Temporary designation of areas as generally infested areas.* The Administrator or an inspector may temporarily designate any area in any State as a generally infested area in accordance with the criteria specified in paragraph (a) of this section. An inspector will give written notice of the designation to the owner or person in possession of the area and thereafter, the interstate movement of any regulated article from such areas is subject to the applicable provisions of this subpart. As soon as practicable, each generally infested area will be added to the list in § 301.45–3 or the designation will be terminated by the Administrator or an authorized inspector, and notice thereof shall be given to the owner or person in possession of the areas.

(d) *Termination of designation as a generally infested area.* The Administrator shall terminate the designation of any area as a generally infested area whenever the Administrator determines that the area no longer requires designation under the criteria specified in paragraph (a) of this section.

[58 FR 39423, July 23, 1993, as amended at 72 FR 70764, Dec. 13, 2007]

§ 301.45–3 Generally infested areas.

(a) The areas described below are designated as generally infested areas:

CONNECTICUT

The entire State.

DELAWARE

The entire State.

DISTRICT OF COLUMBIA

The entire district.

ILLINOIS

Cook County. The entire county.
Du Page County. The entire county.
Lake County. The entire county.
McHenry County. The entire county.

INDIANA

Allen County. The entire county.
De Kalb County. The entire county.
Elkhart County. The entire county.
LaGrange County. The entire county.
LaPorte County. The entire county.
Noble County. The entire county.
Porter County. The entire county.
St. Joseph County. The entire county.
Steuben County. The entire county.

MAINE

Androscoggin County. The entire county.

Aroostook County. The townships of Amity, Bancroft, Benedicta, Cary Plantation, Crystal, Dyer Brook, Forkstown, Glenwood Plantation, Haynesville, Hodgdon, Houlton, Island Falls, Linneus, Macwahoc Plantation, Molunkus, New Limerick, North Yarmouth Academy Grant, Oakfield, Orient, Reed Plantation, Sherman, Silver Ridge, Upper Molunkus, Weston, T1 R5 WELS, T2 R4 WELS, T3 R3 WELS, T3 R4 WELS, T4 R3 WELS, and TA R2 WELS.

Cumberland County. The entire county.

Franklin County. Avon, Carthage, Chesterville, Coplin Plantation, Crockertown, Dallas Plantation, Davis, Eustis, Farmington, Freeman, Industry, Jay, Jerusalem, Kingfield, Lang, Madrid, Mount Abraham, New Sharon, New Vineyard, Perkins, Phillips, Rangeley, Rangeley Plantation, Redington, Salem, Sandy River Plantation, Strong, Temple, Township 6 North of Weld, Township D, Township E, Washington, Weld, Wilton, and Wyman.

Hancock County. The entire county.

Kennebec County. The entire county.

Knox County. The entire county.

Lincoln County. The entire county.

Oxford County. The townships of Adamstown, Albany, Andover, Andover North, Andover West, Batchelders Grant, Bethel, Brownfield, Buckfield, Byron, Canton, Denmark, Dixfield, Fryeburg, Gilead, Grafton, Greenwood, Hanover, Hartford, Hebron, Hiram, Lincoln Plantation, Lovell, Lower Cupsuptic, Magalloway Plantation, Mason Plantation, Mexico, Milton Plantation, Newry, Norway, Oxford, Paris, Parkerstown, Peru, Porter, Richardsontown, Riley, Roxbury, Rumford, Stoneham, Stow, Sumner, Sweden, Upton, Waterford, Woodstock, C, and C Surplus.

Penobscot County. The townships of Alton, Argyle, Bangor City, Bradford, Bradley, Brewer City, Burlington, Carmel, Carroll Plantation, Charleston, Chester, Clifton, Corinna, Corinth, Dexter, Dixmont, Drew Plantation, E. Millinocket, Eddington, Edinburg, Enfield, Etna, Exeter, Garland, Glenburn, Grand Falls Plantation, Greenbush, Greenfield, Grindstone, Hampden, Hermon, Hersey Town, Holden, Hopkins Academy Grant, Howland, Hudson, Indian Purchase, Kenduskeag, Kingman, Lagrange, Lakeville, Lee, Levant, Lincoln, Long A, Lowell, Mattamiscontis, Mattawamkeag, Maxfield, Medway, Milford, Millinocket, Mount Chase, Newburgh, Newport, Old Town City, Orono, Orrington, Passadumkeag, Patten, Plymouth, Prentiss Plantation, Seboeis Plantation, Soldiertown, Springfield, Stacyville, Stetson, Summit, Veazie, Veazie Gore, Webster Plantation, Winn, Woodville, T1 ND, T1 R6 WELS, T1 R8 WELS, T2 R8 NWP, T2 R8 WELS, T2 R9 NWP, T3 R1 NBPP, T3 R9 NWP, T5 R1 NBPP, T5 R8 WELS, T6 R8 WELS, TA R7, TA R8, TA R9, and the portion of T3 R8 within the boundaries of Baxter State Park.

Piscataquis County. The townships of Abbot, Atkinson, Barnard, Blanchard Plantation, Bowerbank, Brownville, Dover-Foxcroft, Elliotsville, Greenville, Guilford, Katahdin Iron Works, Kingsbury Plantation, Lakeview Plantation, Medford, Milo, Monson, Mount Katahdin, Nesourdnahunk, Orneville, Parkman, Sangerville, Sebec, Shirley, Trout Brook, Wellington, Williamsburg, Willimantic, T1 R9 WELS, T1 R10 WELS, T1 R11 WELS, T2 R10 WELS, T2 R9 WELS, T3 R10 WELS, T4 R9 NWP, T4 R9 WELS, T5 R9 NWP, T5 R9 WELS, T6 R10 WELS, T7 R9 NWP, TA R10 WELS, TA R11 WELS, TB R10 WELS, TB R11 WELS, and the portion of T4 R10 WELS within the boundaries of Baxter State Park.

Sagadahoc County. The entire county.

Somerset County. The townships of Anson, Athens, Bald Mountain, Bigelow, Bingham, Bowtown, Brighton Plantation, Cambridge, Canaan, Caratunk, Carrying Place, Carrying Place Town, Concord Plantation, Cornville, Dead River, Detroit, East Moxie, Embden, Fairfield, Harmony, Hartland, Highland Plantation, Lexington Plantation, Lower Enchanted, Madison, Mayfield, Mercer, Moscow, Moxie Gore, New Portland, Norridgewock, Palmyra, Pierce Pond,

Pittsfield, Pleasant Ridge Plantation, Ripley, Skowhegan, Smithfield, Solon, St. Albans, Starks, The Forks Plantation, West Forks Plantation, and T3 R4 BKP WKR.

Waldo County. The entire county.
Washington County. The entire county.
York County. The entire county.

MARYLAND

The entire State.

MASSACHUSETTS

The entire State.

MICHIGAN

The entire State.

MINNESOTA

Cook County. The entire county.
Lake County. The entire county.

NEW HAMPSHIRE

The entire State.

NEW JERSEY

The entire State.

NEW YORK

The entire State.

NORTH CAROLINA

Currituck County. The entire county.
Dare County. The area bounded by a line beginning at the intersection of State Road 1208 and Roanoke Sound; then easterly along this road to its junction with State Road 1206; then southerly along this road to its intersection with U.S. Highway Business 158; then easterly along an imaginary line to its intersection with the Atlantic Ocean; then northwesterly along the coastline to its intersection with the Dare-Currituck County line; then westerly along this county line to its intersection with the Currituck Sound; then southeasterly along this sound to the point of beginning.

OHIO

Ashland County. The entire county.
Ashtabula County. The entire county.
Athens County. The entire county.
Belmont County. The entire county.
Carroll County. The entire county.

Columbiana County. The entire county.
Coshocton County. The entire county.
Crawford County. The entire county.
Cuyahoga County. The entire county.
Defiance County. The entire county.
Delaware County. The entire county.
Erie County. The entire county.
Fairfield County. The entire county.
Franklin County. The entire county.
Fulton County. The entire county.
Geauga County. The entire county.
Guernsey County. The entire county.
Harrison County. The entire county.
Henry County. The entire county.
Hocking County. The entire county.
Holmes County. The entire county.
Huron County. The entire county.
Jefferson County. The entire county.
Knox County. The entire county.
Lake County. The entire county.
Licking County. The entire county.
Lorain County. The entire county.
Lucas County. The entire county.
Mahoning County. The entire county.
Marion County. The entire county.
Medina County. The entire county.
Monroe County. The entire county.
Morgan County. The entire county.
Morrow County. The entire county.
Muskingum County. The entire county.
Noble County. The entire county.
Ottawa County. The entire county.
Perry County. The entire county.
Portage County. The entire county.
Richland County. The entire county.
Sandusky County. The entire county.
Seneca County. The entire county.
Stark County. The entire county.
Summit County. The entire county.
Trumbull County. The entire county.
Tuscarawas County. The entire county.
Vinton County. The entire county.
Washington County. The entire county.
Wayne County. The entire county.
Williams County. The entire county.
Wood County. The entire county.

PENNSYLVANIA

The entire State.

RHODE ISLAND

The entire State.

VERMONT

The entire State.

VIRGINIA

City of Alexandria. The entire city.

City of Bedford. The entire city.
City of Buena Vista. The entire city.
City of Charlottesville. The entire city.
City of Chesapeake. The entire city.
City of Colonial Heights. The entire city.
City of Covington. The entire city.
City of Danville. The entire city.
City of Emporia. The entire city.
City of Fairfax. The entire city.
City of Falls Church. The entire city.
City of Franklin. The entire city.
City of Fredericksburg. The entire city.
City of Hampton. The entire city.
City of Harrisonburg. The entire city.
City of Hopewell. The entire city.
City of Lexington. The entire city.
City of Lynchburg. The entire city.
City of Manassas. The entire city.
City of Manassas Park. The entire city.
City of Newport News. The entire city.
City of Norfolk. The entire city.
City of Petersburg. The entire city.
City of Poquoson. The entire city.
City of Portsmouth. The entire city.
City of Radford. The entire city.
City of Richmond. The entire city.
City of Roanoke. The entire city.
City of Salem. The entire city.
City of South Boston. The entire city.
City of Staunton. The entire city.
City of Suffolk. The entire city.
City of Virginia Beach. The entire city.
City of Waynesboro. The entire city.
City of Williamsburg. The entire city.
City of Winchester. The entire city.
Accomack County. The entire county.
Albemarle County. The entire county.
Alleghany County. The entire county.
Amelia County. The entire county.
Amherst County. The entire county.
Appomattox County. The entire county.
Arlington County. The entire county.
Augusta County. The entire county.
Bath County. The entire county.
Bedford County. The entire county.
Bland County. The entire county.
Botetourt County. The entire county.
Brunswick County. The entire county.
Buckingham County. The entire county.
Campbell County. The entire county.
Caroline County. The entire county.
Charles City County. The entire county.
Charlotte County. The entire county.
Chesterfield County. The entire county.
Clarke County. The entire county.
Craig County. The entire county.
Culpeper County. The entire county.
Cumberland County. The entire county.
Dinwiddie County. The entire county.

Essex County. The entire county.
Fairfax County. The entire county.
Fauquier County. The entire county.
Floyd County. The entire county.
Fluvanna County. The entire county.
Franklin County. The entire county.
Frederick County. The entire county.
Giles County. The entire county.
Gloucester County. The entire county.
Goochland County. The entire county.
Greene County. The entire county.
Greensville County. The entire county.
Halifax County The entire county.
Hanover County. The entire county.
Henrico County. The entire county.
Highland County. The entire county.
Isle of Wight County. The entire county.
James City County. The entire county.
King and Queen County. The entire county.
King George County. The entire county.
King William County. The entire county.
Lancaster County. The entire county.
Loudoun County. The entire county.
Louisa County. The entire county.
Lunenburg County. The entire county.
Madison County. The entire county.
Mathews County. The entire county.
Mecklenburg County The entire county.
Middlesex County. The entire county.
Montgomery County. The entire county.
Nelson County. The entire county.
New Kent County. The entire county.
Northampton County. The entire county.
Northumberland County. The entire county.
Nottoway County. The entire county.
Orange County. The entire county.
Page County. The entire county.
Pittsylvania County. The entire county.
Powhatan County. The entire county.
Prince Edward County. The entire county.
Prince George County. The entire county.
Prince William County. The entire county.
Pulaski County. The entire county.
Rappahannock County. The entire county.
Richmond County. The entire county.
Roanoke County. The entire county.
Rockbridge County. The entire county.
Rockingham County. The entire county.
Shenandoah County. The entire county.
Southampton County. The entire county.

Spotsylvania County. The entire county.
Stafford County. The entire county.
Surry County. The entire county.
Sussex County. The entire county.
Tazewell County. The entire county.
Warren County. The entire county.
Westmoreland County. The entire county.
York County. The entire county.

WEST VIRGINIA

Barbour County. The entire county.
Berkeley County. The entire county.
Braxton County. The entire county.
Brooke County. The entire county.
Calhoun County. The entire county.
Doddridge County. The entire county.
Fayette County. The entire county.
Gilmer County. The entire county.
Grant County. The entire county.
Greenbrier County. The entire county.
Hampshire County. The entire county.
Harrison County. The entire county.
Hancock County. The entire county.
Hardy County. The entire county.
Jackson County. The entire county.
Jefferson County. The entire county.
Lewis County. The entire county.
Marion County. The entire county.
Marshall County. The entire county.
McDowell County. The entire county.
Mercer County. The entire county.
Mineral County. The entire county.
Monongalia County. The entire county.
Monroe County. The entire county.
Morgan County. The entire county.
Nicholas County. The entire county.
Ohio County. The entire county.
Pendleton County. The entire county.
Pleasants County. The entire county.
Pocahontas County. The entire county.
Preston County. The entire county.
Raleigh County. The entire county.
Randolph County. The entire county.
Ritchie County. The entire county.
Summers County. The entire county.
Taylor County. The entire county.
Tucker County. The entire county.
Tyler County. The entire county.
Upshur County. The entire county.
Webster County. The entire county.
Wetzel County. The entire county.
Wirt County. The entire county.
Wood County. The entire county.
Wyoming County. The entire county.

WISCONSIN

Adams County. The entire county.
Ashland County. The entire county.
Bayfield County. The entire county.
Brown County. The entire county.
Calumet County. The entire county.
Clark County. The entire county.
Columbia County. The entire county.
Dane County. The entire county.
Dodge County. The entire county.
Door County. The entire county.
Florence County. The entire county.
Fond du Lac County. The entire county.
Forest County. The entire county.
Green Lake County. The entire county.
Iowa County. The entire county.
Iron County. The entire county.
Jackson County. The entire county.
Jefferson County. The entire county.
Juneau County. The entire county.
Kenosha County. The entire county.
Kewaunee County. The entire county.
Langlade County. The entire county.
Lincoln County. The entire county.
Manitowoc County. The entire county.
Marathon County. The entire county.
Marinette County. The entire county.
Marquette County. The entire county.
Menominee County. The entire county.
Milwaukee County. The entire county.
Monroe County. The entire county.
Oconto County. The entire county.
Oneida County. The entire county.
Outagamie County. The entire county.
Ozaukee County. The entire county.
Portage County. The entire county.
Price County. The entire county.
Racine County. The entire county.
Rock County. The entire county.
Sauk County. The entire county.
Shawano County. The entire county.
Sheboygan County. The entire county.
Vilas County. The entire county.
Walworth County. The entire county.
Washington County. The entire county.
Waukesha County. The entire county.
Waupaca County. The entire county.
Waushara County. The entire county.
Winnebago County. The entire county.
Wood County. The entire county.

[58 FR 39423, July 23, 1993]

EDITORIAL NOTE: For FEDERAL REGISTER citations affecting § 301.45–3, see the List of CFR Sections Affected, which appears in the Finding Aids section of the printed volume and at *www.govinfo.gov.*

§ 301.45-4 Conditions governing the interstate movement of regulated articles and outdoor household articles from generally infested areas.

(a) Regulated articles and outdoor household articles from generally infested areas. (1) A regulated article, except for an article moved in accordance with paragraph (c) of this section, shall not be moved interstate from any generally infested area into or through any area that is not generally infested unless a certificate or permit has been issued and attached to such regulated article in accordance with §§ 301.45-5 and 301.45-8. [3]

(2) An outdoor household article shall not be moved interstate from any generally infested area into or through any area that is not generally infested unless a certificate or OHA document has been issued and attached to such outdoor household article in accordance with §§ 301.45-5 and 301.45-8.

(b) A regulated article originating outside of any generally infested area may be moved interstate directly through any generally infested area without a certificate or permit if the point of origin of the article is clearly indicated by shipping documents, its identity has been maintained, and it has been safeguarded against infestation while in any generally infested area during the months of April through August.

(c) A regulated article originating in a generally infested area may be moved interstate from a generally infested area without a certificate if it complies with (1) or (2) of this paragraph:

(1) The article is moved by the U.S. Department of Agriculture for experimental or scientific purposes, and:

(i) Is moved pursuant to a permit issued for each article by the Administrator;

(ii) Is moved in accordance with conditions specified on the permit and found by the Administrator to be adequate to prevent the dissemination of the gypsy moth, i.e., conditions of treatment, processing, shipment, and disposal; and

[3] Requirements under all other applicable Federal domestic plant quarantines must also be met.

(iii) Is moved with a tag or label securely attached to the outside of the container containing the article or securely attached to the article itself if not in a container, and with such tag or label bearing a permit number corresponding to the number of the permit issued for such article.

(2) The article is logs, pulpwood, or bark and bark products, and the person moving the article has attached a signed accurate statement to the waybill or other shipping documents accompanying the article stating that he or she has inspected the article in accordance with the Gypsy Moth Program Manual no more than 5 days prior to the date of movement and has found no life stages of gypsy moth on the article.

[58 FR 39423, July 23, 1993, as amended at 70 FR 33268, June 7, 2005; 71 FR 40878, July 19, 2006; 72 FR 70764, Dec. 13, 2007; 80 FR 12917, Mar. 12, 2015]

§ 301.45-5 Issuance and cancellation of certificates, limited permits, and outdoor household article documents.

(a) A certificate may be issued by an inspector for the movement of a regulated article or an outdoor household article (OHA) if the inspector determines that it is eligible for certification for movement to any destination under all Federal domestic plant quarantines applicable to such article and:

(1) It has originated in noninfested premises in a generally infested area and has not been exposed to the gypsy moth while within the generally infested area; or

(2) The inspector inspects the article no more than 5 days prior to the date of movement during the months of April through August (14 days prior to the date of movement from September through March) and finds it to be free of the gypsy moth; or

(3) It has been treated under the direction of an inspector to destroy the gypsy moth in accordance with the treatment manual and part 305 of this chapter; or

(4) It has been grown, produced, manufactured, stored, or handled in such a manner that no infestation would be

transmitted thereby as determined by an inspector.

(b) Limited permits may be issued by an inspector to allow interstate movement of any regulated article under this subpart to specified destinations for specified handling, utilization, processing, or treatment in accordance with the treatment manual, when, upon evaluation of all of the circumstances involved in each case, the Administrator determines that such movement will not result in the spread of the gypsy moth because life stages of the moths will be destroyed by such specified handling, utilization, processing or treatment, or the pest will not survive in areas to which shipped, and the requirements of all other applicable Federal domestic plant quarantines have been met.

(c) Certificate and limited permit forms may be issued by an inspector to any person for use for subsequent shipments of regulated articles provided the person is operating under a compliance agreement. Any person operating under a compliance agreement may reproduce the forms as needed to attach them to regulated articles moved under a compliance agreement. Any person operating under a compliance agreement may execute and issue the certificate forms or reproduction of such forms, for the interstate movement of regulated articles from the premises of such person identified in the compliance agreement, if the person has treated such regulated articles as specified in the compliance agreement, and if the regulated articles are eligible for certification for movement to any destination under all applicable Federal domestic plant quarantines. Any person operating under a compliance agreement may execute and issue the limited permit forms, or reproductions of such forms, for the interstate movement of regulated articles to specified destinations when an inspector has made the determinations specified in paragraph (b) of this section.

(d) A certificate may be issued by a qualified certified applicator for the interstate movement of any outdoor household article or mobile home if such qualified certified applicator determines the following:

(1) That the article has been inspected by the qualified certified applicator and found to be free of any life stage of the gypsy moth; or

(2) That the article has been treated by, or treated under the direct supervision of, the qualified certified applicator to destroy any life stage of the gypsy moth in accordance with methods and procedures prescribed in section III of the Gypsy Moth Program Manual.

(e) An OHA document may be issued by the owner of an outdoor household article for the interstate movement of the article if such person has inspected the outdoor household article and has found it to be free of any life stage of gypsy moth.

(f) Any certificate or permit which has been issued or authorized may be withdrawn by an inspector if he determines that the holder thereof has not complied with any condition for the use of such document. The reasons for the withdrawal shall be confirmed in writing as promptly as circumstances permit. Any person whose certificate or permit has been withdrawn may appeal the decision in writing to the Administrator within ten (10) days after receiving the written notification of the withdrawal. The appeal shall state all of the facts and reasons upon which the person relies to show that the certificate or permit was wrongfully withdrawn. The Administrator shall grant or deny the appeal, in writing, stating the reasons for his decision as promptly as circumstances permit. If there is a conflict as to any material fact, a hearing shall be held to resolve such conflict. Rules of practice concerning such a hearing will be adopted by the Administrator.

(Approved by the Office of Management and Budget under control number 0579–0088)

[58 FR 39423, July 23, 1993, as amended at 59 FR 46902, Sept. 13, 1994; 70 FR 33268, June 7, 2005]

§ 301.45–6 Compliance agreement and cancellation thereof.

(a) Any person engaged in the business of growing, handling, or moving regulated articles may enter into a compliance agreement to facilitate the movement of such articles under this subpart. Qualified certified applicators must enter into compliance agreements, in accordance with the definition of qualified certified applicator in § 301.45–1. A compliance agreement shall specify safeguards necessary to prevent spread of the gypsy moth, such as disinfestation practices or application of chemical materials in accordance with the treatment manual and part 305 of this chapter. Compliance agreement forms may be obtained from the Administrator or an inspector.

(b) Any compliance agreement may be canceled by the inspector who is supervising its enforcement, orally or in writing, whenever the inspector finds that such person has failed to comply with the conditions of the agreement. If the cancellation is oral, the decision and the reasons therefore shall be confirmed in writing, as promptly as circumstances permit. Any person whose compliance agreement has been canceled may appeal the decision in writing to the Administrator within ten (10) days after receiving written notification of the cancellation. The appeal shall state all of the facts and reasons upon which the person relies to show that the compliance agreement was wrongfully canceled. The Administrator shall grant or deny the appeal, in writing, stating the reasons for such decision, as promptly as circumstances permit. If there is a conflict as to any material fact, a hearing shall be held to resolve such conflict. Rules of practice concerning such a hearing will be adopted by the Administrator.

(Approved by the Office of Management and Budget under control number 0579–0088)

[58 FR 39423, July 23, 1993, as amended at 59 FR 46902, Sept. 13, 1994; 70 FR 33268, June 7, 2005]

§ 301.45–7 Assembly and inspection of regulated articles and outdoor household articles.

Persons (other than those authorized to use certificates or limited permits, or reproductions thereof, under § 301.45–5(c)) who desire to move interstate a regulated article which must be accompanied by a certificate or permit shall, at least 7 days before the desired movement, request an inspector to examine the article prior to movement. Persons who desire to move interstate an outdoor household article accompanied by a certificate issued in accordance with § 301.45–5 shall, at least 14 days before the desired movement, request an inspector to examine the article prior to movement. Persons who desire to move interstate an outdoor household article or a mobile home accompanied by a certificate issued by a qualified certified applicator in accordance with § 301.45–5(d) shall request a qualified certified applicator to examine the article prior to movement. Such articles shall be assembled at such point and in such manner as the inspector or qualified certified applicator designates to facilitate inspection. An owner who wants to move outdoor household articles interstate may self-inspect the articles and issue an OHA document in accordance with § 301.45–5(e).

[58 FR 39423, July 23, 1993, as amended at 72 FR 70764, Dec. 13, 2007]

§ 301.45–8 Attachment and disposition of certificates, limited permits, and outdoor household article documents.

(a) A certificate, limited permit, or OHA document required for the interstate movement of a regulated article or outdoor household article must at all times during such movement be securely attached to the outside of the container containing the regulated article or outdoor household article, securely attached to the article itself if not in a container, or securely attached to the consignee's copy of the waybill or other shipping document: Provided, however, That the requirements of this section may be met by attaching the certificate, limited permit, or OHA document to the consignee's copy of the waybill or other shipping document only if the regulated article or outdoor household article is sufficiently described on the certificate, limited permit, OHA document or shipping document to identify such article.

(b) The certificate, limited permit, or OHA document for the movement of a regulated article or outdoor household article shall be furnished by the carrier to the consignee at the destination of the shipment.

(c) Any qualified certified applicator who issues a certificate or OHA document shall at the time of issuance send

41

a copy of the certificate or OHA document to the APHIS State Plant Health Director for the State in which the document is issued.

(Approved by the Office of Management and Budget under control number 0579–0088)

[58 FR 39423, July 23, 1993, as amended at 59 FR 46902, Sept. 13, 1994; 72 FR 70764, Dec. 13, 2007]

§ 301.45–9 Inspection and disposal of regulated articles and pests.

Any properly identified inspector is authorized to stop and inspect, and to seize, destroy, or otherwise dispose of, or require disposal of regulated articles, outdoor household articles, and gypsy moths as provided in sections 414, 421, and 434 of the Plant Protection Act (7 U.S.C. 7714, 7731, and 7754).

[58 FR 39423, July 23, 1993, as amended at 66 FR 21050, Apr. 27, 2001]

§ 301.45–10 Movement of live gypsy moths.

Regulations requiring a permit for, and otherwise governing the movement of, live gypsy moths in interstate or foreign commerce are contained in the Federal Plant Pest Regulations in part 330 of this chapter.

§ 301.45–11 Costs and charges.

The services of the inspector shall be furnished without cost. The U.S. Department of Agriculture will not be responsible for any costs or charges incident to inspections or compliance with the provisions of the quarantine and regulations in this subpart, other than for the services of the inspector.

§ 301.45–12 Disqualification of qualified certified applicator to issue certificates.

(a) Any qualified certified applicator may be disqualified from issuing certificates by the Administrator if he determines that one of the following has occurred:

(1) Such person is not certified by a State and/or the Federal government as a commercial certified applicator under the Federal Insecticide, Fungicide, and Rodenticide Act (7 U.S.C. 136i) in a category allowing the application of restricted use pesticides.

(2) Noncompliance with any of the provisions of this subpart or with stipulations agreed on in the compliance agreement between the certified applicator and the Administrator.

(b) The disqualification is effective upon oral or written notification, whichever is earlier. The reasons for the disqualification shall be confirmed in writing as promptly as circumstances permit, unless contained in the written notification. Any qualified certified applicator who is disqualified from issuing certificates may appeal the decision in writing to the Administrator within ten (10) days after receiving written notification of the disqualification. The appeal shall state all of the facts and reasons upon which the person relies to show that the disqualification was a wrongful action. The Administrator shall grant or deny the appeal, in writing, stating the reasons for his decision as promptly as circumstances permit. If there is a conflict as to any material fact, a hearing shall be held to resolve such conflict. Rules of practice concerning such a hearing will be adopted by the Administrator.

(Approved by the Office of Management and Budget under control number 0579–0088)

[58 FR 39423, July 23, 1993, as amended at 59 FR 46902, Sept. 13, 1994; 72 FR 70764, Dec. 13, 2007]

Subpart F—Japanese Beetle

SOURCE: 44 FR 24035, Apr. 24, 1979, unless otherwise noted. Redesignated at 84 FR 2428, Feb. 7, 2019.

QUARANTINE AND REGULATIONS

§ 301.48 Notice of quarantine; quarantine restrictions on interstate movement of regulated articles.

(a) Pursuant to the provisions of sections 411, 412, 414, 431, and 434 of the Plant Protection Act (7 U.S.C. 7711, 7712, 7714, 7751, and 7754), the Secretary of Agriculture heretofore determined after public hearing to quarantine the States of Alabama, Arkansas, Connecticut, Delaware, Georgia, Illinois, Indiana, Iowa, Kentucky, Maine, Maryland, Massachusetts, Michigan, Minnesota, Missouri, New Hampshire, New Jersey, New York, North Carolina,

Ohio, Pennsylvania, Rhode Island, South Carolina, Tennessee, Vermont, Virginia, West Virginia, Wisconsin, and the District of Columbia in order to prevent the spread of the Japanese beetle, a dangerous insect injurious to cultivated crops and not theretofore widely prevalent or distributed within or throughout the United States.

(b) No person shall move any regulated article interstate from any regulated airport destined to any of the following States except in accordance with the conditions prescribed in this subpart: Arizona, California, Colorado, Idaho, Montana, Nevada, Oregon, Utah, and Washington.

[44 FR 24035, Apr. 24, 1979, as amended at 61 FR 32640, June 25, 1996; 66 FR 21050, Apr. 27, 2001; 68 FR 43614, July 24, 2003; 69 FR 40534, July 6, 2004; 71 FR 35493, June 21, 2006]

§ 301.48–1 Definitions.

Terms used in the singular form in this subpart shall be deemed to import the plural and vice versa, as the case may demand. The following terms, when used in this subpart shall be construed, respectively, to mean:

Administrator. The Administrator of the Animal and Plant Health Inspection Service or any person authorized to act for the Administrator.

Animal and Plant Health Inspection Service (APHIS). The Animal and Plant Health Inspection Service of the U.S. Department of Agriculture.

Compliance agreement. A written agreement between the Animal and Plant Health Inspection Service and a person engaged in the business of moving regulated articles interstate, in which the person agrees to comply with the provisions of this subpart.

Inspector. Any employee of the Animal and Plant Health Inspection Service, U.S. Department of Agriculture, or other person, authorized by the Administrator to enforce the provisions of the quarantine and regulations in this subpart.

Interstate. From any State into or through any other State.

Japanese beetle. The live insect known as the Japanese beetle (*Popillia japonica* Newm.) in any stage of development (egg, larva, pupa, or adult).

Person. Any individual, corporation, company, partnership, society, or association, or other organized group of any of the foregoing.

Regulated airport. Any airport or portions of an airport in a quarantined State declared regulated in accordance with provisions in § 301.48–2 of this subpart.

Regulated articles. Aircraft at or from regulated airports.

State. Any State, territory, or district of the United States, including Puerto Rico.

State Plant Regulatory Official. The authorized official of a State who has responsibility for the operation of the State plant regulatory program.

[44 FR 24035, Apr. 24, 1979, as amended at 61 FR 32640, June 25, 1996; 70 FR 33268, June 7, 2005]

§ 301.48–2 Authorization to designate, and terminate designation of, regulated airports.

(a) An inspector may declare any airport within a quarantined State to be a regulated airport when he or she determines that adult populations of Japanese beetle exist during daylight hours at the airport to the degree that aircraft constitute a threat to spread the Japanese beetle and aircraft destined for the States listed in § 301.48(b) may be leaving the airport.

(b) An inspector shall terminate the designation provided for under paragraph (a) of this section when he or she determines that adult populations of Japanese beetle no longer exist at the airport to the degree that the aircraft pose a threat to spread the Japanese beetle.

[44 FR 24035, Apr. 24, 1979, as amended at 61 FR 32640, June 25, 1996]

§ 301.48–3 Notification of designation, and termination of designation, of regulated airports.

Upon designating, or terminating the designation of, an airport as regulated, the inspector shall give written notice to the official in charge of the airport that the airport has been designated as a regulated airport or that the designation has been terminated. The inspector shall also give the same information in writing to the official at the airport in charge of each airline or the operator of any other aircraft, which will move a regulated article to any

State designated in § 301.48(b). The Administrator shall also give the same information to the State Plant Regulatory Official of each State designated in § 301.48(b) to which any regulated article will move.

[44 FR 24035, Apr. 24, 1979, as amended at 61 FR 32640, June 25, 1996]

§ 301.48-4 Conditions governing the interstate movement of regulated articles from quarantined States.

A regulated article may be moved interstate from a regulated airport to any State[1] designated in § 301.48(b) only if:

(a) An inspector, upon visual inspection of the airport and/or the aircraft, determines that the regulated article does not present a threat to spread the Japanese beetle because adult beetle populations are not present; or

(b) The aircraft is opened and loaded only while it is enclosed inside a hangar that an inspector has determined to be free of and safeguarded against Japanese beetle; or

(c) The aircraft is loaded during the hours of 8:00 p.m. to 7:00 a.m. only or lands and departs during those hours and, in either situation, is kept completely closed while on the ground during the hours of 7:00 a.m. to 8:00 p.m.; or

(d) If opened and loaded between the hours of 7:00 a.m. to 8:00 p.m., the aircraft is inspected, treated, and safeguarded. Inspection, treatment, and safeguarding must be done either under a compliance agreement in accordance with § 301.48-8 or under the direct supervision of an inspector. On a case-by-case basis, inspectors will determine which of the following conditions, and any supplemental conditions deemed necessary by the Administrator to prevent the spread of Japanese beetle, are required:

(1) All openings of the aircraft must be closed or safeguarded during the hours of 7:00 a.m. to 8:00 p.m. by exclusionary devices or by other means approved by the Administrator.

(2) All cargo containers that have not been safeguarded in a protected area

must be inspected immediately prior to and during the loading process. All personnel must check their clothing immediately prior to entering the aircraft. All Japanese beetles found must be removed and destroyed.

(3) All areas around doors and hatches or other openings in the aircraft must be inspected prior to removing the exclusionary devices. All Japanese beetles found must be removed and destroyed. All doors and hatches must be closed immediately after the exclusionary devices are moved away from the aircraft.

(4) Aircraft must be treated in accordance with part 305 of this chapter no more than 1 hour before loading. Particular attention should be paid to the ball mat area and the holes around the main entrance. The aircraft must then be aerated under safeguard conditions as required by part 305 of this chapter.

(5) Aircraft treatment records must be maintained by the applicator completing or supervising the treatment for a period of 2 years. These records must be provided upon request for review by an inspector. Treatment records shall include the pesticide used, the date of application, the location where the pesticide was applied (airport and aircraft), the amount of pesticide applied, and the name of the applicator.

(6) When a designated aircraft is replaced with an alternate one just prior to departure (the procedure known as "tail swapping"), the alternate aircraft must be inspected and all Japanese beetles must be removed. The aircraft must be safeguarded by closing all openings and hatches or by equipping the aircraft with exclusionary devices until the aircraft is ready for use. During loading, all treatment and safeguard requirements applicable to regularly scheduled aircraft must be implemented.

(7) Aircraft may be retreated in the noninfested State if live Japanese beetles are found.

(8) Notification of unscheduled commercial flights and of all military flights must be given at least 1 hour before departure to the appropriate person in the destination airport of any

---

[1] Requirements under all other applicable Federal domestic plant quarantines must be met.

of the States listed in §301.48(b). Notification of arriving military flights should also be given to base commanders to facilitate the entrance of Federal and/or State inspectors onto the base if necessary.

[61 FR 32640, June 25, 1996, as amended at 61 FR 56404, Nov. 1, 1996; 70 FR 33268, June 7, 2005]

§301.48–5 Inspection and disposal of regulated articles and pests.

Any properly identified inspector is authorized to stop and inspect, and to seize, destroy, or otherwise dispose of or require disposal of regulated articles and Japanese beetles as provided in sections 414, 421, and 434 of the Plant Protection Act (7 U.S.C. 7714, 7731, and 7754) in accordance with instructions issued by the Administrator.

[44 FR 24035, Apr. 24, 1979, as amended at 61 FR 32641, June 25, 1996; 66 FR 21051, Apr. 27, 2001]

§301.48–6 Movement of live Japanese beetles.

Regulations requiring a permit for and otherwise governing the movement of live Japanese beetles in interstate or foreign commerce are contained in the Federal Plant Pest Regulations in part 330 of this chapter. Applications for permits for the movement of the pest may be made to the Administrator.

[44 FR 24035, Apr. 24, 1979, as amended at 61 FR 32641, June 25, 1996]

§301.48–7 Nonliability of the Department.

The U.S. Department of Agriculture disclaims liability for any costs incident to inspections or compliance with the provisions of the quarantine and regulations in this subpart other than for the services of the inspector.

§301.48–8 Compliance agreements and cancellation.

(a) Any person engaged in the business of moving regulated articles may enter into a compliance agreement to facilitate the movement of such articles under this subpart. Any person who enters into a compliance agreement, and employees or agents of that person, must allow an inspector access to all records regarding treatment of aircraft and to all areas where loading, unloading, and treatment of aircraft occurs.

(b) A compliance agreement may be canceled by an inspector, orally or in writing, whenever he or she determines that the person who has entered into the compliance agreement has failed to comply with the agreement or this subpart. If the cancellation is oral, the cancellation and the reasons for the cancellation will be confirmed in writing within 20 days of oral notification. Any person whose compliance agreement has been canceled may appeal the decision, in writing, to the Administrator within 10 days after receiving written notification of the cancellation. The appeal must state all of the facts and reasons upon which the person relies to show that the compliance agreement was wrongfully canceled. A hearing will be held to resolve any conflict as to any material fact. The Administrator shall adopt rules of practice for the hearing. An appeal shall be granted or denied, in writing, as promptly as circumstances allow, and the reasons for the decision shall be stated. The compliance agreement will remain canceled pending the decision on the appeal.

[61 FR 32641, June 25, 1996]

Subpart G—Pine Shoot Beetle [Reserved]

Subpart H—Asian Longhorned Beetle

SOURCE: 62 FR 10416, Mar. 7, 1997, unless otherwise noted. Redesignated at 84 FR 2428, Feb. 7, 2019.

§301.51–1 Definitions.

Administrator. The Administrator, Animal and Plant Health Inspection Service, or any individual authorized to act for the Administrator.

Animal and Plant Health Inspection Service (APHIS). The Animal and Plant Health Inspection Service of the United States Department of Agriculture.

Asian longhorned beetle. The insect known as Asian longhorned beetle (*Anoplophora glabripennis*) in any stage of development.

Certificate. A document which is issued for a regulated article by an inspector or by a person operating under a compliance agreement, and which represents that such article is eligible for interstate movement in accordance with § 301.51-5(a).

Compliance agreement. A written agreement between APHIS and a person engaged in growing, handling, or moving regulated articles that are moved interstate, in which the person agrees to comply with the provisions of this subpart and any conditions imposed under this subpart.

Infestation. The presence of the Asian longhorned beetle in any life stage.

Inspector. Any employee of the Animal and Plant Health Inspection Service, or other individual authorized by the Administrator to enforce the provisions of this subpart.

Interstate. From any State into or through any other State.

Limited permit. A document in which an inspector affirms that the regulated article not eligible for a certificate is eligible for interstate movement only to a specified destination and in accordance with conditions specified on the permit.

Moved (movement, move). Shipped, offered for shipment, received for transportation, transported, carried, or allowed to be moved, shipped, transported, or carried.

Person. Any association, company, corporation, firm, individual, joint stock company, partnership, society, or any other legal entity.

Quarantined area. Any State, or any portion of a State, listed in § 301.51-3(c) of this subpart or otherwise designated as a quarantined area in accordance with § 301.51-3(b) of this subpart.

Regulated article. Any article listed in § 301.51-2(a) of this subpart or otherwise designated as a regulated article in accordance with § 301.51-2(b) of this subpart.

State. The District of Columbia, Puerto Rico, the Northern Mariana Islands, or any State, territory, or possession of the United States.

§ 301.51-2 Regulated articles.

The following are regulated articles:

(a) Firewood (all hardwood species), and green lumber and other material living, dead, cut, or fallen, inclusive of nursery stock, logs, stumps, roots, branches, and debris of half an inch or more in diameter of the following genera: *Acer* (maple), *Aesculus* (horse chestnut), *Albizia* (mimosa), *Betula* (birch), *Cercidiphyllum* (katsura), *Fraxinus* (ash), *Koelreuteria* (golden rain tree), *Platanus* (sycamore), *Populus* (poplar), *Salix* (willow), *Sorbus* (mountain ash), and *Ulmus* (elm).

(b) Any other article, product, or means of conveyance not covered by paragraph (a) of this section if an inspector determines that it presents a risk of spreading Asian longhorned beetle and notifies the person in possession of the article, product, or means of conveyance that it is subject to the restrictions of this subpart.

[62 FR 10416, Mar. 7, 1997, as amended at 62 FR 60764, Nov. 13, 1997; 68 FR 26985, May 19, 2003; 75 FR 34322, June 17, 2010; 76 FR 52542, Aug. 23, 2011; 81 FR 39176, June 16, 2016]

§ 301.51-3 Quarantined areas.

(a) Except as otherwise provided in paragraph (b) of this section, the Administrator will list as a quarantined area in paragraph (c) of this section, each State or each portion of a State in which the Asian longhorned beetle has been found by an inspector, in which the Administrator has reason to believe that the Asian longhorned beetle is present, or that the Administrator considers necessary to regulate because of its inseparability for quarantine enforcement purposes from localities where Asian longhorned beetle has been found. Less than an entire State will be designated as a quarantined area only if the Administrator determines that:

(1) The State has adopted and is enforcing restrictions on the intrastate movement of regulated articles that are equivalent to those imposed by this subpart on the interstate movement of regulated articles; and

(2) The designation of less than an entire State as a quarantined area will be adequate to prevent the artificial interstate spread of the Asian longhorned beetle.

(b) The Administrator or an inspector may temporarily designate any nonquarantined area as a quarantined area in accordance with the criteria specified in paragraph (a) of this section. The Administrator will give written notice of this designation to the owner or person in possession of the nonquarantined area, or, in the case of publicly owned land, to the person responsible for the management of the nonquarantined area. Thereafter, the interstate movement of any regulated article from an area temporarily designated as a quarantined area is subject to this subpart. As soon as practicable, this area either will be added to the list of designated quarantined areas in paragraph (c) of this section, or the Administrator will terminate the designation. The owner or person in possession of, or, in the case of publicly owned land, the person responsible for the management of, an area for which the designation is terminated will be given written notice of the termination as soon as practicable.

(c) The following areas are designated as quarantined areas:

MASSACHUSETTS

Worcester County. The portion of Worcester County, including portions or all of the municipalities of Worcester, Holden, West Boylston, Boylston, Auburn, and Shrewsbury that is bounded by a line starting at the intersection of Route 9 (Belmont Street) and the eastern boundary of the town of Shrewsbury; then follow the Shrewsbury town boundary northerly until the Boylston town boundary; then follow the entirety of the Boylston town boundary until it comes to the West Boylston town boundary on the Massachusetts Department of Conservation and Recreation Watershed property; then along the West Boylston town boundary until it intersects Manning Street; then southwest on Manning Street in Holden to Wachusett Street (Route 31); then south on Wachusett Street to Highland Street (still Route 31); then southwest on Highland Street to Main Street; then southeast on Main Street to Bailey Road; then south on Bailey Road to Chapin Road; then south on Chapin Road to its end; then continuing in a southeasterly direction to Fisher Road; then southwest on Fisher Road to Stonehouse Hill Road; then south on Stonehouse Hill Road to Reservoir Street; then southeast on Reservoir Street until it intersects the Worcester city boundary; turn south on Oxford Street to Auburn Street; then southeast on Auburn Street crossing under the Massachusetts Turnpike (I-90) and continuing southeast on Millbury Street; at the intersection of Washington Street, turn northeast and continue along Washington Street to the northern boundary of the Massachusetts Turnpike (I-90); then east along the northern boundary of the Massachusetts Turnpike (I-90) to the Auburn town boundary; then follow the Auburn town boundary northerly to the Worcester city boundary; continue along the Worcester city boundary until the Shrewsbury town boundary; then follow the entirety of the Shrewsbury town boundary until the point of beginning.

NEW YORK

New York City. That area in the boroughs of Brooklyn and Queens in the City of New York that is bounded by a line beginning at the point where the Brooklyn Battery Tunnel intersects the Brooklyn shoreline of the East River; then east and north along the shoreline of the East River to its intersection with the City of New York/Nassau County line; then southeast along the City of New York/Nassau County line to its intersection with the Grand Central Parkway; then west on the Grand Central Parkway to the Jackie Robinson Parkway; then west on the Jackie Robinson Parkway to Park Lane; then south on Park Lane to Park Lane South; then south and west on Park Lane South to 112th Street; then south on 112th Street to Atlantic Avenue; then west on Atlantic Avenue to 106th Street; then south on 106th Street to Liberty Avenue; then west on Liberty Avenue to Euclid Avenue; then south on Euclid Avenue to Linden Boulevard; then west on Linden Boulevard to Canton Avenue; then west on Canton Avenue to the Prospect Expressway; then north and west on the Prospect Expressway to the Gowanus Expressway; then north and west on the Gowanus Expressway; then north on Hamilton Avenue to the point of beginning.

Nassau and Suffolk Counties. That area in the villages of Amityville, West Amityville, North Amityville, Babylon, West Babylon, Copiague, Lindenhurst, Massapequa, Massapequa Park, and East Massapequa; in the towns of Oyster Bay and Babylon; in the counties of Nassau and Suffolk that is bounded as follows: Beginning at a point where West Main Street intersects the west shoreline of Carlis Creek; then west along West Main Street to Route 109; then north along Route 109 to Arnold Avenue; then northwest along Arnold Avenue to Albin Avenue; then west along Albin Avenue to East John Street; then west along East John Street to Wellwood Avenue; then north along Wellwood Avenue to the Southern State Parkway; then west along the Southern State Parkway to Broadway; then south along Broadway to Hicksville Road; then south along Hicksville Road to Division Avenue; then south along Division Avenue to South Oyster Bay; then east along the shoreline of South Oyster Bay to Carlis Creek;

47

then along the west shoreline of Carlis Creek to the point of beginning.

OHIO

Clermont County. (1) The portion of Clermont County, including all of the municipalities of Tate and East Fork State Park, and the portions of the Township of Monroe that include the following land parcels: 232609C094, 232609C113, 232609C215, 232609C085, 232609C128, 232609B224, 232609B188, 232609E223, 232609B215, 32609B193, 232609E075, 232609B161, 232609E156, 232609E245, 232609E037, 232609E074, 232609E230, 232609E031, 232609E220, 232609E232, 232609E240, 232609E239, 232609E241, 232609E175, 232609E228, 232609E250, 232609E235, 232609E238, 232609E227, 232609E242, 32609E226, 232609E249, 232609E236, 232609E234, 232609C217, 232609C040, 234715.008, 232609C227, 232609C222, 232609C092, 232609C093, 232609C129, 232609C098, 232609C195, 232609C100, 232609C169, 232609C136, 232609C097, 232609C139, 232609C148, 232609C042, 232609C150, 232609C182, 234715.009, 234715.005, 234715.006, 234715.001, 232609E246, 232609E247, 234715.004, 234715.003, 232609E222, 232609C228, 234425.001, 232609E233, 232609C170, 232609C216, 232609C196, 232609C105, 232609E237, 232609C225, 232609C091, 232609C197, 232609C218, 232609C198, 232609C041, 232609C212, 232609C194, 232609C214, 232609E224, 232609E231, 232609E248, 234715.007, 234715.002, 232609C120, 232609C226, 232609C229, 232609C043; and

(2) The portions of the Townships of Batavia and Stonelick that include the following land parcels: 302909I048, 304436.008, 302909H084, 302909K030, 022003B040, 012003H093, 025503D053, 022003B024, 302909G132, 304436A017, 304436.004, 302909F109, 012003E028, 012003C087, 012003H097, 022003C080, 302909G120, 302909K046, 302909H083, 302909F116, 012003E031, 012003C085, 022003B039, 302909G128, 302909F120, 302909H095, 302909F104, 012003E029, 012003C086, 022003F033, 302912E118, 302909G119, 302909G115, 302909H082, 302909F115, 012003E027, 302909G010, 022003F015, 302909B081, 302909G114, 302909H094, 302909I097, 012003E022, 022004H018, 022004H061, 302909E112, 302909B065, 302909G130, 302909H096, 302909F107, 025503B015, 012003H078, 022004H019, 302909E113, 302909G129, 304436A018, 302909F118, 302909I098, 012003E043, 302909H117, 304436A012, 302909E120, 302909B069, 304436.007, 304436.003, 302909K109, 012003E032, 302909G131, 302909F110, 302909E116, 302909J087, 302909H089, 302909F123, 302912E029, 012003E037, 302909F101, 022003A077, 302909E073, 302909J089, 302909H086, 304436.002, 302909F021, 0155031058, 302909F063, 022003A022, 302909E110, 304436A013, 304436A019, 304436.001, 302909F102, 012003E041, 302909G133, 302909F130, 302909E102, 302909E097, 302909H029, 302909F117, 012005H008, 012003E039, 302909D109, 302909F129, 302909K084, 302909J091, 302909H087, 302909F124, 302909I099, 012003E026, 302909E043, 302909I095, 302909K055, 302909K051, 304436.006, 302909B093, 302909K028P, 012003E040, 302909D049, 022003A023, 302909K107, 302909J086, 304436A020, 302909F128, 302909K110, 025503B013, 302909E090, 302909E114, 302909K054, 304436A014,

302909H090, 302909K047, 302909I025, 012003E034, 302909K044, 304436.010, 302909H093, 302909B040, 012005I010, 025503B012, 302912E135, 302909F132, 302909K032, 304436A015, 302909H085, 302909F112, 022003A078, 012003H102, 022003E020, 302909F131, 302909K059, 302909B015, 302909F119, 302909F125, 302909I096, 012003E023, 022003B038, 302909F022, 302909K108, 302909J041, 302909H088, 302909F113, 302909I024, 012003E033, 012003E035, 302909B083, 302909K031, 304436.009, 304436.005, 302909F114, 012003H101, 012003H100, 012003E011, 302909B070, 302909K099, 302909B103, 302909F122, 302909F127, 022003F034, 012003E024, 025503I058, 302909B076, 302909K028, 304436A016, 302909B064, 302909F111, 022003F018, 012003H104, 022003G023, 304436.011, 302909K042, 302909A039, 302909H091, 302909H097, 302909I023, 012003E036, 022003G022, 302909B038, 302909K052, 302909F121, 302909K071, 022003A076, 012003C027, 025503B016, 022003B043, 302909K094, 302909G116, 302909H081, 302909F126, 012005I005, 012003H094, 025503B017, 022003B044, 302909K092, 302909K106, 302909E106, 302909K053, 022003G014, 012003I072, 060224.026, 302909G110, 302909D107, 302909K061, 302909E115, 302909K058, 022003F019, 022003C062, 060224.009, 302909G109, 302909H018, 302909E016, 302909K057, 012003E038, 012003E030, 012003D009, 302909G121, 302909G127, 302909H017, 302909E014, 302912B150, 012003C061, 012003E047, 012003C078, 302909G108, 302909E013, 302909H080, 302909K027, 302909K066, 012003C028, 012003E046, 302909H099, 022003B042, 302909H020, 022005G014, 302909K068, 012003C088, 022003B025, 302909G011, 302909H098, 302909H092, 012003E042, 012003B021, 012003E045, 012005I007, 302909J067, 302909K105, 302909G122, 302909G123, 302909G124, 302909G125, 302909G126.

[62 FR 10416, Mar. 7, 1997]

EDITORIAL NOTE: For FEDERAL REGISTER citations affecting § 301.51–3, see the List of CFR Sections Affected, which appears in the Finding Aids section of the printed volume and at *www.govinfo.gov.*

§ 301.51–4 Conditions governing the interstate movement of regulated articles from quarantined areas.

(a) Any regulated article may be moved interstate from a quarantined area only if moved under the following conditions:

(1) With a certificate or limited permit issued and attached in accordance with §§ 301.51–5 and 301.51–8;

(2) Without a certificate or limited permit if:

(i) The regulated article is moved by the United States Department of Agriculture for experimental or scientific purposes; or

(ii) The regulated article originates outside the quarantined area and is moved interstate through the quarantined area under the following conditions:

(A) The points of origin and destination are indicated on a waybill accompanying the regulated article; and

(B) The regulated article is moved through the quarantined area without stopping, or has been stored, packed, or handled at locations approved by an inspector as not posing a risk of infestation by Asian longhorned beetle; and

(C) The article has not been combined or commingled with other articles so as to lose its individual identity.

(b) When an inspector has probable cause to believe a person or means of conveyance is moving a regulated article interstate, the inspector is authorized to stop the person or means of conveyance to determine whether a regulated article is present and to inspect the regulated article. Articles found to be infected by an inspector, and articles not in compliance with the regulations in this subpart, may be seized, quarantined, treated, subjected to other remedial measures, destroyed, or otherwise disposed of.

§ 301.51–5 Issuance and cancellation of certificates and limited permits.

(a) An inspector[1] or person operating under a compliance agreement will issue a certificate for the interstate movement of a regulated article if he or she determines that the regulated article:

(1)(i) Is apparently free of Asian longhorned beetle in any stage of development, based on inspection of the regulated article; or

(ii) Has been grown, produced, manufactured, stored, or handled in such a manner that, in the judgment of the inspector, the regulated article does not present a risk of spreading Asian longhorned beetle; and

(2) Is to be moved in compliance with any additional conditions deemed necessary under section 414 of the Plant Protection Act (7 U.S.C. 7714)[2] to prevent the artificial spread of the Asian longhorned beetle; and

(3) Is eligible for unrestricted movement under all other Federal domestic plant quarantines and regulations applicable to the regulated articles.

(b) An inspector or a person operating under a compliance agreement will issue a limited permit for the interstate movement of a regulated article not eligible for a certificate if he or she determines that the regulated article:

(1) Is to be moved interstate to a specified destination for specific processing, handling, or utilization (the destination and other conditions to be listed on the limited permit), and this interstate movement will not result in the spread of Asian longhorned beetle because Asian longhorned beetle will be destroyed by the specific processing, handling, or utilization; and

(2) It is to be moved in compliance with any additional conditions that the Administrator may impose under section 414 of the Plant Protection Act (7 U.S.C. 7714) in order to prevent the spread of the Asian longhorned beetle; and

(3) Is eligible for unrestricted movement under all other Federal domestic plant quarantines and regulations applicable to the regulated article.

(c) An inspector shall issue blank certificates and limited permits to a person operating under a compliance agreement in accordance with §301.51–6 or authorize reproduction of the certificates or limited permits on shipping containers, or both, as requested by the person operating under the compliance agreement. These certificates and limited permits may then be completed and used, as needed, for the interstate movement of regulated articles that have met all of the requirements of paragraph (a) or (b), respectively, of this section.

(d) Any certificate or limited permit may be canceled orally or in writing by

---

[1] Inspectors are assigned to local offices of APHIS, which are listed in local telephone directories. Information concerning such local offices may also be obtained from the Animal and Plant Health Inspection Service, Plant Protection and Quarantine, Domestic and Emergency Operations, 4700 River Road Unit 134, Riverdale, Maryland 20737–1236.

[2] An inspector may hold, seize, quarantine, treat, apply other remedial measures to, destory, or otherwise dispose of plants, plant pests, or other articles in accordance with sections 414, 421, and 434 of the Plant Protection Act (7 U.S.C. 7714, 7731, and 7754).

an inspector whenever the inspector determines that the holder of the certificate or limited permit has not complied with this subpart or any conditions imposed under this subpart. If the cancellation is oral, the cancellation will become effective immediately, and the cancellation and the reasons for the cancellation will be confirmed in writing as soon as circumstances permit. Any person whose certificate or limited permit has been cancelled may appeal the decision in writing to the Administrator within 10 days after receiving the written cancellation notice. The appeal must state all of the facts and reasons that the person wants the Administrator to consider in deciding the appeal. A hearing may be held to resolve a conflict as to any material fact. Rules of practice for the hearing will be adopted by the Administrator. As soon as practicable, the Administrator will grant or deny the appeal, in writing, stating the reasons for the decision.

[62 FR 10416, Mar. 7, 1997, as amended at 66 FR 21051, Apr. 27, 2001]

§ 301.51–6 Compliance agreements and cancellation.

(a) Persons engaged in growing, handling, or moving regulated articles interstate may enter into a compliance agreement[3] if such persons review with an inspector each stipulation of the compliance agreement. Any person who enters into a compliance agreement with APHIS must agree to comply with the provisions of this subpart and any conditions imposed under this subpart.

(b) Any compliance agreement may be canceled orally or in writing by an inspector whenever the inspector determines that the person who has entered into the compliance agreement has not complied with this subpart or any conditions imposed under this subpart. If the cancellation is oral, the cancellation will become effective imme-

diately, and the cancellation and the reasons for the cancellation will be confirmed in writing as soon as circumstances permit. Any person whose compliance agreement has been cancelled may appeal the decision in writing to the Administrator within 10 days after receiving the written cancellation notice. The appeal must state all of the facts and reasons that the person wants the Administrator to consider in deciding the appeal. A hearing may be held to resolve a conflict as to any material fact. Rules of practice for the hearing will be adopted by the Administrator. As soon as practicable, the Administrator will grant or deny the appeal, in writing, stating the reasons for the decision.

§ 301.51–7 Assembly and inspection of regulated articles.

(a) Persons requiring certification or other services must request the services from an inspector[4] at least 48 hours before the services are needed.

(b) The regulated articles must be assembled at the place and in the manner that the inspector designates as necessary to comply with this subpart.

§ 301.51–8 Attachment and disposition of certificates and limited permits.

(a) A regulated article must be plainly marked with the name and address of the consignor and the name and address of the consignee and must have the certificate or limited permit issued for the interstate movement of a regulated article securely attached at all times during interstate movement to:

(1) The outside of the container encasing the regulated article;

(2) The article itself, if it is not in a container; or

(3) The consignee's copy of the accompanying waybill; Provided, that the description of the regulated article on the certificate or limited permit, and on the waybill, are sufficient to identify the regulated article; and

(b) The carrier must furnish the certificate or limited permit authorizing interstate movement of a regulated article to the consignee at the destination of the shipment.

---

[3] Compliance agreements may be initiated by contacting a local office of APHIS. The addresses and telephone numbers of local offices are listed in local telephone directories and may also be obtained from the Animal and Plant Health Inspection Service, Plant Protection and Quarantine, Domestic and Emergency Operations, 4700 River Road Unit 134, Riverdale, Maryland 20737–1236.

[4] See footnote 1 to § 301.51–5.

§ 301.51–9 Costs and charges.

The services of the inspector during normal business hours will be furnished without cost to persons requiring the services. The user will be responsible for all costs and charges arising from inspection and other services provided outside of normal business hours.

Subpart I—Pink Bollworm

SOURCE: 32 FR 16385, Nov. 30, 1967, unless otherwise noted. Redesignated at 84 FR 2428, Feb. 7, 2019.

QUARANTINE AND REGULATIONS

§ 301.52 Quarantine; restriction on interstate movement of specified regulated articles.

(a) *Notice of quarantine.* The following States are quarantined to prevent the spread of the pink bollworm (*Pectinophora gossypiella* (Saund.)): Arizona, California, New Mexico, and Texas.

(b) *Regulated articles.* No common carrier or other person shall move interstate from any quarantined State any regulated article, except in accordance with this subpart. The following are regulated articles:

(1) Cotton and wild cotton, including all parts of these plants.

(2) Seed cotton.

(3) Cottonseed.

(4) American-Egyptian (long-staple) varieties of cotton lint, linters, and lint cleaner waste; except:[1]

(i) American-Egyptian cotton lint, linters, and lint cleaner waste compressed to a density of at least 22 pounds per cubic foot.

(ii) Trade samples of American-Egyptian cotton lint and linters.

(5) Cotton waste produced at cotton gins and cottonseed oil mills.

(6) Cotton gin trash.

(7) Used bagging and other used wrappers for cotton.

(8) Used cotton harvesting equipment and used cotton ginning and used cotton oil mill equipment.

---

[1] The articles hereby exempted remain subject to applicable restrictions under other quarantines and must have not been exposed to pink bollworm infestation after ginning or compression as prescribed.

(9) Kenaf, including all parts of the plants.

(10) Okra, including all parts of these plants, except:

(i) Canned or frozen okra; or

(ii) Okra seed; and

(iii) Fresh, edible fruits of okra:

(A) During December 1 through May 15 if moved interstate, but only during January 1 through March 15 if moved to California.

(B) During May 16 through November 30, if moved interstate to any portion of Illinois, Kentucky, Missouri, or Virginia that is north of the 38th parallel; or to any destination in Colorado, Connecticut, Delaware, District of Columbia, Idaho, Indiana, Iowa, Kansas, Maine, Maryland, Massachusetts, Michigan, Minnesota, Montana, Nebraska, New Hampshire, New Jersey, New York, North Dakota, Ohio, Oregon, Pennsylvania, Rhode Island, South Dakota, Utah, Vermont, Washington, West Virginia, Wisconsin, or Wyoming.

(11) Any other product, article, or means of conveyance not covered by paragraphs (b)(1) through (10) of this section, when an inspector determines that it presents a risk of spread of the pink bollworm and the person in possession of the product, article, or means of conveyance has actual notice that it is subject to the restrictions of this subpart.

[32 FR 16385, Nov. 30, 1967]

EDITORIAL NOTE: For FEDERAL REGISTER citations affecting § 301.52, see the List of CFR Sections Affected, which appears in the Finding Aids section of the printed volume and at *www.govinfo.gov.*

§ 301.52–1 Definitions.

Terms used in the singular form in this subpart shall be deemed to import the plural, and vice versa, as the case may demand. The following terms, when used in this subpart, shall be construed, respectively to mean:

Certificate. A document issued or authorized to be issued under this subpart by an inspector to allow the interstate movement of regulated articles to any destination.

Compliance agreement. A written agreement between a person engaged in growing, handling, or moving regulated articles, and the Plant Protection and

Quarantine Programs, wherein the former agrees to comply with the requirements of this subpart identified in the agreement by the inspector who executes the agreement on behalf of the Plant Protection and Quarantine Programs as applicable to the operations of such person.

Deputy Administrator. The Deputy Administrator of the Plant Protection and Quarantine Programs, Animal and Plant Health Inspection Service, U.S. Department of Agriculture, or any other officer or employee of said Service to whom authority to act in his stead has been or may hereafter be delegated.

Generally infested area. Any part of a regulated area not designated as a suppressive area in accordance with § 301.52–2.

Infestation. The presence of the pink bollworm or the existence of circumstances that make it reasonable to believe that pink bollworm is present.

Inspector. Any employee of the Plant Protection and Quarantine Programs, Animal and Plant Health Inspection Service, U.S. Department of Agriculture, or other person authorized by the Deputy Administrator to enforce the provisions of the quarantine and regulations in this subpart.

Interstate. From any State, territory, or district of the United States into or through any other State, territory, or district of the United States (including Puerto Rico).

Limited permit. A document issued or authorized to be issued by an inspector to allow the interstate movement of noncertified regulated articles to a specified destination for limited handling, utilization, or processing or for treatment.

Moved (movement, move). Shipped, offered for shipment to a common carrier, received for transportation or transported by a common carrier, or carried, transported, moved, or allowed to be moved by any means. "Movement" and "move" shall be construed accordingly.

Person. Any individual, corporation, company, society, or association, or other organized group of any of the foregoing.

Pink bollworm. The live insect known as the pink bollworm of cotton (Pectinophora gossypiella Saund.), in any stage of development.

Regulated area. Any quarantined State, territory, or district, or any portion thereof, listed as a regulated area in § 301.52–2a by the Deputy Administrator in accordance with § 301.52–2(a).

Regulated articles. Any articles described in § 301.52(b).

Restricted destination permit. A document issued or authorized to be issued by an inspector to allow the interstate movement of regulated articles not certified under all applicable Federal domestic plant quarantines to a specified destination for other than scientific purposes.

Scientific permit. A document issued by the Deputy Administrator to allow the interstate movement to a specified destination of regulated articles for scientific purposes.

Suppressive area. That part of a regulated area where eradication of infestation is undertaken as an objective, as designated by the Deputy Administrator under § 301.52–2(a).

[32 FR 16385, Nov. 30, 1967, as amended at 35 FR 2859, Feb. 12, 1970; 36 FR 24917, Dec. 24, 1971; 37 FR 10554, May 25, 1972; 52 FR 26943, July 17, 1987; 67 FR 34818, May 16, 2002; 70 FR 33268, June 7, 2005]

§ 301.52–2 Authorization for Deputy Administrator to list regulated areas and suppressive or generally infested areas.

The Deputy Administrator shall publish and amend from time to time as the facts warrant, the following lists:

(a) *List of regulated areas and suppressive or generally infested areas.* The Deputy Administrator shall list as regulated areas in a supplemental regulation designated as § 301.52–2a, the quarantined States, territories, or districts, or portions thereof, in which pink bollworm has been found or in which there is reason to believe that pink bollworm is present, or which it is deemed necessary to regulate because of their proximity to infestation or their inseparability for quarantine enforcement purposes from infested localities. The Deputy Administrator, in the supplemental regulation, may divide any regulated area into a suppressive area and a generally infested area in accordance with the definitions thereof in § 301.52–1. Less than an entire quarantined State, territory, or district will be designated as a regulated area only if the Deputy Administrator is of the opinion that:

(1) The State, territory, or district has adopted and is enforcing a quarantine or regulations which imposes restrictions on the intrastate movement of the regulated articles which are substantially the same as those which are imposed with respect to the interstate movement of such articles under this subpart; and

(2) The designation of less than the entire State, territory, or district, as a regulated area will otherwise be adequate to prevent the interstate spread of the pink bollworm.

[32 FR 16385, Nov. 30, 1967, as amended at 52 FR 26943, July 17, 1987]

§ 301.52–2a Regulated areas; suppressive and generally infested areas.

The civil divisions and part of civil divisions described below are designated as pink bollworm regulated areas within the meaning of the provisions of this subpart; and such regulated areas are hereby divided into generally infested areas or suppressive areas as indicated below.

ARIZONA

(1) *Generally infested area.* Entire State.
(2) *Suppressive area.* None.

CALIFORNIA

(1) *Generally infested area.*
Imperial County. The entire county.
Inyo County. The entire county.
Los Angeles County. The entire county.
Orange County. The entire county.
Riverside County. The entire county.
San Bernardino County. The entire county.
San Diego County. The entire county.
(2) *Suppressive area.*
Fresno County. The entire county.
Kern County. The entire county.
Kings County. The entire county.
Madera County. The entire county.
Merced County. The entire county.
San Benito County. The entire county.
Tulare County. The entire county.

NEW MEXICO

(1) *Generally infested area.* Entire State.
(2) *Suppressive area.* None.

TEXAS

(1) *Generally infested area.* Entire State.

(2) *Suppressive area.* None.

[42 FR 13533, Mar. 11, 1977]

EDITORIAL NOTE: For FEDERAL REGISTER citations affecting § 301.52–2a, see the List of CFR Sections Affected, which appears in the Finding Aids section of the printed volume and at *www.govinfo.gov.*

§ 301.52–3 Conditions governing the interstate movement of regulated articles from quarantined States. [2]

Any regulated articles may be moved interstate from any quarantined State under the following conditions:

(a) From any regulated area, with certificate or permit issued and attached in accordance with §§ 301.52–4 and 301.52–7 if moved:

(1) From any regulated area into or through any point outside of the regulated areas; or

(2) From any generally infested area into or through any suppressive area; or

(3) Between any noncontiguous suppressive areas; or

(4) Between contiguous suppressive areas when it is determined by the inspector that the regulated articles present a hazard of the spread of the pink bollworm and the person in possession thereof has been so notified; or

(b) From any regulated area, without certificate or permit if moved;

(1) From a generally infested area to a contiguous generally infested area; or

(2) From a suppressive area to a contiguous generally infested area; or

(3) Between contiguous suppressive areas unless the person in possession of the articles has been notified by an inspector that a hazard of spread of the pink bollworm exists; or

(4) Through or reshipped from any regulated area if the articles originated outside of any regulated area and if the point of origin of the articles is clearly indicated, their identity has been maintained and they have been safeguarded against infestation while in the regulated area in a manner satisfactory to the inspector; or

(c) From any area outside the regulated areas, without a certificate or permit if the point of origin of such

---

[2] Requirements under all other applicable Federal domestic plant quarantines must also be met.

movement is clearly indicated on the articles or shipping document which accompanies the articles and if the movement is not made through any regulated area.

[32 FR 16385, Nov. 30, 1967, as amended at 52 FR 26943, July 17, 1987; 70 FR 33268, June 7, 2005]

§ 301.52–4 Issuance and cancellation of certificates and permits.

(a) Certificates may be issued for any regulated articles by any inspector if he determines that they are eligible for certification for movement to any destination under all Federal domestic plant quarantines applicable to such articles and:

(1) Have originated in noninfested premises in a regulated area and have not been exposed to infestation while within the regulated areas; or

(2) Upon examination, have been found to be free of infestation; or

(3) Have been treated to destroy infestation in accordance with part 305 of this chapter; or

(4) Have been grown, produced, manufactured, stored, or handled in such manner that no infestation would be transmitted thereby.

(b) Limited permits may be issued by an inspector to allow interstate movement of regulated articles, not eligible for certification under this subpart, to specified destinations for limited handling, utilization, or processing, or for treatment in accordance with part 305 of this chapter, when upon evaluation of the circumstances involved in each specific case the inspector determines that such movement will not result in the spread of the pink bollworm and requirements of other applicable Federal domestic plant quarantines have been met.

(c) Restricted destination permits may be issued by an inspector to allow the interstate movement of regulated articles to any destination permitted under all applicable Federal domestic plant quarantines (for other than scientific purposes) if such articles are not eligible for certification under all such quarantines but would otherwise qualify for certification under this subpart.

(d) Scientific permits may be issued by the Deputy Administrator to allow the interstate movement of regulated articles for scientific purposes under such conditions as may be prescribed in each specific case by the Deputy Administrator.

(e) Certificate, limited permit, and restricted destination permit forms may be issued by an inspector to any person for use by the latter for subsequent shipments provided such person is operating under a compliance agreement; and any such person may be authorized by an inspector to reproduce such forms on shipping containers or otherwise. Any such person may use the certificate forms, or reproductions of such forms, for the interstate movement of regulated articles from the premises of such person identified in the compliance agreement if such person has made one of the determination specified in paragraph (a) of this section with respect to such articles. Any such person may use the limited permit forms, or reproductions of such forms, for interstate movement of regulated articles to specific destinations authorized by the inspector in accordance with paragraph (b) of this section. Any such person may use the restricted destination permit forms, or reproductions of such forms, for the interstate movement of regulated articles not eligible for certification under all Federal domestic plant quarantines applicable to such articles, under the conditions specified in paragraph (c) of this section.

(f) Any certificate or permit which has been issued or authorized may be withdrawn by the inspector if the inspector determines that the holder thereof has not complied with any condition for the use of such document imposed by this subpart.

[32 FR 16385, Nov. 30, 1967, as amended at 70 FR 33268, June 7, 2005]

§ 301.52–5 Compliance agreements; and cancellation thereof.

(a) Any person engaged in the business of growing, handling, or moving regulated articles may enter into a compliance agreement to facilitate the movement of such articles under this subpart. Compliance agreement forms may be obtained from the Deputy Administrator or an inspector.

(b) Any compliance agreement may be cancelled by the inspector who is supervising its enforcement whenever the inspector finds, after notice and reasonable opportunity to present views has been accorded to the other party thereto, that such other party has failed to comply with the conditions of the agreement.

[32 FR 16385, Nov. 30, 1967, as amended at 70 FR 33268, June 7, 2005]

§ 301.52–6 Assembly and inspection of regulated articles.

Persons (other than those authorized to use certificates, limited permits, or restricted destination permits, or reproductions thereof, under § 301.52–4(e)) who desire to move interstate regulated articles which must be accompanied by a certificate or permit shall, as far in advance as possible, request an inspector to examine the articles prior to movement. Such articles shall be assembled at such points and in such manner as the inspector designates to facilitate inspection.

§ 301.52–7 Attachment and disposition of certificates or permits.

(a) If a certificate or permit is required for the interstate movement of regulated articles, the certificate or permit shall be securely attached to the outside of the container in which such articles are moved, except that, where the certificate or permit is attached to the waybill or other shipping document, and the regulated articles are adequately described on the certificate, permit, or shipping document, the attachment of the certificate or permit to each container of the articles is not required.

(b) In all cases, certificates or permits shall be furnished by the carrier to the consignee at the destination of the shipment.

§ 301.52–8 Inspection and disposal of regulated articles and pests.

Any properly identified inspector is authorized to stop and inspect, and to seize, destroy, or otherwise dispose of, or require disposal of regulated articles and pink bollworms as provided sections 414, 421, and 434 of the Plant Protection Act (7 U.S.C. 7714, 7731, and 7754), in accordance with instructions issued by the Deputy Administrator.

[32 FR 16385, Nov. 30, 1967, as amended at 66 FR 21051, Apr. 27, 2001]

§ 301.52–9 Movement of live pink bollworms.

Regulations requiring a permit for, and otherwise governing the movement of live pink bollworms in interstate or foreign commerce are contained in the Federal Plant Pest regulations in part 330 of this chapter. Applications for permits for the movement of the pest may be made to the Deputy Administrator.

§ 301.52–10 Nonliability of the Department.

The U.S. Department of Agriculture disclaims liability for any costs incident to inspections or compliance with the provisions of the quarantine and regulations in this subpart, other than for the services of the inspector.

Subpart J—Emerald Ash Borer [Reserved]

Subpart K—South American Cactus Moth

SOURCE: 74 FR 27073, June 8, 2009, unless otherwise noted. Redesignated at 84 FR 2428, Feb. 7, 2019.

§ 301.55 Restrictions on interstate movement of regulated articles.

No person may move interstate from any quarantined area any regulated article except in accordance with this subpart.[1]

§ 301.55–1 Definitions.

Administrator. The Administrator, Animal and Plant Health Inspection Service, or any person authorized to act for the Administrator.

---

[1] Any properly identified inspector is authorized, upon probable cause, to stop and inspect persons and means of conveyance moving in interstate commerce and to hold, seize, quarantine, treat, apply other remedial measures to, destroy, or otherwise dispose of regulated articles as provided in sections 414, 421, and 434 of the Plant Protection Act (7 U.S.C. 7714, 7731, and 7754).

Animal and Plant Health Inspection Service (APHIS). The Animal and Plant Health Inspection Service of the United States Department of Agriculture.

Cactus plants. Any of various fleshy-stemmed plants of the botanical family Cactaceae.

Certificate. A document in which an inspector or person operating under a compliance agreement affirms that a specified regulated article is free of South American cactus moth and may be moved interstate to any destination.

Compliance agreement. A written agreement between APHIS and a person engaged in growing, handling, or moving regulated articles, wherein the person agrees to comply with this subpart.

Departmental permit. A document issued by the Administrator in which he or she affirms that interstate movement of the regulated article identified on the document is for scientific or experimental purposes and that the regulated article is eligible for interstate movement in accordance with § 301.55–4(c).

Infestation. The presence of the South American cactus moth or the existence of circumstances that makes it reasonable to believe that the South American cactus moth may be present.

Inspector. Any employee of APHIS or other person authorized by the Administrator to perform the duties required under this subpart.

Interstate. From any State into or through any other State.

Limited permit. A document in which an inspector or person operating under a compliance agreement affirms that the regulated article identified on the document is eligible for interstate movement in accordance with § 301.55–5(b) only to a specified destination and only in accordance with specified conditions.

Moved (move, movement). Shipped, offered for shipment, received for transportation, transported, carried, or allowed to be moved, shipped, transported, or carried.

Person. Any association, company, corporation, firm, individual, joint stock company, partnership, society, or other entity.

Plant Protection and Quarantine (PPQ). The Plant Protection and Quarantine program of the Animal and Plant Health Inspection Service, United States Department of Agriculture.

Quarantined area. Any State, or any portion of a State, listed in § 301.55–3(c) or otherwise designated as a quarantined area in accordance with § 301.55–3(b).

Regulated article. Any article listed in § 301.55–2(a) or (b), or otherwise designated as a regulated article in accordance with § 301.55–2(c).

South American cactus moth. The live insect known as the South American cactus moth, *Cactoblastis cactorum,* in any life stage (egg, larva, pupa, adult).

State. The District of Columbia, Puerto Rico, the Northern Mariana Islands, or any State, territory, or possession of the United States.

§ 301.55–2 Regulated articles.

The following are regulated articles:

(a) The South American cactus moth, in any living stage of its development. [2]

(b) Cactus plants or parts thereof (excluding seeds and canned, preserved, or frozen pads or fruits) of the following genera: *Consolea,* *Cylindropuntia,* *Nopalea,* and *Opuntia.*

(c) Any other product, article, or means of conveyance not listed in paragraphs (a) or (b) of this section that an inspector determines presents a risk of spreading the South American cactus moth, after the inspector provides written notification to the person in possession of the product, article, or means of conveyance that it is subject to the restrictions of this subpart.

§ 301.55–3 Quarantined areas.

(a) Except as otherwise provided in paragraph (b) of this section, the Administrator will list as a quarantined area in paragraph (c) of this section each State, or each portion of a State, in which the South American cactus moth has been found by an inspector, in which the Administrator has reason to believe that the South American cactus moth is present, or that the Administrator considers necessary to

---

[2] Permit and other requirements for the interstate movement of South American cactus moths are contained in part 330 of this chapter.

quarantine because of its inseparability for quarantine enforcement purposes from localities where South American cactus moth has been found. Less than an entire State will be designated as a quarantined area only if the Administrator determines that:

(1) The State has adopted and is enforcing restrictions on the intrastate movement of the regulated articles that are equivalent to those imposed by this subpart on the interstate movement of regulated articles; and

(2) The designation of less than the entire State as a quarantined area will be adequate to prevent the interstate spread of the South American cactus moth.

(b) The Administrator or an inspector may temporarily designate any nonquarantined area in a State as a quarantined area in accordance with the criteria specified in paragraph (a) of this section. The Administrator will give a copy of this regulation along with written notice of the temporary designation to the owner or person in possession of the nonquarantined area, or, in the case of publicly owned land, to the person responsible for the management of the nonquarantined area. Thereafter, the interstate movement of any regulated article from an area temporarily designated as a quarantined area will be subject to this subpart. As soon as practicable, the area will be added to the list in paragraph (c) of this section or the designation will be terminated by the Administrator or an inspector. The owner or person in possession of, or, in the case of publicly owned land, the person responsible for the management of, an area for which designation is terminated will be given written notice of the termination as soon as practicable.

(c) The following areas are designated as quarantined areas: The States of Alabama, Florida, Georgia, Louisiana, Mississippi, and South Carolina.

[74 FR 27073, June 8, 2009, as amended at 75 FR 41074, July 15, 2010]

§ 301.55–4 **Conditions governing the interstate movement of regulated articles from quarantined areas.**

Any regulated article may be moved interstate from a quarantined area[3] only if moved under the following conditions:

(a) With a certificate or limited permit issued and attached in accordance with §§ 301.555 and 301.55–8;

(b) Without a certificate or limited permit if:

(1) The regulated article originated outside the quarantined area and is either moved in an enclosed vehicle or is completely enclosed by a covering (such as canvas, plastic, or closely woven cloth) adequate to prevent access by South American cactus moths while moving through the quarantined area; and

(2) The point of origin of the regulated article is indicated on the waybill, and the enclosed vehicle or the enclosure that contains the regulated article is not opened, unpacked, or unloaded in the quarantined area; and

(3) The regulated article is moved through the quarantined area without stopping except for refueling or for traffic conditions, such as traffic lights or stop signs.

(c) Without a certificate or limited permit if the regulated articles are cactus pads and fruits for consumption from outside the quarantined area that are being moved in accordance with the protocols described in a compliance agreement (see § 301.55–6(a)) to a commercial food warehouse or distribution center within the quarantined area and the regulated articles remain enclosed by a covering (such as canvas, plastic, or closely woven cloth) adequate to prevent access by South American cactus moths while within the quarantined area: and

(d) Without a certificate or limited permit if the regulated article is moved:

(1) By the United States Department of Agriculture for experimental or scientific purposes;

---

[3] Requirements under all other applicable Federal domestic plant quarantines and regulations must also be met.

(2) Pursuant to a departmental permit issued by the Administrator for the regulated article;

(3) Under conditions specified on the departmental permit and found by the Administrator to be adequate to prevent the spread of the South American cactus moth; and

(4) With a tag or label bearing the number of the departmental permit issued for the regulated article attached to the outside of the container of the regulated article or attached to the regulated article itself if not in a container.

§ 301.55–5 Issuance and cancellation of certificates and limited permits.

(a) An inspector [4] may issue a certificate for the interstate movement of a regulated article if the inspector determines that:

(1) The regulated article to be moved and all other regulated articles on the premises have been grown and maintained indoors in a shadehouse or greenhouse and no other cactus moth host material exists on the premises outside of a shadehouse or greenhouse;

(2) The regulated article to be moved and all other regulated articles on the premises are maintained on benches that are kept separate from benches containing non-host material;

(3) The regulated article to be moved and all other regulated articles on the premises have been placed on a 21-day insecticide spray cycle and have been sprayed with *Bacillus thuringiensis* subsp. *kurstaki*, carbaryl, spinosad, or imidaploprid if maintained in the nursery for longer than 21 days;

(4) The regulated article to be moved has been sprayed with *Bacillus thuringiensis* subsp. *kurstaki*, carbaryl, spinosad, or imidaploprid 3 to 5 days prior to shipment and inspected and found free of cactus moth egg sticks and larval damage; and

(5) If the regulated article was moved into the premises from another premises in a quarantined area listed in § 301.55–3, it was immediately placed inside the shadehouse or greenhouse and sprayed with *Bacillus thuringiensis*

---

[4] Services of an inspector may be requested by contacting local offices of Plant Protection and Quarantine, which are listed in telephone directories.

subsp. *kurstaki*, carbaryl, spinosad, or imidaploprid within 24 hours.

(b) An inspector will issue a limited permit for the interstate movement of a regulated article if the inspector determines that:

(1) The regulated article is to be moved interstate to a specified destination for specified handling, processing, or utilization (the destination and other conditions to be listed in the limited permit), and this interstate movement will not result in the spread of the South American cactus moth because life stages of the South American cactus moth will be destroyed by the specified handling, processing, or utilization;

(2) It is to be moved in compliance with any additional conditions that the Administrator may impose under section 414 of the Plant Protection Act (7 U.S.C. 7714) in order to prevent the spread of the South American cactus moth; and

(3) It is eligible for unrestricted movement under all other Federal domestic plant quarantines and regulations applicable to the regulated article.

(c) Certificates and limited permits for the interstate movement of regulated articles may be issued by an inspector or person operating under a compliance agreement. A person operating under a compliance agreement may issue a certificate or limited permit for interstate movement of a regulated article after an inspector has determined that the regulated article is eligible for a certificate or limited permit in accordance with paragraphs (a) or (b) of this section.

(d) Any certificate or limited permit that has been issued may be canceled, either orally or in writing, by an inspector whenever the inspector determines that the holder of the limited permit has not complied with this subpart or any conditions imposed under this subpart. If the cancellation is oral, the cancellation will become effective immediately, and the cancellation and the reasons for the cancellation will be confirmed in writing as soon as circumstances permit. Any person whose certificate or limited permit has been canceled may appeal the decision in writing to the Administrator within 10 days after receiving the written cancellation notice. The appeal must state all of the facts and reasons that the person wants the Administrator to consider in deciding the appeal. A hearing may be held to resolve a conflict as to any material fact. Rules of practice for the hearing will be adopted by the Administrator. As soon as practicable, the Administrator will grant or deny the appeal, in writing, stating the reasons for the decision.

(Approved by the Office of Management and Budget under control number 0579–0337)

§ 301.55–6 Compliance agreements and cancellation.

(a) Any person engaged in growing, handling, or moving regulated articles may enter into a compliance agreement when an inspector determines that the person is aware of this subpart, agrees to comply with its provisions, and agrees to comply with all the provisions contained in the compliance agreement.[5]

. (b) Any compliance agreement may be canceled, either orally or in writing, by an inspector whenever the inspector finds that the person who has entered into the compliance agreement has failed to comply with this subpart or the terms of the compliance agreement. If the cancellation is oral, the cancellation and the reasons for the cancellation will be confirmed in writing as promptly as circumstances allow. Any person whose compliance agreement has been canceled may appeal the decision, in writing, to the Administrator, within 10 days after receiving written notification of the cancellation. The appeal must state all of the facts and reasons upon which the person relies to show that the compliance agreement was wrongfully canceled. As promptly as circumstances allow, the Administrator will grant or deny the appeal, in writing, stating the reasons for the decision. A hearing will be held to resolve any conflict as to any material fact. Rules of practice

concerning a hearing will be adopted by the Administrator.

(Approved by the Office of Management and Budget under control number 0579–0337)

§ 301.55–7 Assembly and inspection of regulated articles.

(a) Any person (other than a person authorized to issue limited permits under § 301.555(c)) who desires a certificate or limited permit to move a regulated article interstate must request an inspector[6] to examine the articles as far in advance of the desired interstate movement as possible, but no less than 48 hours before the desired interstate movement.

(b) The regulated article must be assembled at the place and in the manner the inspector designates as necessary to comply with this subpart.

§ 301.55–8 Attachment and disposition of certificates and limited permits.

(a) A certificate or limited permit required for the interstate movement of a regulated article must, at all times during the interstate movement, be:

(1) Attached to the outside of the container containing the regulated article; or

(2) Attached to the regulated article itself if not in a container; or

(3) Attached to the consignee's copy of the accompanying waybill. If the certificate or limited permit is attached to the consignee's copy of the waybill, the regulated article must be sufficiently described on the certificate or limited permit and on the waybill to identify the regulated article.

(b) The certificate or limited permit for the interstate movement of a regulated article must be furnished by the carrier or the carrier's representative to the consignee listed on the certificate or limited permit upon arrival at the location provided on the certificate or limited permit.

(Approved by the Office of Management and Budget under control number 0579–0337)

§ 301.55–9 Costs and charges.

The services of the inspector during normal business hours (8 a.m. to 4:30 p.m., Monday through Friday, except

---

[5] Compliance agreement forms are available without charge from local Plant Protection and Quarantine offices, which are listed in telephone directories.

[6] *See* footnote 4.

holidays) will be furnished without cost. APHIS will not be responsible for all costs or charges incident to inspections or compliance with the provisions of the quarantine and regulations in this subpart, other than for the services of the inspector.

Subpart L—Plum Pox

Source: 65 FR 35264, June 2, 2000, unless otherwise noted. Redesignated at 84 FR 2428, Feb. 7, 2019.

§ 301.74 Restrictions on interstate movement of regulated articles.

No person may move interstate from any quarantined area any regulated article except in accordance with this subpart.[1]

[65 FR 35264, June 2, 2000, as amended at 66 FR 21051, Apr. 27, 2001]

§ 301.74–1 Definitions.

The following definitions apply to this subpart.

Administrator. The Administrator, Animal and Plant Health Inspection Service, or any person authorized to act for the Administrator.

Animal and Plant Health Inspection Service. The Animal and Plant Health Inspection Service (APHIS) of the United States Department of Agriculture.

Departmental permit. A document issued by the Administrator in which he or she affirms that interstate movement of the regulated article identified on the document is for scientific or experimental purposes and that the regulated article is eligible for interstate movement in accordance with § 301.74–4 of this subpart.

Infestation (infested, infected). The presence of plum pox or circumstances or symptoms that makes it reasonable to believe that plum pox is present.

Inspector. Any employee of the Animal and Plant Health Inspection Serv-

ice, United States Department of Agriculture, or other person authorized by the Administrator to enforce this subpart.

Interstate. From any State into or through any other State.

Moved (move, movement). Shipped, offered for shipment, received for transportation, transported, carried, or allowed to be moved, shipped, transported, or carried.

Person. Any association, company, corporation, firm, individual, joint stock company, partnership, society, or other entity.

Plant Protection and Quarantine. Plant Protection and Quarantine, Animal and Plant Health Inspection Service, United States Department of Agriculture.

Plum pox. A plant disease caused by plum pox potyvirus that can affect many *Prunus* (stone fruit) species, including, but not limited to, almond, apricot, nectarine, peach, plum, and sweet and tart cherry. The strain of plum pox in Pennsylvania does not affect cherry trees.

Quarantined area. Any State, or any portion of a State, listed in § 301.74–3(c) of this subpart or otherwise designated as a quarantined area in accordance with § 301.74–3(b) of this subpart.

Regulated article. Any article listed in § 301.74–2(a) or otherwise designated as a regulated article in accordance with § 301.74–2(b), based on its susceptibility to the form or strain of plum pox detected in the quarantined area.

State. The District of Columbia, Puerto Rico, the Northern Mariana Islands, or any State, territory, or possession of the United States.

§ 301.74–2 Regulated articles.

The following are regulated articles:

(a) All plant material and plant parts of *Prunus* (stone fruit) species other than *P. avium, P. cerasus, P. effusa, P. laurocerasus, P. mahaleb, P. padus, P. sargentii, P. serotina, P. serrula, P. serrulata, P. subhirtella, P. yedoensis,* and *P. virginiana,* except for seeds and fruit that is free of leaves and other plant parts. This includes, but is not limited to, trees, seedlings, root stock, budwood, branches, twigs, and leaves.

(b) Any other product or article that an inspector determines to present a

---

[1] Any properly identified inspector is authorized to stop and inspect persons and means of conveyance and to seize, quarantine, treat, apply other remedial measures to, destroy, or otherwise dispose of regulated articles a provided in sections 414, 421, and 434 of the Plant Protection Act (7 U.S.C. 7714, 7731, and 7754).

risk of spreading plum pox when the inspector notifies the person in possession of the product or article that it is subject to the restrictions in the regulations.

§ 301.74–3 Quarantined areas.

(a) Except as otherwise provided in paragraph (b) of this section, the Administrator will list as a quarantined area in paragraph (c) of this section each State, or each portion of a State, in which plum pox has been detected through inspection and laboratory testing, or in which the Administrator has reason to believe that plum pox is present, or that the Administrator considers necessary to quarantine because of its inseparability for quarantine enforcement purposes from localities in which plum pox has been detected. Less than an entire State will be designated as a quarantined area if the Administrator determines that:

(1) The State has adopted and is enforcing restrictions on the intrastate movement of the regulated articles that are substantially the same as those imposed by this subpart on the interstate movement of regulated articles; and

(2) The designation of less than the entire State as a quarantined area will prevent the interstate spread of plum pox.

(b) The Administrator or an inspector may temporarily designate any nonquarantined area in a State as a quarantined area in accordance with paragraph (a) of this section. The Administrator will give a copy of this regulation along with a written notice for the temporary designation to the owner or person in possession of the nonquarantined area. Thereafter, the interstate movement of any regulated article from an area temporarily designated as a quarantined area will be subject to this subpart. As soon as practicable, this area will be added to the list in paragraph (c) of this section or the designation will be terminated by the Administrator or an inspector. The owner or person in possession of an area for which the quarantine designation is terminated will be given notice of the termination as soon as practicable.

(c) The areas described below are designated as quarantined areas:

NEW YORK

Niagara County. (1) That area of Niagara County in the Towns of Burt, Newfane, and Wilson bordered on the north by Lake Ontario; bordered on the west by Maple Road; then south on Maple Road to Wilson-Burt Road; then east on Wilson-Burt Road to Beebe Road; then south on Beebe Road to Ide Road; then east on Ide Road to Route 78 (Lockport-Olcott Road); then north on Route 78 (Lockport-Olcott Road) to the Lake Ontario shoreline.

(2) That area of Niagara County in the Town of Lewiston bordered on the west by Porter Center Road starting at its intersection with Route 104 (Ridge Road); then north-northeast on Porter Center Road to Langdon Road; then east on Langdon Road to Dickersonville Road; then north on Dickersonville Road to Schoolhouse Road; then east on Schoolhouse Road to Ransomville Road; then south on Ransomville Road to Route 104 (Ridge Road); then northeast on Route 104 (Ridge Road) to Simmons Road; then south on Simmons Road to Albright Road; then east on Albright Road to Townline Road; then south on Townline Road to Lower Mountain Road; then west on Lower Mountain Road to Meyers Hill Road; then south on Meyers Hill Road to Upper Mountain Road; then west on Upper Mountain Road to Indian Hill Road; then northeast on Indian Hill Road to Route 104 (Ridge Road); then east on Route 104 (Ridge Road) to Porter Center Road.

(3) That area of Niagara County bordered on the north by Lake Ontario and on the east by Keg Creek; then south on Keg Creek to Route 18 (Lake Road); then east on Route 18 (Lake Road) to Hess Road; then south on Hess Road to Drake Settlement Road; then west on Drake Settlement Road to Transit Road; then north on Transit Road to Route 18 (Lake Road); then west on Route 10 (Lake Road) to Lockport Olcott Road; then north on Lockport Olcott Road to the Lake Ontario shoreline.

Orleans County. That area of Orleans County in the Towns of Ridgeway and Gaines bordered on the north by Route 104 (Ridge Road) at its intersection with Eagle Harbor Waterport Road; then south on Eagle Harbor Waterport Road to Eagle Habor Knowlesville Road; then west on Eagle Harbor Knowlesville Road to Presbyterian Road; then southwest on Presbyterian Road to Longbridge Road; then south on Longbridge Road to State Route 31; then west on State Route 31 to Wood Road; then south on Wood Road to West County House Road; then west on West County House Road to Maple Ridge Road; then west on Maple Ridge Road to Culvert Road; then north on Culvert Road to Telegraph Road; then west on Telegraph Road to Beales Road; then north on Beales Road to Portage Road; then east on Portage Road to Culvert Rd; then north on Culvert Road to Route 104 (Ridge Road).

Wayne County. (1) That area of Wayne County in the Town of Sodus bordered on the north by Lake Road at its intersection with Redman Road; then east on Lake Road to Maple Avenue; then south on Maple Avenue to Middle Road; then west on Middle Road to Rotterdam Road; then south on Rotterdam Road to State Route 104; then west on State Route 104 to Pratt Road; then south on Pratt Road to Ridge Road; then west on Ridge Road to Richardson Road; then south on Richardson Road to Tripp Road; then south on Tripp Road to Podger Road; then west on Podger Road to East Townline Road; then north on East Townline Road to Everdyke Road; then west on Everdyke Road to Russell Road; then south on Russell Road to Pearsall Road; then west on Pearsall Road to State Route 21; then north on State Route 21 to State Route 104; then east on State Route 104 to East Townline Road; then north on East Townline Road to Van Lare Road; then east on Van Lare Road to Redman Road; then north on Redman Road to Lake Road.

(2) That area of Wayne County in the Towns of Ontario and Williamson bordered on the north by Shepard Road at its intersection with Fisher Road; then east on Shepard Road to Salmon Creek Road; then southwest on Salmon Creek Road to Kenyon Road; then west on Kenyon Road to Furnace Road; then north on Furnace Road to Putnam Road; then east on Putnam Road to Fisher Road; then north on Fisher Road to Shepard Road.

(3) That area of Wayne County in the Town of Sodus beginning on the Sodus Bay shoreline at Ridge Road; then west on Ridge Road to Boyd Road; then north on Boyd Road to Sergeant Road; then north on Sergeant Road to Morley Road; then east on Morley Road to State Route 14; then north on State Route 14 to South Shore Road; then east on South Shore Road and continuing to the shoreline of Sodus Bay.

[65 FR 35264, June 2, 2000, as amended at 75 FR 81089, Dec. 27, 2010; 76 FR 27219, May 11, 2011]

§ 301.74-4 Conditions governing the interstate movement of regulated articles from quarantined areas.

The interstate movement of any regulated article from a quarantined area[2] is prohibited except when:

(a) The regulated article is moved by the United States Department of Agriculture:

(1) For an experimental or scientific purpose;

(2) Pursuant to a Departmental permit issued by the Administrator for the regulated article;

(3) Under conditions specified on the Departmental permit and found by the Administrator to be adequate to prevent the spread of plum pox; and

(4) With a tag or label bearing the number of the Departmental permit issued for the regulated article attached to the outside of the container of the regulated article or attached to the regulated article itself if not in a container; or

(b) The regulated article originated outside the quarantined area and:

(1) Is moved in an enclosed vehicle or is completely enclosed by a covering (such as canvas, plastic, or other closely woven cloth) adequate to prevent access by aphids or other transmission agents of plum pox while in the quarantined area;

(2) The regulated article's point of origin is indicated on the waybill; and

(3) The regulated article must not be uncovered, unpacked, or unloaded while moving through the quarantined area.

§ 301.74-5 Compensation.

(a) *Eligibility.* The following individuals are eligible to receive compensation from the U.S. Department of Agriculture to mitigate losses or expenses incurred because of the plum pox quarantine and emergency actions:

(1) *Owners of commercial stone fruit orchards.* Owners of commercial stone fruit orchards are eligible to receive compensation for losses associated with the destruction of trees in order to control plum pox pursuant to an emergency action notification issued

---

[2] Requirements under all other applicable Federal domestic plant quarantines and regulations must also be met.

by the Animal and Plant Health Inspection Service (APHIS).

(i) *Direct marketers.* Orchard owners eligible for compensation under this paragraph who market all fruit they produce under the conditions described in this paragraph may receive compensation at the rates specified in paragraph (b)(1)(i) of this section. In order to be eligible to receive compensation at the rates specified in paragraph (b)(1)(i) of this section, orchard owners must have marketed fruit produced in orchards subsequently destroyed because of plum pox under the following conditions:

(A) The fruit must have been sold exclusively at farmers markets or similar outlets that require orchard owners to sell only fruit that they produce;

(B) The fruit must not have been marketed wholesale or at reduced prices in bulk to supermarkets or other retail outlets;

(C) The fruit must have been marketed directly to consumers; and

(D) Orchard owners must have records documenting that they have met the requirements of this section, and must submit those records to APHIS as part of their application submitted in accordance with paragraph (c) of this section.

(ii) *All other orchard owners.* Orchard owners eligible for compensation under this paragraph who do not meet the requirements of paragraph (a)(1)(i) of this section are eligible for compensation only in accordance with paragraph (b)(1)(ii) of this section.

(2) *Owners of fruit tree nurseries.* The owner of a fruit tree nursery will be eligible to receive compensation for net revenue losses associated with the prohibition on the movement or sale of nursery stock as a result of the issuance of an emergency action notification by APHIS with respect to regulated articles within the nursery in order to control plum pox.

(3) *Owners of non-fruit-bearing ornamental tree nurseries.* The owner of a non-fruit-bearing ornamental tree nursery will be eligible to receive compensation for net revenue losses associated with the prohibition on the movement or sale of nursery stock as a result of the issuance of an emergency action notification by APHIS with respect to regulated articles within the nursery in order to control plum pox.

(b) *Amount of payment.* Upon approval of a claim submitted in accordance with paragraph (c) of this section, individuals eligible for compensation under paragraph (a) of this section will be paid at the rates indicated in this paragraph.

(1) *Owners of commercial stone fruit orchards*—(i) *Direct marketers.* Owners of commercial stone fruit orchards who APHIS has determined meet the eligibility requirements of paragraph (a)(1)(i) of this section will be compensated according to the following table on a per-acre basis at a rate based on the age of the trees destroyed. If the trees were not destroyed by the date specified on the emergency action notification, the compensation payment will be reduced by 10 percent and by any tree removal costs incurred by the State or the U.S. Department of Agriculture (USDA). The maximum USDA compensation rate is 85 percent of the loss in value, adjusted for any State-provided compensation to ensure total compensation from all sources does not exceed 100 percent of the loss in value.

Age of trees (years)	Maximum compensation rate ($/acre, equal to 85% of loss in value) based on 3-year fallow period	Maximum additional compensation ($/acre, equal to 85% of loss in value) for 4th fallow year	Maximum additional compensation ($/acre, equal to 85% of loss in value) for 5th fallow year
Less than 1	$3,302	$954	$842
1	11,639	1,936	1,721
2	16,327	1,936	1,721
3	20,725	1,936	1,721
4	26,222	1,936	1,721
5	28,820	1,936	1,721
6	29,592	1,936	1,721
7	29,743	1,936	1,721
8	29,196	1,936	1,721
9	28,581	1,936	1,721
10	27,889	1,936	1,721
11	27,110	1,936	1,721

Age of trees (years)	Maximum compensation rate ($/acre, equal to 85% of loss in value) based on 3-year fallow period	Maximum additional compensation ($/acre, equal to 85% of loss in value) for 4th fallow year	Maximum additional compensation ($/acre, equal to 85% of loss in value) for 5th fallow year
12	26,234	1,936	1,721
13	25,248	1,936	1,721
14	24,140	1,936	1,721
15	22,892	1,936	1,721
16	21,489	1,936	1,721
17	20,054	1,936	1,721
18	18,582	1,936	1,721
19	17,070	1,936	1,721
20	15,513	1,936	1,721
21	13,905	1,936	1,721
22	12,382	1,936	1,721
23	10,955	1,936	1,721
24	9,638	1,936	1,721
25	8,442	1,936	1,721

(ii) *All other orchard owners.* Owners of commercial stone fruit orchards who meet the eligibility requirements of paragraph (a)(1)(ii) of this section will be compensated according to the following table on a per-acre basis at a rate based on the age of the trees destroyed. If the trees were not destroyed by the date specified on the emergency action notification, the compensation payment will be reduced by 10 percent and by any tree removal costs incurred by the State or the U.S. Department of Agriculture (USDA). The maximum USDA compensation rate is 85 percent of the loss in value, adjusted for any State-provided compensation to ensure total compensation from all sources does not exceed 100 percent of the loss in value.

Age of trees (years)	Maximum compensation rate ($/acre, equal to 85% of loss in value) based on 3-year fallow period	Maximum additional compensation ($/acre, equal to 85% of loss in value) for 4th fallow year	Maximum additional compensation ($/acre, equal to 85% of loss in value) for 5th fallow year
Less than 1	$3,302	$954	$842
1	6,959	1,072	953
2	10,090	1,072	953
3	12,737	1,072	953
4	16,263	1,072	953
5	17,929	1,072	953
6	18,423	1,072	953
7	18,519	1,072	953
8	18,167	1,072	953
9	17,771	1,072	953
10	17,325	1,072	953
11	16,823	1,072	953
12	16,259	1,072	953
13	15,625	1,072	953
14	14,911	1,072	953
15	14,107	1,072	953
16	13,204	1,072	953
17	12,279	1,072	953
18	11,331	1,072	953
19	10,356	1,072	953
20	9,352	1,072	953
21	8,314	1,072	953
22	7,330	1,072	953
23	6,408	1,072	953
24	5,554	1,072	953
25	4,777	1,072	953

(2) *Owners of fruit tree nurseries.* Owners of fruit tree nurseries who meet the eligibility requirements of paragraph (a)(2) of this section will be compensated for up to 85 percent of the net

revenues lost from their first and second year crops as the result of the issuance of an emergency action notification which will be calculated as follows:

(i) *First year crop.* The net revenue loss for trees that were expected to be sold in the year during which the emergency action notification was issued (*i.e.*, the first year crop) will be calculated as (*expected number of trees to be sold*) × (*average price per tree*) − (*digging, grading, and storage costs*) = net revenue lost for first year crop, where:

(A) The expected number of trees to be sold equals the number of trees in the field minus 2 percent culls minus 3 percent unsold trees; and

(B) The average price per tree is $5.22 for plum and apricot trees and $3.69 for peach and nectarine trees; and

(C) Digging, grading and storage costs are $0.10 per tree.

(ii) *Second year crop.* The net revenue loss for trees that would be expected to be sold in the year following the year during which the emergency action notification was issued (*i.e.*, the second year crop) will be calculated as (*expected number of trees to be sold*) × (*average price per tree*) = net revenue lost for second year crop, where:

(A) The expected number of trees to be sold equals the number of budded trees in the field minus 20 percent death loss minus 2 percent culls; and

(B) The average price per tree is $5.22 for plum and apricot trees and $3.69 for peach and nectarine trees.

(3) *Owners of non-fruit bearing ornamental tree nurseries.* Owners of non-fruit-bearing ornamental tree nurseries who meet the eligibility requirements of paragraph (a)(3) of this section will be compensated for up to 85 percent of the net revenues lost from their crop as the result of the issuance of an emergency action notification. Net revenues will be calculated using an average price of $10.80 per tree or shrub.

(c) *How to apply.* The form necessary to submit a claim for compensation may be obtained from the National Director of the Plum Pox Eradication Program contact listed at *http://www.aphis.usda.gov/plant__health/plant__pest__info/plum__pox/index.shtml.* Claims for trees or nursery stock destroyed on or before February 3, 2012

must be received within 60 days after February 3, 2012. Claims for trees or nursery stock destroyed after February 3, 2012 must be received within 60 days after the destruction of the trees or nursery stock. Claims must be submitted as follows:

(1) *Claims by owners of stone fruit orchards who are direct marketers.* The completed application must be accompanied by:

(i) A copy of the emergency action notification ordering the destruction of the trees and its accompanying inventory that describes the acreage and ages of trees removed;

(ii) Documentation verifying that the destruction of trees has been completed and the date of that destruction; and

(iii) Records documenting that the grower meets the eligibility requirements of paragraph (a)(1)(i) of this section.

(2) *Claims by owners of commercial stone fruit orchards who are not direct marketers.* The completed application must be accompanied by a copy of the emergency action notification ordering the destruction of the trees, its accompanying inventory that describes the acreage and ages of trees removed, and documentation verifying that the destruction of trees has been completed and the date of that destruction.

(3) *Claims by owners of fruit tree nurseries and owners of non-fruit-bearing ornamental tree nurseries.* The completed application must be accompanied by a copy of the order prohibiting the sale or movement of the nursery stock, its accompanying inventory that describes the total number of trees and the age and variety, and documentation describing the final disposition of the nursery stock.

(d) *Replanting.* Trees of susceptible *Prunus* species (*i.e.*, *Prunus* species identified as regulated articles) may not be replanted on premises within a contiguous quarantined area until 3 years from the date the last trees within that area were destroyed because of plum pox pursuant to an emergency action notification issued by APHIS.

(Approved by the Office of Management and Budget under control numbers 0579–0159 and 0579–0251)

[65 FR 55435, Sept. 14, 2000, as amended at 69 FR 30816, June 1, 2004; 77 FR 5383, Feb. 3, 2012]

Subpart M—Citrus Canker

SOURCE: 50 FR 51231, Dec. 13, 1985, unless otherwise noted. Redesignated at 84 FR 2428, Feb. 7, 2019.

NOTICE OF QUARANTINE AND REGULATIONS

§ 301.75–1 Definitions.

ACC coverage. The crop insurance coverage against Asiatic citrus canker (ACC) provided under the Florida Fruit Tree Pilot Crop Insurance Program authorized by the Federal Crop Insurance Corporation.

Administrator. The Administrator of the Animal and Plant Health Inspection Service or any individual authorized to act for the Administrator.

Animal and Plant Health Inspection Service. The Animal and Plant Health Inspection Service of the United States Department of Agriculture.

Budded citrus nursery stock. Liners or rootstock citrus plants that have been grafted with a portion of a stem or branch with a vegetative bud (also known as budwood) that are maintained 1 month after grafting or until the plant reaches marketability.

Budded container/greenhouse grown citrus plants. Individual, budded citrus nursery stock maintained in climate-controlled greenhouses in 4-or 6-inch diameter pots until it is sold for commercial use.

Budded field grown citrus plants. Individual, budded citrus nursery stock maintained in the fields until it is sold for commercial use.

Certificate. An official stamp, form, or other document of the United States Department of Agriculture authorizing the interstate movement of a regulated article from a quarantined area into any area of the United States.

Certified citrus nursery stock. Citrus nursery stock, such as trees or plants, grown at a nursery that is in compliance with State certification requirements and approved for producing citrus nursery stock for commercial sale.

Citrus canker. A plant disease caused by strains of the bacterium *Xanthomonas axonopodis* pv. *citri.*

Commercial citrus grove. An establishment maintained for the primary purpose of producing citrus fruit for commercial sale.

Commercial citrus nursery. An establishment engaged in, but not limited to, the production of certified citrus nursery stock, including plants for planting or replanting in commercial groves or for wholesale or retail sales.

Commercial citrus-producing area. Any area designated as a commercial citrus-producing area in accordance with § 301.75–5 of this subpart.

Commercial packinghouse. An establishment in which space and equipment are maintained for the primary purpose of disinfecting and packing citrus fruit for commercial sale. A commercial packinghouse must also be licensed, registered, or certified for handling citrus fruit with the State in which it operates and meet all the requirements for the license, registration, or certification that it holds.

Compliance agreement. A written agreement between the Animal and Plant Health Inspection Service and a person engaged in the business of growing, maintaining, processing, handling, packing, or moving regulated articles for interstate movement, in which the person pledges to comply with this subpart.

Departmental permit. An official document of the United States Department of Agriculture authorizing the movement of a regulated article from a quarantined area.

Departmental tag or label. An official tag or label of the United States Department of Agriculture, which, attached to a regulated article or its container, indicates that the regulated article is eligible for interstate movement with a Departmental permit.

Exposed. Determined by an inspector to be at risk for developing citrus canker because of proximity during the past 2 years to infected plants, or to personnel, vehicles, equipment, or other articles that may have been contaminated with bacteria that cause citrus canker.

Grove. Any tree or stand of trees maintained to produce fruit and separated from other trees by a boundary, such as a fence, stream, road, canal, irrigation ditch, hedgerow, open space, or sign or marker denoting change of fruit variety.

Infected. Containing bacteria that cause citrus canker.

Infestation. The presence of a plant or plants infected with citrus canker at a particular location, except when the plant or plants contracted the infection at a previous location and the infection has not spread to any other plant at the present location.

Inspector. An individual authorized by the Administrator to perform the specified duties.

Interstate. From any State into or through any other State.

Limited permit. An official stamp, form, or other document of the United States Department of Agriculture authorizing the interstate movement of a regulated article from a quarantined area, but restricting the areas of the United States into which the regulated article may be moved.

Liner or rootstock. Culled seedlings in the growing stage prior to the budding process.

Lot. The inspectional unit for fruit composed of a single variety of fruit that has passed through the entire packing process in a single continuous run not to exceed a single workday (*i.e.*, a run started one day and completed the next is considered two lots).

Move. Ship, carry, transport, offer for shipment, receive for shipment, or allow to be transported by any means.

Movement. The act of shipping, carrying, transporting, offering for shipment, receiving for shipment, or allowing to be transported by any means.

Nursery. Any premises, including greenhouses but excluding any grove, at which nursery stock is grown or maintained.

Nursery stock. Living plants and plant parts intended to be planted, to remain planted, or to be replanted.

Person. Any individual, partnership, corporation, company, society, association, or other organized group.

Public order. Either an "Agreement to Destroy and Covenant Not to Sue" signed by the grove owner and the Florida Department of Food and Consumer Services, Division of Plant Industry (DPI), or an "Immediate Final Order" issued by DPI, both of which identify citrus trees infected with or exposed to citrus canker and order their destruction.

Quarantined area. Any area designated as a quarantined area in accordance with § 301.75–4 of this subpart.

Regulated article. Any article listed in § 301.75–3 (a) or (b) of this subpart or designated as a regulated article in accordance with § 301.75–3(c) of this subpart.

Regulated fruit, regulated nursery stock, regulated plant, regulated seed, regulated tree. Any fruit, nursery stock, plant, seed, or tree defined as a regulated article.

Seedlings. Certified citrus seeds densely planted in seed beds and allowed to germinate and grow until their viability as liners or rootstock can be assessed.

State. Each of the 50 States of the United States, the District of Columbia, Guam, the Northern Mariana Islands, Puerto Rico, the Virgin Islands of the United States, and all other territories and possessions of the United States.

United States. All of the States, the District of Columbia, Guam, the Northern Mariana Islands, Puerto Rico, the Virgin Islands of the United States, and all other territories and possessions of the United States.

[55 FR 37450, Sept. 11, 1990, as amended at 61 FR 1521, Jan. 22, 1996; 65 FR 61080, Oct. 16, 2000; 66 FR 32717, June 18, 2001; 71 FR 33172, June 8, 2006; 72 FR 13427, Mar. 22, 2007; 72 FR 65203, Nov. 19, 2007; 74 FR 54444, Oct. 22, 2009; 76 FR 23457, Apr. 27, 2011]

§ 301.75–2 General prohibitions.

(a) Regulated articles may not be moved interstate from a quarantined area except in accordance with a protocol in §§ 301.75–6, 301.75–7, or 301.75–8, or in accordance with § 301.75–4 if less than an entire State is designated as a quarantined area. Regulated articles may be moved in accordance with the regulations in § 301.75–9 for scientific or experimental purposes only.

(b) Regulated articles moved from a quarantined area with a limited permit may not be moved interstate into any commercial citrus-producing area, except as follows: The regulated articles may be moved through a commercial citrus-producing area if they are covered, or enclosed in containers or in a compartment of a vehicle, while in the commercial citrus-producing area, and are not unloaded in the commercial citrus-producing area without the permission of an inspector.

(c) Regulated articles moved interstate with a limited permit to an area of the United States that is not a commercial citrus-producing area may not subsequently be moved interstate into any commercial citrus-producing area.

[55 FR 37450, Sept. 11, 1990; 55 FR 48208, Nov. 19, 1990; 72 FR 13427, Mar. 22, 2007]

§ 301.75–3   Regulated articles.

(a) Plants or plant parts, including fruit and seeds, or any of the following: All species, clones, cultivars, strains, varieties, and hybrids of the genera *Citrus* and *Fortunella*, and all clones, cultivars, strains, varieties, and hybrids of the species *Clausena lansium* and *Poncirus trifoliata*. The most common of these are: lemon, pummelo, grapefruit, key lime, persian lime, tangerine, satsuma, tangor, citron, sweet orange, sour orange, mandarin, tangelo, ethrog, kumquat, limequat, calamondin, trifoliate orange, and wampi.

(b) Grass, plant, and tree clippings.

(c) Any other product, article, or means of conveyance, of any character whatsoever, not covered by paragraph (a) of this section, when it is determined by an inspector that it presents a risk of spread of citrus canker and the person in possession thereof has actual notice that the product, article, or means of conveyance is subject to the provisions of this subpart.

[50 FR 51231, Dec. 13, 1985, as amended at 54 FR 12180, Mar. 24, 1989. Redesignated and amended at 55 FR 37450, Sept. 11, 1990]

§ 301.75–4   Quarantined areas.

(a) The following States or portions of States are designated as quarantined areas: The State of Florida.

(b) The Administrator may designate any non-quarantined area as a quarantined area in accordance with paragraphs (c) and (d) of this section upon giving written notice of this designation to the owner or persons in possession of the non-quarantined area. Thereafter, regulated articles may be moved interstate from that area only in accordance with this subpart. As soon as practicable, this area will be added to the list in paragraph (a) of this section, or the Administrator will terminate the designation. The owner or person in possession of an area for which designation is terminated will be given written notice as soon as practicable.

(c) Any State or portion of a State where an infestation is detected will be designated as a quarantined area and will remain so until the area has been without infestation for 2 years.

(d) Less than an entire State will be designated as a quarantined area only if all of the following conditions are met:

(1) *Survey.* No area has been designated a survey area.

(2) *Intrastate movement of regulated articles.* The State enforces restrictions on the intrastate movement of regulated articles from the quarantined area that are at least as stringent as those on the interstate movement of regulated articles from the quarantined area, except as follows:

(i) Regulated fruit may be moved intrastate from a quarantined area for processing into a product other than fresh fruit if all of the following conditions are met:

(A) The regulated fruit is accompanied by a document that states the location of the grove in which the regulated fruit was produced, the variety and quantity of regulated fruit being moved intrastate, the address to which the regulated fruit will be delivered for processing, and the date the intrastate movement began,

(B) The regulated fruit and any leaves and litter are completely covered, or enclosed in containers or in a compartment of a vehicle, during the intrastate movement.

(C) The vehicles, covers, and any containers used to carry the regulated fruit intrastate are treated in accordance with part 305 of this chapter before leaving the premises where the regulated fruit is unloaded for processing, and

(D) All leaves, litter, and culls collected from the shipment of regulated fruit at the processing facility are either incinerated at the processing facility or buried at a public landfill that is fenced, prohibits the removal of dumped material, and covers dumped material with dirt at the end of every day that dumping occurs.

(ii) Regulated fruit may be moved intrastate from a quarantined area for packing, either for subsequent interstate movement with a limited permit or for export from the United States, if all of the following conditions are met:

(A) The regulated fruit is accompanied by a document that states the location of the grove in which the regulated fruit was produced, the variety and quantity of regulated fruit being moved intrastate, the address to which the regulated fruit will be delivered for packing, and the date the intrastate movement began.

(B) The regulated fruit and any leaves and litter are completely covered, or enclosed in containers or in a compartment of a vehicle, during the intrastate movement.

(C) The vehicles, covers, and any containers used to carry the regulated fruit intrastate are treated in accordance with part 305 of this chapter before leaving the premises where the regulated fruit is unloaded for packing.

(D) Any equipment that comes in contact with the regulated fruit at the packing plant is treated in accordance with part 305 of this chapter before being used to handle any fruit eligible for interstate movement to commercial citrus-producing areas, and

(E) All leaves and litter collected from the shipment of regulated fruit at the packing plant are either incinerated at the packing plant or buried at a public landfill that is fenced, prohibits the removal of dumped material, and covers dumped material with dirt at the end of every day that dumping occurs. All culls collected from the shipment of regulated fruit are either processed into a product other than fresh fruit, incinerated at the packing plant, or buried at a public landfill that is fenced, prohibits the removal of dumped material, and covers dumped material with dirt at the end of every day that dumping occurs. Any culls moved intrastate for processing must be completely covered, or enclosed in containers or in a compartment of a vehicle, during the intrastate movement, and the vehicles, covers, and any containers used to carry the regulated fruit must be treated in accordance with part 305 of this chapter before leaving the premises where the regulated fruit is unloaded for processing.

(iii) Grass, tree, and plant clippings may be moved intrastate from the quarantined area for disposal in a public landfill or for composting in a recycling facility, if all of the following conditions are met:

(A) The public landfill or recycling facility is located within the survey area described in paragraph (d)(1) of this section,

(B) The grass, tree, or plant clippings are completely covered during the movement from the quarantined area to the public landfill or recycling facility, and

(C) Any public landfill used is fenced, prohibits the removal of dumped material, and covers dumped material with dirt at the end of every day that dumping occurs.

(3) *Inspections.* (i) In the quarantined area, every regulated plant and regulated tree, except indoor houseplants and regulated plants and regulated trees at nurseries, is inspected for citrus canker at least once a year, between May 1 through December 31, by an inspector.

(ii) In the quarantined area, every regulated plant and regulated tree at every nursery containing regulated plants or regulated trees is inspected for citrus canker by an inspector at intervals of no more than 45 days.

(4) *Treatment of personnel, vehicles, and equipment.* In the quarantined area, all vehicles, equipment, and other articles used in providing inspection, maintenance, harvesting, or related services in any grove containing regulated plants or regulated trees, or in providing landscaping or lawn care services on any premises containing regulated plants or regulated trees, must be treated in accordance with part 305 of this chapter upon leaving the grove or premises. All personnel who enter the grove or premises to provide these services must be treated in accordance with part 305 of this chapter upon leaving the grove or premises.

(5) *Destruction of infected plants and trees.* No more than 7 days after a State or Federal laboratory confirms that a regulated plant or regulated tree is infected, the State must provide written

notice to the owner of the infected plant or infected tree that the infected plant or infected tree must be destroyed. The owner must have the infected plant or infected tree destroyed within 45 days after receiving the written notice.

[55 FR 37450, Sept. 11, 1990]

EDITORIAL NOTE: For FEDERAL REGISTER citations affecting § 301.75–4, see the List of CFR Sections Affected, which appears in the Finding Aids section of the printed volume and at *www.govinfo.gov.*

§ 301.75–5 Commercial citrus-producing areas.

(a) The following are designated as commercial citrus-producing areas:

American Samoa	Northern Mariana
Arizona	Islands
California	Puerto Rico
Florida	Texas
Guam	Virgin Islands of the
Hawaii	United States
Louisiana	

(b) The list in paragraph (a) of this section is intended to include jurisdictions which have commercial citrus-producing areas. Less than an entire State may be designated as a commercial citrus-producing area only if the Administrator determines that the area not included as a commercial citrus-producing area does not contain commercial citrus plantings; that the State has adopted and is enforcing a prohibition on the intrastate movement from areas not designated as commercial citrus-producing areas to commercial citrus-producing areas of fruit which are designated as regulated articles and which were moved interstate from a quarantined State pursuant to a limited permit; and that the designation of less than the entire State as a commercial citrus-producing area will otherwise be adequate to prevent the interstate spread of citrus canker.

[50 FR 51231, Dec. 13, 1985, 51 FR 2873, Jan. 22, 1986, as amended at 53 FR 13242, Apr. 22, 1988; 53 FR 44173, Nov. 2, 1988. Redesignated at 55 FR 37450, Sept. 11, 1990]

§ 301.75–6 Interstate movement of regulated nursery stock from a quarantined area.

(a) Regulated nursery stock may not be moved interstate from a quar-

antined area unless such movement is authorized in this section.

(b) Kumquat (*Fortunella* spp.) plants, with or without fruit attached, may be moved interstate from a quarantined area into any area of the United States except commercial citrus-producing areas if all of the following conditions are met:

(1) The plants are own-root-only and have not been grafted or budded;

(2) The plants are started, are grown, and have been maintained solely at the nursery from which they will be moved interstate.

(3) If the plants are not grown from seed, then the cuttings used for propagation of the plants are taken from plants located on the same nursery premises or from another nursery that is eligible to produce kumquat plants for interstate movement under the requirements of this paragraph (b). Cuttings may not be obtained from properties where citrus canker is present.

(4) All citrus plants at the nursery premises have undergone State inspection and have been found to be free of citrus canker no less than three times. The inspections must be at intervals of 30 to 45 days, with the most recent inspection being within 30 days of the date on which the plants are removed and packed for shipment.

(5) All vehicles, equipment, and other articles used in providing inspection, maintenance, or related services in the nursery must be treated in accordance with part 305 of this chapter before entering the nursery to prevent the introduction of citrus canker. All personnel who enter the nursery to provide these services must be treated in accordance with part 305 of this chapter before entering the nursery to prevent the introduction of citrus canker.

(6) If citrus canker is found in the nursery, all regulated plants and plant material must be removed from the nursery and all areas of the nursery's facilities where plants are grown and all associated equipment and tools used at the nursery must be treated in accordance with part 305 of this chapter in order for the nursery to be eligible to produce kumquat plants to be moved interstate under this paragraph (b). Fifteen days after these actions are

completed, the nursery may receive new kumquat seed or cuttings from a nursery that is eligible to produce kumquat plants for interstate movement under this paragraph (b).

(7) The plants, except for plants that are hermetically sealed in plastic bags before leaving the nursery, are completely enclosed in containers or vehicle compartments during movement through the quarantined area.

(8) The kumquat plants or trees are accompanied by a limited permit issued in accordance with §301.75–12. The statement "Limited permit: Not for distribution in AZ, CA, HI, LA, TX, and American Samoa, Guam, Northern Mariana Islands, Puerto Rico, and Virgin Islands of the United States" must be displayed on a plastic or metal tag attached to each plant, or on the box or container if the plant is sealed in plastic. In addition, this statement must be displayed on the outside of any shipping containers used to transport these plants, and the limited permit must be attached to the bill of lading or other shipping document that accompanies the plants.

(c) Regulated nursery stock produced in a nursery within a quarantined area may be moved interstate to any area within the United States, if all of the following conditions are met:

(1) The nursery in which the nursery stock is produced has entered into a compliance agreement in which it agrees to meet the relevant construction standards, sourcing and certification requirements, cleaning, disinfecting, and safeguarding requirements, labeling requirements, and recordkeeping and inspection requirements specified in a PPQ protocol document. The protocol document will be provided to the person at the time he or she enters into the compliance agreement.[1] The compliance agreement may also specify additional conditions determined by APHIS to be necessary in order to prevent the dissemination of citrus canker under which the nursery stock must be grown, maintained, and shipped in order to obtain a certificate for its movement. The compliance agreement will also specify that APHIS may amend the agreement.

(2) An inspector has determined that the nursery has adhered to all terms and conditions of the compliance agreement.

(3) The nursery stock is accompanied by a certificate issued in accordance with §301.75–12.

(4) The nursery stock is completely enclosed in a sealed container that is clearly labeled with the certificate and is moved interstate in that container.

(5) A copy of the certificate is attached to the consignee's copy of the accompanying waybill.

(d) Regulated nursery stock produced in a nursery located in a quarantined area that is not eligible for movement under paragraph (b) or paragraph (c) of this section may be moved interstate only for immediate export. The regulated nursery stock must be accompanied by a limited permit issued in accordance with §301.75–12 and must be moved in a container sealed by APHIS directly to the port of export in accordance with the conditions of the limited permit.

(Approved by the Office of Management and Budget under control number 0579–0369)

[72 FR 13427, Mar. 22, 2007, as amended at 74 FR 16104, Apr. 9, 2009; 75 FR 4240, Jan. 26, 2010; 76 FR 23457, Apr. 27, 2011]

§301.75–7 Interstate movement of regulated fruit from a quarantined area.

(a) Regulated fruit produced in a quarantined area or moved into a quarantined area for packing may be moved interstate with a certificate issued and attached in accordance with §301.75-12 if all of the following conditions are met:

(1) The regulated fruit was packed in a commercial packinghouse whose owner or operator has entered into a compliance agreement with APHIS in accordance with §301.75-13.

(2) The regulated fruit was treated in accordance with part 305 of this chapter.

---

[1] The protocol document is also available on the Internet at *http://www.aphis.usda.gov/plant_health/plant_pest_info/citrus/index.shtml* and may be obtained from local Plant Protection and Quarantine offices, which are listed in telephone directories.

(3) The regulated fruit is free of leaves, twigs, and other plant parts, except for stems that are less than 1 inch long and attached to the fruit.

(4) If the fruit is repackaged after being packed in a commercial packinghouse and before it is moved interstate from the quarantined area, the person that repackages the fruit must enter into a compliance agreement with APHIS in accordance with § 301.75-13 and issue and attach a certificate for the interstate movement of the fruit in accordance with § 301.75-12.

(b) Regulated fruit that is not eligible for movement under paragraph (a) of this section may be moved interstate only for immediate export. The regulated fruit must be accompanied by a limited permit issued in accordance with § 301.75-12 and must be moved in a container sealed by APHIS directly to the port of export in accordance with the conditions of the limited permit.

(Approved by the Office of Management and Budget under control number 0579-0325)

[74 FR 54444, Oct. 22, 2009, as amended at 75 FR 4240, Jan. 26, 2010]

§ 301.75-8 Interstate movement of regulated seed from a quarantined area.

Regulated seed may be moved interstate from a quarantined area into any area of the United States if all of the following conditions are met:

(a) During the 2 years before the interstate movement, no plants or plant parts infected with or exposed to citrus canker were found in the grove or nursery producing the fruit from which the regulated seed was extracted.

(b) The regulated seed was treated in accordance with part 305 of this chapter.

(c) The regulated seed is accompanied by a certificate issued in accordance with § 301.75-12 of this subpart.

[55 FR 37452, Sept. 11, 1990, as amended at 75 FR 4240, Jan. 26, 2010]

§ 301.75-9 Interstate movement of regulated articles from a quarantined area for experimental or scientific purposes.

A regulated article may be moved interstate from a quarantined area if:

(a) Moved by the United States Department of Agriculture for experimental or scientific purposes;

(b) Moved pursuant to a Departmental permit issued for such article by the Administrator;

(c) Moved in accordance with conditions specified on the Departmental permit and determined by the Administrator to be adequate to prevent the spread of citrus canker, i.e., conditions of treatment, processing, growing, shipment, disposal; and

(d) Moved with a Departmental tag or label securely attached to the outside of the container containing the article or securely attached to the article itself if not in a container, with such tag or label bearing a Departmental permit number corresponding to the number of the Departmental permit issued for such article.

[50 FR 51231, Dec. 13, 1985. Redesignated and amended at 55 FR 37450, Sept. 11, 1990]

§ 301.75-10 Interstate movement of regulated articles through a quarantined area.

Any regulated article not produced in a quarantined area may be moved interstate through a quarantined area, without a certificate, limited permit, or Departmental permit, if all of the following conditions are met:

(a) The regulated article is accompanied by either: A receipt showing that the regulated article was purchased outside the quarantined area, or a bill of lading stating the location of the premises where the shipment originated, the type and quantity of regulated articles being moved interstate, and the date the interstate movement began.

(b) The regulated article is moved through the quarantined area without being unloaded, and no regulated article, except regulated fruit that qualifies for interstate movement from the quarantined area in accordance with § 301.75-7 of this subpart, is added to the shipment in the quarantined area.

(c) The regulated article is completely covered, or enclosed in containers or in a compartment of a vehicle, during movement through the quarantined area, except that, covering or enclosure is not required if the regulated article is moved through the quarantined area without stopping, except for refueling or for traffic conditions, such as traffic lights or stop signs.

[55 FR 37452, Sept. 11, 1990]

§301.75–11 [Reserved]

§301.75–12 Certificates and limited permits.

(a) *Issuance and withdrawal.* (1) Certificates and limited permits may be issued for the interstate movement of regulated articles only by an inspector or by persons operating under a compliance agreement.

(2) A certificate or limited permit may be withdrawn by an inspector if the inspector determines that any of the applicable requirements of this subpart have not been met. The decision of the inspector and the reason for the withdrawal must be confirmed in writing as promptly as circumstances allow. Any person whose certificate or limited permit is withdrawn may appeal the decision in writing to the Administrator within 10 days after receiving the written notification. The appeal must state all of the facts and reasons upon which the person relies to show that the certificate or limited permit was wrongfully withdrawn. The Administrator must grant or deny the appeal in writing, stating the reasons for the decision, as promptly as circumstances allow. If there is a conflict as to any material fact, a hearing will be held to resolve the conflict. Rules of practice concerning the hearing will be adopted by the Administrator.

(b) *Attachment and disposition.* (1) Except as provided in §301.75–6(b)(8) for kumquat plants, or in §301.75–6(c)(4) through (c)(5) for any regulated nursery stock, certificates and limited permits accompanying regulated articles interstate must be attached during the interstate movement to one of the following:

(i) The outside of the regulated article, if the regulated article is not packed in a container, or

(ii) The outside of the container in which the regulated article is packed, or

(iii) The consignee's copy of the accompanying waybill, but only if the regulated article is described on the certificate, limited permit, or waybill in a way that allows the regulated article to be identified.

(2) Certificates and limited permits accompanying regulated articles interstate must be given to the consignee at the point of destination.

[55 FR 37453, Sept. 11, 1990, as amended at 72 FR 13428, Mar. 22, 2007; 74 FR 16104, Apr. 9, 2009; 76 FR 23457, Apr. 27, 2011]

§301.75–13 Compliance agreements.

(a) *Eligibility.* Any person engaged in the business of growing or handling regulated articles for interstate movement may enter into a compliance agreement with the Animal and Plant Health Inspection Service to facilitate the interstate movement of regulated articles in accordance with this subpart. Compliance agreements may be arranged by contacting a local office of Plant Protection and Quarantine, Animal and Plant Health Inspection Service (listed in local telephone directories), or by contacting the Animal and Plant Health Inspection Service, Plant Protection and Quarantine, Domestic and Emergency Operations, 4700 River Road Unit 134, Riverdale, Maryland 20737–1236.

(b) *Cancellation.* Any compliance agreement may be cancelled orally or in writing by an inspector if the inspector finds that the person who entered into the compliance agreement has failed to comply with this subpart, or any term or condition of the compliance agreement itself. If the person is given notice of cancellation orally, written confirmation of the decision and the reasons for it must be provided as promptly as circumstances allow. Any person whose compliance agreement is cancelled may appeal the decision in writing to the Administrator within 10 days after receiving the written notification. The appeal must state all of the facts and reasons upon which the person relies to show that the compliance agreement was wrongfully cancelled. The Administrator must grant or deny the appeal, in writing, stating the reasons for the decision, as promptly as circumstances allow. If there is a conflict as to any material fact, a hearing will be held to resolve the conflict. Rules of practice concerning the hearing will be adopted by the Administrator.

[55 FR 37453, Sept. 11, 1990, as amended at 59 FR 67608, Dec. 30, 1994; 76 FR 23457, Apr. 27, 2011]

§ 301.75–14 Costs and charges.

The services of the inspector shall be furnished without cost. The United States Department of Agriculture will not be responsible for any costs or charges incident to inspections or compliance with the provisions in this subpart, other than for the services of the inspector.

[50 FR 51231, Dec. 13, 1985. Redesignated at 55 FR 37450, Sept. 11, 1990]

§ 301.75–15 Funds for the replacement of commercial citrus trees.

Subject to the availability of appropriated funds, the owner of a commercial citrus grove may be eligible to receive funds to replace commercial citrus trees in accordance with the provisions of this section.

(a) *Eligibility.* The owner of a commercial citrus grove may be eligible to receive funds to replace commercial citrus trees removed to control citrus canker if the trees were removed pursuant to a public order between 1986 and 1990 or on or after September 28, 1995.

(b) *Tree replacement payments.* The owner of a commercial citrus grove who is eligible under paragraph (a) of this section to receive funds to replace commercial citrus trees will, upon approval of an application submitted in accordance with paragraph (c) of this section, receive a payment of $26 per tree up to the following per-acre maximum payments:

Variety	Maximum payment per acre
Grapefruit, red seedless	$2,704
Orange, Valencia	3,198
Orange, early/midseason/navel	3,068

Variety	Maximum payment per acre
Tangelo	2,964
Lime	4,004
Other or mixed citrus	2,704

(c) *How to apply for tree replacement funds.* The form necessary to apply for funds to replace commercial citrus trees may be obtained from any local citrus canker eradication program office in Florida, or from the USDA Citrus Canker Eradication Program, 6901 West Sunrise Boulevard, Plantation, FL 33313. The completed application should be accompanied by a copy of the public order directing the destruction of the trees and its accompanying inventory that describes the number and the variety of trees removed. Your completed application must be sent to the USDA Citrus Canker Eradication Program, Attn: Commercial Tree Replacement Program, c/o Division of Plant Industry, 3027 Lake Alfred Road, Winter Haven, FL 33881. Claims for trees destroyed on or before October 16, 2000, must have been received on or before December 15, 2000. Claims for trees destroyed after October 16, 2000, must be received within 60 days after the destruction of the trees. The Administrator may, on a case-by-case basis, approve the consideration of late claims when it appears that the claim was late through no fault of the owner of the trees, in the opinion of the Administrator. However, any request for consideration of a late claim must be submitted to the Administrator on or before August 19, 2002 for trees destroyed on or before August 17, 2001, and within 1 year after the destruction of the trees for trees destroyed after August 17, 2001.

(Approved by the Office of Management and Budget under control number 0579–0163)

[65 FR 61080, Oct. 16, 2000, as amended at 66 FR 43066, Aug. 17, 2001]

§ 301.75–16 Payments for the recovery of lost production income.

Subject to the availability of appropriated funds, the owner of a commercial citrus grove may be eligible to receive payments in accordance with the provisions of this section to recover income from production that was lost as the result of the removal of commercial citrus trees to control citrus canker.

(a) *Eligibility.* The owner of a commercial citrus grove may be eligible to receive payments to recover income from production that was lost as the result of the removal of commercial citrus trees to control citrus canker if the trees were removed pursuant to a public order between 1986 and 1990 or on or after September 28, 1995.

(b) *Calculation of payments.* (1) The owner of a commercial citrus grove who is eligible under paragraph (a) of this section to receive payments to recover lost production income will, upon approval of an application submitted in accordance with paragraph (c) of this section, receive a payment calculated using the following rates:

Citrus variety	Payment (per acre)
Grapefruit	$3,342
Orange, Valencia, and tangerine	6,446
Orange, navel (includes early and midseason oranges)	6,384
Tangelo	1,989
Lime	6,503
Other or mixed citrus	3,342

(2) *Payment adjustments.* (i) In cases where the owner of a commercial citrus grove had obtained ACC coverage for trees in his or her grove and received crop insurance payments following the destruction of the insured trees, the payment provided for under paragraph (b)(1) of this section will be reduced by the total amount of the crop insurance payments received by the commercial citrus grove's owner for the insured trees.

(ii) In cases where ACC coverage was available for trees in a commercial citrus grove but the owner of the grove had not obtained ACC coverage for his or her insurable trees, the per-acre payment provided for under paragraph (b)(1) of this section will be reduced by 5 percent.

(c) *How to apply for lost production payments.* The form necessary to apply for lost production payments may be obtained from any local citrus canker eradication program office in Florida, or from the USDA Citrus Canker Eradication Program, 6901 West Sunrise Boulevard, Plantation, FL 33313. The completed application should be accompanied by a copy of the public order directing the destruction of the trees and its accompanying inventory that describes the acreage, number, and the variety of trees removed. Your completed application must be sent to the USDA Citrus Canker Eradication Program, Attn: Lost Production Payments Program, c/o Division of Plant Industry, 3027 Lake Alfred Road, Winter Haven, FL 33881. Claims for losses attributable to the destruction of trees on or before the effective date of this rule must be received on or before September 17, 2001. Claims for losses attributable to the destruction of trees after the effective date of this rule must be received within 60 days after the destruction of the trees. The Administrator may, on a case-by-case basis, approve the consideration of late claims when the circumstances appear, in the opinion of the Administrator, to warrant such consideration. However, any request for consideration of a late claim must be submitted to the Administrator on or before July 18, 2002 for trees destroyed on or before July 18, 2001, and within 1 year after the destruction of the trees for trees destroyed after July 18, 2001.

[66 FR 32717, June 18, 2001; 66 FR 33740, June 25, 2001; 71 FR 33172, June 8, 2006]

§301.75–17 Funds for the replacement of certified citrus nursery stock.

Subject to the availability of appropriated funds, a commercial citrus nursery may be eligible to receive funds to replace certified citrus nursery stock in accordance with the provisions of this section.

(a) *Eligibility.* A commercial citrus nursery may be eligible to receive funds to replace certified citrus nursery stock removed to control citrus canker if the nursery stock was removed pursuant to a public order after September 30, 2001, and before January 10, 2006.

(b) *Certified citrus nursery stock payments.* A commercial citrus nursery that is eligible under paragraph (a) of this section to receive funds to replace certified citrus nursery stock will, upon approval of an application submitted in accordance with paragraph (c) of this section, receive a payment calculated using the following rates:

Type of certified nursery stock	Payment (dollars)
Seedlings	0.18/plant.
Liners or rootstock	1.50/plant.
Budded field grown citrus plants	4.00/plant.
Budded container/greenhouse citrus plants.	4.50/plant.
Citrus nursery stock in containers for wholesale or retail sale:	
1 gallon	5.00/container.
3 gallon	10.00/container.
5 gallon	15.00/container.
7 gallon	20.00/container.
Larger than 7 gallon	26.00/container.

(c) *How to apply for certified nursery stock replacement funds.* The form necessary to apply for funds to replace certified nursery stock may be obtained from any local citrus canker eradication program office in Florida, or from the USDA Citrus Canker Eradication Program, 6901 West Sunrise Boulevard, Plantation, FL 33313. The completed application should be accompanied by a copy of the public order directing the destruction of the trees and its accompanying inventory that describes the number and type of the certified nursery stock removed. If the certified nursery stock was planted in pots, the inventory should specify the size of the container. If the certified nursery stock was bare root plants or in a temporary container, the inventory should specify whether the plant was non-budded or budded. The completed application must be sent to the USDA Citrus Canker Eradication Program, Attn: Commercial Compensation, 10300 Sunset Dr., Suite 150, Miami, FL 33173. Claims for certified nursery stock must be received by August 7, 2006.

[71 FR 33172, June 8, 2006]

Subpart N—Citrus Greening and Asian Citrus Psyllid

SOURCE: 75 FR 34332, June 17, 2010, unless otherwise noted. Redesignated at 84 FR 2428, Feb. 7, 2019.

§ 301.76 Restrictions on the interstate movement of regulated articles.

No person may move interstate from any quarantined area any articles regulated for citrus greening and Asian citrus psyllid, except in accordance with this subpart.[1]

§ 301.76–1 Definitions.

Administrator. The Administrator of the Animal and Plant Health Inspection Service or any individual authorized to act for the Administrator.

Animal and Plant Health Inspection Service (APHIS). The Animal and Plant Health Inspection Service of the United States Department of Agriculture.

Asian citrus psyllid. The insect known as Asian citrus psyllid (*Diaphorina citri* Kuwayama) in any stage of development.

Certificate. A document, stamp, or other means of identification approved by APHIS and issued by an inspector or person operating under a compliance agreement when he or she finds that, because of certain conditions, a regulated article can be moved safely from an area quarantined for Asian citrus psyllid and/or citrus greening without spreading the psyllid or the disease.

Citrus greening. A plant disease, also commonly referred to as Huanglongbing disease of citrus, that is caused by several strains of the uncultured, phloem-limited bacterial pathogen "*Candidatus* Liberibacter asiaticus".

Commercial citrus grove. A solid-set planting of trees maintained for the primary purpose of producing citrus fruit for commercial sale.

Compliance agreement. A written agreement between APHIS and a person engaged in the business of growing, maintaining, processing, handling, packing, or moving regulated articles for interstate movement, in which the person agrees to comply with this subpart. For the purposes of this subpart, a memorandum of understanding is considered a compliance agreement.

EPA. The U.S. Environmental Protection Agency.

Established population. Presence of Asian citrus psyllid within an area

---

[1] In order to enforce this section, any properly identified inspector is authorized to stop and inspect persons and means of conveyance and to seize, quarantine, treat, apply other remedial measures to, destroy, or otherwise dispose of host articles as provided in sections 414, 421, and 434 of the Plant Protection Act (7 U.S.C. 7714, 7731, and 7754).

that the Administrator determines is likely to persist for the foreseeable future.

Inspector. An individual authorized by the Administrator to perform the duties required under this subpart.

Interstate. From any State into or through any other State.

Limited permit. A document issued by an inspector or person operating under a compliance agreement to allow the interstate movement of regulated articles to a specified destination, for specified handling, processing, or utilization.

Moved (move, movement). Shipped, offered for shipment, received for transportation, transported, carried (whether on one's person or by any other means of conveyance), or allowed to be moved, shipped, transported, or carried. For the purposes of this subpart, movements include any type of shipment, including mail and Internet commerce.

Nursery. Any commercial location where nursery stock is grown, propagated, stored, maintained, or sold, or any location from which nursery stock is distributed.

Nursery stock. Any plants or plant parts, excluding fruit, intended to be planted, to remain planted, or to be replanted. Nursery stock includes, but is not limited to, trees, shrubs, cuttings, grafts, scions, and buds.

Person. Any association, company, corporation, firm, individual, joint stock company, partnership, society, or other entity.

Port. Any place designated by the President, Secretary of the Treasury, or Congress at which a Customs officer is assigned with authority to accept entries of merchandise, to collect duties, and to enforce the various provisions of the Customs and Navigation laws in force at that place.

Quarantined area. Any State or portion of a State designated as a quarantined area for Asian citrus psyllid or citrus greening in accordance with §301.76–3.

Regulated article. Any article listed in §301.76–2 or otherwise designated as a regulated article in accordance with §301.76–2(c).

State. The District of Columbia, Puerto Rico, the Northern Mariana Islands, or any State, territory, or possession of the United States.

[75 FR 34332, June 17, 2010, as amended at 77 FR 59712, Oct. 1, 2012]

§301.76–2 Regulated articles for Asian citrus psyllid and citrus greening.

The following are regulated articles for Asian citrus psyllid and citrus greening:

(a) All plants and plant parts (including leaves), except fruit, of: *Aegle marmelos, Aeglopsis chevalieri, Afraegle gabonensis, A. paniculata, Amyris madrensis, Atalantia* spp. (including *Atalantia monophylla), Balsamocitrus dawei, Bergera (=Murraya) koenigii, Calodendrum capense, Choisya ternate, C. arizonica,* X *Citroncirus webberi, Citropsis articulata, Citropsis gilletiana, Citrus madurensis (=* X *Citrofortunella microcarpa), Citrus* spp., *Clausena anisum-olens, C. excavata, C. indica, C. lansium, Eremocitrus glauca, Eremocitrus* hybrid, *Esenbeckia berlandieri, Fortunella* spp., *Limonia acidissima, Merrillia caloxylon, Microcitrus australasica, M. australis, M. papuana,* X *Microcitronella* spp., *Murraya* spp., *Naringi crenulata, Pamburus missionis, Poncirus trifoliata, Severinia buxifolia, Swinglea glutinosa, Tetradium ruticarpum, Toddalia asiatica, Triphasia trifolia, Vepris (=Toddalia) lanceolata,* and *Zanthoxylum fagara.*

(b) Propagative seed of the species listed in paragraph (a) of this section is considered a host of citrus greening but not a host of Asian citrus psyllid. Therefore, notwithstanding the other provisions of this subpart, the movement of propagative seed of these species from an area quarantined for citrus greening is prohibited, while the movement of such seed from an area quarantined only for Asian citrus psyllid, but not for citrus greening, is allowed without restriction.

(c) Any other product, article, or means of conveyance may be designated a regulated article for Asian citrus psyllid or citrus greening, if an inspector determines that it presents a risk of spreading these pests, and after the inspector provides written notification to the person in possession of the product, article, or means of conveyance that it is subject to the restrictions of this subpart.

(d) Plant parts of the species listed in paragraph (a) of this section may be exempted from the regulations in this subpart, provided that the parts have been processed such that an inspector determines they no longer present a risk of spreading Asian citrus psyllid or citrus greening.

§ 301.76–3 Quarantined areas; citrus greening and Asian citrus psyllid.

(a) The Administrator will designate an area as a quarantined area for citrus greening or as a quarantined area for Asian citrus psyllid in accordance with the criteria listed in paragraph (c) of this section. The Administrator will publish a description of all areas quarantined for citrus greening or Asian citrus psyllid on the Plant Protection and Quarantine (PPQ) Web site: (*http:// www.aphis.usda.gov/plant_health/ plant_pest_info/citrus_greening/ index.shtml*). The description of each quarantined area will include the date the description was last updated and a description of any changes that have been made to the quarantined area. Lists of all quarantined areas may also be obtained by request from any local office of PPQ; local offices are listed in telephone directories and on the Internet at (*http://www.aphis.usda.gov/services/report_pest_disease/report_pest_disease.shtml*). After a change is made to the description of quarantined areas, we will publish a notice in the FEDERAL REGISTER informing the public that the change has occurred and describing the change to the quarantined areas.

(b) *Designation of an area less than an entire State as a quarantined area.* Less than an entire State will be designated as a quarantined area for citrus greening or the Asian citrus psyllid only if the Administrator determines that:

(1) The State has adopted and is enforcing restrictions on the intrastate movement of regulated articles that are equivalent to those imposed by this subpart on the interstate movement of regulated articles; and

(2) The designation of less than the entire State as a quarantined area will prevent the interstate spread of citrus greening or Asian citrus psyllid.

(c) *Criteria for designation of a State, or a portion of a State, as a quarantined area for citrus greening or Asian citrus psyllid.* (1) A State, or portion of a State, will be designated as a quarantined area for citrus greening when the presence of citrus greening is confirmed within the area by an APHIS-administered test.

(2) A State, or portion of a State, will be designated as a quarantined area for Asian citrus psyllid in which an established population of Asian citrus psyllids has been detected.

(3) A State, or portion of a State, will be designated as a quarantined area for either citrus greening or Asian citrus psyllid if the Administrator considers it necessary to quarantine the area because of its inseparability for quarantine enforcement purposes from localities in which citrus greening or an established population of Asian citrus psyllids has been found.

§ 301.76–4 Labeling requirements for regulated nursery stock produced within an area quarantined for citrus greening.

(a) Effective September 15, 2010, except as provided in paragraphs (b) and (c) of this section, all regulated nursery stock offered for commercial sale within an area quarantined for citrus greening must have an APHIS-approved plastic or metal tag on which a statement alerting consumers to Federal prohibitions regarding the interstate movement of the article is prominently and legibly displayed. Alternatively, if the article is destined for commercial sale in a box or container, the statement may be printed on the box or container, or printed on a label permanently affixed to the box or container, provided that, in either case, the statement is prominently and legibly displayed. The operator of the site of propagation of the nursery stock and the person offering the plants for commercial sale are jointly responsible for all such labeling.

(b) Nursery stock produced within a quarantined area for planting in a commercial citrus grove within that same area and moved directly to that grove, without movement outside of the quarantined area, may be moved without being labeled in accordance with paragraph (a) of this section.

(c) Nursery stock that will be moved interstate in accordance with §301.76–7 may be moved without being labeled in accordance with paragraph (a) of this section.

(Approved by the Office of Management and Budget under control number 0579–0363)

[75 FR 34332, June 17, 2010, as amended at 76 FR 23457, Apr. 27, 2011]

§301.76–5 General conditions governing the issuance of any certificate or limited permit; provisions for cancellation of a certificate or limited permit.

(a) *Certificates.* In addition to all other relevant conditions within this subpart, an inspector or person operating under a compliance agreement will issue a certificate only if a regulated article:

(1) Will be moved in compliance with any additional emergency conditions that the Administrator may impose under section 414 of the Plant Protection Act (7 U.S.C. 7714)[2] to prevent the spread of Asian citrus psyllid; and

(2) Is eligible for unrestricted movement under all other Federal domestic plant quarantines and regulations applicable to the article.

(b) *Limited permits.* In addition to all other relevant conditions within this subpart, an inspector or person operating under a compliance agreement may issue a limited permit for the interstate movement of a regulated article only if the regulated article:

(1) Is to be moved interstate to a specified destination for specified handling, processing, or utilization (the destination and other conditions to be listed in the limited permit) and this movement of the regulated article will not result in the spread of citrus greening or the Asian citrus psyllid;

(2) Is to be moved in compliance with any additional emergency conditions the Administrator may impose under section 414 of the Plant Protection Act (7 U.S.C. 7714) to prevent the spread of citrus greening and the Asian citrus psyllid; and

---

[2] An inspector may hold seize, quarantine, treat, apply other remedial measures to, destroy, or otherwise dispose of plants, plant pests, or other articles in accordance with sections 414, 421, and 423 of the Plant Protection Act (7 U.S.C. 7714, 7731, and 7754).

(3) Is eligible for interstate movement under all other Federal domestic plant quarantines and regulations applicable to the article.

(c) Certificates and limited permits for the interstate movement of a regulated article may be issued by an inspector or person operating under a compliance agreement. A person operating under a compliance agreement may issue a certificate for the interstate movement of a regulated article after he or she has determined that the article is eligible for a certificate in accordance with paragraph (a) of this section and all other relevant conditions of this subpart. A person operating under a compliance agreement may issue a limited permit for interstate movement of a regulated article after he or she has determined that the article is eligible for a limited permit in accordance with paragraph (b) of this section and all other relevant conditions of this subpart.

(d) Any certificate or limited permit that has been issued may be withdrawn, either orally or in writing, by an inspector if he or she determines that the holder of the certificate or limited permit has not complied with all of the provisions in this subpart or has not complied with all the conditions contained in the certificate or limited permit. If the withdrawal is oral, the withdrawal and the reasons for the withdrawal will be confirmed in writing as soon as circumstances allow. Any person whose certificate or limited permit has been withdrawn may appeal the decision in writing to the Administrator within 10 days after receiving the written notification of the withdrawal. The appeal must state all of the facts and reasons upon which the person relies to show that the certificate or limited permit was wrongfully withdrawn. As promptly as circumstances allow, the Administrator will grant or deny the appeal, in writing, stating the reasons for the decision. A hearing will be held to resolve any conflict as to any material fact. Rules of practice concerning a hearing will be adopted by the Administrator.

(e) Unless specific provisions exist in §301.76–6 or §301.76–7 of this subpart to allow the interstate movement of a

certain regulated article, the interstate movement of that article is prohibited.

(Approved by the Office of Management and Budget under control number 0579–0363)

§ 301.76–6 Additional conditions for issuance of certificates and limited permits for regulated articles moved interstate from areas quarantined for Asian citrus psyllid, but not for citrus greening. .

(a) *Additional conditions for issuance of a certificate; any regulated article.* In addition to the general conditions for issuance of a certificate contained in § 301.76–5(a), an inspector or person operating under a compliance agreement may issue a certificate for the interstate movement of any regulated article to any State if:

(1) The article is treated with methyl bromide[3] in accordance with 7 CFR part 305 of this chapter. ·

(2) The article is shipped in a container that has been sealed with an agricultural seal placed by an inspector.

(3) The container that will be moved interstate is clearly labeled with the certificate.

(4) A copy of the certificate will be attached to the consignee's copy of the accompanying waybill.

(b) *Additional conditions for issuance of a certificate; regulated nursery stock.* In addition to the general conditions for issuance of a certificate contained in § 301.76–5(a), an inspector or person operating under a compliance agreement may issue a certificate for interstate movement of regulated nursery stock to any State if:

(1) The nursery in which the nursery stock is produced has entered into a compliance agreement with APHIS in which it agrees to meet the relevant construction standards, sourcing and certification requirements, cleaning, disinfecting, and safeguarding requirements, labeling requirements, and recordkeeping and inspection requirements specified in a PPQ protocol document. The protocol document will be provided to the person at the time he or she enters into the compliance

agreement.[4] The compliance agreement may also specify additional conditions determined by APHIS to be necessary in order to prevent the spread of Asian citrus psyllid under which the nursery stock must be grown, maintained, and shipped in order to obtain a certificate for its movement. The compliance agreement will also specify that APHIS may amend the agreement.

(2) An inspector determines that the nursery has adhered to all terms and conditions of the compliance agreement.

(3) The nursery stock is completely enclosed in a sealed container that is clearly labeled with the certificate and is moved interstate in that container.

(4) A copy of the certificate is attached to the consignee's copy of the accompanying waybill.

(c) *Additional conditions for issuance of a limited permit; regulated nursery stock.* (1) *Nursery stock that will not be moved through American Samoa, Arizona, California, Florida, Guam, Hawaii, the Northern Mariana Islands, Puerto Rico, Texas, or the U.S. Virgin Islands.* In addition to the general conditions for the issuance of a limited permit contained in § 301.76–5(b), an inspector or person operating under a compliance agreement, other than the operator of the nursery in which the nursery stock was produced and his or her employees, may issue a limited permit for the interstate movement of regulated nursery stock through areas of the United States other than American Samoa, Arizona, California, Florida, Guam, Hawaii, Louisiana, the Northern Mariana Islands, Puerto Rico, Texas, or the U.S. Virgin Islands, and to areas of the United States other than those portions of Arizona and California that are not quarantined due to the presence of Asian citrus psyllid or citrus greening, if:

(i) The nursery in which the nursery stock is produced has entered into a compliance agreement with APHIS in accordance with § 301.76–8;

---

[3] EPA and State and local environmental authorities may restrict the use of methyl bromide on certain articles.

[4] The protocol document is also available on the Internet at *http://www.aphis.usda.gov/plant_health/plant_pest_info/citrus/index.shtml* and may be obtained from local Plant Protection and Quarantine offices, which are listed in telephone directories.

(ii) All citrus nursery stock at the nursery has been inspected by an inspector every 30 days, and any findings of Asian citrus psyllid during an inspection have been reported to APHIS immediately;

(iii) The nursery stock is treated for Asian citrus psyllid with an APHIS-approved soil drench or in-ground granular application no more than 90 days and no fewer than 30 days before shipment, followed by an APHIS-approved foliar spray no more than 10 days before shipment. All treatments must be applied according to their EPA label, including directions on application, restrictions on place of application and other restrictions, and precautions, and including statements pertaining to Worker Protection Standards;

(iv) The nursery stock is affixed prior to movement with a plastic or metal tag on which the statement "Limited permit: USDA–APHIS–PPQ. Not for distribution in those portions of AZ and CA not quarantined due to the presence of Asian citrus psyllid or citrus greening" is prominently and legibly displayed on the obverse, and adequate information as determined by APHIS regarding the identity of the nursery stock and its source of production to conduct traceback to the nursery in which the nursery stock was produced is prominently and legibly printed on the reverse. If the nursery stock is destined for movement or sale in boxes or containers, the statement and the identifying information may be printed on the box or container, or printed on a label permanently affixed to the box or container, provided that, in either case, the statement and the identifying information are prominently and legibly displayed;

(v) A copy of the limited permit will be attached to the consignee's copy of the accompanying waybill; and

(vi) The nursery stock is shipped in accordance with the conditions specified on the limited permit to the destination specified on the permit.

(2) *Nursery stock that will be moved through American Samoa, Arizona, California, Florida, Guam, Hawaii, Louisiana, the Northern Mariana Islands, Puerto Rico, Texas, or the U.S. Virgin Islands.* In addition to the general conditions for the issuance of a limited permit contained in § 301.76–5(b), an inspector or person operating under a compliance agreement may issue a permit for the interstate movement of regulated nursery stock through American Samoa, Arizona, California, Florida, Guam, Hawaii, Louisiana, the Northern Mariana Islands, Puerto Rico, Texas, or the U.S. Virgin Islands, and to areas of the United States other than those portions of Arizona and California that are not quarantined due to the presence of Asian citrus psyllid or citrus greening, if:

(i) All conditions for movement of regulated nursery stock in paragraphs (c)(1)(i) through (c)(1)(vi) of this section are fulfilled;

(ii) The nursery stock is inspected by an inspector on the date of shipment and found free of Asian citrus psyllid;

(iii) The nursery stock is completely enclosed in a container sealed with an agricultural seal and is moved interstate in that container;

(iv) The container prominently and legibly displays the statement and identifying information specified in paragraph (c)(1)(iv) of this section;

(v) The agricultural seal remains intact throughout movement to the destination specified on the limited permit; and

(vi) The agricultural seal is removed at the destination specified on the limited permit by an inspector.

(d) *Additional conditions for issuance of a limited permit; regulated articles intended for consumption, as apparel or as a similar personal accessory, or for other decorative use.*[5] In addition to the general conditions for issuance of a limited permit contained in § 301.76–5(b), an inspector or person operating under a compliance agreement may issue a limited permit for the interstate movement of regulated articles intended for consumption, as apparel or as a similar personal accessory, or for other decorative use if:

(1) The articles are treated with irradiation in accordance with 7 CFR part 305 of this chapter at an irradiation facility that is not located in an area quarantined for citrus greening.

---

[5] Examples of such articles include *Bergera* (=*Murraya*) *koenigii* leaves, as well as *Murraya paniculata* flowers or foliage.

(2) The container that will be used to move the articles interstate is clearly labeled with the limited permit, which must contain the name of the State or portion of a State where the articles were produced and a statement that the articles were treated in accordance with 7 CFR part 305 of this chapter.

(3) A copy of the limited permit is attached to the consignee's copy of the accompanying waybill.

(Approved by the Office of Management and Budget under control number 0579–0369)

[75 FR 34332, June 17, 2010, as amended at 76 FR 23457, Apr. 27, 2011; 77 FR 59712, Oct. 1, 2012; 78 FR 63367, Oct. 24, 2013]

§ 301.76-7 Additional conditions for issuance of certificates and limited permits for regulated articles moved interstate from areas quarantined for citrus greening.

(a) *Additional conditions for the issuance of a certificate; regulated nursery stock produced within a nursery located in the quarantined area.* In addition to the general conditions for issuance of a certificate contained in § 301.76–5(a), an inspector or person operating under a compliance agreement may issue a certificate for interstate movement of regulated nursery stock to any State if all of the following conditions are met:

(1) The nursery in which the nursery stock is produced has entered into a compliance agreement with APHIS in which it agrees to meet the relevant construction standards, sourcing and certification requirements, cleaning, disinfecting, and safeguarding requirements, labeling requirements, and recordkeeping and inspection requirements specified in a PPQ protocol document. The protocol document will be provided to the person at the time he or she enters into the compliance agreement.[6] The compliance agreement may also specify additional conditions determined by APHIS to be necessary in order to prevent the dissemination of citrus greening under which the nursery stock must be

grown, maintained, and shipped in order to obtain a certificate for its movement. The compliance agreement will also specify that APHIS may amend the agreement.

(2) An inspector has determined that the nursery has adhered to all terms and conditions of the compliance agreement.

(3) The nursery stock is completely enclosed in a sealed container that is clearly labeled with the certificate and is moved interstate in that container.

(4) A copy of the certificate is attached to the consignee's copy of the accompanying waybill.

(b) *Additional conditions for issuance of a limited permit; regulated nursery stock grown, produced, or maintained at a nursery or other facility located in the quarantined area.* In addition to the general conditions for issuance of a limited permit contained in § 301.76–5(b), an inspector or person operating under a compliance agreement may issue a limited permit for the interstate movement for immediate export of regulated nursery stock grown, produced, or maintained at a nursery or other facility located in the quarantined area if:

(1) The nursery stock is treated for Asian citrus psyllid with an APHIS-approved soil drench or in-ground granular application, followed by an APHIS-approved foliar spray, in accordance with § 301.76–6(b)(1), or with methyl bromide, in accordance with 7 CFR part 305 of this chapter.

(2) The nursery stock is inspected by an inspector in accordance with § 301.76–9 and found free of Asian citrus psyllid, if treated in accordance with § 301.76–6(b)(1).

(3) The nursery stock is affixed prior to movement with a plastic or metal tag on which the statement "Limited permit: USDA-APHIS-PPQ. For immediate export only" is prominently and legibly displayed. If the nursery stock is destined for movement or sale in a box or container, the statement may be printed on the box or container, or printed on a label permanently affixed to the box or container, provided that, in either case, the statement is prominently and legibly displayed.

---

[6] The protocol document is also available on the Internet at *http://www.aphis.usda.gov/ plant_health/plant_pest_info/citrus/ index.shtml* and may be obtained from local Plant Protection and Quarantine offices, which are listed in telephone directories.

(4) The nursery stock is accompanied by a copy of this limited permit attached to the consignee's copy of the waybill.

(5) The nursery stock is moved in accordance with the conditions specified on the limited permit directly to the port of export specified on the limit permit, in a container sealed with an agricultural seal placed by an inspector.

(6) A copy of the limited permit is attached to or legibly printed on this container.

(7) The nursery stock remains in this container, and the container remains sealed, as long as the plants are within the United States.

(c) Except for nursery stock for which a limited permit has been issued in accordance with the conditions of paragraph (a) or (b) of this section, no other regulated article may be moved interstate from an area quarantined for citrus greening.

(Approved by the Office of Management and Budget under control number 0579–0369)

[75 FR 34332, June 17, 2010, as amended at 76 FR 23458, Apr. 27, 2011; 77 FR 59712, Oct. 1, 2012]

§301.76–8 Compliance agreements and cancellation.

(a) Any person involved in the growing, maintaining, processing, handling, packing, treating, or moving of regulating articles from areas quarantined for citrus greening or Asian citrus psyllid may enter into a compliance agreement when an inspector determines that the person understands this subpart, agrees to comply with its provisions, and agrees to comply with all the provisions contained in the compliance agreement. The person must also agree to maintain and offer for inspection such records as are necessary to demonstrate continual adherence to the requirements of the regulations and the provisions of the compliance agreement.[7]

---

[7] Compliance agreement forms are available without charge from the Animal and Plant Health Inspection Service, Plant Protection and Quarantine, Domestic and Emergency Operations, 4700 River Road Unit 134, Riverdale, MD 20737–1236, and from local offices of the Plant Protection and Quarantine

(b) Any compliance agreement may be canceled, either orally or in writing, by an inspector whenever the inspector finds that the person who has entered into the compliance agreement has failed to comply with this subpart, or any term or condition of the compliance agreement itself. If the cancellation is oral, the cancellation and the reasons for the cancellation will be confirmed in writing as promptly as circumstances allow. Any person whose compliance agreement has been canceled may appeal the decision, in writing, within 10 days after receiving written notification of the cancellation. The appeal must state all of the facts and reasons upon which the person relies to show that the compliance agreement was wrongly canceled. As promptly as circumstances allow, the Administrator will grant or deny the appeal, in writing, stating the reasons for the decision. A hearing will be held to resolve any conflict as to any material fact. Rules of practice concerning a hearing will be adopted by the Administrator.

(Approved by the Office of Management and Budget under control numbers 0579–0363 and 0579–0369)

[75 FR 34332, June 17, 2010, as amended at 76 FR 23457, 23459, Apr. 27, 2011; 77 FR 59712, Oct. 1, 2012]

§301.76–9 Inspection of regulated nursery stock.

All nursery stock intended for interstate movement for immediate export from an area quarantined for citrus greening, must be inspected by an inspector[8] no more than 72 hours prior to movement. The person who desires to move the articles interstate must notify the inspector as far in advance of the desired interstate movement as possible. The articles must be inspected at the place and in the manner the inspector designates as necessary

---

offices, which are listed in telephone directories.

[8] Inspectors are assigned to local offices of APHIS, which are listed in local telephone directories. Information concerning local offices may also be obtained from the Animal and Plant Health Inspection Service, Plant Protection and Quarantine, Domestic and Emergency Operations, 4700 River Road Unit 134, Riverdale, MD 20737–1236.

to comply with this subpart. If the inspector has reason to believe that the interstate movement of the articles may lead to the artificial spread of citrus greening or Asian citrus psyllid, he or she may deny issuance of a limited permit for interstate movement of the article or take other remedial measures to prohibit such spread.

(Approved by the Office of Management and Budget under control number 0579–0363)

[75 FR 34332, June 17, 2010, as amended at 76 FR 23457, 23459, Apr. 27, 2011; 77 FR 59712, Oct. 1, 2012]

§ 301.76–10 Attachment and disposition of certificates and limited permits.

(a) A certificate or limited permit required for the interstate movement of a regulated article, or a copy thereof, must, at all times during the interstate movement, be:

(1) Attached to or legibly printed on the outside of the container containing the regulated article or attached to the regulated article itself, if the article is not packed in a container; and

(2) Attached to or legibly printed on the sealed container in which the article is shipped; and

(3) Attached to the consignee's copy of the accompanying waybill. The host article must be sufficiently described on the certificate or limited permit and on the waybill to identify the article.

(b) The certificate or limited permit for the interstate movement of a host article must be furnished by the carrier or the carrier's representative to the consignee listed on the certificate or limited permit upon arrival at the location provided on the certificate or limited permit.

§ 301.76–11 Costs and charges.

The services of the inspector during normal business hours (8 a.m. to 4:30 p.m., Monday through Friday, except holidays) will be furnished without cost. APHIS will not be responsible for any costs or charges incident to inspections or compliance with the provisions of the quarantine and regulations in this subpart, other than for the services of the inspector.

Subpart O—Witchweed

QUARANTINE AND REGULATIONS

§ 301.80 Quarantine; restriction on interstate movement of specified regulated articles.

(a) *Notice of quarantine.* Under the authority of sections 411, 412, 414, and 434 of the Plant Protection Act (7 U.S.C. 7711, 7712, 7714, and 7754), the Secretary of Agriculture quarantines the States of North Carolina and South Carolina in order to prevent the spread of witchweed (*Striga* spp.), a parasitic plant that causes a dangerous disease of corn, sorghum, and other crops of the grass family and is not widely prevalent or distributed within and throughout the United States. Through the aforementioned authorities, the Secretary imposes a quarantine on the States of North Carolina and South Carolina with respect to the interstate movement from those States of articles described in paragraph (b) of this section, issues regulations in this subpart governing the movement of such articles, and gives notice of this quarantine action.

(b) *Quarantine restrictions on interstate movement of specified regulated articles.* No common carrier or other person shall move interstate from any quarantined State any of the following articles (defined in § 301.80–1(p) as regulated articles), except in accordance with the conditions prescribed in this subpart:

(1) Soil, compost, peat, humus, muck, and decomposed manure, separately or with other things; sand; and gravel.

(2) Plants with roots.

(3) Grass sod.

(4) Plant crowns and roots for propagation.

(5) True bulbs, corms, rhizomes, and tubers of ornamental plants.

(6) Root crops, except those from which all soil has been removed.

(7) Peanuts in shells and peanut shells, except boiled or roasted peanuts.

(8) Small grains and soybeans.

(9) Hay, straw, fodder, and plant litter of any kind.

(10) Seed cotton and gin trash.

(11) Stumpwood.

(12) Long green cucumbers, cantaloupes, peppers, squash, tomatoes, and

watermelons, except those from which all soil has been removed.

(13) Pickling cucumbers, string beans, and field peas.

(14) Cabbage, except firm heads with loose outer leaves removed.

(15) Leaf tobacco, except flue-cured leaf tobacco.

(16) Ear corn, except shucked ear corn.

(17) Sorghum.

(18) Used crates, boxes, burlap bags, and cotton-picking sacks, and other used farm products containers.

(19) Used farm tools.

(20) Used mechanized cultivating equipment and used harvesting equipment.

(21) Used mechanized soil-moving equipment.

(22) Any other products, articles, or means of conveyance, of any character whatsoever, not covered by paragraphs (b) (1) through (20) of this section, when it is determined by an inspector that they present a hazard of spread of witchweed, and the person in possession thereof has been so notified.

[35 FR 10553, June 30, 1970, as amended at 36 FR 24917, Dec. 24, 1971; 41 FR 27372, July 2, 1976; 58 FR 216, Jan. 5, 1993; 66 FR 21052, Apr. 27, 2001]

§ 301.80–1 Definitions.

Terms used in the singular form in this subpart shall be deemed to import the plural, and vice versa, as the case may demand. The following terms, when used in this subpart, shall be construed, respectively, to mean:

Certificate. A document issued or authorized to be issued under this subpart by an inspector to allow the interstate movement of regulated articles to any destination.

Compliance agreement. A written agreement between a person engaged in growing, handling, or moving regulated articles, and the Plant Protection and Quarantine Programs, wherein the former agrees to comply with the requirements of this subpart identified in the agreement by the inspector who executes the agreement on behalf of the Plant Protection and Quarantine Programs as applicable to the operations of such person.

Deputy Administrator. The Deputy Administrator of the Plant Protection

and Quarantine Programs, Animal and Plant Health Inspection Service, U.S. Department of Agriculture, or any other officer or employee of said Service to whom authority to act in his stead has been or may hereafter be delegated.

Farm tools. An instrument worked or used by hand, e.g., hoes, rakes, shovels, axes, hammers, and saws.

Generally infested area. Any part of a regulated area not designated as a suppressive area in accordance with § 301.80–2.

Infestation. The presence of witchweed or the existence of circumstances that make it reasonable to believe that witchweed is present.

Inspector. Any employee of the Plant Protection and Quarantine Programs, Animal and Plant Health Inspection Service, U.S. Department of Agriculture, or other person, authorized by the Deputy Administrator to enforce the provisions of the quarantine and regulations in this subpart.

Interstate. From any State into or through any other State.

Limited permit. A document issued, or authorized to be issued by an inspector to allow the interstate movement of noncertifiable regulated articles to a specified destination for limited handling, utilization, or processing, or for treatment.

Mechanized cultivating equipment; and mechanized harvesting equipment. Mechanized equipment used for soil tillage, including tillage attachments for farm tractors, e.g., tractors, disks, plows, harrows, planters, and subsoilers; mechanized equipment used for harvesting purposes, e.g., mechanical cotton harvesters, hay balers, corn pickers, and combines.

Mechanized soil-moving equipment. Mechanized equipment used to move or transport soil, e.g., draglines, bulldozers, road scrapers, and dumptrucks.

Moved (movement, move). Shipped, offered for shipment to a common carrier, received for transportation or transported by a common carrier, or carried, transported, moved or allowed to be moved by any means. "Movement" and "move" shall be construed accordingly.

Person. Any individual, corporation, company, society, or association, or other organized group of any of the foregoing.

Plant Protection and Quarantine Programs. The organizational unit with the Animal and Plant Health Inspection Service, U.S. Department of Agriculture, delegated responsibility for enforcing provisions of the Plant Protection Act and related legislation, quarantines, and regulations.

Regulated area. Any quarantined State, or any portion thereof, designated as a regulated area in § 301.80–2a or otherwise designated as a regulated area in accordance with § 301.80–2(b).

Regulated articles. Any articles described in § 301.80(b).

Restricted destination permit. A document issued or authorized to be issued by an inspector to allow the interstate movement of regulated articles not certifiable under all applicable Federal domestic plant quarantines to a specified destination for other than scientific purposes.

Scientific permit. A document issued by the Deputy Administrator to allow the interstate movement to a specified destination of regulated articles for scientific purposes.

Soil. That part of the upper layer of earth in which plants can grow.

State. Any State, territory, or district of the United States, including Puerto Rico.

Suppressive' area. That portion of a regulated area where eradication of infestation is undertaken as an objective, as designated by the Deputy Administrator under § 301.80–2(a).

Treatment Manual. The provisions currently contained in the "Manual of Administratively Authorized Procedures to be Used Under the Witchweed Quarantine" and the "Fumigation Procedures Manual" and any amendments thereto.[1]

Witchweed. Parasitic plants of the genus *Striga* and reproductive parts thereof, including seeds.

[41 FR 27372, July 2, 1976, as amended at 66 FR 21052, Apr. 27, 2001]

§ 301.80–2 Authorization to designate, and terminate designation of, regulated areas and suppressive or generally infested areas; and to exempt articles from certification, permit, or other requirements.

(a) *Regulated areas and suppressive or generally infested areas.* The Deputy Administrator shall designate as regulated areas, in a supplemental regulation designated as § 301.80–2a, each quarantined State, or each portion thereof in which witchweed has been found or in which there is reason to believe that witchweed is present or which it is deemed necessary to regulate because of its proximity to infestation or its inseparability for quarantine enforcement purposes from infested localities. The Deputy Administrator, in the supplemental regulation, may designate any regulated area or portion thereof, as a suppressive area or a generally infested area in accordance with the definitions thereof in § 301.80–1. Less than an entire quarantined State will be designated as a regulated area only if the Deputy Administrator is of the opinion that:

(1) The State has adopted and is enforcing a quarantine or regulation which imposes restrictions on the intrastate movement of the regulated articles which are substantially the same as those which are imposed with respect to the interstate movement of such articles under this subpart; and

(2) The designation of less than the entire State as a regulated area will otherwise be adequate to prevent the interstate spread of witchweed.

(b) *Temporary designation of regulated areas and suppressive or generally infested areas.* The Deputy Administrator or an authorized inspector may temporarily designate any other premises in a quarantined State as a regulated area and may designate the regulated area or portions thereof as a suppressive or generally infested area, in accordance with the criteria specified in paragraph (a) of this section for designating such area, by serving written notice thereof

---

[1] Pamphlets containing such provisions are available upon request to the Deputy Administrator, Plant Protection and Quarantine Programs, Animal and Plant Health Inspection Service, U.S. Department of Agriculture, Washington, DC 20250, or from an inspector.

on the owner or person in possession of such premises, and thereafter the interstate movement of regulated articles from such premises by any person having notice of the designation shall be subject to the applicable provisions of this subpart. As soon as practicable, such premises shall be added to the list in §301.80–2a if a basis then exists for their designation.

(c) *Termination of designation as a regulated area and a suppressive or generally infested area.* The Deputy Administrator shall terminate the designation provided for under paragraph (a) of this section of any area designated as a regulated area or a suppressive or a generally infested area when he determines that such designation is no longer required under the criteria specified in paragraph (a) of this section. The Deputy Administrator or an inspector shall terminate the designation provided for under paragraph (b) of this section of any premises designated as a regulated area or a suppressive or a generally infested area when he determines that such designation is no longer required under the criteria specified in paragraph (a) of this section, and notice thereof shall be given to the owner or person in possession of the premises.

(d) *Exemption of articles from certification, permit, or other requirements.* The Deputy Administrator may, in a supplemental regulation designated as §301.80–2b, list regulated articles or movements of regulated articles which shall be exempt from the certification, permit, or other requirements of this subpart under such conditions as he may prescribe, if he finds that facts exist as to the pest risk involved in the movement of such regulated articles which make it safe to so relieve such requirements.

[41 FR 27372, July 2, 1976]

§301.80–2a Regulated areas; generally infested and suppressive areas.

The civil divisions and parts of civil divisions described below are designated as witchweed regulated areas within the meaning of this subpart.

NORTH CAROLINA

(1) *Generally infested areas.* None.
(2) *Suppressive areas.*

Bladen County. That area located north and east of the Cape Fear River.

The Hardison, H.B., farm located on a field road 0.25 mile northwest of its intersection with State Secondary Road 1719 and 0.2 mile west of its intersection with State Secondary Road 1797.

Cumberland County. That area bounded on the west by the Cape Fear River, then by a line running east and northeast along the Fayetteville city limits to U.S. Highway 301, then northeast on U.S. Highway 301 to Interstate 95, then northeast on Interstate 95 to U.S. Highway 13, then east and northeast on U.S. Highway 13 to the Cumberland-Sampson County line.

The Barker, P.R., farm located on the south side of State Secondary Road 2242, 0.2 mile south of Interstate 95 on State Secondary Road 2252.

The Jackson, Ellis, farm located on the west side of State Secondary Road 1003 and 0.4 mile south of its intersection with N.C. Highway 59.

The Lovick, Eugene, farm located on the north side of State Secondary Road 1732 and 0.9 mile west of its intersection with U.S. Highway 301.

The McLaughlin, Cornell, farm located on the south side of State Secondary Road 2221 and 0.2 mile east of its intersection with State Secondary Road 2367.

The Thigpen, William, farm located on the south side of State Secondary Road 2212 and 1 mile west of its intersection with N.C. Highway 87.

Pender County. The Cones Folly farm located along a farm road 2.3 miles south of its intersection with State Secondary Road 1201 and 2.2 miles southeast of its intersection with State Secondary Road 1200.

Robeson County. That area bounded on the west by the Robeson County/Scotland County line; then by a line running east along the Robeson County/Hoke County line to N.C. Highway 211; then southeast on N.C. Highway 211 to the Robeson County/Bladen County line; then south along the Robeson County/Bladen County line and the Robeson County/Columbus County line to U.S. Highway 74; then northwest on U.S. Highway 74 to N.C. Highway 41; then south on N.C. Highway 41 to the South Carolina State line; and then northwest along the South Carolina State line to the Robeson County/Scotland County line. (This area may be more generally described as that part of Robeson County lying south and west of N.C. Highway 211, bounded by U.S. Highway 74 east of N.C. Highway 41 and by the South Carolina line west of N.C. Highway 41.)

The Brown, James, farm located on the south side of a private road known as Reola Drive, 0.1 mile east of its intersection with State Secondary Road 1823, which intersection is 0.7 mile south of the intersection of

State Secondary Road 1823 with State Secondary Road 1774.

The Buie, Joshua, farm located on a farm road 0.8 mile south of its intersection with State Secondary Road 1529 and 0.3 mile southwest of the right of way of Interstate Highway 95.

The Lewis, Knox, farm located on the south side of State Secondary Road 1752, 0.5 mile east of its intersection with State Secondary Road 1318.

Sampson County. That area bounded on the north by N.C. Highway 24 and on the east by U.S. Highway 701.

The Brady-Johnson, William, property located on a private road in the town of Salemburg, 0.1 mile north of its intersection with Church Street and 0.1 mile west of its intersection with N.C. Highway 242.

The Carter, Raeford, farm located on the west side of State Secondary Road 1144, 0.2 mile north of its intersection with State Secondary Road 1143.

The Lucas, June, estate located at the end of State Secondary Road 1496, 1.0 mile northwest of its intersection with State Secondary Road 1233.

The Parker, David, farm located on the northwest side of the intersection of a private road known as David Parker Lane and State Secondary Road 1301, 0.5 mile north of the intersection of State Secondary Road 1301 with N.C. Highway 24.

The Riley, Troy Lee, property located 0.05 mile west of the end of a private road known as Stage Coach Lane, 0.2 mile north of the intersection of Stage Coach Lane and N.C. Highway 24, in the town of Autryville.

SOUTH CAROLINA

(1) *Generally infested areas.* None.

(2) *Suppressive areas.*

Horry County. The Bell, Richard, farm located on the east side of State Highway 90, 5.7 miles north of its intersection with State Highway 22.

The Chestnut, Jacob T., farm located on the west side of an unpaved road known as Short Cut Road, 0.2 mile north of its junction with an unpaved road known as Pint Circle Road, 0.4 mile east of its junction with and 0.8 mile north of its junction with State Highway 90.

The Cribbs, L.V., farm located on the west side of an unpaved road known as Causey Road, 3.3 miles north of its intersection with a secondary road known as Sandplant Road and 2.1 miles west of its intersection with State Highway 76.

The Cribbs, L.V., farm located on the east side of an unpaved road known as Causey Road, 2.8 miles north of its intersection with a secondary road known as Sandplant Road and 2.1 miles west of its intersection with State Highway 76.

The Gerald, Kenneth, farm located on the south side of a secondary highway known as Lake Swamp Road, 0.4 mile east of its intersection with a secondary highway known as Nichols Highway South and 1.6 miles south of its intersection with State Highway 917.

The Gerald, Ravenell, farm located on the north side of an unpaved road known as Farming Dale Road, 0.6 mile north of its junction with State Highway 917 and 1.1 miles east of its intersection with a secondary highway known as Nichols Highway.

The Hammonds, Austin J., farm located on the north side of a secondary road known as Sandplant Road, 1.5 miles west of its intersection with State Highway 76 and 1.7 miles north of its intersection with State Highway 9.

The Livingston, Pittman, farm located on the east side of State Highway 90, 2.2 miles north of its junction with State Highway 22.

The Mae, Blossie, farm located on the west side of an unpaved road known as Dela Road, 0.3 mile south of its intersection with a secondary road known as Pint Circle Road, 0.2 mile west of its intersection with State Highway 90, and 3.5 miles north of its intersection with State Highway 22.

The McDaniel, Ellis, farm located on the south side of State Highway 917, 1.4 miles west of its intersection with a secondary highway known as Nichols Highway.

The Smith, Tommy G., farm located on the south side of a secondary road known as Old Chesterfield Road, 1.5 miles east of its intersection with State Highway 90 and 2.7 miles north of its intersection with State Highway 22.

The Strickland, Quincy, farm located on the north side of State Highway 917, 1.2 miles west of its intersection with a secondary highway known as Nichols Highway.

The Stroud, J.B., farm located on the east side of an unpaved road known as Providence Drive, 1.3 miles north of its junction with an unpaved road known as Tranquil Road, 0.5 mile west of its junction with a secondary highway known as Nichols Highway North and 2.3 miles north of its intersection with State Highway 917.

The Vault, Bennie, farm located on the west side of an unpaved road known as Strawberry Road, 0.5 mile south of its junction with State Highway 90.

Vereen, Isiah, farm located on the west side of an unpaved road known as West Shore Road, 1.6 miles south of its junction with State Highway 90.

Vereen, Lula, farm located on the north side of a secondary road known as Dogwood Road, 1.6 miles north of its intersection with State Highway 22, then 0.7 mile east of its intersection with State Highway 90.

The Willoughby, Shane, farm located on the north side of an unpaved road known as Farming Dale Road, 0.4 mile north of its junction with State Highway 917 and 1.1 miles east of its intersection with a secondary highway known as Nichols Highway.

The Worley, Floyd C., farm located on both sides of a secondary road known as Sandplant Road, 1.1 miles west of its intersection with State Highway 76 and 1.7 miles north of its intersection with State Highway 9.

Marion County. The Baxley, Warner, farm located on the west side of Penderboro Road, 1.6 miles north of its intersection with the State Highway 501 Bypass.

The Best Woods Road and Bubba Road farm located on both sides of Best Woods Road, 1.4 miles south of its intersection with State Highway 76.

The Erwin, Harold, farm located on the west side of the State secondary road known as Laughin Road, 1 mile north of its intersection with State Highway 76.

The Gerald, Issaic, farm located on the west side of a secondary road known as Foxworth Road, 0.3 mile northwest of its intersection with Secondary Road 9.

The Holmes, Issaic, farm located on the east side of an unpaved road known as Phill Road, 0.5 mile south of its junction with State Highway 9 and 5 miles east of its intersection with State Highway 41–A.

The Johnson, J. D., farm located on the west side of an unpaved road known as Harold Road, 0.6 mile north of its intersection with Old Mullins Road and 1.3 miles west of its intersection with North Main Street in Nichols.

The Keen, Davis, Estate farm located on the south side of an unpaved road known as Frazier Road, 0.7 mile northwest of its intersection with Secondary Road 9.

The Richardson, Billy, farm located on the east side of Secondary Road 908, 0.8 mile north of its intersection with State Highway 378.

The Rogers, Paul, farm located on the north side of an unpaved road known as Tobacco Barn Road, 0.8 mile west of its intersection with a State secondary road known as E. Sellers Road and 1.7 miles north of its intersection with State Highway 41–A.

[68 FR 6604, Feb. 10, 2003, as amended at 68 FR 51876, Aug. 29, 2003; 72 FR 7925, Feb. 22, 2007; 72 FR 44951, Aug. 10, 2007]

§ 301.80-2b Exempted articles. [1]

(a) The following articles are exempt from the certification and permit and other requirements of this subpart if they meet the applicable conditions prescribed in paragraphs (a) (1) through (5) of this section and have not been exposed to infestation after cleaning or

other handling as prescribed in said paragraph:

(1) Small grains, if harvested in bulk or into new or treated containers, and if the grains and containers for the grains have not come in contact with the soil or if they have been cleaned at a designed facility. [2]

(2) Soybeans, when determined by an inspector that the soybeans were grown, harvested, and handled in a manner to prevent contamination from witchweed seed.

(3) Pickling cucumbers, string beans, and field peas, if washed free of soil with running water.

(4) Used farm tools, if cleaned free of soil.

(5) Used mechanized cultivating equipment and used mechanized soil-moving equipment, if cleaned free of soil.

(b) The following article is exempt from the certification and permit requirements of § 301.80–4 under the applicable conditions as prescribed in paragraph (b)(1) of this section:

(1) Seed cotton, if moving to a designated gin. [2]

[42 FR 56334, Oct. 25, 1977, as amended at 53 FR 24924, July 1, 1988]

§ 301.80-3 Conditions governing the interstate movement of regulated articles from quarantined States. [3]

(a) Any regulated articles, except soil samples for processing, testing, or analysis, may be moved interstate from any quarantined State under the following conditions:

(1) With certificate or permit issued and attached in accordance with §§ 301.80–4 and 301.80–7, if moved:

(i) From any generally infested area or any suppressive area into or through any point outside of the regulated areas; or

---

[1] The articles hereby exempted remain subject to applicable restrictions under other quarantines.

[2] Information as to designated facilities, gins, oil mills, and processing plants may be obtained from an inspector. Any facility, gin, oil mill, or processing plant is eligible for designation under this subpart if the operator thereof enters into a compliance agreement (as defined in § 301.80–1(b)).

[3] Requirements under all other applicable Federal domestic plant quarantines must also be met.

(ii) From any generally infested area into or through any suppressive area; or

(iii) Between any noncontiguous suppressive areas; or

(iv) Between contiguous suppressive areas when it is determined by an inspector that the regulated articles present a hazard of the spread of the witchweed and the person in possession thereof has been so notified; or

(v) Through or reshipped from any regulated area when such movement is not authorized under paragraph (a)(2)(v) of this section; or

(2) Without certificate or permit if moved:

(i) From any regulated area under the provisions of § 301.80–2b which exempts certain articles from certificate and permit requirements; or

(ii) From a generally infested area to a contiguous generally infested area; or

(iii) From a suppressive area to a contiguous generally infested area; or

(iv) Between contiguous suppressive areas unless the person in possession of the articles has been notified by an inspector that a hazard of spread of the witchweed exists; or

(v) Through or reshipped from any regulated area if the articles originated outside of any regulated area and if the point of origin of the articles is clearly indicated, their identity has been maintained, and they have been safeguarded against infestation while in the regulated area in a manner satisfactory to the inspector; or

(3) From any area outside the regulated areas, if moved:

(i) With a certificate or permit attached; or

(ii) Without a certificate or permit, if:

(A) The regulated articles are exempt from certification and permit requirements under the provisions of § 301.80–2b; or

(B) The point of origin of such movement is clearly indicated on the articles or shipping document which accompanies the articles and if the movement is not made through any regulated area.

(b) Unless specifically authorized by the Deputy Administrator in emergency situations, soil samples for processing, testing, or analysis may be moved interstate from any regulated area only to laboratories approved [4] by the Deputy Administrator and so listed by him in a supplemental regulation. [5] A certificate or permit will not be required to be attached to such soil samples except in those emergency situations where the Deputy Administrator has authorized such movement to another destination with a certificate or permit issued and attached in accordance with §§ 301.80–4(d) and 301.80–7. Soil samples originating in areas outside of the regulated areas will not require such a certificate or permit and their movement is not restricted to approved laboratories if the point of origin of such samples is clearly indicated on the articles or shipping document which accompanies the articles and if the movement is not made through any regulated area.

[41 FR 27373, July 2, 1976]

§ 301.80–4 Issuance and cancellation of certificates and permits.

(a) Certificates may be issued for any regulated articles (except soil samples for processing, testing, or analysis) by an inspector if he determines that they are eligible for certification for movement to any destination under all Federal domestic plant quarantines applicable to such articles and:

(1) Have originated in noninfested premises in a regulated area and have not been exposed to infestation while within the regulated areas; or

(2) Have been treated to destroy infestation in accordance with the treatment manual; or

(3) Have been grown, produced, manufactured, stored, or handled in such a manner that no infestation would be transmitted thereby.

(b) Limited permits may be issued by an inspector to allow interstate movement of regulated articles (except soil samples for processing, testing, or analysis) not eligible for certification

---

[4] Pamphlets containing provisions for laboratory approval may be obtained from the Deputy Administrator, Plant Protection and Quarantine Programs, APHIS, U.S. Department of Agriculture, Washington, DC 20250.

[5] For list of approved laboratories, see (41 FR 4615 and amendments thereof).

under this subpart, to specified destinations for limited handling, utilization, or processing, or for treatment in accordance with the treatment manual, when upon evaluation of the circumstances involved in each specific case he determines that such movement will not result in the spread of witchweed and requirements of other applicable Federal domestic plant quarantines have been met.

(c) Restricted destination permits may be issued by an inspector to allow the interstate movement (for other than scientific purposes) of regulated articles (except soil samples for processing, testing, or analysis) to any destination permitted under all applicable Federal domestic plant quarantines if such articles are not eligible for certification under all such quarantines but would otherwise qualify for certification under this subpart.

(d) Scientific permits to allow the interstate movement of regulated articles, and certificates or permits to allow the movement of soil samples for processing, testing, or analysis in emergency situations, may be issued by the Deputy Administrator under such conditions as may be prescribed in each specific case by the Deputy Administrator to prevent the spread of witchweed.

(e) Certificate, limited permit, and restricted destination permit forms may be issued by an inspector to any person for use by the latter for subsequent shipments of regulated articles (except soil samples for processing, testing, or analysis) provided such person is operating under a compliance agreement; and any such person may be authorized by an inspector to reproduce such forms on shipping containers or otherwise. Any such person may execute and issue the certificate forms, or reproductions of such forms, for the interstate movement of regulated articles from the premises of such person identified in the compliance agreement if such person has treated such regulated articles to destroy infestation in accordance with the treatment manual, and if such regulated articles are eligible for certification for movement to any destination under all Federal domestic plant quarantines applicable to such articles. Any such person may

execute and issue the limited permit forms, or reproductions of such forms, for interstate movement of regulated articles to specified destinations when the inspector has made the determinations specified in paragraph (b) of this section. Any such person may execute and issue the restricted destination permit forms, or reproductions of such forms, for the interstate movement of regulated articles not eligible for certification under all Federal domestic plant quarantines applicable to such articles, under the conditions specified in paragraph (c) of this section.

(f) Any certificate or permit which has been issued or authorized may be withdrawn by the inspector or the Deputy Administrator if he determines that the holder thereof has not complied with any condition for the use of such document imposed by this subpart. As soon as possible after such withdrawal, the holder of the certificate or permit shall be notified in writing by the Deputy Administrator or an inspector of the reason therefor and afforded reasonable opportunity to present his views thereon, and if there is a conflict as to any material fact, a hearing shall be held to resolve such conflict.

[35 FR 10553, June 30, 1970, as amended at 36 FR 24917, Dec. 24, 1971; 41 FR 27374, July 2, 1976]

§ 301.80-5 Compliance agreements; and cancellation thereof.

(a) Any person engaged in the business of growing, handling, or moving regulated articles may enter into a compliance agreement to facilitate the movement of such articles under this subpart. Compliance agreement forms may be obtained from the Deputy Administrator or an inspector.

(b) Any compliance agreement may be canceled by the inspector who is supervising its enforcement whenever he finds that such other party has failed to comply with the conditions of the agreement. As soon as possible after such cancellation, such party shall be notified in writing by the Deputy Administrator or an inspector of the reason therefor and afforded reasonable opportunity to present views thereon, and if there is a conflict as to any material fact, a hearing shall be held to resolve such conflict.

[35 FR 10553, June 30, 1970, as amended at 36 FR 24917, Dec. 24, 1971; 41 FR 27374, July 2, 1976]

§ 301.80–6 Assembly and inspection of regulated articles.

Persons (other than those authorized to use certificates, limited permits, or restricted destination permits, or reproductions thereof, under § 301.80–4(e)) who desire to move interstate regulated articles which must be accompanied by a certificate or permit shall, as far in advance as possible, request an inspector to examine the articles prior to movement. Such articles shall be assembled at such points and in such a manner as the inspector designates to facilitate inspection.

[35 FR 10553, June 30, 1970, as amended at 36 FR 24917, Dec. 24, 1971; 41 FR 27374, July 2, 1976]

§ 301.80–7 Attachment and disposition of certificates or permits.

(a) If a certificate or permit is required for the interstate movement of regulated articles, the certificates or permit shall be securely attached to the outside of the container in which such articles are moved except that, where the certificate or permit is attached to the waybill or other shipping document, and the regulated articles are adequately described on the certificate, permit or shipping document, the attachment of the certificate or permit to each container of the articles is not required.

(b) In all cases, certificates or permits shall be furnished by the carrier to the consignee at the destination of the shipment.

[35 FR 10553, June 30, 1970, as amended at 36 FR 24917, Dec. 24, 1971]

§ 301.80–8 Inspection and disposal of regulated articles and pests.

Any properly identified inspector is authorized to stop and inspect, and to seize, destroy, or otherwise dispose of, or require disposal of regulated articles and witchweed as provided in sections 414, 421, and 434 of the Plant Protection Act (7 U.S.C. 7714, 7731, and 7754), in ac-

cordance with instructions issued by the Deputy Administrator.

[35 FR 10553, June 30, 1970, as amended at 36 FR 24917, Dec. 24, 1971; 41 FR 27374, July 2, 1976; 66 FR 21052, Apr. 27, 2001]

§ 301.80–9 Movement of witchweed.

Regulations requiring a permit for, and otherwise governing the movement of witchweed in interstate or foreign commerce are contained in the Federal plant pest regulations in part 330 of this chapter. Applications for permits for the movement of the pest may be made to the Deputy Administrator.

[35 FR 10553, June 30, 1970, as amended at 36 FR 24917, Dec. 24, 1971]

§ 301.80–10 Nonliability of the Department.

The U.S. Department of Agriculture disclaims liability for any costs incident to inspections or compliance with the provisions of the quarantine and regulations in this subpart, other than for the services of the inspector.

[35 FR 10553, June 30, 1970, as amended at 36 FR 24917, Dec. 24, 1971]

Subpart P—Imported Fire Ant

SOURCE: 57 FR 57327, Dec. 4, 1992, unless otherwise noted. Redesignated at 84 FR 2428, Feb. 7, 2019.

QUARANTINE AND REGULATIONS

§ 301.81 Restrictions on interstate movement of regulated articles.

No person may move interstate from any quarantined area any regulated article except in accordance with this subpart.

§ 301.81–1 Definitions.

Administrator. The Administrator, Animal and Plant Health Inspection Service, or any person authorized to act for the Administrator.

Animal and Plant Health Inspection Service. The Animal and Plant Health Inspection Service of the U.S. Department of Agriculture (APHIS).

Certificate. A document in which an inspector or a person operating under a compliance agreement affirms that a specified regulated article meets the requirements of this subpart and may be moved interstate to any destination.

Compliance agreement. A written agreement between APHIS and a person engaged in growing, handling, or moving regulated articles that are moved interstate, in which the person agrees to comply with the provisions of this subpart and any conditions imposed under this subpart.

Imported fire ant. Living imported fire ants of the species *Solenopsis invicta* Buren and *Solenopsis richteri* Forel, and hybrids of these species.

Infestation (infested). The presence of an imported fire ant queen or a reproducing colony of imported fire ants, except that on grass sod and plants with roots and soil attached, an infestation is the presence of any life form of the imported fire ant.

Inspector. An APHIS employee or other person authorized by the Administrator to enforce the provisions of this subpart.

Interstate. From any State into or through any other State.

Limited permit. A document in which an inspector affirms that a specified regulated article not eligible for a certificate is eligible for interstate movement only to a specified destination and in accordance with conditions specified on the permit.

Movement (moved). The act of shipping, transporting, delivering, or receiving for movement, or otherwise aiding, abetting, inducing or causing to bo movod.

Noncompacted soil. Soil that can be removed from an article by brisk brushing or washing with water under normal city water pressure (at least 4 gallons per minute at 40 to 50 pounds per square inch through a ½-inch orifice).

Person. Any association, company, corporation, firm, individual, joint stock company, partnership, society, or any other legal entity.

Reproducing colony. A combination of one or more imported fire ant workers and one or more of the following immature imported fire ant forms: Eggs, larvae, or pupae.

Soil. Any non-liquid combination of organic and/or inorganic material in which plants can grow.

Soil-moving equipment. Equipment used for moving or transporting soil, including, but not limited to, bulldozers, dump trucks, or road scrapers.

State. The District of Columbia, Puerto Rico, the Northern Mariana Islands, or any State, territory, or possession of the United States.

§301.81–2 Regulated articles.

The following are regulated articles:

(a) Imported fire ant queens and reproducing colonies of imported fire ants.[1]

(b) Soil,[2] separately or with other articles, except potting soil that is shipped in original containers in which the soil was placed after commercial preparation.

(c) Baled hay and baled straw stored in direct contact with the ground;

(d) Plants and sod with roots and soil attached, except plants maintained indoors in a home or office environment and not for sale;

(e) Used soil-moving equipment, unless removed of all noncompacted soil; and

(f) Any other article or means of conveyance when:

(1) An inspector determines that it presents a risk of spread of the imported fire ant due to its proximity to an infestation of the imported fire ant; and

(2) The person in possession of the product, article, or means of conveyance has been notified that it is regulated under this subpart.

(Approved by the Office of Management and Budget under control number 0579–0102)

[57 FR 57327, Dec. 4, 1992, as amended at 59 FR 67133, Dec. 29, 1994]

§301.81–3 Quarantined areas.

(a) The Administrator will quarantine each State or each portion of a State that is infested.

(b) Less than an entire State will be listed as a quarantine area only if the Administrator determines that:

---

[1] Permit and other requirements for the interstate movement of imported fire ants are contained in part 330 of this chapter.

[2] The movement of soil from Puerto Rico is subject to additional provisions in part 330 of this chapter.

(1) The State has adopted and is enforcing restrictions on the intrastate movement of the regulated articles listed in § 301.81-2 that are equivalent to the interstate movement restrictions imposed by this subpart; and

(2) Designating less than the entire State as a quarantined area will prevent the spread of the imported fire ant.

(c) The Administrator may include uninfested acreage within a quarantined area due to its proximity to an infestation or inseparability from the infested locality for quarantine purposes, as determined by:

(1) Projections of spread of imported fire ant around the periphery of the infestation, as determined by previous years' surveys;

(2) Availability of natural habitats and host materials, within the uninfested acreage, suitable for establishment and survival of imported fire ant populations; and

(3) Necessity of including uninfested acreage within the quarantined area in order to establish readily identifiable boundaries.

(d) The Administrator or an inspector may temporarily designate any nonquarantined area as a quarantined area in accordance with the criteria specified in paragraphs (a), (b), and (c) of this section. The Administrator will give written notice of this designation to the owner or person in possession of the nonquarantined area, or, in the case of publicly owned land, to the person responsible for the management of the nonquarantined area; thereafter, the interstate movement of any regulated article from an area temporarily designated as a quarantined area is subject to this subpart. As soon as practicable, this area either will be added to the list of designated quarantined areas in paragraph (e) of this section, or the Administrator will terminate the designation. The owner or person in possession of, or, in the case of publicly owned land, the person responsible for the management of, an area for which the designation is terminated will be given written notice of the termination as soon as practicable.

(e) The areas described below are designated as quarantined areas:

ALABAMA

The entire State.

ARKANSAS

Ashley County. The entire county.
Bradley County. The entire county.
Calhoun County. The entire county.
Chicot County. The entire county.
Clark County. The entire county.
Cleveland County. The entire county.
Columbia County. The entire county.
Dallas County. The entire county.
Desha County. The entire county.
Drew County. The entire county.
Faulkner County. The entire county.
Garland County. The entire county.
Grant County. The entire county.
Hempstead County. The entire county.
Hot Spring County. The entire county.
Howard County. The entire county.
Jefferson County. The entire county.
Lafayette County. The entire county.
Lincoln County. The entire county.
Little River County. The entire county.
Lonoke County. The entire county.
Miller County. The entire county.
Montgomery County. The entire county.
Nevada County. The entire county.
Ouachita County. The entire county.
Perry County. The entire county.
Pike County. The entire county.
Polk County. The entire county.
Pulaski County. The entire county.
Saline County. The entire county.
Sevier County. The entire county.
Union County. The entire county.
Yell County. The entire county.

CALIFORNIA

Los Angeles County. That portion of Los Angeles County in the Cerritos area bounded by a line beginning at the intersection of Artesia Boulevard and Marquardt Avenue; then south along Marquardt Avenue to the Los Angeles/Orange County Line; then south and west along the Los Angeles/Orange County Line to Carson Street; then west along Carson Street to Norwalk Boulevard; then north along Norwalk Boulevard to Centralia Street; then west along Centralia Street to Pioneer Boulevard; then north along Pioneer Boulevard to South Street; then east along South Street to Norwalk Boulevard; then north along Norwalk Boulevard to 183rd Street; then east along 183rd Street to Bloomfield Avenue; then north along Bloomfield Avenue to Artesia Boulevard; then east along Artesia Boulevard to the point of beginning.

That portion of Los Angeles County in the Azusa area bounded by a line beginning at the intersection of Irwindale Avenue and Foothill Boulevard; then east along Foothill Boulevard to Azusa Avenue; then south along Azusa Avenue to East Fifth Street; then east along East Fifth Street to North

Cerritos Avenue; then south along North Cerritos Avenue to Arrow Highway; then west along Arrow Highway to Azusa Avenue, then south along Azusa Avenue to Covina Boulevard; then west along an imaginary line to the intersection of Martinez Street and Irwindale Avenue; then north along Irwindale Avenue to the point of beginning. Orange County. The entire county.

Riverside County. That portion of Riverside County in the Indio area bounded by a line beginning at the intersection of Avenue 50 and Jackson Street; then south along Jackson Street to 54th Avenue; then west along 54th Avenue to Madison Street; then north along Madison Street to Avenue 50; then east along Avenue 50 to the point of beginning.

That portion of Riverside County in the Moreno Valley area bounded by a line beginning at the intersection of Reche Vista Drive and Canyon Ranch Road; then southeast along Canyon Ranch Road to Valley Ranch Road; then east along Valley Ranch Road to Michael Way; then south along Michael Way to Casey Court; then east along Casey Court to the Moreno Valley City Limits; then south and east along the Moreno Valley City Limits to Pico Vista Way; then southwest along Pico Vista Way to Los Olivos Drive; then south along Los Olivos Drive to Jaclyn Avenue; then west along Jaclyn Avenue to Perris Boulevard; then south along Perris Boulevard to Kalmia Avenue; then west along Kalmia Avenue to Hubbard Street; then north along Hubbard Street to Nightfall Way; then west and south along Nightfall Way to Sundial Way; then west along Sundial Way to Indian Avenue; then south along Indian Avenue to Ebbtide Lane; then west along Ebbtide Lane to Ridgecrest Lane; then south along Ridgecrest Lane to Moonraker Lane; then west along Moonraker Lane to Davis Street; then south along Davis Street to Gregory Lane; then west along Gregory Lane to Heacock Street; then northwest along an imaginary line to the intersection of Lake Valley Drive and Breezy Meadow Drive; then north along Breezy Meadow Drive to its intersection with Stony Creek; then north along an imaginary line to the intersection of Old Lake Drive and Sunnymead Ranch Parkway; then northwest along Sunnymead Ranch Parkway to El Granito Street; then east along El Granito Street to Lawless Road; then east along an imaginary line to the intersection of Heacock Street and Reche Vista Drive; then north along Reche Vista Drive to the point of beginning.

That portion of Riverside County in the Bermuda Dunes, Palm Desert, and Rancho Mirage areas bounded by a line beginning at the intersection of Ramon Road and Bob Hope Drive; then south along Bob Hope Drive to Dinah Shore Drive; then east along Dinah Shore Drive to Key Largo Avenue; then south along Key Largo Avenue to Gerald Ford Drive; then west along Gerald Ford Drive to Bob Hope Drive; then south along Bob Hope Drive to Frank Sinatra Drive; then east along Frank Sinatra Drive to Vista Del Sol; then south along Vista Del Sol to Country Club Drive; then east along Country Club Drive to Adams Street; then south along Adams Street to 42nd Avenue; then east along 42nd Avenue to Tranquillo Place; then south along Tranquillo Place to its intersection with Harbour Court; then southwest along an imaginary line to the intersection of Granada Drive and Caballeros Drive; then southeast along Caballeros Drive to Kingston Drive; then west along Kingston Drive to Mandeville Road; then east along Mandeville Road to Port Maria Road; then south along Port Maria Road to Fred Waring Drive; then west along Fred Waring Drive to its intersection with Dune Palms Road; then southwest along an imaginary line to the intersection of Adams Street and Miles Avenue; then west along Miles Avenue to Washington Street; then northwest along Washington Street to Fred Waring Drive; then west along Fred Waring Drive to Joshua Road; then north along Joshua Road to Park View Drive; then west along Park View Drive to State Highway 111; then northwest along State Highway 111 to Magnesia Fall Drive; then west along Magnesia Fall Drive to Gardess Road; then northwest along Gardess Road to Dunes View Road; then northeast along Dunes View Road to Halgar Road; then northwest along Halgar Road to Indian Trail Road; then northeast along Indian Trail Road to Mirage Road; then north along Mirage Road to State Highway 111; then northwest along State Highway 111 to Frank Sinatra Drive; then west along Frank Sinatra Drive to Da Vall Drive; then north along Da Vall Drive to Ramon Road; then east along Ramon Road to the point of beginning.

That portion of Riverside County in the Palm Springs area bounded by a line beginning at the intersection of Tramway Road, State Highway 111, and San Rafael Drive; then east along San Rafael Drive to Indian Canyon Drive; then south along Indian Canyon Drive to Francis Drive; then east along Francis Drive to North Farrell Drive; then south along North Farrell Drive to Verona Road; then east along Verona Road to Whitewater Club Drive; then east along an imaginary line to the intersection of Verona Road and Ventura Drive; then east along Verona Road to Avenida Maravilla; then east and south along Avenida Maravilla to 30th Avenue; then west along 30th Avenue to its end; then due west along an imaginary line to the Whitewater River; then southeast along the Whitewater River to Dinah Shore Drive; then west along an imaginary line to the east end of 34th Avenue; then west along 34th Avenue to Golf Club Drive; then south along

Golf Club Drive to East Palm Canyon Drive; then south along an imaginary line to the intersection of Desterto Vista and Palm Hills Drive; then south along Palm Hills Drive to its end; then southwest along an imaginary line to the intersection of Murray Canyon and Palm Canyon Drive; then northwest along Palm Canyon Drive to the Palm Springs city limits; then west and north along Palm Springs city limits to Tahquitz Creek; then due north along an imaginary line to Tramway Road; then northeast along Tramway Road to the point of beginning.

FLORIDA

The entire State.

GEORGIA

The entire State.

LOUISIANA

The entire State.

MISSISSIPPI

The entire State.

NEW MEXICO

Dona Ana County. The entire county.

NORTH CAROLINA

Anson County. The entire county.

Beaufort County. The entire county.

Bertie County. That portion of the county bounded by a line beginning at the intersection of State Highway 11/42 and the Hertford/Bertie County line; then east along the Hertford/Bertie County line to the Bertie/Chowan County line; then south along the Bertie/Chowan County line to the Bertie/Martin County line; then west along the Bertie/Martin County line to State Highway 11/42; then north along State Highway 11/42 to the point of beginning.

Bladen County. The entire county.

Brunswick County. The entire county.

Cabarrus County. The entire county.

Camden County. That portion of the county bounded by a line beginning at the intersection of State Road 1112 and State Highway 343; then east along State Highway 343 to State Road 1107; then south along State Road 1107 to the Camden/Pasquotank County line; then north along the Camden/Pasquotank County line to State Road 1112; then north along State Road 1112 to the point of beginning.

Carteret County. The entire county.

Chatham County. The entire county.

Cherokee County. That portion of the county lying south and west of a line beginning at the intersection of the Cherokee/Clay County line and the North Carolina/Georgia State line; then north to U.S. Highway 64; then northwest along the southern shoreline of Hiwassee Lake to the Tennessee State line.

Chowan County. That portion of the county bounded by a line beginning at the intersection of the Chowan/Gates County line and State Highway 32; then south along State Highway 32 to State Highway 37; then east along State Highway 37 to the Chowan/Perquimans County line; then south along the Chowan/Perquimans County line to the shoreline of the Albemarle Sound; then west along the shoreline of the Albemarle Sound to the Chowan/Bertie County line; then north along the Chowan/Bertie County line to the Chowan/Hertford County line; then north along the Chowan/Hertford County line to the Chowan/Gates County line; then east along the Chowan/Gates County line to the point of beginning.

Clay County. That portion of the county lying southwest of State Highway 69 and the North Carolina/Georgia State line; then north along Interstate 70 to its intersection with U.S. Highway 64; then west along U.S. Highway 64 to the Clay/Cherokee County boundary.

Cleveland County. The entire county.

Columbus County. The entire county.

Craven County. The entire county.

Cumberland County. The entire county.

Currituck County. That portion of the county bounded by a line beginning at the intersection of the Currituck/Camden County line and State Road 1112; then east along State Road 1112 to U.S. Highway 158; then south along U.S. Highway 158 to State Road 1111; then east along State Road 1111 to the shoreline of the Atlantic Ocean; then south along the shoreline of the Atlantic Ocean to the Currituck/Duck County line; then south and west along the Currituck/Duck County line to the Currituck/Camden County line; then north along the Currituck/Camden County line to the point of beginning.

Dare County. The entire county, excluding the portion of the barrier islands south of Oregon Inlet.

Duplin County. The entire county.

Durham County. That portion of the county lying south of Interstate 85.

Edgecombe County. That portion of the county lying south of a line beginning at the intersection of State Highway 111 and the Martin/Edgecombe County line; then southwest on State Highway 111 to U.S. Highway 64 Alternate; then west on U.S. Highway 64 Alternate to County Route 1252; then west of this northerly line to County Route 1408; then west on County Route 1408 to County Route 1407; then south on County Route 1407 to the Edgecombe/Nash County line.

Gaston County. The entire county.

Greene County. The entire county.

Harnett County. The entire county.

Hertford County. That portion of the county lying south and east of a line beginning at the intersection of State Highway 11 and the

Bertie/Hertford county line; then northeast on State Highway 11 to the U.S. Highway 13 Bypass; then northeast on U.S. Highway 13 to the Hertford/Gates County line.

Hoke County. The entire county.

Hyde County. The entire county.

Iredell County. That portion of the county lying south of State Highway 150.

Johnston County. The entire county.

Jones County. The entire county.

Lee County. The entire county.

Lenoir County. The entire county.

Lincoln County. That portion of the county lying east of State Highway 321.

Martin County. That portion of the county lying south of a line beginning at the intersection of State Highway 111 and the Edgecombe/Martin County line; then north and east on State Highway 111 to State Highway 11/42; then northeast along State Highway 11/42 to the Martin/Bertie County line.

Mecklenburg County. The entire county.

Montgomery County. The entire county.

Moore County. The entire county.

Nash County. That portion of the county lying south and east of the line beginning at the intersection of U.S. Highway 64 and the Franklin/Nash County line; then northeast on U.S. Highway 64 to Interstate 95; then north on Interstate 95 to State Highway 4; then east on State Highway 4 to U.S. Highway 301; then east along a straight line from the intersection of State Highway 64 and U.S. Highway 301 to the Nash/Edgecombe County line.

New Hanover County. The entire county.

Onslow County. The entire county.

Orange County. The portion of the county that lies south of Interstate 85.

Pamlico County. The entire county.

Pasquotank County. That portion of the county bounded by a line beginning at the intersection of the Pasquotank/Perquimans County line and U.S. Highway 17; then east along U.S. Highway 17 to the Pasquotank/Camden County line; then south along the Pasquotank/Camden County line to the shoreline of the Albemarle Sound; then west along the shoreline of the Albemarle Sound to the Pasquotank/Perquimans County line; then north along the Pasquotank/Perquimans County line to the point of beginning.

Pender County. The entire county.

Perquimans County. That portion of the county bounded by a line beginning at the intersection of the Perquimans/Chowan County line and State Road 1118; then east along State Road 1118 to State Road 1200; then north along State Road 1200 to State Road 1213; then east along State Road 1213 to State Road 1214; then southeast along State Road 1214 to State Road 1221; then northeast along State Road 1221 to the Perquimans/Pasquotank County line; then south along the Perquimans/Pasquotank County line to the shoreline of the Albemarle Sound; then west along the shoreline of the Albemarle Sound to the Perquimans/Chowan County line; then north along the Perquimans/Chowan County line to the point of beginning.

Pitt County. The entire county.

Polk County. The entire county.

Randolph County. That portion of the county lying south of the line beginning at the intersection of State Highway 49 and the Davidson/Randolph County line; then east on State Highway 49 to U.S. Highway 64; then east on U.S. Highway 64 to its intersection with the Randolph/Chatham County line.

Richmond County. The entire county.

Robeson County. The entire county.

Rutherford County. That portion of the county lying south of State Highway 74.

Sampson County. The entire county.

Scotland County. The entire county.

Stanly County. The entire county.

Tyrrell County. The entire county.

Union County. The entire county.

Wake County. The entire county.

Washington County. The entire county.

Wayne County. The entire county.

Wilson County. The entire county.

OKLAHOMA

Bryan County. The entire county.

Carter County. The entire county.

Choctaw County. The entire county.

Comanche County. The entire county.

Johnston County. The entire county.

Love County. The entire county.

Marshall County. The entire county.

McCurtain County. The entire county.

PUERTO RICO

The entire State.

South Carolina

The entire State.

TENNESSEE

Anderson County. That portion of the county lying east and south of a line beginning at the intersection of the Roane/Anderson County line and Tennessee Highway 330; then northeast on Tennessee Highway 330 to Tennessee Highway 116; then north on Tennessee Highway 116 to Interstate 75; then southeast on Interstate 75 to the Anderson/Knox County line.

Bedford County. The entire county.

Benton County. The entire county.

Bledsoe County. The entire county.

Blount County. The entire county.

Bradley County. The entire county.

Carroll County. The entire county.

Chester County. The entire county.

Coffee County. That portion of the county lying west and south of a line beginning at the intersection of the Cannon/Coffee County line and Tennessee Highway 53; then south on Tennessee Highway 53 to Riddle Road; then southeast on Riddle Road to Keele

Road; then northeast on Keele Road to Tennessee Highway 55; then northeast on Tennessee Highway 55 to Swann Road; then east on Swann Road to Wiser Road; then north on Wiser Road to Rock Road; then east on Rock Road to Pleasant Knoll Road; then north on Pleasant Knoll Road to Marcrom Road; then east on Marcrom Road to the Coffee/Warren County line.

Crockett County. That portion of the county lying east of a line beginning at the intersection of the Haywood/Crockett County line and U.S. Highway 70A/79; then northeast on U.S. Highway 70A/79 to Tennessee Highway 88; then north on Tennessee Highway 88 to Tennessee Highway 54; then northeast on Tennessee Highway 54 to the Crockett/Gibson County line.

Cumberland County. That portion of the county lying southeast of a line beginning at the intersection of the White/Cumberland County line and U.S. Highway 70; then east on U.S. Highway 70 to Market Street (in Crab Orchard); then north on Market Street to Main Street; then west on Main Street to Chestnut Hill Road; then northeast on Chestnut Hill Road to Westchester Drive; then north on Westchester Drive to Peavine Road; then east on Peavine Road to Hebbertsburg Road; then northeast on Hebbertsburg Road to the Cumberland/Morgan County line.

Davidson County. That portion of the county lying southeast of a line beginning at the intersection of the Williamson/Davidson County line and U.S. Highway 431; then northeast on U.S. Highway 431 to Tennessee Highway 254; then east on Tennessee Highway 254 to U S. Highway 31A/41A; then north on U.S. Highway 31A/41A to Tennessee Highway 255; then northeast on Tennessee Highway 255 to Interstate 40; then east on Interstate 40 to the Davidson/Wilson County line.

Decatur County. The entire county.

Fayette County. The entire county.

Franklin County. The entire county.

Gibson County. That portion of the county lying southeast of a line beginning at the intersection of the Madison/Gibson County line and U.S. Highway 45W; then northwest on U.S. Highway 45W to U.S. Highway 45 Bypass (Tennessee Highway 366); then north on U.S. Highway 45 Bypass to U.S. Highway 79/70A; then northeast on U.S. Highway 79/70A to the Gibson/Carroll County line.

Giles County. The entire county.

Grundy County. The entire county.

Hamilton County. The entire county.

Hardeman County. The entire county.

Hardin County. The entire county.

Haywood County. That portion of the county lying southeast of Tennessee Highway 54.

Henderson County. The entire county.

Hickman County. The entire county.

Humphreys County. That portion of the county lying south of a line beginning at the intersection of the Benton/Humphreys County line and the line of latitude 36°; then continuing east along the line of latitude 36° to Forks River Road; then south on Forks River Road to Old Highway 13; then southeast on Old Highway 13 to Tennessee Highway 13; then south on Tennessee Highway 13 to Interstate 40; then east on Interstate 40 to the Humphreys/Hickman County line.

Knox County. That portion of the county lying southwest of a line beginning at the intersection of the Union/Knox County line and Tennessee Highway 33; then south on Tennessee Highway 33 to the Tennessee River; then northeast along the Tennessee River to the French Broad River; then east along the French Broad River to the Knox/Sevier County line.

Lawrence County. The entire county.

Lewis County. The entire county.

Lincoln County. The entire county.

Loudon County. The entire county.

Madison County. The entire county.

Marion County. The entire county.

Marshall County. That portion of the county lying south of a line beginning at the intersection of the Maury/Marshall County line and Moses Road; then northeast on Moses Road to Wilson School Road; then southeast on Wilson School Road to Lunns Store Road; then south on Lunns Store Road to Tennessee Highway 99; then east on Tennessee Highway 99 to U.S. Highway 31A; then south on U.S. Highway 31A to James Shaw Road; then south on James Shaw Road to Clay Hill Road; then east on Clay Hill Road to Warner Road; then south on Warner Road to Batten Road; then southeast on Batten Road to the Marshall/Bedford County line.

Maury County. That portion of the county lying south of a line beginning at the intersection of the Hickman/Maury County line and Jones Valley Road; then east on Jones Valley Road to Leipers Creek Road; then south on Leipers Creek Road to Tennessee Highway 247; then northeast on Tennessee Highway 247 to Tennessee Highway 246; then north on Tennessee Highway 246 to the Maury/Williamson County line.

McMinn County. The entire county.

McNairy County. The entire county.

Meigs County. The entire county.

Monroe County. The entire county.

Moore County. The entire county.

Morgan County. That portion of the county lying south of a line beginning at the intersection of the Cumberland/Morgan County line and Tennessee Highway 298; then northeast on Tennessee Highway 298 to Tennessee Highway 62; then southeast on Tennessee Highway 62 to the Morgan/Roane County line.

Perry County. The entire county.

Polk County. The entire county.

Rhea County. The entire county.

Roane County. The entire county.

Rutherford County. The entire county.

Sequatchie County. The entire county.

Shelby County. The entire county.

Tipton County. That portion of the county lying south of a line beginning at the intersection of the Shelby/Tipton County line and Tennessee Highway 14; then northeast on Tennessee Highway 14 to Tennessee Highway 179; then southeast on Tennessee Highway 179 to the Tipton/Haywood County line.

Van Buren County. The entire county.

Warren County. That portion of the county lying southeast of a line beginning at the intersection of the Coffee/Warren County line and Marcrom Road; then east on Marcrom Road to Fred Hoover Road; then north on Fred Hoover Road to Tennessee Highway 287; then northwest on Tennessee Highway 287 to Vervilla Road; then northeast on Vervilla Road to Swan Mill Road; then east on Swan Mill Road to Grove Road; then southeast on Grove Road to Tennessee Highway 108/127; then northeast on Tennessee Highway 108/127 to the split between Tennessee Highway 108 and Tennessee Highway 127; then northeast on Tennessee Highway 127 to Tennessee Highway 56; then southeast on Tennessee Highway 56 to Fairview Road; then northeast on Fairview Road to Tennessee Highway 8; then southeast on Tennessee Highway 8 to Dark Hollow Road; then north on Dark Hollow Road to Tennessee Highway 30; then northeast on Tennessee Highway 30 to the Warren/Van Buren County line.

Wayne County. The entire county.

Williamson County. That portion of the county lying northeast of a line beginning at the intersection of the Davidson/Williamson County line and U.S. Highway 31; then southwest on U.S. Highway 31 to U.S. Highway Business 431; then southeast on U.S. Highway Business 431 to Mack Hatcher Parkway; then north on Mack Hatcher Parkway to South Royal Oaks Boulevard; then northeast on South Royal Oaks Boulevard to Tennessee Highway 96; then east on Tennessee Highway 96 to Clovercroft Road; then northeast on Clovercroft Road to Wilson Pike; then north on Wilson Pike to Clovercroft Road; then northeast on Clovercroft Road to Rocky Fork Road; then east on Rocky Fork Road to the Rutherford/Williamson County line. Also, that portion of the county enclosed by a line beginning at the intersection of the Maury/Williamson County line and Tennessee Highway 246; then north on Tennessee Highway 246 to Thompson Station Road West; then east on Thompson Station Road West to Thompson Station Road East; then east on Thompson Station Road East to Interstate 65; then south on Interstate 65 to the Williamson/Maury County line.

TEXAS

Anderson County. The entire county.
Angelina County. The entire county.
Aransas County. The entire county.
Atascosa County. The entire county.
Austin County. The entire county.
Bandera County. The entire county.
Bastrop County. The entire county.
Bee County. The entire county.
Bell County. The entire county.
Bexar County. The entire county.
Blanco County. The entire county.
Bosque County. The entire county.
Bowie County. The entire county.
Brazoria County. The entire county.
Brazos County. The entire county.
Brooks County. The entire county.
Brown County. The entire county.
Burleson County. The entire county.
Burnet County. The entire county.
Caldwell County. The entire county.
Calhoun County. The entire county.
Cameron County. The entire county.
Camp County. The entire county.
Cass County. The entire county.
Chambers County. The entire county.
Cherokee County. The entire county.
Collin County. The entire county.
Colorado County. The entire county.
Comal County. The entire county.
Comanche County. The entire county.
Cooke County. The entire county.
Coryell County. The entire county.
Dallas County. The entire county.
Delta County. The entire county.
Denton County. The entire county.
De Witt County. The entire county.
Dimmit County. The entire county.
Duval County. The entire county.
Eastland County. The entire county.
Ector County. The entire county.
Edwards County. The entire county.
Ellis County. The entire county.
Erath County. The entire county.
Falls County. The entire county.
Fannin County. The entire county.
Fayette County. The entire county.
Fort Bend County. The entire county.
Franklin County. The entire county.
Freestone County. The entire county.
Frio County. The entire county.
Galveston County. The entire county.
Gillespie County. The entire county.
Goliad County. The entire county.
Gonzales County. The entire county.
Grayson County. The entire county.
Gregg County. The entire county.
Grimes County. The entire county.
Guadalupe County. The entire county.
Hamilton County. The entire county.
Hardin County. The entire county.
Harris County. The entire county.
Harrison County. The entire county.
Hays County. The entire county.
Henderson County. The entire county.
Hidalgo County. The entire county.
Hill County. The entire county.
Hood County. The entire county.
Hopkins County. The entire county.
Houston County. The entire county.
Hunt County. The entire county.
Jack County. The entire county.

Jackson County. The entire county.
Jasper County. The entire county.
Jefferson County. The entire county.
Jim Wells County. The entire county.
Johnson County. The entire county.
Jones County. The entire county.
Karnes County. The entire county.
Kaufman County. The entire county.
Kendall County. The entire county.
Kenedy County. The entire county.
Kerr County. The entire county.
Kimble County. The entire county.
Kinney County. The entire county.
Kleberg County. The entire county.
La Salle County. The entire county.
Lamar County. The entire county.
Lampasas County. The entire county.
Lavaca County. The entire county.
Lee County. The entire county.
Leon County. The entire county.
Liberty County. The entire county.
Limestone County. The entire county.
Live Oak County. The entire county.
Llano County. The entire county.
Madison County. The entire county.
Marion County. The entire county.
Mason County. The entire county.
Matagorda County. The entire county.
Maverick County. The entire county.
McCulloch County. The entire county.
McLennan County. The entire county.
McMullen County. The entire county.
Medina County. The entire county.
Midland County. The entire county.
Milam County. The entire county.
Montague County. The entire county.
Montgomery County. The entire county.
Morris County. The entire county.
Nacogdoches County. The entire county.
Navarro County. The entire county.
Newton County. The entire county.
Nueces County. The entire county.
Orange County. The entire county.
Palo Pinto County. The entire county.
Panola County. The entire county.
Parker County. The entire county.
Polk County. The entire county.
Rains County. The entire county.
Real County. The entire county.
Red River County. The entire county.
Refugio County. The entire county.
Robertson County. The entire county.
Rockwall County. The entire county.
Rusk County. The entire county.
Sabine County. The entire county.
San Augustine County. The entire county.
San Jacinto County. The entire county.
San Patricio County. The entire county.
San Saba County. The entire county.
Shelby County. The entire county.
Smith County. The entire county.
Somervell County. The entire county.
Stephens County. The entire county.
Tarrant County. The entire county.
Taylor County. The entire county.
Titus County. The entire county.
Tom Green County. The entire county.

Travis County. The entire county.
Trinity County. The entire county.
Tyler County. The entire county.
Upshur County. The entire county.
Uvalde County. The entire county.
Val Verde County. The entire county.
Van Zandt County. The entire county.
Victoria County. The entire county.
Walker County. The entire county.
Waller County. The entire county.
Washington County. The entire county.
Webb County. The entire county.
Wharton County. The entire county.
Wichita County. The entire county.
Willacy County. The entire county.
Williamson County. The entire county.
Wilson County. The entire county.
Wise County. The entire county.
Wood County. The entire county.
Young County. The entire county.
Zavala County. The entire county.

[57 FR 57327, Dec. 4, 1992]

EDITORIAL NOTE: For FEDERAL REGISTER citations affecting § 301.81–3, see the List of CFR Sections Affected, which appears in the Finding Aids section of the printed volume and at *www.govinfo.gov.*

§ 301.81–4 Interstate movement of regulated articles from quarantined areas.

(a) Any regulated article may be moved interstate from a quarantined area into or through an area that is not quarantined only if moved under the following conditions:

(1) With a certificate or limited permit issued and attached in accordance with §§ 301.81–5 and 301.81–9 of this subpart;

(2) Without a certificate or limited permit, provided that each of the following conditions is met:

(i) The regulated article was moved into the quarantined area from an area that is not quarantined;

(ii) The point of origin is indicated on a waybill accompanying the regulated article;

(iii) The regulated article is moved through the quarantined area (without stopping except for refueling, or for traffic conditions, such as traffic lights or stop signs), or has been stored, packed, or parked in locations inaccessible to the imported fire ant, or in locations that have been treated in accordance with part 305 of this chapter, while in or moving through any quarantined area; and

(iv) The article has not been combined or commingled with other articles so as to lose its individual identity; or

(3) Without a certificate or limited permit provided the regulated article is a soil sample being moved to a laboratory approved by the Administrator[3] to process, test, or analyze soil samples.

(b) Inspectors are authorized to stop any person or means of conveyance moving in interstate commerce they have probable cause to believe is moving regulated articles, and to inspect the articles being moved and the means of conveyance. Articles found to be infested by an inspector, and articles not in compliance with the regulations in this subpart, may be seized, quarantined, treated, subjected to other remedial measures, destroyed, or otherwise disposed of. Any treatments will be in accordance with part 305 of this chapter.

(Approved by the Office of Management and Budget under control number 0579–0102)

[57 FR 57327, Dec. 4, 1992, as amended at 59 FR 67133, Dec. 29, 1994; 59 FR 67609, Dec. 30, 1994; 67 FR 8464, Feb. 25, 2002; 70 FR 33268, June 7, 2005; 75 FR 4240, Jan. 26, 2010]

§ 301.81–5 Issuance of a certificate or limited permit.

(a) An inspector[4] or person operating under a compliance agreement will issue a certificate for the interstate movement of a regulated article approved under such compliance agreement if he or she determines that the regulated article:

(1) Is eligible for unrestricted movement under all other applicable Federal domestic plant quarantines and regulations;

(2) Is to be moved interstate in compliance with any additional conditions deemed necessary under section 414 of the Plant Protection Act (7 U.S.C. 7714) to prevent the spread of the imported fire ant;[5] and

(3)(i) Is free of an imported fire ant infestation, based on his or her visual examination of the article;

(ii) Has been grown, produced, manufactured, stored, or handled in a manner that would prevent infestation or destroy all life stages of the imported fire ant;

(iii) Has been treated in accordance with part 305 of this chapter; or.

(iv) If the article is containerized nursery stock, it has been produced in accordance with § 301.81–11.

(b) An inspector will issue a limited permit for the interstate movement of a regulated article not eligible for a certificate if the inspector determines that the regulated article:

(1) Is to be moved interstate to a specified destination for specified handling, utilization, or processing (the destination and other conditions to be listed in the limited permit), and this interstate movement will not result in the spread of the imported fire ant because the imported fire ant will be destroyed by the specified handling, utilization, or processing;

(2) Is to be moved interstate in compliance with any conditions that the Administrator may impose under section 414 of the Plant Protection Act (7 U.S.C. 7714) to prevent the spread of the imported fire ant; and

(3) Is eligible for interstate movement under all other Federal domestic plant quarantines and regulations applicable to the regulated article.

(c) An inspector shall issue blank certificates to a person operating under a compliance agreement (in accordance with § 301.81–6 of this subpart) or authorize reproduction of the certificates on shipping containers, or both, as requested by the person operating under

---

[3] Criteria that laboratories must meet to become approved to process, test, or analyze soil, and the list of currently approved laboratories, may be obtained from the Animal and Plant Health Inspection Service, Plant Protection and Quarantine, Domestic and Emergency Operations, 4700 River Road Unit 134, Riverdale, Maryland 20737–1236.

[4] Inspectors are assigned to local offices of APHIS, which are listed in local telephone directories. Information on local offices may also be obtained from the Animal and Plant Health Inspection Service, Plant Protection and Quarantine, Domestic and Emergency Operations, 4700 River Road Unit 134, Riverdale, Maryland 20737–1236.

[5] An inspector may hold, seize, quarantine, treat, apply other remedial measures to, destory, or otherwise dispose of plants, plant pests, or other articles in accordance with sections 414, 421, and 434 of the Plant Protection Act (7 U.S.C. 7714, 7731, and 7754).

the compliance agreement. These certificates may then be completed and used, as needed, for the interstate movement of regulated articles that have met all of the requirements of paragraph (a) of this section.

(Approved by the Office of Management and Budget under control number 0579–0088)

[57 FR 57327, Dec. 4, 1992, as amended at 59 FR 67133, Dec. 29, 1994; 59 FR 67609, Dec. 30, 1994; 66 FR 21052, Apr. 27, 2001; 75 FR 4240, Jan. 26, 2010]

§ 301.81-6 Compliance agreements.

Persons who grow, handle, or move regulated articles interstate may enter into a compliance agreement[6] if such persons review with an inspector each stipulation of the compliance agreement, have facilities and equipment to carry out disinfestation procedures or application of chemical materials in accordance with part 305 of this chapter, and meet applicable State training and certification standards as authorized by the Federal Insecticide, Fungicide, and Rodenticide Act (86 Stat. 983; 7 U.S.C. 136b). Any person who enters into a compliance agreement with APHIS must agree to comply with the provisions of this subpart and any conditions imposed under this subpart.

[57 FR 57327, Dec. 4, 1992, as amended at 59 FR 67609, Dec. 30, 1994; 75 FR 4240, Jan. 26, 2010]

§ 301.81-7 Cancellation of a certificate, limited permit, or compliance agreement.

Any certificate, limited permit, or compliance agreement may be canceled orally or in writing by an inspector whenever the inspector determines that the holder of the certificate or limited permit, or the person who has entered into the compliance agreement, has not complied with this subpart or any conditions imposed under

---

[6] Compliance agreements may be initiated by contacting a local office of Plant Protection and Quarantine, which are listed in telephone directories. The addresses and telephone numbers of local offices of Plant Protection and Quarantine may also be obtained from the Animal and Plant Health Inspection Service, Plant Protection and Quarantine, Domestic and Emergency Operations, 4700 River Road Unit 134, Riverdale, Maryland 20737–1236.

this subpart. If the cancellation is oral, the cancellation will become effective immediately and the cancellation and the reasons for the cancellation will be confirmed in writing as soon as circumstances allow within 20 days after oral notification of the cancellation. Any person whose certificate, limited permit, or compliance agreement has been canceled may appeal the decision, in writing, within 10 days after receiving the written cancellation notice. The appeal must state all of the facts and reasons that the person wants the Administrator to consider in deciding the appeal. A hearing may be held to resolve any conflict as to any material fact. Rules of practice for the hearing will be adopted by the Administrator. As soon as practicable, the Administrator will grant or deny the appeal, in writing, stating the reasons for the decision.

(Approved by the Office of Management and Budget under control number 0579–0102)

[57 FR 57327, Dec. 4, 1992, as amended at 59 FR 67133, Dec. 29, 1994]

§ 301.81-8 Assembly and inspection of regulated articles.

(a) Persons requiring certification or other services must request the services from an inspector[7] at least 48 hours before the services are needed.

(b) The regulated articles must be assembled at the place and in the manner the inspector designates as necessary to comply with this subpart.

(Approved by the Office of Management and Budget under control number 0579–0088)

[57 FR 57327, Dec. 4, 1992, as amended at 59 FR 67133, Dec. 29, 1994]

§ 301.81-9 Attachment and disposition of certificates and limited permits.

(a) The consignor must ensure that the certificate or limited permit authorizing interstate movement of a regulated article is, at all times during interstate movement, attached to:

(1) The outside of the container encasing the regulated article;

(2) The article itself, if it is not in a container; or

(3) The consignee's copy of the accompanying waybill: *Provided,* that the

---

[7] See footnote 4 to § 301.81–5(a).

descriptions of the regulated article on the certificate or limited permit, and on the waybill, are sufficient to identify the regulated article; and

(b) The carrier must furnish the certificate or limited permit authorizing interstate movement of a regulated article to the consignee at the shipment's destination.

(Approved by the Office of Management and Budget under control number 0579–0088)

[57 FR 57327, Dec. 4, 1992, as amended at 59 FR 67133, Dec. 29, 1994]

§ 301.81-10 Costs and charges.

The services of the inspector during normal business hours will be furnished without cost to persons requiring the services. The United States Department of Agriculture will not be responsible for any other costs or charges.

§ 301.81-11 Imported fire ant detection, control, exclusion, and enforcement program for nurseries producing containerized plants.

This detection, control, exclusion, and enforcement program is designed to keep nurseries free of the imported fire ant and provides a basis to certify containerized nursery stock for interstate movement. Participating regulated establishments must be operating under a compliance agreement in accordance with § 301.81-6. Such compliance agreements shall state the specific requirements that a shipper agrees to follow to move plants in accordance with the requirements of the program. Certificates and a nursery identification number may be issued to the nursery for use on shipments of regulated articles.

(a) *Detection.* (1) Nursery owners are required to visually survey their entire premises twice monthly for the presence of imported fire ants.

(2) Nurseries participating in this program will be inspected by Federal or State inspectors at least twice per year. More frequent inspections may be necessary depending upon imported fire ant infestation levels immediately surrounding the nursery, the thoroughness of nursery management in maintaining imported-fire-ant-free premises, and the number of previous detections of imported fire ants in or near containerized plants. Inspections by Federal and State inspectors should be more frequent just before and during the peak shipping season. Any nurseries determined during nursery inspections to have imported fire ant colonies must be immediately treated to the extent necessary to eliminate the colonies.

(b) *Control.* Nursery plants that are shipped under this program must originate in a nursery that meets the requirements of this section. Nursery owners must implement a treatment program with registered bait and contact insecticides. The premises, including growing and holding areas, must be maintained free of the imported fire ant. As part of this treatment program, all exposed soil surfaces (including sod and mulched areas) on property where plants are grown, potted, stored, handled, loaded, unloaded, or sold must be treated in accordance with part 305 of this chapter at least once every 6 months. The first application must be performed early in the spring. Followup treatments with a contact insecticide in accordance with part 305 of this chapter must be applied to eliminate all remaining colonies.

(c) *Exclusion.* (1) For plants grown on the premises, treatment of soil or potting media in accordance with part 305 of this chapter prior to planting is required.

(2) For plants received from outside sources, to prevent the spread into a nursery free of the imported fire ant by newly introduced, infested nursery plants, all plants must be:

(i) Obtained from nurseries that comply with the requirements of this section and that operate under a compliance agreement in accordance with § 301.81-6; or

(ii) Treated upon delivery in accordance with part 305 of this chapter, and within the specified number of days be either:

(A) Repotted in treated potting soil media;

(B) Retreated in accordance with part 305 of this chapter at the specified interval; or

(C) Shipped.

(d) *Enforcement.* (1) The nursery owner must maintain records of the nursery's surveys and treatments for the imported fire ant. These records must be made available to State and Federal inspectors upon request.

(2) If imported fire ants are detected in nursery stock during an inspection by a Federal or State inspector, issuance of certificates for movement will be suspended until necessary treatments are applied and the plants and nursery premises are determined to be free of the imported fire ant. A Federal or State inspector may declare a nursery to be free of the imported fire ant upon reinspection of the premises. This inspection must be conducted no sooner than 30 days after treatment. During this period, certification may be based upon treatments for plants in accordance with part 305 of this chapter.

(3) Upon notification by the department of agriculture in any State of destination that a confirmed imported fire ant infestation was found on a shipment from a nursery considered free of the imported fire ant, the department of agriculture in the State of origin must cease its certification of shipments from that nursery. An investigation by Federal or State inspectors will commence immediately to determine the probable source of the problem and to ensure that the problem is resolved. If the problem is an infestation, issuance of certification for movement on the basis of imported-fire-ant-free premises will be suspended until treatment and elimination of the infestation is completed. Reinstatement into the program will be granted upon determination that the nursery premises are free of the imported fire ant, and that all other provisions of this subpart are being followed.

(4) In cases where the issuance of certificates is suspended through oral notification, the suspension and the reasons for the suspension will be confirmed in writing within 20 days of the oral notification of the suspension. Any person whose issuance of certificates has been suspended may appeal the decision, in writing, within 10 days after receiving the written suspension notice. The appeal must state all of the facts and reasons that the person wants the Administrator to consider in deciding the appeal. A hearing may be held to resolve any conflict as to any material fact. Rules of practice for the hearing will be adopted by the Adminis-

trator. As soon as practicable, the Administrator will grant or deny the appeal, in writing, stating the reasons for the decision.

[75 FR 4240, Jan. 26, 2010]

Subpart Q—Unshu Oranges [Reserved]

Subpart R—Golden Nematode

Source: 37 FR 24330, Nov. 16, 1972, unless otherwise noted. Redesignated at 84 FR 2428, Feb. 7, 2019.

QUARANTINE AND REGULATIONS

§ 301.85 **Quarantine; restriction on interstate movement of specified regulated articles.**

(a) *Notice of quarantine.* Under the authority of sections 411, 412, 414, and 434 of the Plant Protection Act (7 U.S.C. 7711, 7712, 7714, and 7754), the Secretary of Agriculture quarantines the State of New York in order to prevent the spread of the golden nematode (*Globodera rostochiensis*), which causes a dangerous disease of potatoes and certain other plants and is not widely prevalent or distributed within and throughout the United States. Through the aforementioned authorities, the Secretary imposes a quarantine on the State of New York with respect to the interstate movement from that State of the articles described in paragraph (b) of this section, issues regulations in this subpart governing the movement of such articles, and gives notice of this quarantine action.

(b) *Quarantine restrictions on interstate movement of specified regulated articles.* No common carrier or other person shall move interstate from any quarantined State any of the following articles (defined in § 301.85–1 as regulated articles), except in accordance with the conditions prescribed in this subpart:

(1) Soil, compost, humus, muck, peat, and decomposed manure, separately or with other things.

(2) Plants with roots, except soil-free aquatic plants.

(3) Grass sod.

(4) Plant crowns and roots for propagation.

(5) True bulbs, corms, rhizomes, and tubers of ornamental plants.

(6) Irish potatoes included within any one or more of the following paragraph (b)(6)(i), (ii), or (iii) of this section:

(i) Irish potatoes for seed; and

(ii) Irish potatoes unless—

(A) Each is at least 1½ inches in diameter based on measurement by a sizing screen or sizing chain, each is substantially free of soil as a result of grading (a method of removing soil mechanically) under a compliance agreement in accordance with §301.85–5(b), and they are moved in an approved container; or

(B) Each is substantially free of soil as a result of washing or fluming under a compliance agreement in accordance with §301.85–5(b), and they are moved in an approved container; or

(iii) Irish potatoes harvested from a field tested and found by an inspector to contain an identifiable population of viable golden nematodes, unless such field had been subsequently treated in accordance with paragraph (b)(6)(iii) (A), (B), or (C) of this section under the supervision of an inspector and in accordance with any additional conditions found necessary by the inspector to assure effective application of the pesticide used; and unless headlands and farm roads are treated in accordance with paragraph (b)(6)(iii)(D) of this section:

(A) Applications of 140.3 liters of Vorlex (1,3 dichloropropene; 1,2 dichloropropane, and other related compounds, 80 percent; plus methyl isothiocyanate, 20 percent active ingredients) per hectare (15 gallons per acre); two applications 5 to 10 days apart with a third application 5 to 10 days after the second application to areas in which the inspector finds upon microscopic examination of soil samples that viable golden nematodes may still exist; soil to be from 3 °C to 29 °C (38 °F to 84 °F).

(B) Applications of 280.6 liters of D-D (1,3 dichloropropene; 1,2 dichloropropane, and other related compounds, 100 percent active ingredients) per hectare (30 gallons per acre); two applications 5 to 10 days apart with a third application 5 to 10 days after the second application to areas in which the inspector finds upon microscopic examination of soil samples that viable golden nematodes may still

exist (consult product label for heavier dosage in muck or peat soils); soil to be from 4.5 °C to 29 °C (40 °F to 84 °F).

(C) Applications of 168.4 liters of Telone II (1,3 dichloropropene, 92 percent active ingredient) per hectare (18 gallons per acre); two applications 5 to 10 days apart with a third application 5 to 10 days after the second application to areas in which the inspector finds upon microscopic examination of soil samples that viable golden nematodes may still exist (consult product label for heavier dosage in muck or peat soils); soil to be from 4.5 °C to 32 °C (40 °F to 90 °F).

(D) Application of Vapam (sodium-N-methyl dithiocarbamate, 32.7 percent active ingredient) mixed with water at the rate of 1 part Vapam to 60 parts water and applied as a drench at the rate of 14.96 cubic meters per hectare (1600 gallons per acre); soil to be from 4.5 °C to 32 °C (40 °F to 90 °F).

(7) Root crops other than Irish potatoes.

(8) Small grains and soybeans.

(9) Hay, straw, fodder, and plant litter, of any kind.

(10) Ear corn, except shucked ear corn.

(11) Used crates, boxes, and burlap bags, and other used farm products containers.

(12) Used farm tools.

(13) Used mechanized cultivating equipment and used harvesting equipment.

(14) Used mechanized soil-moving equipment.

(15) Any other products, articles, or means of conveyance of any character whatsoever, not covered by paragraphs (b) (1) through (14) of this section, when it is determined by an inspector that they present a hazard of spread of golden nematode, and the person in possession thereof has been so notified.

[37 FR 24330, Nov. 16, 1972, as amended at 47 FR 12331, Mar. 23, 1982; 66 FR 21052, Apr. 27, 2001; 67 FR 8465, Feb. 25, 2002; 69 FR 21040, Apr. 20, 2004]

§301.85–1 Definitions.

Terms used in the singular form in this subpart shall be deemed to import the plural and vice versa, as the case may demand. The following terms, when used in this subpart shall be construed respectively to mean:

Certificate. A document issued or authorized to be issued under this subpart by an inspector to allow the interstate movement of regulated articles to any destination.

Compliance agreement. A written agreement between a person engaged in growing, handling, or moving regulated articles, and the Plant Protection and Quarantine Programs, wherein the former agrees to comply with the requirements of this subpart identified in the agreement by the inspector who executes the agreement on behalf of the Plant Protection and Quarantine Programs as applicable to the operations of such person.

Deputy Administrator. The Deputy Administrator of the Plant Protection and Quarantine Programs, Animal and Plant Health Inspection Service, U.S. Department of Agriculture, or any other officer or employee of said service to whom authority to act in his stead has been or may hereafter be delegated.

Farm tools. An instrument worked or used by hand, e.g., hoes, rakes, shovels, axes, hammers, and saws.

Generally infested area. Any part of a regulated area not designated as a suppressive area in accordance with §301.85-2.

Golden nematode. The nematode known as the golden nematode (*Globodera rostochiensis*), in any stage of development.

Infestation. The presence of the golden nematode or the existence of circumstances that make it reasonable to believe that the golden nematode is present.

Inspector. Any employee of the Plant Protection and Quarantine Programs, Animal and Plant Health Inspection Service, U.S. Department of Agriculture, or other person, authorized by the Deputy Administrator to enforce the provisions of the Quarantine and regulations in this subpart.

Interstate. From any State into or through any other State.

Limited permit. A document issued or authorized to be issued by an inspector to allow the interstate movement of noncertifiable regulated articles to a specified destination for limited handling, utilization or processing or for treatment.

Mechanized cultivating equipment; and mechanized harvesting equipment. Mechanized equipment used for soil tillage, including tillage attachments for farm tractors, e.g., tractors, disks, plows, harrows, planters, and subsoilers; mechanized equipment used for harvesting purposes, e.g., combines, potato conveyors, and harvesters and hay balers.

Mechanized soil-moving equipment. Equipment used for moving or transporting soil, e.g., draglines, bulldozers, dump trucks, road scrapers, etc.

Moved (movement, move). Shipped, deposited for transmission in the mail, otherwise offered for shipment, received for transportation, carried, or otherwise transported, or moved, or allowed to be moved, by mail or otherwise. "Movement" and "move" shall be construed in accordance with this definition.

Person. Any individual, corporation, company, society, or association, or other organized group of any of the foregoing.

Plant Protection and Quarantine Programs. The organizational unit within the Animal and Plant Health Inspection Service, U.S. Department of Agriculture, delegated responsibility for enforcing provisions of the Plant Protection Act and related legislation, quarantines, and regulations.

Regulated area. Any quarantined State, or any portion thereof, listed as a regulated area in §301-85-2a, or otherwise designated as a regulated area in accordance with §301.85-2(b).

Regulated article. Any articles as described in §301.85(b).

Restricted destination permit. A document issued or authorized to be issued by an inspector to allow the interstate movement of regulated articles not certifiable under all applicable Federal domestic plant quarantines to a specified destination for other than scientific purposes.

Scientific permit. A document issued by the Deputy Administrator to allow the interstate movement to a specified destination of regulated articles for scientific purposes.

Soil. That part of the upper layer of earth in which plants can grow.

State. Any State, territory, or district of the United States, including Puerto Rico.

Suppressive area. That portion of a regulated area where eradication of infestation is undertaken as an objective, as designated under §301.85–2(a).

[37 FR 24330, Nov. 16, 1972, as amended at 47 FR 12331, Mar. 23, 1982; 66 FR 21052, Apr. 27, 2001; 67 FR 8465, Feb. 25, 2002; 70 FR 33268, June 7, 2005]

§301.85–2 Authorization to designate, and terminate designation of, regulated areas and suppressive or generally infested areas; and to exempt articles from certification, permit, or other requirements.

(a) *Regulated areas and suppressive or generally infested areas.* The Deputy Administrator shall list as regulated areas, in a supplemental regulation designated as §301.85–2a, each quarantined State; or each portion thereof in which golden nematode has been found or in which there is reason to believe that golden nematode is present, or which it is deemed necessary to regulate because of their proximity to infestation or their inseparability for quarantine enforcement purposes from infested localities. The Deputy Administrator, in the supplemental regulation, may divide any regulated area into a suppressive area or a generally infested area in accordance with the definitions thereof in §301.85–1. Less than an entire quarantined State will be designated as a regulated area only if the Deputy Administrator is of the opinion that:

(1) The State has adopted and is enforcing a quarantine or regulation which imposes restrictions on the intrastate movement of the regulated articles which are substantially the same as those which are imposed with respect to the interstate movement of such articles under this subpart; and

(2) The designation of less than the entire State as a regulated area will otherwise be adequate to prevent the interstate spread of the golden nematode.

(b) *Temporary designation of regulated areas and suppressive or generally infested areas.* The Deputy Administrator or an authorized inspector may temporarily designate any other premises in a quarantined State as a regulated area and a suppressive or generally infested area, in accordance with the criteria specified in paragraph (a) of this section for listing such area, by serving written notice thereof on the owner or person in possession of such premises, and thereafter the interstate movement of regulated articles from such premises by any person having notice of the designation shall be subject to the applicable provisions of this subpart. As soon as practicable, such premises shall be added to the list in §301.85–2a if a basis then exists for their designation; otherwise the designation shall be terminated by the Deputy Administrator or an authorized inspector and notice thereof shall be given to the owner or person in possession of the premises.

(c) *Termination of designation as a regulated area and a suppressive or generally infested area.* The Deputy Administrator shall terminate the designation provided for under paragraph (a) of this section of any area listed as a regulated area and suppressive or generally infested area when he determines that such designation is no longer required under the criteria specified in paragraph (a) of this section.

(d) *Exemption of articles from certification, permit, or other requirements.* The Deputy Administrator may, in a supplemental regulation designated as §301.85–2b, list regulated articles or movements of regulated articles which shall be exempt from the certification, permit, or other requirements of this subpart under such conditions as he or she may prescribe, if he or she finds that facts exist as to the pest risk involved in the movement of such regulated articles which make it safe to so relieve such requirements.

[37 FR 24330, Nov. 16, 1972, as amended at 70 FR 33268, June 7, 2005]

§301.85–2a Regulated areas; suppressive and generally infested areas.

The civil divisions and parts of civil divisions described below are designated as golden nematode regulated areas within the meaning of the provisions of this subpart; and such regulated areas are hereby divided into generally infested areas or suppressive areas as indicated below:

NEW YORK

(1) *Generally infested area:*

Cayuga County. (A) The Town of Montezuma;

(B) That portion of land within the Town of Mentz owned or operated by Martens Farm which lies in an area bounded as follows: Beginning at the intersection of Tow Path Road and Maiden Lane; then west along Tow Path Road to its intersection with the Town of Mentz boundary; then north along the Town of Mentz boundary to its intersection with Maiden Lane; then east along Maiden Lane to the point of beginning.

Livingston County. (A) That portion of land in the area of South Lima North Muck in the town of Lima bounded as follows: Beginning at a point along the north side of South Lima Rd. marked by latitude/longitude coordinates 42.8553, −77.6738; then north along a farm road to coordinates 42.8588, −77.6712; then east along a farm road to coordinates 42.8596, −77.6678; then north along a farm road to coordinates 42.8624, −77.6683; then east along a farm road to coordinates 42.8624, −77.6648; then north along a farm road to coordinates number 42.8735, −77.6651; then west along a farm road to coordinates number 42.8735, −77.6684; then south along Little Conesus Creek to coordinates 42.8712, −77.6693; then west to include a portion of an access road and gravel clean off site to coordinates 42.8712, −77.6705; then south to coordinates 42.8711, −77.6704; then east to coordinates 42.8711, −77.6699; then north to coordinates 42.8712, −77.6698; then east to coordinates 42.8711, −77.6693; then south along Little Conesus Creek to coordinates 42.8688, −77.6702; then west along a farm road to coordinates 42.8688, −77.6713; then south along a farm road to coordinates 42.8659, −77.6733; then south along a farm road to coordinates 42.8642, −77.6740; then west along a farm road to coordinates 42.8643, −77.6761; then south along a farm road to coordinates 42.8567, −77.6802; then east to coordinates 42.8564, −77.6741; then south along Little Conesus Creek to coordinates 42.8553, −77.6745; then east to point of beginning at coordinates 42.8553, −77.6738;

(B) That portion of land in the area of South Lima South Muck in the town of Lima bounded as follows: Beginning at a point along the south side of South Lima Rd. marked by latitude/longitude coordinates 42.8552, −77.6774; then south to coordinates 42.8548, −77.6774; then east to coordinates 42.8548, −77.6767; then south to coordinates 42.8509, −77.6770; then south to coordinates 42.8447, −77.6772; then east to coordinates 42.8446, −77.6739; then north along a farm road to coordinates 42.8477, −77.6728; then east along a farm road to coordinates 42.8488, −77.6700; then north along a farm road to coordinates 42.8512, −77.6701; then west along a farm road to coordinates 42.8512, −77.6720;

then north along a farm road to coordinates 42.8516, −77.6720; then west along a farm road to coordinates 42.8518, −77.6740; then north to coordinates 42.8541, −77.6740; then west to coordinates 42.8545, −77.6766; then north to coordinates 42.8552, −77.6765; then west to point of beginning at coordinates 42.8552, −77.6774;

(C) That portion of land in the area of Wiggle Muck in the town of Livonia bounded as follows: Beginning at a point along the west side of Plank Rd. (State highway 15A) marked by latitude/longitude coordinates 42.8489, −77.6136; then west to coordinates 42.8491, −77.6203; then south along a farm road to coordinates 42.8468, −77.6192; then south along a farm road to coordinates 42.8419, −77.6188; then east to coordinates 42.8422, −77.6161; then north along a farm road to coordinates 42.8487, −77.6168; then east to the west side of Plank Rd. marked by coordinates 42.8487, −77.6135; then north to point of beginning at coordinates 42.8489, −77.6136; and

(D) That portion of land in the town of Avon bounded as follows: Beginning at a point marked by latitude/longitude coordinates 42.9056, −77.6872; then east along a farm road to coordinates 42.9054, −77.6850; then east along a farm road to coordinates 42.9060, −77.6825; then north along a drainage ditch to coordinates 42.9069, −77.6823; then north along a drainage ditch to coordinates 42.9079, −77.6847; then north to coordinates 42.9103, −77.6844; then west along the south side of a farm road to coordinates 42.9103, −77.6857; then south along a farm road to point of beginning at coordinates 42.9056, −77.6872.

Nassau County. (A) That portion of land in the town of Oyster Bay and the village of Old Brookville bounded as follows: Beginning at a point marked by latitude-longitude coordinates 40.8312 −73.5917; then proceeding north-northwest 152′ along Hegemans Lane to coordinates 40.8317, −73.5917; continuing north-northwest 581′ along Hegemans Lane to coordinates 40.8332, −73.5920; then heading north 473′ along Hegemans Lane to coordinates 40.8345, −73.5921; continuing north 461′ along Hegemans Lane to coordinates 40.8358, −73.5921; continuing north 182′ along Hegemans Lane to coordinates 40.8363, −73.5921; continuing north 1227′ along Hegemans Lane to coordinates 40.8397, −73.5920; continuing north 718′ along Hegemans Lane to coordinates 40.8416, −73.5919; then heading west 188′ to Chicken Valley Road at coordinates 40.8416, −73.5926: then heading southwest 70′ along Chicken Valley Road to coordinates 40.8415, −73.5928; continuing southwest 297′ along Chicken Valley Road to coordinates 40.8409, −73.5936; continuing southwest 106′ along Chicken Valley Road to coordinates 40.8407, −73.5938; continuing southwest 155′ along Chicken Valley Road to coordinates 40.8404, −73.5914;

continuing southwest 142' along Chicken Valley Road to coordinates 40.8400, −73.5944; continuing southwest 125' along Chicken Valley Road to coordinates 40.8397, −73.5946; continuing southwest 36' along Chicken Valley Road to coordinates 40.8397, −73.5947: continuing southwest 38' along Chicken Valley Road to coordinates 40.8396, −73.5948; continuing southwest 201' along Chicken Valley Road to coordinates 40.8392, −73.5953; continuing southwest 98' along Chicken Valley Road to coordinates 40.8390, −73.5955; continuing southwest 61' along Chicken Valley Road to coordinates 40.8389, −73.5957; continuing southwest 187' along Chicken Valley Road to coordinates 40.8384, −73.5960; continuing southwest 62' along Chicken Valley Road to coordinates 40.8383, −73.5962; continuing southwest 142' along Chicken Valley Road to coordinates 40.8380, −73.5965; continuing southwest 252' along Chicken Valley Road to coordinates 40.8375, −73.5971; then heading southeast 254' along a wood line to coordinates 40.8370, −73.5965; then heading south-southeast 79' along a wood line to coordinates 40.8363, −73.5963; continuing south-southeast 182' along a wood line to coordinates 40.8358, −73.5962; then heading south 303' along a wood line to coordinates 40.8350, −73.5961; then heading south-southeast 175' along a wood line to coordinates 40.8346, −73.5958; then heading south 329' along a farm road to coordinates 40.8337, −73.5957; then heading southwest 76' along a farm road to coordinates 40.8335, −73.5958; then heading west-southwest 104' along a farm road to coordinates 40.8334, −73.5962; then heading south-southwest 278' along a farm road to coordinates 40.8327, −73.5965; then heading southeast 1441' along the south boundary of the Young's Farm to the starting point at coordinates 40.8312, −73.5917;

(B) That portion of land in the town of Oyster Bay and in the village of Old Brookville bounded as follows: Beginning at a point marked by latitude-longitude coordinates 40.8343, −73.5921; then heading south 300' along Hogemans Lane to coordinates 40.8332, −73.5920; then heading 403' northeast along a tree line to coordinates 40.8335, −73.5906; then heading 263' north to coordinates 40.8342, −73.5906; then heading northwest 6' along Linden Lane to coordinates 40.8342, −73.5906; continuing northwest 98' along Linden Lane to coordinates 40.8343, −73.5909; continuing northwest 52' along Linden Lane to coordinates 40.8344, −73.5911; then continuing 54' northwest along Linden Lane to coordinates 40.8344, −73.5913; then heading southwest 43' along Linden Lane to coordinates 40.8344, −73.5915; then continuing southwest 170' to the starting point at coordinates 40.8343, −73.5921;

(C) That portion of land in the town of Oyster Bay and the village of Old Brookville bounded as follows: Beginning at a point marked by latitude-longitude coordinates

40.8365, −73.5874; then heading south 452' to coordinates 40.8352, −73.5873; then proceeding northeast 583' to coordinates 40.8356, −73.5853; then heading north-northwest 400' to coordinates 40.8367, −73.5855; then heading 529' southwest to the starting point at coordinates 40.8365, −73.5874;

(D) That portion of land in the town of Oyster Bay in the villages of Upper Brookville and Matinecock bounded as follows: Beginning at a point marked by latitude-longitude coordinates 40.8631, −73.5671; proceeding south 240' along Chicken Valley Road to coordinates 40.8625, −73.5670; continuing south 293' along Chicken Valley Road to coordinates 40.8617, −73.5670; continuing south 131' along Chicken Valley Road to coordinates 40.8613, −73.5670; proceeding south-southwest 116' along Chicken Valley Road to coordinates 40.8610, −73.5671; continuing south-southwest 113' along Chicken Valley Road to coordinates 40.8670, −73.5672; proceeding southwest 116' along Chicken Valley Road to coordinates 40.8604, −3.5674; continuing southwest 337' along Chicken Valley Road to coordinates 40.8596, −73.5679; continuing southwest 82' along Chicken Valley Road to coordinates 40.8594, −73.5680; then heading east-northeast 819' through a wooded area to coordinates 40.8596, −73.5651; then heading southeast 1462' through a wooded area to coordinates 40.8558, −73.5634; continuing southeast 1065' through a wooded area to coordinates 40.8531, −73.5618; then heading northeast 1167' along the border with a golf course to coordinates 40.8547, −73.5581; continuing northeast 317' along the border with the golf course to coordinates 40.8552, −73.5573; then heading east-northeast 1278' along the border with a golf course to coordinates 40.8557, −73.5527; then heading south 275' along the border with a golf course to coordinates 40.8549, −73.5526; then heading northeast 873' along the golf course boundary to coordinates 40.8564, −73.5501; then heading northwest 1463' through a wooded area to coordinates 40.8602, −73.5519; then heading 615' southwest along the border of a residential area to coordinates 40.8599, −73.5541; then heading 280' northwest through a wooded area to coordinates 40.8604, −73.5548; then heading north-northwest 188' along a residential driveway to coordinates 40.8609, −73.5549; then heading west 337' along a residential driveway to coordinates 40.8609, −73.5561; then heading northwest 309' along a residential driveway to coordinates 40.8617, −73.5565; then heading north-northwest 103' along a residential driveway to coordinates 40.8620, −73.5566; continuing north-northwest 38' along a residential driveway to coordinates 40.8621, −73.5566; then heading north 108' along a residential driveway to coordinates 40.8624, −73.5566; then heading northwest 135' along a residential driveway to coordinates 40.8627, −73.5567; continuing northwest 95' along a residential driveway to coordinates

40.8630, −73.5568; then heading east 106′ to a residential driveway at coordinates 40.8630, −73.5565; then heading northeast 172′ along a residential driveway to coordinates 40.8631, −73.5558; continuing northeast 160′ along a residential driveway to coordinates 40.8633, −73.5553; continuing northeast 20′ along a residential driveway to coordinates 40.8633, −73.5552; continuing northeast 141′ along a residential driveway to coordinates 40.8636, −73.5549; continuing northeast 106′ along a residential driveway to coordinates 40.8639, −73.5547; then heading north-northeast 20′ along a residential driveway to coordinates 40.8639, −73.5547; continuing north-northeast 107′ along a residential driveway to coordinates 40.8642, −73.5546; continuing north-northeast 111′ along a residential driveway to coordinates 40.8645, −73.5545; continuing north-northeast 207′ along a residential driveway to coordinates 40.8650, −73.5545; then heading north 53′ along a residential driveway to coordinates 40.8652, −73.5545; then heading northeast 153′ along a paved driveway to coordinates 40.8655, −73.5540; continuing northeast 173′ along a paved driveway to coordinates 40.8657, −73.5535; continuing northeast 71′ along a paved driveway to coordinates 40.8658, −73.5533; then heading north-northeast 47′ along a paved driveway to coordinates 40.8659, −73.5532; continuing north-northeast 70′ along a paved driveway to coordinates 40.8661, −73.5531; continuing north-northeast 79′ along a paved driveway to coordinates 40.8663, −73.5530; then heading north-northeast 109′ along a paved driveway to coordinates 40.8666, −73.5530; then heading north 129′ along a paved driveway to coordinates 40.8669, −73.5529; continuing north 122′ along a paved driveway to coordinates 40.8673, −73.5529; then heading north-northeast 182′ along a paved driveway to coordinates 40.8678, −73.5527; then heading north 23′ along a paved driveway to coordinates 40.8678, −73.5527; then heading north-northwest 36′ along a paved driveway to Planting Fields Road at coordinates 40.8679, −73.5528; then heading southwest 6′ along Planting Fields Road to coordinates 40.8679, −73.5528; continuing southwest 48′ along Planting Fields Road to coordinates 40.8679, −73.5530; then heading west 339′ along Planting Fields Road to coordinates 40.8679, −73.5542; then heading west-southwest 93′ along Planting Fields Road to coordinates 40.8679, −73.5545; then heading southwest 69′ along Planting Fields Road to coordinates 40.8678, −73.5548; continuing southwest 79′ along Planting Fields Road to coordinates 40.8677, −73.5550; continuing southwest 133′ along Planting Fields Road to coordinates 40.8675, −73.5554; continuing southwest 60′ along Planting Fields Road to coordinates 40.8674, −73.5556; continuing southwest 71′ along Planting Fields Road to coordinates 40.8674, −73.5558; heading west-southwest 225′ along Planting Fields

Road to coordinates 40.8672, −73.5566; continuing west-southwest 89′ along Planting Fields Road to coordinates 40.8672, −73.5570; continuing west-southwest 132′ along Planting Fields Road to coordinates 40.8671, −73.5574, then heading southwest 91′ along Planting Fields Road to coordinates 40.8670, −73.5577; continuing southwest 59′ along Planting Fields Road to coordinates 40.8669, −73.5579; continuing southwest 240′ along Planting Fields Road to coordinates 40.8665, −73.5586; continuing southwest 219′ along Planting Fields Road to coordinates 40.8661, −73.5592; continuing southwest 89′ along Planting Fields Road to coordinates 40.8660, −73.5594; continuing southwest 132′ along Planting Fields Road to coordinates 40.8658, −73.5598; continuing southwest 93′ along Planting Fields Road to coordinates 40.8657, −73.5602; continuing southwest 97′ along Planting Fields Road to coordinates 40.8655, −73.5605; continuing southwest 90′ along Planting Fields Road to coordinates 40.8654, −73.5608; continuing southwest 131′ along Planting Fields Road to coordinates 40.8653, −73.5612; continuing southwest 80′ along Planting Fields Road to coordinates 40.8652, −73.5614; continuing southwest 257′ along Planting Fields Road to coordinates 40.8649, −73.5623; continuing southwest 133′ along Planting Fields Road to coordinates 40.8647, −73.5627; continuing southwest 59′ along Planting Fields Road to coordinates 40.8646, −73.5629; continuing southwest 85′ along Planting Fields Road to coordinates 40.8645, −73.5632; then heading west-southwest 177′ along Planting Fields Road to coordinates 40.8644, −73.5638; continuing west-southwest 213′ along Planting Fields Road to coordinates 40.8643, −73.5646; then heading southwest 89′ along Planting Fields Road to coordinates 40.8642, −73.5649; continuing southwest 58′ along Planting Fields Road to coordinates 40.8642, −73.5651; continuing southwest 133′ along Planting Fields Road to coordinates 40.8639, −73.5654; continuing southwest 325′ along Planting Fields Road to coordinates 40.8635, −73.5664; continuing southwest 116′ along Planting Fields Road to coordinates 40.8633, −73.5668; continuing southwest 100′ along Planting Fields Road to the intersection with Chicken Valley Road at starting point coordinates 40.8631, −73.5671; and

(E) That portion of land in the town of Oyster Bay and in the hamlet of Old Bethpage bounded as follows: Beginning at a point marked by latitude-longitude coordinates 40.7703, −73.4460; then proceeding southwest 1997′ through a wood line to Winding Road at coordinates 40.7655, −73.4493; then heading southeast 102′ along Winding Road to coordinates 40.7653, −73.4490; continuing southeast 57′ along Winding Road to coordinates 40.7652, −73.4488; then heading south-southeast 52′ along Winding Road to coordinates 40.7650, −73.4488; continuing south-southeast 99' along Winding Road to coordinates 40.7648, −73.4487; then heading south 654' along Winding Road to coordinates 40.7630, −73.4485; then heading southeast 24' along a ramp to coordinates 40.7629, −73.4485; then heading northeast 2134' through a wood line to coordinates 40.7644, −73.4411; then heading north-northwest 1197' along the Nassau County-Suffolk County Border to coordinates 40.7677, −73.4417; then heading southwest 250' through a wood line to coordinates 40.7676, −73.4426; then heading 132' northnorthwest along Restoration Road to coordinates 40.7679, −73.4426; then heading northeast 311' along Restoration Road to coordinates 40.7687, −73.4422; then heading northwest 96' along Restoration Road to coordinates 40.7689, −73.44245; continuing northwest 262' along Restoration Road to coordinates 40.7695, −73.4430; continuing northwest 173' along Restoration Road to coordinates 40.7699, −73.4433; continuing northwest 33' along Restoration Road to coordinates 40.7699, −73.4434; then heading southwest 275' along Restoration Road to coordinates 40.76942, −73.4440; continuing southwest 194' along Restoration Road to coordinates 40.76900, −73.4445; then heading north-northwest 334' along a gravel path to coordinates 40.7698, −73.4448; then heading northwest 364' along a gravel path to the starting point at coordinates 40.7703, −73.4460.

Orleans County. (A) That portion of land in the town of Barre bounded as follows: Beginning at a point on the north side of Spoil Bank Road marked by latitude-longitude coordinates 43.1327, −78.1234; then east along a farm road running parallel to Spoil Bank Road to coordinates 43.1327, −78.1191; then north along a willow hedge row to coordinates 43.1354, −78.1191; then west along a drainage ditch to coordinates 43.1353, −78.1227; then northwest along a drainage ditch to coordinates 43.1354, −78.1230; then northwest to coordinates 43.1355, −78.1232; then west to coordinates 43.1355, −78.1233; then southwest to coordinates 43.1354, 78.1234; then south along a drainage ditch to the point of beginning at coordinates 43.1327, −78.1234;

(B) That portion of land in the town of Barre bounded as follows: Beginning at a point marked by latitude-longitude coordinates 43.1548, −078.1199; then east along a farm road to coordinates 43.1548, −078.1166; then south to coordinates 43.1524, −078.1167; then west to coordinates 43.1524, −078.1199; then north along a willow hedgerow to the point beginning at coordinates 43.1548, −078.1199; and

(C) That portion of land in the town of Barre bounded as follows: Beginning at a point marked by latitude-longitude coordinates 43.1551, −78.1240; then west to coordinates 43.1551, −78.1244; then southwest to coordinates 43.1550, −78.1245; then southwest to coordinates 43.1550, −78.1245; then south to

coordinates 43.1549, −78.1245; then west along a drainage ditch to coordinates 43.1548, −78.1264, then north to coordinates 43.1596, −78.1266; then east along a farm road to coordinates 43.1597, −78.1243; then south along a willow hedge to the point of beginning at coordinates 43.1551, −78.1240.

Seneca County. The town of Tyre.

Steuben County. (A) The towns of Prattsburgh and Wheeler.

(B) The area known as "Arkport Muck North" located in the town of Dansville bounded as follows: Beginning at a point along the west bank of the Marsh Ditch that intersects a farm road marked by latitude/longitude coordinates 42.4230, −77.7121; then north along the Marsh Ditch to coordinates 42.4314, −77.7158; then west along a farm road to coordinates 42.4307, −77.7204; then south along the edge of a forest to coordinates 42.4284, −77.7194; then west along a farm road to coordinates 42.4282, −77.7201; then south along a farm road to coordinates 42.4255, −77.7189; then east along a tree line to coordinates 42.4254, −77.7180; then south along a tree line to coordinates 42.4230, −77.7157; then east to point of beginning at coordinates 42.4230, −77.7121;

(C) The area known as "Arkport Muck South" located in the town of Dansville bounded as follows: Beginning at a point along the west side of New York Route 36 marked by latitude/longitude coordinates 42.4034, −77.6986; then north along the west side of New York Route 36 to coordinates 42.4145, −77.6999; then west along a farm road to coordinates 42.4145, −77.7029; then north along a farm road to coordinates 42.4160, −77.7036; then west along a farm road to coordinates 42.4162, −77.7083; then north along the west bank of the Marsh Ditch to coordinates 42.4186, −77.7097; then west along a farm road to coordinates 42.4181, −77.7121; then north along a farm road to coordinates 42.4214, −77.7140; then west along a farm road to coordinates 42.4211, −77.7198; then south along the east side of the Conrail right-of-way (Erie Lackawanna Railroad) to coordinates 42.4050, −77.7107; then east along a farm road to coordinates 42.4049, −77.7038; then south along a farm road to coordinates 42.4034, −77.7030; then east to point of beginning at coordinates 42.4034, −77.6986;

(D) That portion of land in the town of Cohocton (formerly known as the "Werthwhile Farm") on the north side of County Road 5 (known as Brown Hill Road), and 0.2 mile west of the junction of County Road 5 with County Road 58 (known as Wager Road); and

(E) That portion of land in the town of Fremont that is bounded as follows: Beginning at a point on Babcock Road that intersects a farm road marked by latitude/longitude coordinates 42.4368, −77.5751; then west along the farm road to coordinates 42.4367, −77.5780; then south to coordinates 42.4360,

−77.5780; then west to coordinates 42.4359, −77.5807; then south to coordinates 42.4335, −77.5806; then east to coordinates 42.4333, −77.5778; then south to coordinates 42.4318, −77.5777; then east to coordinates 42.4323, −77.5771; then north to coordinates 42.4330, −77.5763; then east to coordinates 42.4330, −77.5761; then north to coordinates 42.4349, −77.5756; then east to coordinates 42.4349, −77.5749; then north to the point of beginning at coordinates 42.4368, −77.5751.

Suffolk County. (A) Towns of Riverhead, East Hampton, Southampton, Southold, and Shelter Island in their entirety in Suffolk County;

(B) That portion of land in the town of Huntington and the hamlet of Melville bounded as follows: Beginning at a point marked by latitude-longitude coordinates 40.7767, −72.4202; then proceeding southwest 788′ along Broad Hollow Rd. to coordinates 40.7746, −72.4210; then heading east 2354′ along Huntington Quadrangle Rd. to coordinates 40.7748, −72.4125; then heading south 2095′ parallel to Maxess Rd. to coordinates 40.7691, −72.4121; then heading southeast 250′ along Bayliss Rd. to coordinates 40.7689, −72.4113; then heading 2734′ north to coordinates 40.7764, −72.4114; then heading east 1820′ to coordinates 40.7767, −72.4049; then heading north 233′ along Pinelawn Rd. to coordinates 40.7773, −72.4048; then heading west 4267′ to the starting point at coordinates 40.7767, −72.4202;

(C) That portion of land in the town of Huntington and the hamlet of Melville bounded as follows: Beginning at a point marked by latitude-longitude coordinates 40.7954, −72.4080; then proceeding south 1645′ along the west boundary of White Post Farms to coordinates 40.7909, −72.4073; then proceeding east 1110′ along the south boundary of White Post Farms and a housing development to coordinates 40.7910, −72.4033; then heading north 2033′ parallel to Bedell Place to coordinates 40.7965, −72.4041; then heading southwest 1170′ along Old Country Road to the starting point at coordinates 40.7954, −72.4080;

(D) That portion of land in the town of Huntington and the hamlet of Dix Hills bounded as follows: Beginning at a point marked by latitude-longitude coordinates 40.7904, −72.3410; then proceeding southeast 306′ along Deer Park Road to coordinates 40.7896, −72.3407; continuing southeast 272′ along Deer Park Road to coordinates 40.7888, −72.3404; continuing southeast 530′ along Deer Park Road to coordinates 40.7874, −72.3399; then proceeding northeast 1002′ along the south boundary of the DeLalio Sod Company to coordinates 40.7883, −72.3364; then proceeding northwest 541′ along the east boundary of the DeLalio Sod Company and the Garden Depot to coordinates 40.7897, −72.3371; continuing northwest 554′ along the east boundary of field 15–C–21 to coordinates

40.7911, −72.3377; then proceeding southwest 952′ along the north boundary of field 15–C–21 to the starting point at coordinates 40.7904, −72.3410; and

(E) That portion of land in the town of Brookhaven and the hamlet of Manorville bounded as follows: Beginning at a point marked by latitude-longitude coordinates 40.8542, −72.8240; then proceeding northeast 442′ along South Street to coordinates 40.8545, −72.8225; continuing northeast 1086′ along South Street to coordinates 40.8550, −72.8186; then proceeding east 413′ to coordinates 40.8551, −72.8171 at the intersection of South Street and Wading River Rd.; then proceeding northwest 714′ along Wading River Road to coordinates 40.8568, −72.8183; then continuing northwest 695′ along Wading River Road to coordinates 40.8586, −72.8194; continuing northwest 497′ along Wading River Road to coordinates 40.8598, −72.8202; continuing northwest 221′ along Wading River Road to coordinates 40.8603, −72.8205; continuing northwest 203′ along Wading River Road to coordinates 40.8608, −72.8209; continuing 194′ along Wading River Road to coordinates 40.8613, −72.8212; continuing 212′ along Wading River Road to coordinates 40.8618, −72.8215; proceeding northwest 30′ to coordinates 40.8618, −72.8216; then heading 45′ west to coordinates 40.8618, −72.8218; then heading 183′ southwest along the south ramp of the Long Island Expressway to coordinates 40.8617, −72.8224; then heading west 179′ parallel with the south ramp of the Long Island Expressway to coordinates 40.8617, −72.8231; then continuing west 182′ to coordinates 40.8617, −72.8237; continuing west 299′ parallel with the south ramp of the Long Island Expressway to coordinates 40.8618, −72.8248; then proceeding 201′ southeast to coordinates 40.8617, −72.8255; continuing southwest 88′ to coordinates 40.8615, −72.8257; then south 83′ along a wood line to coordinates 40.8613, −72.8257; continuing south 116′ along a wood line to coordinates 40.8610, −72.8257; continuing southeast 96′ along a wood line to coordinates 40.8607, −72.8256; then heading 92′ southwest along the wood line to coordinates 40.8605, −72.8257; then heading 47′ south along the wood line to coordinates 40.8603, −72.8257; then heading southeast 194′ along the wood line to coordinates 40.8599, −72.8261; continuing 87′ southwest along the wood line to coordinates 40.8597, −72.8262; continuing 200′ southwest along the wood line to coordinates 40.8592, −72.8265; then heading southeast 112′ along the wood line to coordinates 40.8589, −72.8264; then heading east 232′ along the wood line to coordinates 40.8589, −72.8256; then heading south 828′ along the wood line to coordinates 40.8566, −72.8253; then heading east 246′ along the northern boundary of a horse farm to coordinates 40.8567, −72.8244; then heading south 940′ along the boundary of a horse farm to the starting point at coordinates 40.8542, −72.8240.

Wayne County. The town of Savannah.

(2) *Suppressive area:* None.

[51 FR 30050, Aug. 22, 1986, as amended at 69 FR 249, Jan. 5, 2004; 69 FR 64640, Nov. 8, 2004; 76 FR 60358, Sept. 29, 2011; 78 FR 1714, Jan. 9, 2013; 78 FR 3827, Jan. 17, 2013; 80 FR 59553, Oct. 2, 2015]

§ 301.85–2b Exempted articles. [1]

(a) The following articles are exempt from the certification and permit requirements of this subpart if they meet the applicable conditions prescribed in paragraphs (a) (1) through (4) of this section and have not been exposed to infestation after cleaning or other handling as prescribed in said paragraphs:

(1) Small grains, if harvested in bulk or directly into approved containers, and if the small grains and containers thereof have not come into contact with the soil; or, if they have been cleaned to meet State seed sales requirements.

(2) Soybeans (other than for seed), if harvested in bulk or directly into approved containers, and if the soybeans and containers thereof have not come into contact with the soil.

(3) Unshucked ear corn, if harvested in bulk or directly into approved containers, and if the corn and containers thereof have not come into contact with the soil.

(4) Used farm tools, if cleaned free of soil.

(b) The following articles are exempt from the certification and permit requirements of this subpart if they meet the applicable conditions prescribed in paragraphs (b) (1) through (3) of this section and have not been exposed to infestation after cleaning or other handling as prescribed in said paragraphs: *Provided,* That this exemption shall not apply to any class of regulated articles specified by an inspector in a written notification to the owner or person in possession of the premises that the movement of such articles from such premises under this exemption would involve a hazard of spread of the golden nematode:

(1) Root crops (other than Irish potatoes and sugar beets), if moved in approved containers.

(2) Hay, straw, fodder, and plant litter, if moved in approved containers.

(c) Containers of the following types are approved for the purposes of this section:

(1) New paper bags; and consumer packages of any material except cloth or burlap.

(2) Crates, pallet boxes, trucks, and boxcars, if free of soil.

[35 FR 4692, Mar. 18, 1970, as amended at 47 FR 12331, Mar. 23, 1982; 67 FR 8465, Feb. 25, 2002]

§ 301.85–3 Conditions governing the interstate movement of regulated articles from quarantined States. [2]

(a) Any regulated articles except soil samples for processing, testing, or analysis may be moved interstate from any quarantined State under the following conditions:

(1) With certificate or permit issued and attached in accordance with §§ 301.85–4 and 301.85–7 if moved:

(i) From any generally infested area or any suppressive area into or through any point outside of the regulated areas; or

(ii) From any generally infested area into or through any suppressive area; or

(iii) Between any noncontiguous suppressive areas; or

(iv) Between contiguous suppressive areas when it is determined by an inspector that the regulated articles present a hazard of the spread of the golden nematode and the person in possession thereof has been so notified; or

(v) Through or reshipped from any regulated area when such movement is not authorized under paragraph (a)(2)(v) of this section; or

(2) From any regulated area, without certificate or permit if moved:

(i) Under the provisions of § 301.85–2b which exempts certain articles from certificate and permit requirements; or

(ii) From a generally infested area to a contiguous generally infested area; or

---

[1] The articles hereby exempted remain subject to applicable restrictions under other quarantines and other provisions of this subpart.

[2] Requirements under all other applicable Federal domestic plant quarantines must also be met.

(iii) From a suppressive area to a contiguous generally infested area; or

(iv) Between contiguous suppressive areas unless the person in possession of the articles has been notified by an inspector that a hazard of spread of the golden nematode exists; or

(v) Through or reshipped from any regulated area if the articles originated outside of any regulated area and if the point of origin of the articles is clearly indicated, their identity has been maintained, and they have been safeguarded against infestation while in the regulated area in a manner satisfactory to the inspector; or

(3) From any area outside the regulated areas, if moved:

(i) With a certificate or permit attached; or

(ii) Without a certificate or permit, if:

(a) The regulated articles are exempt from certification and permit requirements under the provisions of § 301.85–2b; or

(b) The point of origin of such movement is clearly indicated on the articles or shipping document which accompanies the articles and if the movement is not made through any regulated area.

(b) Unless specifically authorized by the Deputy Administrator in emergency situations, soil samples for processing, testing or analysis may be moved interstate from any regulated area only to laboratories approved[3] by the Deputy Administrator and so listed by him in a supplemental regulation.[4] A certificate or permit is not required to be attached to such soil samples except in those situations where the Deputy Administrator has authorized such movement only with a certificate or permit issued and attached in accordance with §§ 301.85–4 and 301.85–7. A certificate or permit is not required to be attached to soil samples originating in areas outside of the regulated areas if the point of origin of such movement is

clearly indicated on the articles or shipping document which accompanies the articles and if the movement is not made through any regulated area.

[37 FR 24330, Nov. 16, 1972, as amended at 67 FR 8465, Feb. 25, 2002]

§ 301.85–4 Issuance and cancellation of certificates and permits.

(a) Certificates may be issued for any regulated articles (except soil samples for processing, testing, or analysis) by an inspector if the inspector determines that they are eligible for certification for movement to any destination under all Federal domestic plant quarantines applicable to such articles and:

(1) Have originated in noninfested premises in a regulated area and have not been exposed to infestation while within the regulated areas; or

(2) Have been treated to destroy infestation in accordance with part 305 of this chapter; or

(3) Have been grown, produced, manufactured, stored, or handled in such a manner that no infestation would be transmitted thereby.

(b) Limited permits may be issued by an inspector to allow interstate movement of regulated articles (except soil samples for processing, testing or analysis) not eligible for certification under this subpart, to specified destinations for limited handling, utilization, or processing, or for treatment in accordance with part 305 of this chapter, when, upon evaluation of the circumstances involved in each specific case he determines that such movement will not result in the spread of the golden nematode and requirements of other applicable Federal domestic plant quarantines have been met.

(c) Restricted destination permits may be issued by an inspector to allow the interstate movement (for other than scientific purposes) of regulated articles (except soil samples for processing, testing, or analysis) to any destination permitted under all applicable Federal domestic plant quarantines if such articles are not eligible for certification under all such quarantines but would otherwise qualify for certification under this subpart.

---

[3] Pamphlets containing provisions for laboratory approval may be obtained from the Deputy Administrator, Plant Protection and Quarantine Programs, APHIS, U.S. Department of Agriculture, Washington, DC 20250.

[4] For list of approved laboratories, see PP 639 (37 FR 7813, 15525, and amendments thereof).

(d) Scientific permits to allow the interstate movement of regulated articles and certificates or permits to allow the movement of soil samples for processing, testing, or analysis in emergency situations may be issued by the Deputy Administrator under such conditions as may be prescribed in each specific case by the Deputy Administrator to prevent the spread of the golden nematode.

(e) Certificate, limited permit, and restricted destination permit forms may be issued by an inspector to any person for use for subsequent shipments of regulated articles (except for soil samples for processing, testing, or analysis) provided such person is operating under a compliance agreement; and any such person may be authorized by an inspector to reproduce such forms on shipping containers or otherwise. Any such person may execute and issue the certificate forms, or reproductions of such forms, for the interstate movement of regulated articles from the premises of such person identified in the compliance agreement if such person has treated such regulated articles to destroy infestation in accordance with part 305 of this chapter, and if such regulated articles are eligible for certification for movement to any destination under all Federal domestic plant quarantines applicable to such articles. Any such person may execute and issue the limited permit forms, or reproductions of such forms, for interstate movement of regulated articles to specified destinations when the inspector has made the determinations specified in paragraph (b) of this section. Any such person may execute and issue the restricted destination permit forms, or reproductions of such forms, for the interstate movement of regulated articles not eligible for certification under all Federal domestic plant quarantines applicable to such articles, under the conditions specified in paragraph (c) of this section.

(f) Any certificate or permit which has been issued or authorized may be withdrawn by the inspector or the Deputy Administrator if he or she determines that the holder thereof has not complied with any condition for the use of such document imposed by this subpart. Prior to such withdrawal, the holder of the certificate of permit shall be notified of the proposed action and the reason therefor and afforded reasonable opportunity to present his or her views thereon.

[37 FR 24330, Nov. 16, 1972, as amended at 70 FR 33268, June 7, 2005]

§ 301.85-5 Compliance agreement and cancellation thereof.

(a) Any person engaged in the business of growing, handling, or moving regulated articles may enter into a compliance agreement to facilitate the movement of such articles under this subpart. Compliance agreement forms may be obtained from the Deputy Administrator or an inspector.

(b) Any person engaged in the business of removing soil from Irish potatoes by the process of grading, washing, or fluming may enter into a compliance agreement concerning such operations. The compliance agreement shall be a written agreement between the person conducting such operations and Plant Protection and Quarantine wherein such person agrees to conduct such operations in a manner which, in the judgment of the inspector supervising enforcement of the quarantine and regulations, will substantially remove the soil from the potatoes.

(c) Any compliance agreement may be canceled by the inspector who is supervising its enforcement whenever the inspector finds, after notice and reasonable opportunity to present views has been accorded to the other party thereto, that such other party has failed to comply with the conditions of the agreement.

[37 FR 24330, Nov. 16, 1972, as amended at 47 FR 12332, Mar. 23, 1982; 70 FR 33268, June 7, 2005]

§ 301.85-6 Assembly and inspection of regulated articles.

Persons (other than those authorized to use certificates, limited permits, or restricted destination permits, or reproductions thereof, under § 301.85-4(e)) who desire to move interstate regulated articles which must be accompanied by a certificate or permit shall, as far in advance as possible, request an inspector to examine the articles prior to movement. Such articles shall be assembled at such points and in such manner as the inspector designates to facilitate inspection.

§ 301.85–7 Attachment and disposition of certificates and permits.

(a) If a certificate or permit is required for the interstate movement of regulated articles, the certificate or permit shall be securely attached to the outside of the container in which such articles are moved, except that, where the certificate or permit is attached to the waybill or other shipping document, and the regulated articles are adequately described on the certificate, permit, or shipping document, the attachment of the certificate or permit to each container of the articles is not required.

(b) In all cases, certificates or permits shall be furnished by the carrier to the consignee at the destination of the shipment.

§ 301.85–8 Inspection and disposal of regulated articles and pests.

Any properly identified inspector is authorized to stop and inspect, and to seize, destroy, or otherwise dispose of, or require disposal of regulated articles and golden nematodes as provided in sections 414, 421, and 434 of the Plant Protection Act (7 U.S.C. 7714, 7731, and 7754) in accordance with instructions issued by the Deputy Administrator.

[37 FR 24330, Nov. 16, 1972, as amended at 66 FR 21052, Apr. 27, 2001]

§ 301.85–9 Movement of live golden nematodes.

Regulations requiring a permit for and otherwise governing the movement of live golden nematodes in interstate or foreign commerce are contained in the Federal Plant Pest Regulations in part 330 of this chapter. Applications for permits for the movement of the pest may be made to the Deputy Administrator.

§ 301.85–10 Nonliability of the Department.

The U.S. Department of Agriculture disclaims liability for any costs incident to inspections or compliance with the provisions of the quarantine and regulations in this subpart, other than for the services of the inspector.

Subpart S—Pale Cyst Nematode

SOURCE: 72 FR 51984, Sept. 12, 2007, unless otherwise noted. Redesignated at 84 FR 2428, Feb. 7, 2019.

§ 301.86 Restrictions on interstate movement of regulated articles.

No person may move interstate from any quarantined area any regulated article except in accordance with this subpart.[1]

§ 301.86–1 Definitions.

Administrator. The Administrator, Animal and Plant Health Inspection Service, or any person authorized to act for the Administrator.

Animal and Plant Health Inspection Service. The Animal and Plant Health Inspection Service (APHIS) of the United States Department of Agriculture.

Associated field. A field that has been found to be at risk for infestation with pale cyst nematode in accordance with § 301.86–3(c)(2).

Certificate. A document in which an inspector or person operating under a compliance agreement affirms that a specified regulated article is free of pale cyst nematode and may be moved interstate to any destination.

Compliance agreement. A written agreement between APHIS and a person engaged in growing, handling, or moving regulated articles, wherein the person agrees to comply with this subpart.

Departmental permit. A document issued by the Administrator in which he or she affirms that interstate movement of the regulated article identified on the document is for scientific or experimental purposes and that the regulated article is eligible for interstate movement in accordance with § 301.86–4.

Field. A defined production site that is managed separately from surrounding areas for phytosanitary purposes.

---

[1] Any properly identified inspector is authorized to stop and inspect persons and means of conveyance and to seize, quarantine, treat, apply other remedial measures to, destroy, or otherwise dispose of regulated articles as provided in section 414 of the Plant Protection Act (7 U.S.C. 7714).

Infestation (infested). The presence of the pale cyst nematode or the existence of circumstances that makes it reasonable to believe that the pale cyst nematode is present.

Infested field. A field that has been found to be infested with pale cyst nematode in accordance with §301.86-3(c)(1).

Inspector. Any employee of APHIS or other person authorized by the Administrator to perform the duties required under this subpart.

Interstate. From any State into or through any other State.

Limited permit. A document in which an inspector or person operating under a compliance agreement affirms that the regulated article identified on the document is eligible for interstate movement in accordance with §301.86-5(b) only to a specified destination and only in accordance with specified conditions.

Moved (move, movement). Shipped, offered for shipment, received for transportation, transported, carried, or allowed to be moved, shipped, transported, or carried.

Nursery stock. Living plants and plant parts intended to be planted, to remain planted, or to be replanted.

Pale cyst nematode. The pale cyst nematode (*Globodera pallida*), in any stage of development.

Person. Any association, company, corporation, firm, individual, joint stock company, partnership, society, or other entity.

Plant Protection and Quarantine. The Plant Protection and Quarantine program of the Animal and Plant Health Inspection Service, United States Department of Agriculture.

Quarantined area. Any State or portion of a State designated as a quarantined area in accordance with the provisions in §301.86-3.

Regulated article. Any article listed in §301.86-2 or otherwise designated as a regulated article in accordance with §301.86-2(i).

State. The District of Columbia, Puerto Rico, the Northern Mariana Islands, or any State, territory, or possession of the United States.

[72 FR 51984, Sept. 12, 2007, as amended at 74 FR 19381, Apr. 29, 2009]

§301.86-2 Regulated articles.

The following are regulated articles:
(a) Pale cyst nematodes. [2]
(b) The following pale cyst nematode host crops:

Eggplant (*Solanum melongena* L.)
Pepper (*Capsicum* spp.)
Potato (*Solanum tuberosum* L.)
Tomatillo (*Physalis philadelphica*)
Tomato (*Lycopersicon esculentum* L.)

(c) Root crops.
(d) Garden and dry beans (*Phaseolus* spp.) and peas (*Pisum* spp.).
(e) All nursery stock.
(f) Soil, compost, humus, muck, peat, and manure, and products on or in which soil is commonly found, including grass sod and plant litter.
(g) Hay, straw, and fodder.
(h) Any equipment or conveyance used in an infested or associated field that can carry soil if moved out of the field.
(i) Any other product, article, or means of conveyance not listed in paragraphs (a) through (h) of this section that an inspector determines presents a risk of spreading the pale cyst nematode, after the inspector provides written notification to the person in possession of the product, article, or means of conveyance that it is subject to the restrictions of this subpart.

[72 FR 51984, Sept. 12, 2007, as amended at 74 FR 19381, Apr. 29, 2009]

§301.86-3 Quarantined areas.

(a) *Designation of quarantined areas.* In accordance with the criteria listed in paragraph (c) of this section, the Administrator will designate as a quarantined area each field that has been found to be infested with pale cyst nematode, each field that has been found to be associated with an infested field, and any area that the Administrator considers necessary to quarantine because of its inseparability for quarantine enforcement purposes from infested or associated fields. The Administrator will publish the description of the quarantined area on the Plant Protection and Quarantine Web site, *https://www.aphis.usda.gov/*

---

[2] Permit and other requirements for the interstate movement of pale cyst nematodes are contained in part 330 of this chapter.

planthealth/pcn''. The description of the quarantined area will include the date the description was last updated and a description of the changes that have been made to the quarantined area. The description of the quarantined area may also be obtained by request from any local office of PPQ; local offices are listed in telephone directories. After a change is made to the quarantined area, we will publish a notice in the FEDERAL REGISTER informing the public that the change has occurred and describing the change to the quarantined area.

(b) *Designation of an area less than an entire State as a quarantined area.* Less than an entire State will be designated as a quarantined area only if the Administrator determines that:

(1) The State has adopted and is enforcing restrictions on the intrastate movement of the regulated articles that are equivalent to those imposed by this subpart on the interstate movement of regulated articles; and

(2) The designation of less than the entire State as a quarantined area will prevent the interstate spread of the pale cyst nematode.

(c) *Criteria for designation of fields as infested fields and associated fields.* (1) *Infested fields.* A field will be designated as an infested field for pale cyst nematode upon a determination that viable pale cyst nematode is present in the field. The determination will be made in accordance with the criteria established by the Administrator for the designation of infested fields. The criteria are presented in a protocol document that may be viewed at *https://www.aphis.usda.gov/planthealth/pcn.* The protocol may also be obtained by request from any local office of Plant Protection and Quarantine; local offices are listed in telephone directories. Any substantive changes we propose to make to the protocol will be published for comment in the FEDERAL REGISTER. After we review the comments received, we will publish another notice in the FEDERAL REGISTER informing the public of any changes to the protocol.

(2) *Associated fields.* The Administrator will designate a field as an associated field when pale cyst nematode host crops, as listed in § 301.86-2(b),

have been grown in the field in the last 10 years and

(i) The field shares a border with an infested field; or

(ii) The field came into contact with a regulated article listed in § 301.86-2 from an infested field within the last 10 years; or

(iii) Within the last 10 years, the field shared ownership, tenancy, seed, drainage or runoff, farm machinery, or other elements of shared cultural practices with an infested field that could allow spread of the pale cyst nematode, as determined by the Administrator.

(d) *Removal of fields from quarantine.* (1) *Infested fields.* An infested field will be removed from quarantine for pale cyst nematode upon a determination that no viable pale cyst nematode is detected in the field. The determination will be made in accordance with criteria established by the Administrator and sufficient to support removal of infested fields from quarantine. The criteria are presented in a protocol document as provided in paragraph (d)(4) of this section along with information for viewing the protocol.

(2) *Associated fields.* An associated field will be removed from quarantine for pale cyst nematode once surveys are completed and pale cyst nematode is not detected in the field. The determination will be made in accordance with criteria established by the Administrator and sufficient to support removal of associated fields from quarantine. The criteria are presented in a protocol document as provided in paragraph (d)(4) of this section along with information for viewing the protocol.

(3) *Removal of other areas from quarantine.* If the Administrator has quarantined any area other than infested or associated fields because of its inseparability for quarantine enforcement purposes from infested or associated fields, as provided in paragraph (a) of this section, that area will be removed from quarantine when the relevant infested or associated fields are removed from quarantine.

(4) *Protocol for removal of fields from quarantine.* The Administrator will remove infested and associated fields, and other areas as provided in this section, from quarantine for pale cyst

nematode in accordance with the protocols published on the Plant Protection and Quarantine website at *https://www.aphis.usda.gov/planthealth/pcn.* The protocols may also be obtained by request from any local office of Plant Protection and Quarantine; local offices are listed in telephone directories. Any substantive changes we propose to make to the protocolswill be published for comment in the FEDERAL REGISTER. After we review the comments received, we will publish another notice in the FEDERAL REGISTER informing the public of any changes to the protocols.

[72 FR 51984, Sept. 12, 2007, as amended at 74 FR 19381, Apr. 29, 2009; 85 FR 85503, Dec. 29, 2020]

§ 301.86–4 Conditions governing the interstate movement of regulated articles from quarantined areas.

(a) Any regulated article may be moved interstate from a quarantined area only if moved under the following conditions:

(1) With a certificate or limited permit issued and attached in accordance with §§ 301.86–5 and 301.86–8;

(2) Without a certificate or limited permit if:

(i) The regulated article is moved by the United States Department of Agriculture for experimental or scientific purposes; or

(ii) The regulated article originates outside the quarantined area and is moved interstate through the quarantined area under the following conditions:

(A) The points of origin and destination are indicated on a waybill accompanying the regulated article; and

(B) The regulated article is moved through the quarantined area without stopping (except for refueling and for traffic conditions such as traffic lights and stop signs); and

(C) The regulated article is not unpacked or unloaded in the quarantined area; and

(D) The article has not been combined or commingled with other articles so as to lose its individual identity.

(b) When an inspector has probable cause to believe a person or means of conveyance is moving a regulated article interstate, the inspector is authorized to stop the person or means of conveyance to determine whether a regulated article is present and to inspect the regulated article. Articles found to be infested by an inspector, and articles not in compliance with the regulations in this subpart, may be seized, quarantined, treated, subjected to other remedial measures, destroyed, or otherwise disposed of.

§ 301.86–5 Issuance and cancellation of certificates and limited permits.

(a) *Certificates.* An inspector[3] or person operating under a compliance agreement may issue a certificate for the interstate movement of a regulated article if the inspector determines that the regulated article satisfies the general requirements for a certificate in paragraph (a)(1) of this section and any requirements that may apply to the regulated article under paragraphs (a)(2) through (a)(7) of this section.

(1) *Certification requirements for all regulated articles.* The regulated article must be moved in compliance with any additional emergency conditions the Administrator may impose under section 414 of the Plant Protection Act (7 U.S.C. 7714)[4] to prevent the spread of the pale cyst nematode. In addition, the regulated article must be eligible for unrestricted movement under all other Federal domestic plant quarantines and regulations applicable to the regulated article.

---

[3] Inspectors are assigned to local offices of APHIS, which are listed in local telephone directories. Information concerning such local offices may also be obtained from the Animal and Plant Health Inspection Service, Plant Protection and Quarantine, Domestic and Emergency Operations, 4700 River Road Unit 134, Riverdale, Maryland 20737–1236.

[4] Section 414 of the Plant Protection Act (7 U.S.C. 7714) provides that the Secretary of Agriculture may, under certain conditions, hold, seize, quarantine, treat, apply other remedial measures to destroy or otherwise dispose of any plant, plant pest, plant product, article, or means of conveyance that is moving, or has moved into or through the United States or interstate if the Secretary has reason to believe the article is a plant pest or is infested with a plant pest at the time of movement.

(2) *Certification requirements for nursery stock*—(i) *Potatoes.* Potatoes intended for use as nursery stock (*i.e.*, seed potatoes) are prohibited from being moved interstate from the quarantined area.

(ii) *Nursery stock of other host crops.* An inspector may issue a certificate for the interstate movement of nursery stock of pale cyst nematode host crops other than potatoes, as listed in § 301.86-2(b), if the nursery stock was grown in a field that meets the following requirements:

(A) The field has been surveyed by an inspector for pale cyst nematode at least once in the last 3 years;

(B) The pale cyst nematode has not been found in the field; and

(C) No more than one pale cyst nematode host crop, as listed in § 301.86-2(b), has been grown in the field in the last 3 years.

(iii) *Nursery stock of non-host crops*— (A) *With soil.* An inspector may issue a certificate for the interstate movement of nursery stock of non-host crops moved with soil if the nursery stock was grown in a field that meets the following requirements:

(1) The field has been surveyed by an inspector for pale cyst nematode at least once in the last 3 years;

(2) The pale cyst nematode has not been found in the field; and

(3) No more than one pale cyst nematode host crop, as listed in § 301.86-2(b), has been grown in the field in the last 3 years.

(B) *Without soil (bare-rooted).* An inspector may issue a certificate for the interstate movement of nursery stock of non-host crops moved without soil if the inspector finds the nursery stock to be free of soil on its roots and on all other parts of the plant.

(3) *Certification requirements for potatoes for consumption, root crops for consumption, garden or dry beans, and peas.* An inspector may issue a certificate for the movement of potatoes intended for consumption, root crops intended for consumption, garden or dry beans, or peas from the quarantine area only if the field in which the potatoes, root crops, garden or dry beans, or peas were grown meets the following requirements:

(i) The field has been surveyed by an inspector for pale cyst nematode at least once in the last 3 years and prior to the planting of the potatoes or root crops;

(ii) Pale cyst nematode has not been found in the field; and

(iii) No more than one pale cyst nematode host crop, as listed in § 301.86-2(b), has been grown in the field in the last 3 years.

(4) *Certification requirements for soil and associated products.* An inspector may issue a certificate for the interstate movement of a regulated article listed in § 301.86-2(e) only if the article originated in a field that meets the following requirements:

(i) The field has been surveyed by an inspector for pale cyst nematode at least once in the last 3 years;

(ii) The pale cyst nematode has not been found in the field; and

(iii) No more than one pale cyst nematode host crop, as listed in § 301.86-2(b), has been grown in the last 3 years.

(5) *Certification requirements for hay, straw, and fodder.* An inspector may issue a certificate for the movement of hay, straw, or fodder from the quarantined area only if:

(i) The field where the hay, straw, or fodder was produced meets the following requirements:

(A) The field has been surveyed by an inspector for pale cyst nematode at least once in the last 3 years;

(B) The pale cyst nematode has not been found in the field; and

(C) No more than one pale cyst nematode host crop, as listed in § 301.86-2(b), has been grown in the field in the last 3 years; or

(ii) The hay, straw, or fodder is produced according to procedures judged by an inspector to be sufficient to isolate it from soil throughout its production.

(6) *Certification requirements for equipment used in infested or associated fields.* An inspector may issue a certificate for the interstate movement of equipment that has been used in an infested or associated field and that can carry soil if moved out of the field only after the equipment has been pressure-washed under the supervision of an inspector to remove all soil or steam-treated in accordance with part 305 of this chapter.

(b) *Limited permits*—(1) *General conditions.* An inspector[5] may issue a limited permit for the interstate movement of a regulated article if the inspector determines that:

(i) The regulated article is to be moved interstate to a specified destination for specified handling, processing, or utilization (the destination and other conditions to be listed in the limited permit), and this interstate movement will not result in the spread of the pale cyst nematode because life stages of the pale cyst nematode will be destroyed by the specified handling, processing, or utilization;

(ii) The regulated article is to be moved in compliance with any additional emergency conditions the Administrator may impose under section 414 of the Plant Protection Act (7 U.S.C. 7714) to prevent the spread of the pale cyst nematode; and

(iii) The regulated article is eligible for interstate movement under all other Federal domestic plant quarantines and regulations applicable to the regulated article.

(2) *Specific conditions for potatoes for consumption.* An inspector may issue a limited permit to allow the interstate movement of potatoes from the quarantined area for processing or packing only if:

(i) The potatoes are transported in a manner that prevents the potatoes and soil attached to the potatoes from coming into contact with agricultural premises outside the quarantined area; and

(ii) The potatoes are processed or packed at facilities that handle potatoes, waste, and waste water in a manner approved by APHIS to prevent the spread of pale cyst nematode.

(c) Certificates and limited permits for the interstate movement of regulated articles may be issued by an inspector or person operating under a compliance agreement. A person operating under a compliance agreement may issue a certificate for the interstate movement of a regulated article after an inspector has determined that the regulated article is eligible for a certificate in accordance with paragraph (a) of this section. A person oper-

ating under a compliance agreement may issue a limited permit for interstate movement of a regulated article after an inspector has determined that the regulated article is eligible for a limited permit in accordance with paragraph (b) of this section.

(d) Any certificate or limited permit that has been issued may be withdrawn, either orally or in writing, by an inspector if he or she determines that the holder of the certificate or limited permit has not complied with all provisions in this subpart for the use of the certificate or limited permit or has not complied with all the conditions contained in the certificate or limited permit. If the withdrawal is oral, the withdrawal and the reasons for the withdrawal will be confirmed in writing as promptly as circumstances allow. Any person whose certificate or limited permit has been withdrawn may appeal the decision in writing to the Administrator within 10 days after receiving the written notification of the withdrawal. The appeal must state all of the facts and reasons upon which the person relies to show that the certificate or limited permit was wrongfully withdrawn. As promptly as circumstances allow, the Administrator will grant or deny the appeal, in writing, stating the reasons for the decision. A hearing will be held to resolve any conflict as to any material fact. Rules of practice concerning a hearing will be adopted by the Administrator.

(Approved by the Office of Management and Budget under control number 0579–0322)

[72 FR 51984, Sept. 12, 2007, as amended at 74 FR 19381, Apr. 29, 2009]

§301.86–6 Compliance agreements and cancellation.

(a) Any person engaged in growing, handling, or moving regulated articles may enter into a compliance agreement when an inspector determines that the person is aware of this subpart, agrees to comply with its provisions, and agrees to comply with all

---

[5] See footnote 3 to §301.86–5(a).

the provisions contained in the compliance agreement.[6]

(b) Any compliance agreement may be canceled, either orally or in writing, by an inspector whenever the inspector finds that the person who has entered into the compliance agreement has failed to comply with any of the provisions of this subpart. If the cancellation is oral, the cancellation and the reasons for the cancellation will be confirmed in writing as promptly as circumstances allow. Any person whose compliance agreement has been canceled may appeal the decision, in writing, to the Administrator, within 10 days after receiving written notification of the cancellation. The appeal must state all of the facts and reasons upon which the person relies to show that the compliance agreement was wrongfully canceled. As promptly as circumstances allow, the Administrator will grant or deny the appeal, in writing, stating the reasons for the decision. A hearing will be held to resolve any conflict as to any material fact. Rules of practice concerning a hearing will be adopted by the Administrator.

§ 301.86–7 Assembly and inspection of regulated articles.

(a) Any person (other than a person authorized to issue certificates or limited permits under § 301.86–5(c)) who desires a certificate or limited permit to move a regulated article interstate must notify an inspector[7] as far in advance of the desired interstate movement as possible, but no less than 48 hours before the desired interstate movement.

(b) The regulated article must be assembled at the place and in the manner the inspector designates as necessary to comply with this subpart.

§ 301.86–8 Attachment and disposition of certificates and limited permits.

(a) A certificate or limited permit required for the interstate movement of a regulated article must, at all times during the interstate movement, be:

[6] Compliance agreement forms are available without charge from local Plant Protection and Quarantine offices, which are listed in telephone directories.

[7] See footnote 3 to § 301.86–5(a).

(1) Attached to the outside of the container containing the regulated article; or

(2) Attached to the regulated article itself if not in a container; or

(3) Attached to the consignee's copy of the accompanying waybill. If the certificate or limited permit is attached to the consignee's copy of the waybill, the regulated article must be sufficiently described on the certificate or limited permit and on the waybill to identify the regulated article.

(b) The certificate or limited permit for the interstate movement of a regulated article must be furnished by the carrier or the carrier's representative to the consignee listed on the certificate or limited permit upon arrival at the location provided on the certificate or limited permit.

(Approved by the Office of Management and Budget under control number 0579–0322)

§ 301.86–9 Costs and charges.

The services of the inspector during normal business hours (8 a.m. to 4:30 p.m., Monday through Friday, except holidays) will be furnished without cost. APHIS will not be responsible for any costs or charges incident to inspections or compliance with the provisions of the quarantine and regulations in this subpart, other than for the services of the inspector.

Subpart T—Sugarcane Diseases

SOURCE: 48 FR 50059, Oct. 31, 1983, unless otherwise noted. Redesignated at 84 FR 2428, Feb. 7, 2019.

QUARANTINE AND REGULATIONS

§ 301.87 Quarantine; restrictions on interstate movement of specified articles.[1][2]

(a) *Notice of quarantine.* Under the authority of sections 411, 412, 414, and 434

[1] Any inspector is authorized to stop and inspect persons and means of conveyance, and to hold, seize, quarantine, treat, apply other remedial measures to, destroy, or otherwise dispose of plants, plant pests, or other articles in accordance with sections 414, 421, and 434 of the Plant Protection Act (7 U.S.C. 7714, 7731, and 7754).

[2] Regulations concerning the movement of gummosis bacteria and leaf scald bacteria in

of the Plant Protection Act (7 U.S.C. 7711, 7712, 7714, and 7754), the Secretary of Agriculture quarantines Hawaii to prevent the artificial spread of leaf scald disease and quarantines Puerto Rico to prevent the artificial spread of gummosis disease and leaf scald disease. The regulations in this subpart govern the interstate movement from Hawaii and Puerto Rico of the regulated articles described in § 301.87-2.

(b) *Quarantine restrictions on interstate movement of regulated articles.* No common carrier or other person shall move interstate from any regulated area any regulated article except in accordance with the conditions prescribed in this subpart.

[48 FR 50059, Oct. 31, 1983, as amended at 66 FR 21052, Apr. 27, 2001]

§ 301.87-1 Definitions.

Terms used in the singular form in this subpart shall be construed as the plural and vice versa, as the case may demand. The following terms, when used in this subpart, shall be construed, respectively, to mean:

Certificate. A document which is issued for a regulated article by an inspector or by a person operating under a compliance agreement, and which represents that the article is eligible for interstate movement in accordance with § 301.87-5(a) of this subpart.

Compliance agreement. A written agreement between Plant Protection and Quarantine and a person engaged in the business of growing, handling, or moving regulated articles, in which the person agrees to comply with the provisions of this subpart and any conditions imposed pursuant to such provisions.

Deputy Administrator. The Deputy Administrator of the Animal and Plant Health Inspection Service, U.S. Department of Agriculture for Plant Protection and Quarantine, or any officer or employee of the Department to whom authority to act in his or her stead has been or may hereafter be delegated.

Gummosis disease. A dangerous plant disease of sugarcane which is caused by the highly infectious bacterium, *Xanthomonas vasculorum* (Cobb)

Dowson, and which is not widely prevalent or distributed within and throughout the United States.

Inspector. Any employee of Plant Protection and Quarantine, Animal and Plant Health Inspection Service, U.S. Department of Agriculture, or other person, authorized by the Deputy Administrator in accordance with law to enforce the provisions of the quarantine and regulations in this subpart.

Interstate. From any State into or through any other State.

Leaf scald disease. A dangerous plant disease of sugarcane which is caused by the highly infectious bacterium, *Xanthomonas albilineans* (Ashby) Dowson, and which is not widely prevalent or distributed within and throughout the United States.

Limited permit. A document which is issued for a regulated article by an inspector or by a person operating under a compliance agreement, and which represents that the regulated article is eligible for interstate movement in accordance with § 301.87-5(b) of this subpart.

Moved (movement, move). Shipped, offered for shipment to a common carrier, received for transportation or transported by a common carrier, or carried, transported, moved, or caused or allowed to be moved by any means. "Movement" and "move" shall be construed in accordance with this definition.

Person. Any individual, partnership, corporation, company, society, association, or other organized group.

Plant Protection and Quarantine. The organizational unit within the Animal and Plant Health Inspection Service, U.S. Department of Agriculture, delegated responsibility for enforcing provisions of the Plant Protection Act and related legislation, quarantines, and regulations.

Regulated area. Any quarantined State, or any portion thereof, listed as a regulated area in § 301.87-3(c) of this subpart, or otherwise designated as a regulated area in accordance with § 301.87-3(b) of this subpart.

Regulated article. Any article listed in § 301.87-2(a), (b), (c), (d), or otherwise designated as a regulated article in accordance with § 301.87-2(e).

interstate or foreign commerce are contained in part 330 of this chapter.

123

State. Any State, Territory, or District of the United States, including the Commonwealth of Puerto Rico.

Sugarcane disease. This means leaf scald disease with respect to activities in Hawaii, and means gummosis disease or leaf scald disease with respect to activities in Puerto Rico.

[48 FR 50059, Oct. 31, 1983, as amended at 52 FR 31374, Aug. 20, 1987; 66 FR 21052, Apr. 27, 2001]

§ 301.87–2 Regulated articles.

(a) Sugarcane plants, whole or in part, including true seed and bagasse, but not including pieces of cane boiled for a minimum of 30 minutes during processing into sugarcane chews;

(b) Used sugarcane processing equipment (sugarcane mill equipment, such as equipment used for extracting and refining sugarcane juice; and experimental devices, such as devices used for extracting sugarcane juice);

(c) Used sugarcane field equipment (equipment used for sugarcane field production purposes, e.g. planters, tractors, discs, cultivators, and vehicles);

(d) Sugarcane juice; and

(e) Any other product, article, or means of conveyance, of any character whatsoever, not covered by paragraph (a), (b), (c), or (d) of this section, when it is determined by an inspector that it presents a risk of spread of a sugarcane disease and the person in possession of it has actual notice that the product, article, or means of conveyance is subject to the restrictions of this section.

[48 FR 50059, Oct. 31, 1983, as amended at 52 FR 31374, Aug. 20, 1987]

§ 301.87–3 Regulated areas.

(a) Except as otherwise provided in paragraph (b) of this section, the Deputy Administrator shall list as a regulated area in paragraph (c) of this section, each quarantined State, or each portion thereof, in which a sugarcane disease has been found by an inspector or in which the Deputy Administrator has reason to believe that a sugarcane disease is present, or each portion of a quarantined State which the Deputy Administrator deems necessary to regulate because of its proximity to a sugarcane disease or its inseparability for quarantine enforcement purposes from localities in which a sugarcane disease occurs. Less than an entire quarantined State will be designated as a regulated area only if the Deputy Administrator is of the opinion that:

(1) The State has adopted and is enforcing a quarantine or regulation which imposes restrictions on the intrastate movement of the regulated articles which are substantially the same as those which are imposed with respect to the interstate movement of such articles under this subpart; and

(2) The designation of less than the entire State as a regulated area will otherwise be adequate to prevent the artificial interstate spread of a sugarcane disease.

(b) The Deputy Administrator or an inspector may temporarily designate any nonregulated area in a quarantined State as a regulated area in accordance with the criteria specified in paragraph (a) of this section for listing such an area. Written notice of the designation shall be given to the owner or person in possession of the nonregulated area and, thereafter, the interstate movement of any regulated article from the area shall be subject to the applicable provisions of this subpart. As soon as practicable, the area shall be added to the list in paragraph (c) of this section or the designation shall be terminated by the Deputy Administrator or an inspector, and notice thereof shall be given to the owner or person in possession of the area.

(c) The areas described below are designated as regulated areas;

Hawaii

All of Hawaii.

Puerto Rico.

All of Puerto Rico.

§ 301.87–4 Conditions governing the interstate movement of regulated articles from regulated areas in quarantined States. [3]

Any regulated article may be moved interstate from any regulated area in a quarantined State if moved under the following conditions:

---

[3] Requirements under all other applicable Federal domestic plant quarantines must also be met.

(a) With a certificate or limited permit issued and attached in accordance with §§301.87–5 and 301.87–8 of this subpart, or

(b) Without a certificate or limited permit, if

(1) Moved directly through any regulated area, and

(2) The article originated outside of any regulated area, and

(3) The point of origin of the article is clearly indicated by shipping documents, its identity has been maintained, and it has not been used for the production of sugarcane while in the regulated area.

§301.87–5 Issuance and cancellation of certificates and limited permits.

(a) A certificate shall be issued by an inspector for the movement of a regulated article if the inspector:

(1)(i) Determines that it has been treated under the direction of an inspector[4] in accordance with part 305 of this chapter, or

(ii) Determines based on inspection of the article and the premises of origin that it is free from sugarcane diseases;[5]

(2) Determines that it is to be moved in compliance with any additional conditions deemed necessary under section 414 of the Plant Protection Act (7 U.S.C. 7714)[6] to prevent the spread of sugarcane diseases; and

(3) Determines that it is eligible for unrestricted movement under all other Federal domestic plant quarantines applicable to the article.

(b) A limited permit shall be issued by an inspector for the movement of a regulated article if the inspector:

(1) Determines, in consultation with the Deputy Administrator, that it is to be moved:

(i) For a specified purpose (such as for consumption or manufacturing) stated on the limited permit, other than for processing or harvesting sugarcane; and

(ii) To a specified destination stated on the limited permit, which is not in a county or parish where sugarcane is produced, and which is not within 10 miles of a sugarcane field;

(2) Determines that it is to be moved in compliance with any additional conditions deemed necessary under section 414 of the Plant Protection Act (7 U.S.C. 7714)[6] to prevent the spread of sugarcane diseases; and

(3) Determines that it is eligible for such movement under all other Federal domestic plant quarantines applicable to the article.

(c) Certificates and limited permits for shipments of regulated articles may be issued by an inspector or by any person engaged in the business of growing, handling, or moving regulated articles provided such person is operating under a compliance agreement. Any such person may execute and issue a certificate for the interstate movement of a regulated article if the person has treated the regulated article to destroy infection in accordance with the provisions of §301.87–10 of this subpart and the inspector has made the determination that the article is otherwise eligible for a certificate in accordance with paragraph (a) of this section; or if the inspector has made the determination that the article is eligible for a certificate in accordance with paragraph (a) of this section without such treatment. Any such person may execute and issue a limited permit for interstate movement of a regulated article when the inspector has made the determination that the article is eligible for a limited permit in accordance with paragraph (b) of this section.

(d) Any certificate or limited permit which has been issued or authorized may be withdrawn by an inspector if the inspector determines that its holder has not complied with any condition under the regulations for its use. The reasons for the withdrawal shall be confirmed in writing as promptly as

---

[4] Treatments shall be monitored by inspectors in order to assure compliance with requirements in this subpart.

[5] The term *sugarcane diseases* means leaf scald disease with respect to movement of regulated articles from Hawaii and means gummosis disease and leaf scald disease with respect to movements of regulated articles from Puerto Rico.

[6] An inspector may hold, seize, quarantine, treat, apply other remedial measures to, destroy, or otherwise dispose of plants, plant pests, or other articles in accordance with sections 414, 421, and 434 of the Plant Protection Act (7 U.S.C. 7714, 7731, and 7754).

circumstances allow. Any person whose certificate or limited permit has been withdrawn may appeal the decision in writing to the Deputy Administrator within ten days after receiving the written notification of the withdrawal. The appeal shall state all of the facts and reasons upon which the person relies to show that the certificate or limited permit was wrongfully withdrawn. The Deputy Administrator shall grant or deny the appeal in writing, stating the reasons for the decision as promptly as circumstances allow. If there is a conflict as to any material fact, a hearing shall be held to resolve the conflict under rules of practice which shall be adopted by the Administrator of the Animal and Plant Health Inspection Service, USDA, for the proceeding.

[48 FR 50059, Oct. 31, 1983, as amended at 66 FR 21053, Apr. 27, 2001; 75 FR 4241, Jan. 26, 2010]

§ 301.87–6 Compliance agreement; cancellation.

(a) Any person engaged in the business of growing, handling, or moving regulated articles may enter into a compliance agreement to facilitate the movement of regulated articles under this subpart.[7] The compliance agreement shall be a written agreement between a person engaged in such a business and Plant Protection and Quarantine, in which the person agrees to comply with the provisions of this subpart and any conditions imposed pursuant to such provisions.

(b) Any compliance agreement may be canceled orally or in writing by the inspector who is supervising its enforcement whenever the inspector finds that such person has failed to comply with the provisions of this subpart or any conditions imposed pursuant to such provisions. If the cancellation is oral, the decision and the reasons for the cancellation shall be confirmed in writing as promptly as circumstances

allow. Any person whose compliance agreement has been canceled may appeal the decision, in writing, to the Deputy Administrator within ten days after receiving written notification of the cancellation. The appeal shall state all of the facts and reasons upon which the person relies to show that the compliance agreement was wrongfully cancelled. The Deputy Administrator shall grant or deny the appeal, in writing, stating the reasons for the decision, as promptly as circumstances allow. If there is a conflict as to any material fact, a hearing shall be held to resolve the conflict under rules of practice which shall be adopted by the Administrator of the Animal and Plant Health Inspection Service, USDA, for the proceeding.

[48 FR 50059, Oct. 31, 1983, as amended at 59 FR 67609, Dec. 30, 1994]

§ 301.87–7 Assembly and inspection of regulated articles.

(a) Any person (other than a person authorized to issue certificates or limited permits under § 301.87–5(c) of this subpart) who desires to move interstate a regulated article accompanied by a certificate or limited permit shall, as far in advance as possible (should be no less than 48 hours before the desired movement), request an inspector[8] to take any necessary action under this subpart prior to movement of the regulated article.

(b) The regulated article shall be assembled at whatever point and in whatever manner the inspector designates as necessary to comply with the requirements of this subpart.

[48 FR 50059, Oct. 31, 1983, as amended at 59 FR 67609, Dec. 30, 1994]

§ 301.87–8 Attachment and disposition of certificates and limited permits.

(a) A certificate or limited permit required for the interstate movement of

---

[7] Compliance Agreement forms are available without charge from the Animal and Plant Health Inspection Service, Plant Protection and Quarantine, Domestic and Emergency Operations, 4700 River Road Unit 134, Riverdale, Maryland 20737–1236, and from local offices of Plant Protection and Quarantine. (Local offices are listed in telephone directories.)

[8] Inspectors are assigned to local offices of Plant Protection and Quarantine, which are listed in telephone directories. Information concerning local offices may also be obtained from the Animal and Plant Health Inspection Service, Plant Protection and Quarantine, Domestic and Emergency Operations, 4700 River Road Unit 134, Riverdale, Maryland 20737–1236.

a regulated article, at all times during such movement, shall be securely attached to the outside of the container containing the regulated article, securely attached to the article itself if not in a container, or securely attached to the consignee's copy of the accompanying waybill or other shipping document; provided however, that the requirements of this section may be met by attaching the certificate or limited permit to the consignee's copy of the waybill or other shipping document only if the regulated article is sufficiently described on the certificate, limited permit, or shipping document to identify the article.

(b) The certificate or limited permit for the movement of a regulated article shall be furnished by the carrier to the consignee at the destination of the shipment.

§301.87-9 Costs and charges.

The services of the inspector shall be furnished without cost. The U.S. Department of Agriculture will not be responsible for any costs or charges incident to inspections or compliance with the provisions of the quarantine and regulations in this subpart, other than for the services of the inspector.

§301.87-10 [Reserved]

Subpart U—Karnal Bunt

SOURCE: 61 FR 52207, Oct. 4, 1996, unless otherwise noted. Redesignated at 84 FR 2428, Feb. 7, 2019.

§301.89-1 Definitions.

Actual price received. The net price after adjustment for any premiums or discounts stated on the sales receipt.

Administrator. The Administrator, Animal and Plant Health Inspection Service, or any person authorized to act for the Administrator.

Animal and Plant Health Inspection Service (APHIS). The Animal and Plant Health Inspection Service of the U.S. Department of Agriculture.

Certificate. A document in which an inspector or a person operating under a compliance agreement affirms that a specified regulated article meets the requirements of this subpart and may be moved to any destination.

Compliance agreement. A written agreement between APHIS and a person engaged in growing, handling, or moving regulated articles, in which the person agrees to comply with the provisions of this subpart and any conditions imposed under this subpart.

Contaminated seed. Seed from sources in which the Karnal bunt pathogen (*Tilletia indica* (Mitra) Mundkur) has been determined to exist by the presence of bunted kernels or teliospores.

Contract price. The net price after adjustment for any premiums or discounts stated in the contract.

Conveyances. Containers used to move wheat, durum wheat, or triticale, or their products, including trucks, trailers, railroad cars, bins, and hoppers.

Distinct definable area. A commercial wheat production area of contiguous fields that is separated from other wheat production areas by desert, mountains, or other nonagricultural terrain as determined by an inspector, based on survey results.

Grain. Wheat, durum wheat, and triticale used for consumption or processing.

Grain storage facility. That part of a grain handling operation or unit or a grain handling operation, consisting or structures, conveyances, and equipment that receive, unload, and store, grain, and that is able to operate as an independent unit from other units of the grain handling operation. A grain handling operation may be one grain storage facility or may be comprised of many grain storage facilities on a single premises.

Hay. Host crops cut and dried for feeding to livestock. Hay cut after reaching the dough stage may contain mature kernels of the host crop.

Host crops. Plants or plant parts, including grain, seed, or hay, of wheat, durum wheat, and triticale.

Infestation (infected). The presence of Karnal bunt, or any identifiable stage of development (*i.e.,* bunted kernels in grain, bunted kernels or teliospores in seed) of the fungus *Tilletia indica* (Mitra) Mundkur, or the existence of circumstances that make it reasonable to believe that Karnal bunt is present.

Inspector. An APHIS employee or designated cooperator/collaborator authorized by the Administrator to enforce the provisions of this subpart.

Karnal bunt. A plant disease caused by the fungus *Tilletia indica* (Mitra) Mundkur.

Limited permit. A document in which an inspector affirms that a specified regulated article not eligible for a certificate is eligible for movement only to a specified destination and in accordance with conditions specified on the permit.

Mechanized cultivating equipment and mechanized harvesting equipment. Mechanized equipment used for soil tillage, including tillage attachments for farm tractors—*e.g.,* tractors, disks, plows, harrows, planters, and subsoilers; mechanized equipment used for harvesting purposes—*e.g.,* combines, grain buggies, trucks, swathers, and hay balers.

Movement (moved). The act of shipping, transporting, delivering, or receiving for movement, or otherwise aiding, abetting, inducing or causing to be moved.

Person. Any association, company, corporation, firm, individual, joint stock company, partnership, society, or any other legal entity.

Plant. Any plant (including any plant part) for or capable of propagation, including a tree, a tissue culture, a plantlet culture, pollen, a shrub, a vine, a cutting, a graft, a scion, a bud, a bulb, a root, and a seed.

Seed. Wheat, durum wheat, and triticale used for propagation.

Soil. The loose surface material of the earth in which plants grow, in most cases consisting of disintegrated rock with an admixture of organic material.

Soil-moving equipment. Equipment used for moving or transporting soil, including, but not limited to, bulldozers, dump trucks, or road scrapers.

State. The District of Columbia, Puerto Rico, the Northern Mariana Islands, or any State, territory, or possession of the United States.

Straw. The vegetative material left after the harvest of host crops. Straw is generally used as animal feed, bedding, mulch, or for erosion control.

Tilling. The turning of a minimum of the top 6 inches of soil.

[61 FR 52207, Oct. 4, 1996, as amended at 62 FR 23624, May 1, 1997; 62 FR 24751, May 6, 1997; 63 FR 31599, June 10, 1998; 64 FR 23752, May 4, 1999; 69 FR 8095, Feb. 23, 2004]

§ 301.89-2 Regulated articles.

The following are regulated articles:

(a) Conveyances, including trucks, railroad cars, and other containers used to move host crops produced in a regulated area that have tested positive for Karnal bunt through the presence of bunted kernels;

(b) Grain elevators/equipment/structures used for storing and handling host crops produced in a regulated area that have tested positive for Karnal bunt through the presence of bunted kernels;

(c) Seed conditioning equipment and storage/handling equipment/structures that have been used in the production of wheat, durum wheat, and triticale found to contain the spores of *Tilletia indica;*

(d) Plants or plant parts (including grain, seed, and straw) and hay cut after reaching the dough stage of all varieties of wheat (*Triticum aestivum*), durum wheat (*Triticum durum*), and triticale (*Triticum aestivum* × *Secale cereale*) that are produced in a regulated area, except for straw/stalks/seed heads for decorative purposes that have been processed or manufactured prior to movement and are intended for use indoors;

(e) *Tilletia indica* (Mitra) Mundkur;

(f) Mechanized harvesting equipment that has been used in the production of wheat, durum wheat, or triticale that has tested positive for Karnal bunt through the presence of bunted kernels; and

(g) Any other product, article, or means of conveyance when:

(1) An inspector determines that it presents a risk of spreading Karnal bunt based on appropriate testing and the intended use of the product, article, or means of conveyance; and

(2) The person in possession of the product, article, or means of conveyance has been notified that it is regulated under this subpart.

[69 FR 8095, Feb. 23, 2004]

§301.89-3 Regulated areas.

(a) The Administrator will regulate each State or each portion of a State that is infected.

(b) Less than an entire State will be listed as a regulated area only if the Administrator:

(1)(i) Determines that the State has adopted and is enforcing restrictions on the intrastate movement of the regulated articles listed in §301.89-2 that are equivalent to the movement restrictions imposed by this subpart; and

(ii) Determines that designating less than the entire State as a regulated area will prevent the spread of Karnal bunt; or

(2) Exercises his or her extraordinary emergency authority under 7 U.S.C. 150dd.

(c) The Administrator may include noninfected acreage within a regulated area due to its proximity to an infestation or inseparability from the infected locality for regulatory purposes, as determined by:

(1) Projections of the spread of Karnal bunt along the periphery of the infestation;

(2) The availability of natural habitats and host materials within the non-infected acreage that are suitable for establishment and survival of Karnal bunt; and

(3) The necessity of including uninfected acreage within the regulated area in order to establish readily identifiable boundaries.

(d) The Administrator or an inspector may temporarily designate any nonregulated area as a regulated area in accordance with the criteria specified in paragraphs (a), (b), and (c) of this section. The Administrator will give written notice of this designation to the owner or person in possession of the nonregulated area, or, in the case of publicly owned land, to the person responsible for the management of the nonregulated area. Thereafter, the movement of any regulated article from an area temporarily designated as a regulated area is subject to this subpart. As soon as practicable, this area either will be added to the list of designated regulated areas in paragraph (g) of this section, or the Administrator will terminate the designation. The owner or person in possession of,

or, in the case of publicly owned land, the person responsible for the management of, an area for which the designation is terminated will be given written notice of the termination as soon as practicable.

(e) The Administrator will classify a field or area as a regulated area when:

(1) It is a field planted with seed from a lot found to contain a bunted wheat kernel; or

(2) It is a distinct definable area that contains at least one field that was found during survey to contain a bunted wheat kernel (the distinct definable area may include an area where Karnal bunt is not known to exist but where intensive surveys are required because of the area's proximity to a field found during survey to contain a bunted kernel); or

(3) It is a distinct definable area that contains at least one field that has been determined to be associated with grain at a handling facility containing a bunted kernel of a host crop (the distinct definable area may include an area where Karnal bunt is not known to exist but where intensive surveys are required because of the area's proximity to the field associated with the bunted kernel at the handling facility).

(f) A field known to have been infected with Karnal bunt, as well as any non-infected acreage surrounding the field, will be released from regulation if:

(1) The field has been permanently removed from crop production; or

(2) The field is tilled at least once per year for a total of 5 years (the years need not be consecutive). After tilling, the field may be planted with a crop or left fallow. If the field is planted with a host crop, the crop must test negative, through the absence of bunted kernels, for Karnal bunt.

(g) The following areas or fields are designated as regulated areas (maps of the regulated areas may be obtained by contacting the Animal and Plant Health Inspection Service, Plant Protection and Quarantine, 4700 River Road Unit 98, Riverdale, MD 20737-1236):

ARIZONA

La Paz County. Beginning at the northeast corner of sec. 24, T. 8 N., R. 21 W.; then south to the southeast corner of sec. 1, T. 7 N., R.

21 W.; then east to the northeast corner of sec. 7, T. 7 N., R. 20 W.; then south to the southeast corner of sec. 19, T. 7 N., R. 20 W.; then west to the southwest corner of sec. 19, T. 7 N., R. 20 W.; then south to the southeast corner of sec. 36, T. 7 N., R. 21 W.; then west to the southwest corner of sec. 36, T. 7 N., R. 21 W.; then south to the southeast corner of sec. 2, T. 6 N., R. 21 W.; then west to the southwest corner of sec. 2, T. 6 N., R. 21 W.; then south to the southeast corner of sec. 10, T. 6 N., R. 21 W.; then west to the southwest corner of sec. 8, T. 6 N., R. 21 W.; then north to the southwest corner of sec. 5, T. 6 N., R. 21 W.; then west to the southwest corner of sec. 6, T. 6 N., R. 21 W.; then north to the northwest corner of sec. 6, T. 6 N., R. 21 W.; then west to the southwest corner of sec. 36, T. 7 N., R. 22 W., then north to the northwest corner of sec. 24, T. 7 N., R. 22 W.; then east to the northeast corner of sec. 24, T. 7 N., R. 22 W.; then north from that point to the Colorado River; then northeast along the Colorado River to the northern boundary of sec. 16, T. 8 N., R. 21 W.; then east to the northeast corner of sec. 14, T. 8 N., R. 21 W.; then south to the southeast corner of sec. 14, T. 8 N., R. 21 W.; then east to the point of beginning.

Maricopa County. (1) Beginning at the southeast corner of sec. 8, T. 1 S., R. 2 E.; then west to the southwest corner of sec. 8, T. 1 S., R. 2 E.; then south to the southeast corner of sec. 18, T. 1 S., R. 2 E.; then west to the southwest corner of sec. 14, T. 1 S., R. 1 E.; then north to the northwest corner of sec. 14, T. 1 S., R. 1 E.; then west to the southwest corner of sec. 9, T. 1 S., R. 1 E.; then north to the northwest corner of sec. 9, T. 1 S., R. 1 E.; then west to the southwest corner of sec. 5, T. 1 S., R. 1 E.; then north to the northwest corner of sec. 5, T. 1 S., R. 1 E.; then west to the northeast corner of sec. 6, T. 1 S., R. 1 W.; then south to the southeast corner of sec. 7, T. 1 S., R. 1 W.; then west to the northeast corner of sec. 14, T. 1 S., R. 2 W.; then south to the southeast corner of sec. 14, T. 1 S., R. 2 W.; then west to the northeast corner of sec. 20, T. 1 S., R. 2 W.; then south to the southeast corner of sec. 20, T. 1 S., R. 2 W.; then west to the northeast corner of sec. 29, T. 1 S., R. 3 W.; then south to the southeast corner of sec. 29, T. 1 S., R. 3 W.; then west to the southwest corner of sec. 26, T. 1 S., R. 5 W.; then north to the northwest corner of sec. 14, T. 1 N., R. 5 W.; then east to the southwest corner of sec. 7, T. 1 N., R. 2 W.; then north to the northwest corner of sec. 7, T. 1 N., R. 2 W.; then east to the northeast corner of sec. 7, T. 1 N., R. 2 W.; then north to the northwest corner of sec. 5, T. 1 N., R. 2 W.; then east to the northeast corner of sec. 5, T. 1 N., R. 2 W.; then north to the northwest corner of sec. 33, T. 2 N., R. 2 W.; then east to the northeast corner of sec. 33, T. 2 N., R. 2 W.; then north to the northwest corner of sec. 3,

T. 3 N., R. 2 W.; then east to the northeast corner of sec. 1, T. 3 N., R. 1 W.; then south to the northwest corner of sec. 19, T. 3 N., R. 1 E.; then east to the northeast corner of sec. 20, T. 3 N., R. 1 E.; then south to the northeast corner of sec. 29, T. 3 N., R. 1 E.; then east to the northeast corner of sec. 27, T. 3 N., R. 1 E.; then south to the southeast corner of sec. 27, T. 3 N., R. 1 E.; then east to the northeast corner of sec. 35, T. 3 N., R. 1 E.; then south to the southeast corner of sec. 35, T. 3 N., R. 1 E.; then east to the northeast corner of sec. 1, T. 2 N., R. 1 E.; then south to the northeast corner of sec. 1, T. 1 N., R. 1 E.; then east to the northeast corner of sec. 4, T. 1 N., R. 2 E.; then south to the northwest corner of sec. 15, T. 1 N., R. 2 E.; then east to the northeast corner of sec. 14, T. 1 N., R. 2 E.; then south to the southeast corner of sec. 35, T. 1 N., R. 2 E.; then west to the northeast corner of sec. 3, T. 1 S., R. 2 E.; then south to the southeast corner of sec. 3, T. 1 S., R. 2 E.; then west to the southwest corner of sec. 4, T. 1 S., R. 2 E.; then south to the point of beginning.

(2) Beginning at the intersection of the Maricopa/Pinal County line and the southeast corner of sec. 36, T. 2 S., R. 7 E.; then west along the Maricopa/Pinal County line to the southwest corner of sec. 33, T. 2 S., R. 5 E.; then north to the northwest corner of sec. 33; then west to the southwest corner of sec. 30, T. 2 S., R. 5 E.; then north to the southeast corner of sec. 25, T. 2 S., R. 4 E.; then west to the southwest corner of sec. 25, T. 2 S., R. 4 E.; then north to the southwest corner of sec. 13, T. 2 S., R. 4 E.; then west to the southwest corner of sec. 15, T. 2 S., R. 4 E.; then north to the northwest corner of sec. 3, T. 2 S., R. 4 E.; then east to the southwest corner of sec. 35, T. 1 S., R. 4 E.; then north to the northwest corner of sec. 35, T. 1 S., R. 4 E.; then east to the northeast corner of sec. 33, T. 1 S., R. 5 E.; then north to the northwest corner of sec. 27, T. 1 S., R. 5 E.; then east to the northeast corner of sec. 27, T. 1 S., R. 5 E.; then north to the northwest corner of sec. 23, T. 1 S., R. 5 E.; then east to the northeast corner of sec. 21, T. 1 S., R. 6 E.; then south to the southeast corner of sec. 21, T. 1 S., R. 6 E.; then east to the northeast corner of sec. 27, T. 1 S., R. 6 E.; then south to the southeast corner of sec. 27, T. 1 S., R. 6 E.; then east to the northeast corner of sec. 31, T. 1 S., R. 7 E.; then south to the northwest corner of sec. 5, T. 2 S., R. 7 E.; then east to the northeast corner of sec. 3, T. 2 S., R. 7 E.; then north to the northwest corner of sec. 35, T. 1 S., R. 7 E.; then east to the northeast corner of sec. 36, T. 1 S., R. 7 E. and the Maricopa/Pinal County line; then south along the Maricopa/Pinal County line to the point of beginning.

(3) Beginning at the southeast corner of sec. 30, T. 6 S., R. 5 W.; the west to the northeast corner of sec. 33, T. 6 S., R. 6 W.; then south to the southeast corner of sec. 33, T. 6

S., R. 6 W.; then west to the southwest corner of sec. 36, T. 6 S., R. 7 W.; then north to the northwest corner of sec. 36, T. 6 S., R. 7 W.; then west to the southwest corner of sec. 26, T. 6 S., R. 7 W.; then north to the northwest corner of sec. 23, T. 6 S., R. 7 W.; then west to the southeast corner of sec. 18, T. 6 S., R. 7 W.; then north to the northeast corner of sec. 6, T. 6 S., R. 7 W.; then west to the southeast corner of sec. 31, T. 5 S., R. 7 W.; then north to the northwest corner of sec. 29, T. 5 S., R. 7 W.; then east to the northeast corner of sec. 29, T. 5 S., R.7 W.; then east to the southwest corner of sec. 22, T. 5 S., R. 7 W.; then north to northwest corner of sec. 22, T. 5 S., R. 7 W.; then to the southwest corner of sec. 14, T. 5 S., R. 7 W.; then north to the northwest corner of sec. 14, T. 5 S., R. 7 W.; then east to the northeast corner of sec. 13, T. 5 S., R. 6 W.; then south to the southeast corner of sec. 24, T. 5 S., R. 6 W.; then east to the northeast corner of sec. 30, T. 5 S., R. 5 W.; then south to the southeast corner of sec. 30, T. 5 S., R. 5 W.; then east to the northeast corner of sec. 32, T. 5 S., R. 5 W.; then south to the southeast corner of sec. 32, T. 5 S., R. 5 W.; then east to the northeast corner of sec. 5, T. 6 S., R. 5 W.; then south to the southeast corner of sec. 20, T. 6 S., R. 5 W.; then west to the northeast corner of sec. 30, T. 6 S., R. 5 W.; then south to the point of beginning.

(4) Beginning at the southeast corner of sec. 36, T. 2 N., R. 5 E.; then west to the northeast corner of sec. 4, T. 1 N., R. 5 E.; then south to the southeast corner of sec. 4, T. 1 N., R. 5 E.; then west to the southwest corner of sec. 4, T. 1 N., R. 5 E.; then south to the southeast corner of sec. 17, T. 1 N., R. 5 E.; then west to the south west corner of sec. 17, T. 1 N., R. 5 E.; then north to the northwest corner of sec. 27, T. 1 N., R. 5 E.; then west to the southwest corner of sec. 12, T. 1 N., R. 4 E.; then north to the northwest corner of sec. 12, T. 1 N., R. 4 E.; then east to northeast corner of sec. 12, T. 1 N., R. 4 E.; then north to the northwest corner of sec. 7, T. 2 N., R. 5 E.; then east to the northeast corner of sec. 12, T. 2 N., R. 5 E.; then south to the point of beginning.

Pinal County: (1) Beginning at the intersection of the Maricopa/Pinal County line and the northwest corner of sec. 31, T. 1 S., R. 8 E.; then east to the northeast corner of sec. 32, T. 1 S., R. 8 E.; then south to the northwest corner of sec. 4, T. 2 S., R. 8 E.; then east to the northeast corner of sec. 4, T. 2 S., R. 8 E.; then south to the southeast corner of sec. 4, T. 3 S., R. 8 E.; then west to the northeast corner of sec. 8, T. 3 S., R. 8 E.; then south to the southeast corner of sec. 8, T. 3 S., R. 8 E.; then west to the southwest corner of sec. 12, T 3 S., R. 7 E.; then north to the southeast corner of sec. 2, T. 3 S., R. 7 E.; then west to the northeast corner of sec. 9, T. 3 S., R. 6 E.; then south to the southeast corner of sec. 28, T. 3 S., R. 6 E.; then west to the southwest corner of sec. 28, T. 3 S., R. 6 E.; then south to the southeast corner of sec. 32, T. 3 S., R. 6 E.; then west to the southwest corner of sec. 35, T. 3 S., R. 5 E.; then north to the northwest corner of sec. 35, T. 3 S., R. 5 E.; then west to the southwest corner of sec. 27, T. 3 S., R. 5 E.; then north to the northwest corner of sec. 10, T. 3 S., R. 5 E.; then west to the southwest corner of sec. 4, T. 3 S., R. 5 E.; then north to the northwest corner of sec. 4, T. 3 S., R. 5 E. and the intersection of the Maricopa/Pinal County line; then east along the Maricopa/Pinal County line to the northwest corner of sec. 6, T. 3 S., R. 8 E.; then north along the Maricopa/Pinal County line to the point of beginning.

(2) Beginning at the southeast corner of sec. 5, T. 6 S., R. 4 E.; then west to the southwest corner of sec. 1, T. 6 S., R. 3 E.; then south to the southeast corner of sec. 14, T. 6 S., R. 3 E.; then west to the southwest corner of sec. 14, T. 6 S., R. 3 E.; then south to the southeast corner of sec. 22, T. 6 S., R. 3 E.; then west to the northeast corner of sec. 30, T. 6 S., R. 3 E.; then south to the southeast corner of sec. 30, T. 6 S., R. 3 E.; then west to the southwest corner of sec. 30, T. 6 S., R. 3 E.; then north to the southeast corner of sec. 25, T. 6 S., R. 2 E.; then west to the southwest corner of sec. 25, T. 6 S., R. 2 E.; then north to the southeast corner of sec. 11, T. 6 S., R. 2 E.; then west to the southwest corner of sec. 11, T. 6 S., R. 2 E.; then north to the northwest corner of sec. 35, T. 4 S., R. 2 E.; then east to the northeast corner of sec. 35, T. 4 S., R. 2 E.; then north to the northwest corner of sec. 25, T. 4 S., R. 2 E.; then east to the southwest corner of sec. 20, T. 4 S., R. 3 E.; then north to the northwest corner of sec. 20, T. 4 S., R. 3 E.; then east to the northeast corner of sec. 24, T. 4 S., R. 3 E.; then south to the southeast corner of sec. 24, T. 4 S., R. 3 E.; then east to the northeast corner of sec. 28, T. 4 S., R. 4 E.; then south to the northwest corner of sec. 34, T. 4 S., R. 4 E.; then east to the northeast corner of sec. 35, T. 4 S., R. 4 E.; then south to the northwest corner of sec. 1, T. 5 S., R. 4 E.; then east to the northeast corner of sec. 1, T. 5 S., R. 4 E.; then south to the southeast corner of sec. 1, T. 5 S., R. 4 E.; then west to the northeast corner of sec. 12, T. 5 S., R. 4 E.; then south to the southeast corner of sec. 24, T. 5 S., R. 4 E.; then west to the southwest corner of sec. 24, T. 5 S., R. 4 E.; then south to the northeast corner of sec. 35, T. 5 S., R. 4 E.; then west to the northwest corner of sec. 35, T. 5 S., R. 4 E.; then south to the southeast corner of sec. 37, T. 5 S., R. 4 E.; then west to the northwest corner of sec. 50, T. 5 S., R. 4 E.; then south to the southeast corner of sec. 49, T. 6 S., R. 4 E.; then west to the northeast corner of sec. 5, T. 6 S., R. 4 E.; then south to the point of beginning.

[61 FR 52207, Oct. 4, 1996]

EDITORIAL NOTE: For FEDERAL REGISTER citations affecting § 301.89–3, see the List of CFR Sections Affected, which appears in the Finding Aids section of the printed volume and at www.govinfo.gov.

§ 301.89–4　Planting.

Any wheat, durum wheat, or triticale that originates within a regulated area must be tested and found free from bunted wheat kernels and spores before it may be used as seed within or outside a regulated area.

[69 FR 8096, Feb. 23, 2004]

§ 301.89–5　Movement of regulated articles from regulated areas.

(a) Any regulated article may be moved from a regulated area into or through an area that is not regulated only if moved under the following conditions:

(1) With a certificate or limited permit issued and attached in accordance with §§ 301.89–6 and 301.89–10;

(2) Without a certificate or limited permit, provided that each of the following conditions is met:

(i) The regulated article was moved into the regulated area from an area that is not regulated;

(ii) The point of origin is indicated on a waybill accompanying the regulated article;

(iii) The regulated article is moved through the regulated area without stopping, or has been stored, packed, or handled at locations approved by an inspector as not posing a risk of contamination with Karnal bunt, or has been treated in accordance with part 305 of this chapter while in or moving through any regulated area; and

(iv) The article has not been combined or commingled with other articles so as to lose its individual identity;

(b) When an inspector has probable cause to believe a person or means of conveyance is moving a regulated article, the inspector is authorized to stop the person or means of conveyance to determine whether a regulated article is present and to inspect the regulated article. Articles found to be infected by an inspector, and articles not in compliance with the regulations in this subpart, may be seized, quarantined, treated, subjected to other remedial measures, destroyed, or otherwise disposed of. Any treatments will be in accordance with part 305 of this chapter.

[61 FR 52207, Oct. 4, 1996, as amended at 62 FR 23627, May 1, 1997; 63 FR 50751, Sept. 23, 1998; 69 FR 8096, Feb. 23, 2004; 75 FR 4241, Jan. 26, 2010; 75 FR 68945, Nov. 10, 2010]

§ 301.89–6　Issuance of a certificate or limited permit.

(a) An inspector[1] or person operating under a compliance agreement will issue a certificate for the movement of a regulated article outside a regulated area if he or she determines that the regulated article:

(1) Is eligible for unrestricted movement under all other applicable Federal domestic plant quarantines and regulations;

(2) Is to be moved in compliance with any conditions deemed necessary under section 414 of the Plant Protection Act (7 U.S.C. 7714)[2] to prevent the artificial spread of Karnal bunt; and

(3)(i) Is free of Karnal bunt infestation, based on laboratory results of testing, and history of previous infestation;

(ii) Has been grown, produced, manufactured, stored, or handled in a manner that would prevent infestation or destroy all life stages of Karnal bunt; or

(iii) Has been treated in accordance with part 305 of this chapter.

(b) To be eligible for movement under a certificate, hay cut after the dough stage or grain from a field within a regulated area must be tested prior to its movement from the field or before it is commingled with similar commodities and must be found free from bunted kernels. If bunted kernels are found, the grain or hay will be eligible for

---

[1] Inspectors are assigned to local offices of APHIS, which are listed in local telephone directories. Information concerning such local offices may also be obtained from the Animal and Plant Health Inspection Service, Plant Protection and Quarantine, Surveillance and Emergency Programs Planning and Coordination, 4700 River Road Unit 98, Riverdale, Maryland 20737–1236.

[2] An inspector may hold, seize, quarantine, treat, apply other remedial measures to, destroy, or otherwise dispose of plants, plant pests, or other articles in accordance with sections 414, 421, and 431 of the Plant Protection Act (7 U.S.C. 7714, 7731, and 7754).

movement only under a limited permit issued in accordance with paragraph (c) of this section, and the field of production will be considered positive for Karnal bunt.

(c) An inspector or a person operating under a compliance agreement will issue a limited permit for the movement outside the regulated area of a regulated article not eligible for a certificate if the inspector determines that the regulated article:

(1) Is to be moved to a specified destination for specified handling, utilization, or processing (the destination and other conditions to be listed in the limited permit and/or compliance agreement), and this movement will not result in the artificial spread of Karnal bunt because Karnal bunt will be destroyed or the risk mitigated by the specified handling, utilization, or processing;

(2) Is to be moved in compliance with any additional conditions the Administrator may impose under section 414 of the Plant Protection Act (7 U.S.C. 7714) to prevent the artificial spread of Karnal bunt; and

(3) Is eligible for movement under all other Federal domestic plant quarantines and regulations applicable to the regulated article.

(d) An inspector shall issue blank certificates and limited permits to a person operating under a compliance agreement in accordance with §301.89-7 or authorize reproduction of the certificates or limited permits on shipping containers, or both, as requested by the person operating under the compliance agreement. These certificates and limited permits may then be completed and used, as needed, for the movement of regulated articles that have met the applicable requirements of paragraphs (a) and (b) of this section for the issuance of certificates or of paragraph (c) of this section for the issuance of limited permits.

[61 FR 52207, Oct. 4, 1996, as amended at 62 FR 23627, May 1, 1997; 63 FR 50751, Sept. 23, 1998; 64 FR 23754, May 4, 1999; 66 FR 21053, Apr. 27, 2001; 67 FR 21161, Apr. 30, 2002; 69 FR 8096, Feb. 23, 2004; 75 FR 4241, Jan. 26, 2010; 75 FR 68945, Nov. 10, 2010]

§301.89-7 Compliance agreements.

Persons who grow, handle, or move regulated articles may enter into a compliance agreement[3] if such persons review with an inspector each provision of the compliance agreement, have facilities and equipment to carry out disinfestation procedures or application of chemical materials in accordance with part 305 of this chapter, and meet applicable State training and certification standards under the Federal Insecticide, Fungicide, and Rodenticide Act, as amended (7 U.S.C. 136b). Any person who enters into a compliance agreement with APHIS must agree to comply with the provisions of this subpart and any conditions imposed under this subpart.

[61 FR 52207, Oct. 4, 1996, as amended at 62 FR 23628, May 1, 1997; 69 FR 8096, Feb. 23, 2004; 75 FR 4241, Jan. 26, 2010; 75 FR 68945, Nov. 10, 2010]

§301.89-8 Cancellation of a certificate, limited permit, or compliance agreement.

Any certificate, limited permit, or compliance agreement may be canceled orally or in writing by an inspector whenever the inspector determines that the holder of the certificate or limited permit, or the person who has entered into the compliance agreement, has not complied with this subpart or any conditions imposed under this subpart. If the cancellation is oral, the cancellation will become effective immediately and the cancellation and the reasons for the cancellation will be confirmed in writing as soon as circumstances allow, but within 20 days after oral notification of the cancellation. Any person whose certificate, limited permit, or compliance agreement has been canceled may appeal the decision, in writing, within 10 days

_____
[3] Compliance agreements may be initiated by contacting a local office of Plant Protection and Quarantine, which are listed in telephone directories. The addresses and telephone numbers of local offices of Plant Protection and Quarantine may also be obtained from the Animal and Plant Health Inspection Service, Plant Protection and Quarantine, Surveillance and Emergency Program Planning and Coordination, 4700 River Road Unit 98, Riverdale, Maryland 20737-1236.

after receiving the written cancellation notice. The appeal must state all of the facts and reasons that the person wants the Administrator to consider in deciding the appeal. A hearing may be held to resolve any conflict as to any material fact. Rules of practice for the hearing will be adopted by the Administrator. As soon as practicable, the Administrator will grant or deny the appeal, in writing, stating the reasons for the decision.

§ 301.89–9 Assembly and inspection of regulated articles.

(a) Persons requiring certification or other services must request the services of an inspector[4] at least 24 hours before the services are needed.

(b) The regulated articles must be assembled at the place and in the manner the inspector designates as necessary to comply with this subpart.

[61 FR 52207, Oct. 4, 1996, as amended at 62 FR 23628, May 1, 1997; 64 FR 29550, June 2, 1999; 75 FR 68945, Nov. 10, 2010]

§ 301.89–10 Attachment and disposition of certificates and limited permits.

(a) The consignor must ensure that the certificate or limited permit authorizing movement of a regulated article is, at all times during movement, attached to:

(1) The outside of the container encasing the regulated article;

(2) The article itself, if it is not in a container; or

(3) The consignee's copy of the accompanying waybill: Provided, that the descriptions of the regulated article on the certificate or limited permit, and on the waybill, are sufficient to identify the regulated article; and

(b) The carrier must furnish the certificate or limited permit authorizing movement of a regulated article to the consignee at the shipment's destination.

§ 301.89–11 Costs and charges.

The services of the inspector during normal business hours will be furnished without cost to persons requiring the services.

The user will be responsible for all costs and charges arising from inspection and other services provided outside of normal business hours.

§ 301.89–12 Cleaning, disinfection, and disposal.

(a) Mechanized harvesting equipment that has been used to harvest host crops that test positive for Karnal bunt based on the presence of bunted kernels must be cleaned and, if disinfection is determined to be necessary by an inspector, disinfected in accordance with part 305 of this chapter prior to movement from a regulated area.

(b) Seed conditioning equipment that was used in the conditioning of seed that was tested and found to contain spores or bunted kernels of *Tilletia indica* must be cleaned and disinfected in accordance with part 305 of this chapter prior to being used in the conditioning of seed that has tested negative for the spores of *Tilletia indica* or to being moved from a regulated area.

(c) Any grain storage facility, including on-farm storage, that is used to store seed that has tested bunted-kernel or spore positive or grain that has tested bunted-kernel positive must be cleaned and, if disinfection is determined to be necessary by an inspector, disinfected in accordance with part 305 of this chapter if the facility will be used to store grain or seed in the future.

(d) Conveyances used to move bunted-kernel-positive host crops, including trucks, railroad cars, and other containers, that have sloping metal sides leading directly to a bottom door or slide chute, are self cleaning, and will not be required to be cleaned and disinfected.

(e) Spore-positive wheat, durum wheat, or triticale seed that has been treated with any chemical that renders it unfit for human or animal consumption must be disposed of by means of burial under a minimum of 24 inches of soil in a nonagricultural area that will not be cultivated or in an approved landfill.

[69 FR 8096, Feb. 23, 2004, as amended at 75 FR 4241, Jan. 26, 2010]

---

[4] See footnote 1.

§ 301.89–15 **Compensation for growers, handlers, and seed companies in the 1999–2000 and subsequent crop seasons.**

Growers, handlers, and seed companies are eligible to receive compensation from the United States Department of Agriculture (USDA) for the 1999–2000 and subsequent crop seasons to mitigate losses or expenses incurred because of the Karnal bunt regulations and emergency actions, as follows:

(a) *Growers, handlers, and seed companies in areas under first regulated crop season.* Growers, handlers, and seed companies are eligible to receive compensation for the loss in value of their wheat in accordance with paragraphs (a)(1) and (a)(2) of this section if: The wheat was grown in a State where the Secretary has declared an extraordinary emergency; and the wheat was grown in an area of that State that became regulated for Karnal bunt after the crop was planted, or for which an Emergency Action Notification (PPQ Form 523) was issued after the crop was planted; and the wheat was grown in an area that remained regulated or under Emergency Action Notification at the time the wheat was sold. Growers and handlers of wheat grown in Oklahoma during the 2000–2001 growing season are eligible to receive compensation if the wheat was commingled in storage with wheat that meets the above requirements of this paragraph. Growers, handlers, and seed companies in areas under the first regulated crop season are eligible for compensation for 1999–2000 or subsequent crop season wheat and for wheat inventories in their possession that were unsold at the time the area became regulated. The compensation provided in this paragraph is for wheat grain, certified wheat seed, wheat held back from harvest by a grower in the 2000–2001 growing season for use as seed in the next growing season, and wheat grown with the intention of producing certified wheat seed.

(1) *Growers.* Growers of wheat in an area under the first regulated crop season, who sell wheat that was tested by APHIS and found positive for Karnal bunt prior to sale, or that was tested by APHIS and found positive for Karnal bunt after sale and the price received by the grower is contingent on the test results, are eligible to receive compensation as described in paragraphs (a)(1)(i) and (a)(1)(ii) of this section. However, compensation for positive-testing wheat will not exceed $1.80 per bushel under any circumstances.

(i) If the wheat was grown under contract and a price was determined in the contract before the area where the wheat was grown became regulated, compensation will equal the contract price minus the actual price received by the grower.

(ii) If the wheat was not grown under contract or a price was determined in the contract after the area where the wheat was grown became regulated, compensation will equal the estimated market price for the relevant class of wheat (meaning type of wheat, such as durum or hard red winter) minus the actual price received by the grower. The estimated market price will be calculated by APHIS for each class of wheat, taking into account the prices offered by relevant terminal markets (animal feed, milling, or export) during the harvest months for the area, with adjustments for transportation and other handling costs. Separate estimated market prices will be calculated for certified wheat seed and wheat grown with the intention of producing certified wheat seed, and wheat grain.

(2) *Handlers and seed companies.* Handlers and seed companies who sell wheat grown in an area under the first regulated crop season are eligible to receive compensation only if the wheat was not tested by APHIS prior to purchase by the handler or seed company, but was tested by APHIS and found positive for Karnal bunt after purchase by the handler or seed company, as long as the price to be paid is not contingent on the test results. Compensation will equal the estimated market price for the relevant class of wheat (meaning type of wheat, such as durum or hard red winter) minus the actual price received by the handler or seed company. The estimated market price will be calculated by APHIS for each class of wheat, taking into account the prices offered by relevant terminal markets (animal feed, milling, or export) during the harvest months for the

area, with adjustments for transportation and other handling costs. Separate estimated market prices will be calculated for certified wheat seed and wheat grown with the intention of producing certified wheat seed, and wheat grain. However, compensation will not exceed $1.80 per bushel under any circumstances.

(b) *Growers, handlers, and seed companies in previously regulated areas.* For the 1999–2000 crop season and the 2000–2001 crop season only, growers, handlers, and seed companies are eligible to receive compensation for the loss in value of their wheat in accordance with paragraphs (b)(1) and (b)(2) of this section if: The wheat was grown in a State where the Secretary has declared an extraordinary emergency; and the wheat was grown in an area of that State that became regulated for Karnal bunt before the crop was planted, or for which an Emergency Action Notification (PPQ Form 523) was issued before the crop was planted; and the wheat was grown in an area that remained regulated or under Emergency Action Notification at the time the wheat was sold. Growers, handlers, and seed companies in previously regulated areas will not be eligible for compensation for wheat from the 2001–2002 and subsequent crop seasons; except that, for growers or handlers of wheat harvested in any field in the Texas counties of Archer, Baylor, Throckmorton, and Young during the 2000–2001 crop season that has not been found to contain a bunted wheat kernel, this requirement applies to compensation for wheat from the 2002–2003 and subsequent crop seasons. The compensation provided in this paragraph is for wheat grain, certified wheat seed, and wheat grown with the intention of producing certified wheat seed.

(1) *Growers.* Growers of wheat in a previously regulated area who sell wheat that was tested by APHIS and found positive for Karnal bunt prior to sale, or that was tested by APHIS and found positive for Karnal bunt after sale and the price received by the grower is contingent on the test results, are eligible to receive compensation at the rate of $.60 per bushel of positive testing wheat.

(2) *Handlers and seed companies.* Handlers and seed companies who sell wheat grown in a previously regulated area are eligible to receive compensation only if the wheat was not tested by APHIS prior to purchase by the handler, but was tested by APHIS and found positive for Karnal bunt after purchase by the handler or seed company, as long as the price to be paid by the handler or seed company is, not contingent on the test results. Compensation will be at the rate of $.60 per bushel of positive testing wheat.

(c) *To claim compensation.* Compensation payments to growers, handlers, and seed companies under paragraphs (a) and (b) of this section will be issued by the Farm Service Agency (FSA). Claims for compensation for the 1999–2000 crop season must be received by FSA on or before December 4, 2001. Claims for compensation for subsequent crop seasons must be received by FSA on or before March 1 of the year following that crop season. The Administrator may extend the deadline, upon request in specific cases, when unusual and unforeseen circumstances occur that prevent or hinder a claimant from requesting compensation on or before these dates. To claim compensation, a grower, handler, or seed company must complete and submit to the local FSA county office the following documents:

(1) *Growers, handlers, and seed companies.* A grower, handler, or seed company must submit a Karnal Bunt Compensation Claim form, provided by FSA. If the wheat was grown in an area that is not a regulated area, but for which an Emergency Action Notification (PPQ Form 523) (EAN) has been issued, the grower, handler, or seed company must submit a copy of the EAN. Growers, handlers, and seed companies must also submit a copy of the Karnal bunt certificate issued by APHIS that shows the Karnal bunt test results, and verification as to the actual (not estimated) weight of the wheat that tested positive (such as a copy of a facility weigh ticket, or other verification). For compensation claims for wheat seed, a grower or seed company must submit documentation showing that the wheat is either certified seed or was grown with the intention of producing certified seed

(this documentation may include one or more of the following types of documents: an application to the State seed certification agency for field inspection; a bulk sale certificate; certification tags or labels issued by the State seed certification agency; or a document issued by the State seed certification agency verifying that the wheat is certified seed);

(2) *Growers.* In addition to the documents required in paragraph (c)(1) of this section, growers must submit a copy of the receipt for the final sale of the wheat, showing the total bushels sold and the total price received by the grower. Growers compensated under paragraph (b)(1) of this section (previously regulated areas) whose wheat was not tested prior to sale must submit documentation showing that the price paid to the grower was contingent on test results (such as a copy of the receipt for the final sale of the wheat or a copy of the contract the grower has for the wheat, if this information appears on those documents).

(3) *Handlers and seed companies.* In addition to the documents required in paragraph (c)(1) of this section, handlers and seed companies must submit a copy of the receipt for the final sale of the wheat, showing the total bushels sold and the total price received by the handler or seed company. The handler or seed company must also submit documentation showing that the price paid or to be paid to the grower is not contingent on the test results (such as a copy of the receipt for the purchase of the wheat or a copy of the contract the handler or seed company has with the grower, if this information appears on those documents).

(d) *Special allowance for negative wheat grown in Archer, Baylor, Throckmorton, and Young Counties, TX, in the 2000–2001 growing season.* Notwithstanding any other provision of this section, wheat that was harvested from fields in Archer, Baylor, Throckmorton, or Young Counties, TX, in the 2000–2001 growing season, and that tested negative for Karnal bunt after harvest, is eligible for compensation in accordance with paragraph (a) of this section.

(e) *Special allowance for disposal costs for treated uncertified wheat seed in Ar-cher, Baylor, Throckmorton, and Young Counties, TX, in the 2000–2001 growing season.* Notwithstanding any other provision of this section, growers in Archer, Baylor, Throckmorton, or Young Counties, TX, who own treated uncertified wheat seed that tested positive for Karnal bunt spores during the 2000–2001 growing season are eligible for compensation in accordance with this paragraph. The grower is eligible for compensation for the costs of disposing of such wheat seed, by burial on the grower's premises, by burial at a landfill, or through another means approved by APHIS. The compensation for disposing of wheat seed by burial on the grower's premises is $1.00 per bushel. The compensation for disposing of wheat seed by burial at a landfill, or through another means approved by APHIS, is the actual cost of disposal, up to $1.20 per bushel, as verified by receipts for disposal costs. To apply for this compensation, the grower must submit a Karnal Bunt Compensation Claim form, provided by FSA, and must also submit a copy of the Karnal bunt certificate issued by APHIS that shows the Karnal bunt test results, and verification as to the actual (not estimated) weight of the uncertified wheat seed that tested positive for spores (such as a copy of a facility weigh ticket, or other verification). For seed disposed of by burial at a landfill the grower must also submit one or more receipts for the disposal costs of the uncertified wheat seed, showing the total bushels destroyed and the total disposal costs (landfill fees, transportation costs, etc.).

[63 FR 31599, June 10, 1998, as amended at 64 FR 34113, June 25, 1999; 66 FR 40842, Aug. 6, 2001; 67 FR 21566, May 1, 2002]

§ **301.89–16 Compensation for grain storage facilities, flour millers, National Survey participants, and certain custom harvesters and equipment owners or lessees for the 1999–2000 and subsequent crop seasons.**

Owners of grain storage facilities, flour millers, and participants in the National Karnal Bunt Survey are eligible to receive compensation from the United States Department of Agriculture (USDA) for the 1999–2000 and subsequent crop seasons to mitigate losses or expenses incurred because of the Karnal bunt regulations and emergency actions, as follows:

(a) *Decontamination of grain storage facilities.* Owners of grain storage facilities that are in States where the Secretary has declared an extraordinary emergency, and who have decontaminated their grain storage facilities pursuant to either an Emergency Action Notification (PPQ Form 523) issued by an inspector or a letter issued by an inspector ordering decontamination of the facilities, are eligible to be compensated, on a one time only basis for each facility for each covered crop year wheat, for up to 50 percent of the direct cost of decontamination. However, compensation will not exceed $20,000 per grain storage facility (as defined in § 301.89–1). General clean-up, repair, and refurbishment costs are excluded from compensation. Compensation payments will be issued by APHIS. To claim compensation, the owner of the grain storage facility must submit to an inspector records demonstrating that decontamination was performed on all structures, conveyances, or materials ordered by APHIS to be decontaminated. The records must include a copy of the Emergency Action Notification or the letter from an inspector ordering decontamination, contracts with individuals or companies hired to perform the decontamination, receipts for equipment and materials purchased to perform the decontamination, time sheets for employees of the grain storage facility who performed activities connected to the decontamination, and any other documentation that helps show the cost to the owner and that decontamination has been completed. Claims for compensation for the 1999–2000 crop season must be received by APHIS on or before December 4, 2001. Claims for compensation for the 2000–2001 crop season and beyond must be received by March 1 of the year following that crop season. The Administrator may extend these deadlines upon written request in specific cases, when unusual and unforeseen circumstances occur that prevent or hinder a claimant from requesting compensation on or before these dates.

(b) *Flour millers.* Flour millers who, in accordance with a compliance agreement with APHIS, heat treat millfeed that is required by APHIS to be heat treated are eligible to be compensated at the rate of $35.00 per short ton of millfeed. The amount of millfeed compensated will be calculated by multiplying the weight of wheat from the regulated area received by the miller by 25 percent (the average percent of millfeed derived from a short ton of grain). Compensation payments will be issued by APHIS. To claim compensation, the miller must submit to an inspector verification as to the actual (not estimated) weight of the wheat (such as a copy of a facility weigh ticket or a copy of the bill of lading for the wheat, if the actual weight appears on those documents, or other verification). Flour millers must also submit verification that the millfeed was heat treated (such as a copy of the limited permit under which the wheat was moved to a treatment facility and a copy of the bill of lading accompanying that movement; or a copy of PPQ Form 700 (which includes certification of processing) signed by the inspector who monitors the mill). Claims for compensation for the 1999–2000 crop season must be received by APHIS on or before December 4, 2001. Claims for compensation for the 2000–2001 crop season and beyond must be received by March 1 of the year following that crop season. The Administrator may extend these deadlines upon written request in specific cases, when unusual and unforeseen circumstances occur that prevent or hinder a claimant from requesting compensation on or before these dates.

(c) *National Karnal Bunt Survey participants.* If a grain storage facility participating in the National Karnal Bunt Survey tests positive for Karnal bunt, the facility will be regulated, and may be ordered decontaminated, pursuant to either an Emergency Action Notification (PPQ Form 523) issued by an inspector or a letter issued by an inspector ordering decontamination of the facility. If the Secretary has declared an extraordinary emergency in the State in which the grain storage facility is located, the owner will be eligible for compensation as follows:

(1) *Loss in value of positive wheat.* The owner of the grain storage facility will

be compensated for the loss in value of positive wheat. Compensation will equal the estimated market price for the relevant class of wheat minus the actual price received for the wheat. The estimated market price will be calculated by APHIS for each class of wheat, taking into account the prices offered by relevant terminal markets (animal feed, milling, or export) during the relevant time period for that facility, with adjustments for transportation and other handling costs. However, compensation will not exceed $1.80 per bushel under any circumstances. Compensation payments for loss in value of wheat will be issued by the Farm Service Agency (FSA). To claim compensation, the owner of the facility must submit to the local FSA office a Karnal Bunt Compensation Claim form, provided by FSA. The owner of the facility must also submit to FSA a copy of the Emergency Action Notification or letter from an inspector under which the facility is or was quarantined; verification as to the actual (not estimated) weight of the wheat (such as a copy of a facility weigh ticket or a copy of the bill of lading for the wheat, if the actual weight appears on those documents, or other verification); and a copy of the receipt for the final sale of the wheat, showing the total bushels sold and the total price received by the owner of the grain storage facility. Claims for compensation for the 1999–2000 crop season must be received by APHIS on or before December 4, 2001. Claims for compensation for the 2000–2001 crop season and beyond must be received by March 1 of the year following that crop season. The Administrator may extend these deadlines upon written request in specific cases, when unusual and unforeseen circumstances occur that prevent or hinder a claimant from requesting compensation on or before these dates.

(2) *Decontamination of grain storage facilities.* The owner of the facility will be compensated on a one time only basis for each grain storage facility for each covered crop year wheat for the direct costs of decontamination of the facility at the same rate described under paragraph (a) of this section (up to 50 per cent of the direct costs of decon-

tamination, not to exceed $20,000 per grain storage facility). Compensation payments for decontamination of grain storage facilities will be issued by APHIS, and claims for compensation must be submitted in accordance with the provisions in paragraph (a) of this section. Claims for compensation for the 1999–2000 crop season must be received by APHIS on or before December 4, 2001. Claims for compensation for the 2000–2001 crop season and beyond must be received by March 1 of the year following that crop season. The Administrator may extend these deadlines upon written request in specific cases, when unusual and unforeseen circumstances occur that prevent or hinder a claimant from requesting compensation on or before these dates.

(d) *Special allowances for custom harvesters and equipment owners or lessees for costs related to cleaning and disinfection of mechanized harvesting and other equipment in Archer, Baylor, Throckmorton, and Young Counties, TX, in the 2000–2001 crop season.* All claims for compensation under this paragraph §301.89-16(d) must be received by APHIS on or before September 6, 2005. The Administrator may extend this deadline upon written request in specific cases, when unusual and unforeseen circumstances occur that prevent or hinder a claimant from requesting compensation on or before this date. All compensation payments made under this paragraph §301.89-16(d) will be issued by APHIS. Claims for compensation should be sent to Plant Protection and Quarantine, APHIS, USDA, 304 West Main Street, Olney, TX 76374.

(1) *Custom harvesters.* (i) *Cleaning and disinfection of mechanized harvesting equipment.* Custom harvesters who harvested host crops that an inspector determined to be infected with Karnal bunt and that were grown in Archer, Baylor, Throckmorton, or Young Counties, TX, during the 2000–2001 crop season are eligible to receive compensation for the cost of cleaning and disinfecting their mechanized harvesting equipment as required by §301.89-12(a). Compensation for the cost of cleaning and disinfection mechanized harvesting equipment used to harvest Karnal bunt-infected host crops will be either

the actual cost or $750 per cleaned machine, whichever is less. To claim compensation, a custom harvester must provide copies of a contract or other signed agreement for harvesting in Archer, Baylor, Throckmorton, or Young County during the 2000–2001 crop season, signed on a date prior to the designation of the county as a regulated area for Karnal bunt, or an affidavit stating that the custom harvester entered into an agreement to harvest in Archer, Baylor, Throckmorton, or Young County during the 2000–2001 crop season prior to the designation of the county as a regulated area for Karnal bunt, signed by the customer with whom the custom harvester entered into the agreement; a copy of the PPQ–540 certificate issued to allow the movement of mechanized harvesting equipment from a regulated area after it had been used to harvest host crops that an inspector determined to be infected with Karnal bunt and had been subsequently cleaned and disinfected; and a receipt showing the cost of the cleaning and disinfection.

(ii) *Contracts lost due to cleaning and disinfection.* Custom harvesters who harvested host crops that an inspector determined to be infected with Karnal bunt and that were grown in Archer, Baylor, Throckmorton, or Young Counties, TX, during the 2000–2001 crop season are also eligible to be compensated for the revenue lost if they lost one contract due to downtime necessitated by cleaning and disinfection, if the contract to harvest Karnal bunt-infected host crops in a previously nonregulated area was signed before the area was declared a regulated area for Karnal bunt. Compensation will only be provided for one contract lost due to cleaning and disinfection. Compensation for any contract that was lost due to cleaning and disinfection will be either the full value of the contract or $23.48 for each acre that was to have been harvested under the contract, whichever is less. To claim compensation, a custom harvester must provide copies of a contract or other signed agreement for harvesting in Archer, Baylor, Throckmorton, or Young County during the 2000–2001 crop season, signed on a date prior to the designation of the county as a regulated area

for Karnal bunt, or an affidavit stating that the custom harvester entered into an agreement to harvest in Archer, Baylor, Throckmorton, or Young County during the 2000–2001 crop season prior to the designation of the county as a regulated area for Karnal bunt, signed by the customer with whom the custom harvester entered into the agreement; a copy of the PPQ–540 certificate issued to allow the movement of mechanized harvesting equipment from a regulated area after it has been used to harvest host crops that an inspector determined to be infected with Karnal bunt and had been subsequently cleaned and disinfected; and the contract for harvesting in an area not regulated for Karnal bunt that had been lost due to time lost to cleaning and disinfecting harvesting equipment, signed on a date prior to the designation of the relevant county as a regulated area for Karnal bunt, for which the custom harvester will receive compensation, or an affidavit stating that the custom harvester entered into an agreement to harvest in an area not regulated for Karnal bunt prior to the designation of the county as a regulated area for Karnal bunt and stating the number of acres that were to have been harvested and the amount the custom harvester was to have been paid under the agreement, signed by the customer with whom the custom harvester entered into the agreement.

(iii) *Fixed costs incurred during cleaning and disinfection.* Custom harvesters who harvested host crops that an inspector determined to be infected with Karnal bunt and that were grown in Archer, Baylor, Throckmorton, or Young Counties, TX, during the 2000–2001 crop season who do not apply for compensation for a contract lost due to cleaning and disinfection as described in paragraph (d)(1)(ii) of this section are eligible for compensation for fixed costs incurred during cleaning and disinfection. Compensation for fixed costs incurred during cleaning and disinfection will be $2,000. To claim compensation, a custom harvester must provide copies of a contract or other signed agreement for harvesting in Archer, Baylor, Throckmorton, or Young County during the 2000–2001 crop season, signed on a date prior to the designation of the county as a regulated area for Karnal bunt, or an affidavit stating that the custom harvester entered into an agreement to harvest in Archer, Baylor, Throckmorton, or Young County during the 2000–2001 crop season prior to the designation of the county as a regulated area for Karnal bunt, signed by the customer with whom the custom harvester entered into the agreement; and a copy of the PPQ–540 certificate issued to allow the movement of mechanized harvesting equipment from a regulated area after it has been used to harvest host crops that an inspector determined to be infected with Karnal bunt and has been subsequently cleaned and disinfected.

(2) *Other equipment; cleaning and disinfection.* Owners or lessees of equipment other than mechanized harvesting equipment and seed conditioning equipment that came into contact with host crops that an inspector determined to be infected with Karnal bunt in Archer, Baylor, Throckmorton, or Young Counties, TX, during the 2000–2001 crop season and that was required by an inspector to be cleaned and disinfected are eligible for compensation for the cost of cleaning and disinfection. Compensation for the cleaning and disinfection of such equipment will be $100. To receive this compensation, owners or lessees must submit a copy of the PPQ–540 certificate issued to allow the movement of the equipment from a regulated area after it had been in contact with host crops that an inspector determined to be infected with Karnal bunt and had been subsequently cleaned and disinfected.

(Approved by the Office of Management and Budget under control number 0579–0248)

[63 FR 31600, June 10, 1998, as amended at 64 FR 34113, June 25, 1999; 66 FR 40842, Aug. 6, 2001; 69 FR 24915, May 5, 2004; 69 FR 41181, July 8, 2004; 70 FR 24302, May 9, 2005]

Subpart V—Corn Cyst Nematode [Reserved]

Subpart W—European Larch Canker

SOURCE: 49 FR 18992, May 4, 1984, unless otherwise noted. Redesignated at 84 FR 2428, Feb. 7, 2019.

QUARANTINE AND REGULATIONS

§ 301.91 Quarantine and regulations; restrictions on interstate movement of regulated articles. [1]

(a) *Quarantines and regulations.* The secretary of agriculture hereby quarantines the State of Maine in order to prevent the artificial spread of European larch canker, *Lachnellula willkommi* (Dasycypha), a dangerous plant disease of trees of the *Larix* and *Pseudolarix* species not hereto fore widely prevalent or distributed within and throughout the United States; and hereby establishes regulations governing the interstate movement of regulated articles specified in § 301.91–2

(b) *Restrictions on interstate movement of regulated articles.* No common carrier or other person shall move interstate from any regulated area any regulated article except in accordance with the conditions prescribed in this subpart.

[49 FR 18992, May 4, 1984, as amended at 66 FR 21053, Apr. 27, 2001]

§ 301.91–1 Definitions.

Terms used in the singular form in this subpart shall be construed as the plural and vice versa, as the case may demand. The following terms, when used in this subpart, shall be construed, respectively, to mean:

Certificate. A document which is issued for a regulated article by an inspector or by a person operating under a compliance agreement, and which represents that such article is eligible for interstate movement in accordance with § 301.91–5(a).

Compliance agreement. A written agreement between Plant Protection

---

[1] Any properly identified inspector is authorized to stop and inspect persons and means of conveyance, and to seize, quarantine, treat, apply other remedial measures to, destroy, or otherwise dispose of regulated articles as provided in sections 414, 421, and 434 of the Plant Protection Act (7 U.S.C. 7714, 7731, and 7754).

and Quarantine and a person engaged in the business of growing, handling, or moving regulated articles, wherein the person agrees to comply with the provisions of this subpart and any conditions imposed pursuant thereto.

Deputy Administrator. The Deputy Administrator of the Animal and Plant Health Inspection Service for Plant Protection and Quarantine, or any officer or employee of the Department to whom authority to act in his/her stead has been or may hereafter be delegated.

European larch canker. The plant disease known as European larch canker, *Lachnellula willkommi* (Dasycypha), in any stage of development.

Infestation. The presence of European larch canker or the existence of circumstances that make it reasonable to believe that the European larch canker is present.

Inspector. Any employee of Plant Protection and Quarantine, Animal and Plant Health Inspection Service, U.S. Department of Agriculture, or other person, authorized by the Deputy Administrator in accordance with law to enforce the provisions of the quarantines and regulations in this subpart.

Interstate. From any State into or through any other State.

Limited permit. A document which is issued for a regulated article by an inspector or by a person operating under a compliance agreement, and which represents that such regulated article is eligible for interstate movement in accordance with § 301.91–5(b).

Moved (movement, move). Shipped, offered for shipment to a common carrier, received for transportation or transported by a common carrier, or carried, transported, moved, or allowed to be moved or caused to be moved by any means. "Movement" and "move" shall be construed accordingly.

Person. Any individual, partnership, corporation, company, society, association, or other organized group.

Plant Protection and Quarantine. The organizational unit within the Animal and Plant Health Inspection Service, U.S. Department of Agriculture, delegated responsibility for enforcing provisions of the Plant Protection Act and related legislation, quarantines, and regulations.

Regulated area. Any State, or any portion thereof, listed in § 301.91–3(c) or otherwise designated as a regulated area in accordance with § 301.91–3(b).

Regulated article. Any article listed in § 301.91–2(a) or otherwise designated as a regulated article in accordance with § 301.91–2(b).

State. Each of the several States of the United States, the District of Columbia, Guam, the Northern Mariana Islands, Puerto Rico, the Virgin Islands of the United States and all other Territories and Possessions of the United States.

[49 FR 18992, May 4, 1984, as amended at 66 FR 21053, Apr. 27, 2001]

§ 301.91–2 Regulated articles.

The following are regulated articles:

(a) Logs, pulpwood, branches, twigs, plants, scion and other propagative material of the *Larix* or *Pseudolarix* spp. except seeds;

(b) Any other product, article, or means of conveyance, of any character whatsoever, not covered by paragraph (a) of this section, when it is determined by an inspector that it presents a risk of spread of European larch canker and the person in possession thereof has actual notice that the product, article or means of conveyance is subject to the restrictions in the quarantine and regulations.

§ 301.91–3 Regulated areas.

(a) Except as otherwise provided in paragraph (b) of this section, the Deputy Administrator shall list as a regulated area in paragraph (c) of this section, the State, or any portion thereof, in which European larch canker has been found by an inspector or in which the Deputy Administrator has reason to believe that European larch canker is present, or any portion of a quarantined State which the Deputy Administrator deems necessary to regulate because of its proximity to a European larch canker infestation or its inseparability for quarantine enforcement purpose from localities in which European larch canker occurs. Less than an entire quarantined State will be designated as a regulated area only if the Deputy Administrator determines that:

(1) The State has adopted and is enforcing a quarantine or regulation which imposes restrictions on the intrastate movement of the regulated articles which are substantially the same as those which are imposed with respect to the interstate movement of such articles under this subpart; and

(2) The designation of less than the entire State as a regulated area will otherwise be adequate to prevent the artifical interstate spread of European larch canker.

(b) The Deputy Administrator or an inspector may temporarily designate any nonregulated area in a quarantined State as a regulated area in accordance with the criteria specified in paragraph (a) of this section for listing such area. Written notice of such designation shall be given to the owner or person in possession of such nonregulated area, and, thereafter, the interstate movement of any regulated article from such area shall be subject to the applicable provisions of this subpart. As soon as practicable, such area shall be added to the list in paragraph (c) of this section or such designation shall be terminated by the Deputy Administrator or an inspector, and notice thereof shall be given to the owner or person in possession of the area.

(c) The areas described below are designated as regulated areas:

MAINE

Hancock County. The entire townships of Gouldsboro, Sorrento, Sullivan, T7 SD, T9 SD, T10 SD, and T16 MD, and Winter Harbor.

Knox County. The entire townships of Appleton, Camden, Cushing, Friendship, Hope, Owls Head, Rockland, Rockport, Saint George, South Thomaston, Thomaston, Union, Warren, and Washington.

Lincoln County. The entire townships of Alna, Boothbay, Boothbay Harbor, Bremen, Bristol, Damariscotta, Edgecomb, Jefferson, Newcastle, Nobleboro, Somerville, South Bristol, Southport, Waldoboro, Westport Island, and Wiscasset.

Waldo County. The entire townships of Lincolnville and Searsmont.

Washington County. The entire townships of Addison, Baring Plantation, Beals, Beddington, Berry Township, Calais, Cathance Township, Centerville Township, Charlotte, Cherryfield, Columbia, Columbia Falls, Cooper, Cutler, Deblois, Dennysville, East Machias, Eastport, Edmunds Township, Harrington, Jonesboro, Jonesport, Lubec, Machias, Machiasport, Marion Township,

Marshfield, Meddybemps, Milbridge, Northfield, Pembroke, Perry, Robbinston, Roque Bluffs, Steuben, T18 MD BPP, T19 MD BPP, T24 MD BPP, T25 MD BPP, Trescott Township, Whiting, and Whitneyville.

[49 FR 18992, May 4, 1984, as amended at 49 FR 36817, Sept. 20, 1984; 50 FR 7033, Feb. 20, 1985; 50 FR 13178, Apr. 3, 1985; 76 FR 52544, Aug. 23, 2011]

§301.91–4 Conditions governing the interstate movement of regulated articles from regulated areas in quarantined States.[2]

Any regulated article may be moved interstate from any regulated area in a quarantined State only if moved under the following conditions:

(a) With a certificate or limited permit issued and attached in accordance with §§301.91–5 and 301.91–8 of this subpart; or

(b) Without a certificate or limited permit;

(1) If moved to a contiguous regulated area; or

(2)(i) If moved directly through (moved without stopping except under normal traffic conditions such as traffic lights or stop signs) any regulated area in an enclosed vehicle or in an enclosed container on a vehicle to prevent the introduction of European larch canker;

(ii) If the article originated outside of any regulated area; and

(iii) If the point of origin of any article is clearly indicated by shipping documents and its identity has been maintained.

§301.91–5 Issuance and cancellation of certificates and limited permits.

(a) A certificate shall be issued by an inspector, except as provided in paragraph (c) of this section, for the movement of a regulated article if such inspector:

(1)(i) Determines based on inspection of the premises of origin that the premises are free from European larch canker; or

(ii) Determines that it has been grown, processed, stored, or handled in

_____

[2] Requirements under all other applicable Federal domestic plant quarantines must also be met.

such a manner that the regulated article is free of European larch canker; and

(2) Determines that it is to be moved in compliance with any additional conditions deemed necessary under section 414 of the Plant Protection Act (7 U.S.C. 7714)[3] to prevent the spread of European larch canker; and

(3) Determines that it is eligible for unrestricted movement under all other Federal domestic plant quarantines and regulations applicable to such article.

(b) A limited permit shall be issued by an inspector, except as provided in paragraph (c) of this section, for the movement of a regulated article if such inspector:

(1) Determines, in consultation with the Deputy Administrator, that it is to be moved to a specified destination for specified handling, utilization, or processing (such destination and other conditions to be specified on the limited permit), when, upon evaluation of all of the circumstances involved in each case, it is determined that such movement will not result in the spread of European larch canker because the disease will be destroyed by such specified handling, utilization, or processing;

(2) Determines that it is to be moved in compliance with any additional conditions deemed necessary under section 414 of the Plant Protection Act (7 U.S.C. 7714)[3] to prevent the spread of European larch canker; and

(3) Determines that it is eligible for such movement under all other Federal domestic plant quarantines and regulations applicable to such article.

(c) Certificates and limited permits may be issued by any person engaged in the business of growing, handling, or moving regulated articles provided such person has entered into and is operating under a compliance agreement. Any such person may execute and issue a certificate or limited permit for the interstate movement of a regulated article if an inspector has previously made the determination that the arti-

cle is eligible for a certificate in accordance with § 301.91–5(a) or is eligible for a limited permit in accordance with § 301.91–5(b).

(d) Any certificate or limited permit which has been issued or authorized may be withdrawn by an inspector if such inspector determines that the holder thereof has not complied with any conditions under the regulations for the use of such document. The reasons for the withdrawal shall be confirmed in writing as promptly as circumstances permit. Any person whose certificate or limited permit has been withdrawn may appeal the decision in writing to the Deputy Administrator within ten (10) days after receiving the written notification of the withdrawal. The appeal shall state all of the facts and reasons upon which the person relies to show that the certificate or limited permit was wrongfully withdrawn. The Deputy Administrator shall grant or deny the appeal, in witing, stating the reasons for such decision, as promptly as circmstances permit. If there is a conflict as to any material fact, a hearing shall be held to resolve such conflict. Rules of Practice concerning such a hearing will be adopted by the Deputy Administrator.

[49 FR 18992, May 4, 1984, as amended at 66 FR 21053, Apr. 27, 2001]

§ 301.91–6 Compliance agreement and cancellation thereof.

(a) Any person engaged in the business of growing, handling, or moving regulated articles may enter into a compliance agreement to facilitate the movement of regulated articles under this subpart.[4] The compliance agreement shall be a written agreement between a person engaged in such a business and Plant Protection and Quarantine, wherein the person agrees to

---

[3] An inspector may hold, seize, quarantine, treat, apply other remedial measures to, destroy, or otherwise dispose of plants, plant pests, or other articles in accordance with sections 414, 421, and 434 of the Plant Protection Act (7 U.S.C. 7714, 7731, and 7754).

[4] Compliance agreement forms are available without charge from the Animal and Plant Health Inspection Service, Plant Protection and Quarantine, Domestic and Emergency Operations, 4700 River Road Unit 134, Riverdale, Maryland 20737–1236, and from local offices of the Plant Protection and Quarantine. (Local offices are listed in telephone directories).

comply with the provisions of this subpart and any conditions imposed pursuant thereto.

(b) Any compliance agreement may be cancelled orally or in writing by the inspector who is supervising its enforcement whenever the inspector finds that such person has failed to comply with the provisions of this subpart or any conditions imposed pursuant thereto. If the cancellation is oral, the decision and the reasons therefor shall be confirmed in writing, as promptly as circumstances permit. Any person whose compliance agreement has been cancelled may appeal the decision, in writing, to the Deputy Administrator within ten (10) days after receiving written notification of the cancellation. The appeal shall state all of the facts and reasons upon which the person relies to show that the compliance agreement was wrongfully cancelled. The Deputy Administrator shall grant or deny the appeal, in writing, stating the reasons for such decision, as promptly as circumstances permit. If there is a conflict as to any material fact, a hearing shall be held to resolve such conflict. Rules of Practice concerning such a hearing will be adopted by the Deputy Administrator.

[49 FR 18992, May 4, 1984, as amended at 59 FR 67609, Dec. 30, 1994]

§301.91–7 Assembly and inspection of regulated articles.

(a) Any person (other than a person authorized to issue certificates or limited permits under §301.91–5(c)), who desires to move interstate a regulated article accompanied by a certificate or limited permit shall, as far in advance as possible (should be no less than 48 hours before the desired movement), request an inspector[5] to take any necessary action under this subpart prior to movement of the regulated article.

(b) Such articles shall be assembled at such point and in such manner as

[5]Inspectors are assigned to local offices of Plant Protection and Quarantine which are listed in telephone directories. Information concerning such local offices may also be obtained from the Animal and Plant Health Inspection Service, Plant Protection and Quarantine, Domestic and Emergency Operations, 4700 River Road Unit 134, Riverdale, Maryland 20737–1236.

the inspector designates as necessary to comply with the requirements of this subpart.

[49 FR 18992, May 4, 1984, as amended at 59 FR 67609, Dec. 30, 1994]

§301.91–8 Attachment and disposition of certificates and limited permits.

(a) A certificate or limited permit required for the interstate movement of a regulated article, at all times during such movement, shall be securely attached to the outside of the containers containing the regulated article, securely attached to the article itself if not in a container, or securely attached to the consignee's copy of the accompanying waybill or other shipping document; *Provided, however,* That the requirements of this section may be met by attaching the certificate or limited permit to the consignee's copy of the waybill or other shipping documents only if the regulated article is sufficiently described on the certificate, limited permit, or shipping document to identify such article.

(b) The certificate or limited permit for the movement of a regulated article shall be furnished by the carrier to the consignee at the destination of the shipment.

§301.91–9 Costs and charges.

The services of the inspector shall be furnished without cost, except as provided in 7 CFR part 354. The U.S. Department of Agriculture will not be responsible for any costs or charges incident to inspections or compliance with the provisions of the quarantine and regulations in this subpart, other than for the services of the inspector.

Subpart X—Phytophthora Ramorum

SOURCE: 72 FR 8597, Feb. 27, 2007, unless otherwise noted. Redesignated at 84 FR 2428, Feb. 7, 2019.

§301.92 Restrictions on interstate movement.

(a) No person may move interstate from any quarantined area any regulated, restricted, or associated article or any other nursery stock except in accordance with this subpart.[1]

(b) No person may move interstate from any regulated establishment any regulated, restricted, or associated articles except in accordance with this subpart.

(c) No person may move interstate from any quarantined area or regulated establishment any regulated restricted, or associated article or nursery stock that has been tested with a test approved by APHIS and found infected with *Phytophthora ramorum*, or that is part of a plant that was found infected with *Phytophthora ramorum*, unless such movement is in accordance with part 330 of this chapter.

[72 FR 8597, Feb. 27, 2007, as amended at 84 FR 16192, Apr. 18, 2019]

§ 301.92–1 Definitions.

Administrator. The Administrator, Animal and Plant Health Inspection Service, or any person authorized to act for the Administrator.

Animal and Plant Health Inspection Service. The Animal and Plant Health Inspection Service (APHIS) of the United States Department of Agriculture.

Associated article. Any article listed in § 301.92–2(c).

Bark chips. Bark fragments broken or shredded from a log or tree.

Certificate. A document, stamp, or imprint by which an inspector or person operating under a compliance agreement affirms that a specified regulated or associated article meets applicable requirements of this subpart and may be moved interstate to any destination.

Compliance agreement. A written agreement between APHIS and a person engaged in growing, processing, handling, or moving regulated or associated articles, wherein the person agrees to comply with this subpart.

Duff. Decaying plant matter that includes leaf litter, green waste, stem material, bark, and any other plant material that, upon visual inspection, does not appear to have completely decomposed.

Firewood. Wood that has been cut, sawn, or chopped into a shape and size commonly used for fuel, or other wood intended for fuel.

Forest stock. All flowers, trees, shrubs, vines, scions, buds, or other plants that are wild-grown, backyard-grown, or naturally occurring.

From. An article is considered to be "from" a specific site or location for the purposes of this subpart if it was grown or propagated in, stored or sold, or distributed from the site or location.

Growing media. Any material in which plant roots are growing or intended for that purpose.

Inspector. Any employee of APHIS, the U.S. Department of Agriculture, or other person authorized by the Administrator to perform the duties required under this subpart.

Interstate. From any State into or through any other State.

Log. The bole of a tree; trimmed timber that has not been sawn further than to form cants.

Lot. A contiguous block of plants of the same species or cultivar, of the same container size and from the same source, if known.

Lumber. Logs that have been sawn into boards, planks, or structural members such as beams.

Moved (move, movement). Shipped, offered for shipment, received for transportation, transported, carried, or allowed to be moved, shipped, transported, or carried.

Mulch. Bark chips, wood chips, wood shavings, or sawdust, or a mixture thereof, that could be used as a protective or decorative ground cover or as part of a growing media mixture.

Non-host nursery stock. Any taxa of nursery stock not listed in § 301.92–2 as a regulated or associated article.

Nursery. Any location where nursery stock is grown, propagated, stored, or sold, or any location from which nursery stock is distributed. Locations that grow trees for sale without roots (*e.g.*, as Christmas trees) are considered to be nurseries for the purposes of this subpart.

---

[1] Any properly identified inspector is authorized to stop and inspect persons and means of conveyance and to seize, quarantine, treat, apply other remedial measures to, destroy, or otherwise dispose of regulated or restricted articles as provided in sections 414, 421, and 434 of the Plant Protection Act (7 U.S.C. 7714, 7731, and 7754).

Nursery stock. All plants for planting, including houseplants, propagative material that is grown in a nursery, and tree seedlings for reforestation, except the following: Seeds; turf or sod; bulbs, tubers, corms, or rhizomes;[2] greenhouse grown cactus, succulents, and orchids; aquarium grown aquatic plants; greenhouse, container, or field grown palms; greenhouse, container, or field grown cycads, and tissue culture plants grown in vitro; and plants meeting the definition of forest stock.

Permit. A written authorization issued by APHIS to allow the interstate movement of restricted articles in accordance with part 330 of this chapter.

Person. Any association, company, corporation, firm, individual, joint stock company, partnership, society, or other entity.

Plant Protection and Quarantine. The Plant Protection and Quarantine program of the Animal and Plant Health Inspection Service, United States Department of Agriculture.

Quarantined area. Any State, or any portion of a State, listed in §301.92–3(a)(3) of this subpart or otherwise designated as a quarantined area in accordance with §301.92–3(a)(2) of this subpart.

Regulated article. Any article listed in §301.92–2(b) of this subpart.

Restricted article. Any article listed in §301.92–2(a) of this subpart.

Regulated establishment. Any nursery regulated by APHIS pursuant to §301.92–3(b).

Soil. The loose surface material of the earth in which plants grow, in most cases consisting of disintegrated rock with an admixture of organic material.

State. The District of Columbia, Puerto Rico, the Northern Mariana Islands, or any State, territory, or possession of the United States.

[72 FR 8597, Feb. 27, 2007, as amended at 84 FR 16192, Apr. 18, 2019]

[2] Bulbs, tubers, corms, or rhizomes are only considered nursery stock (and therefore, regulated under this subpart) if they are of plant taxa listed in §301.92–2 as regulated articles or associated articles.

§301.92–2 Restricted, regulated, and associated articles; lists of proven hosts and associated plant taxa.

(a) *Restricted articles.* The following are restricted articles:

(1) Bark chips or mulch[3] located in a quarantined area and that are proven host plant taxa listed in paragraph (d) of this section.

(2) Forest stock located or grown in a quarantined area and that are proven host plant taxa or associated plant taxa listed in paragraph (d) or (e) of this section.

(3) Any other product or article that an inspector determines to present a risk of spreading *Phytophthora ramorum*, if an inspector notifies the person in possession of the product or article that it is a restricted article.

(b) *Regulated articles.* The following are regulated articles:

(1) Nursery stock, decorative trees without roots, unprocessed wood and wood products, and plant products, including firewood, logs, lumber,[4] wreaths, garlands, and greenery of proven host plant taxa listed in paragraph (d) of this section.

(2) Soil and growing media.

(3) Any other product or article that an inspector determines to present a risk of spreading *Phytophthora ramorum* if an inspector notifies the person in possession of the product or article that it is subject to the restrictions in the regulations.

(c) *Associated articles.* The following are associated articles: Nursery stock of associated plant taxa listed in paragraph (e) of this section.

(d) *Proven host plant taxa.* The following are proven hosts of *Phytophthora ramorum:*

Acer macrophyllum Bigleaf maple

Acer pseudoplatanus Planetree maple

Adiantum aleuticum Western maidenhair fern

Adiantum jordanii California maidenhair fern

Aesculus californica California buckeye

[3] Bark chips or mulch of species listed in paragraph (d) of this section and that are marked with an asterisk (*) are not restricted articles.

[4] Firewood, logs, and lumber of species listed in paragraph (d) of this section and that are marked with an asterisk (*) are not regulated articles.

Aesculus hippocastanum horse chestnut
*Arbutus menziesii Madrone
*Arctostaphylos manzanita Manzanita
*Calluna vulgaris Scotch heather
*Camellia spp. Camellia—all species, hybrids and cultivars
*Castanea sativa Sweet chestnut
*Cinnamomum camphora Camphor tree
Fagus sylvatica European beech
*Frangula californica (≡*Rhamnus californica*) California coffeeberry
*Frangula purshiana (≡*Rhamnus purshiana*) Cascara
Fraxinus excelsior European ash
*Gaultheria procumbens, Eastern teaberry
*Griselinia littoralis Griselinia
*Hamamelis virginiana Witch hazel
*Heteromeles arbutifolia Toyon
*Kalmia spp. Kalmia—includes all species, hybrids, and cultivars
*Laurus nobilis Bay laurel
Lithocarpus densiflorus Tanoak
*Lonicera hispidula California honeysuckle
*Maianthemum racemosum (=*Smilacina racemosa*) False Solomon's seal
*Michelia doltsopa Michelia
*Parrotia persica Persian ironwood
*Photinia fraseri Red tip photinia
*Pieris spp. Pieris—includes all species, hybrids, and cultivars
*Pseudotsuga menziesii var. *menziesii* and all nursery-grown *P. menziesii* Douglas fir
Quercus agrifolia Coast live oak
Quercus cerris European turkey oak
Quercus chrysolepis Canyon live oak
Quercus falcata Southern red oak
*Quercus ilex Holm oak
Quercus kelloggii California black oak
Quercus parvula var. *shrevei* and all nursery grown *Q. parvula* Shreve's oak
*Rhododendron spp. Rhododendron (including azalea)—includes all species, hybrids, and cultivars
*Rosa gymnocarpa Wood rose
*Salix caprea Goat willow
*Sequoia sempervirens Coast redwood
*Syringa vulgaris Lilac
*Taxus baccata European yew
*Trientalis latifolia Western starflower
*Umbellularia californica California bay laurel, pepperwood, Oregon myrtle
*Vaccinium ovatum Evergreen huckleberry
*Viburnum spp. Viburnum—all species, hybrids, and cultivars

(e) *Associated plant taxa.* The following plant taxa are considered to be associated with *Phytophthora ramorum:*

Abies concolor White fir
Abies grandis Grand fir
Abies magnifica Red fir
Acer circinatum Vine maple
Acer davidii Striped bark maple
Acer laevigatum Evergreen maple
Arbutus unedo Strawberry tree
Arctostaphylos columbiana Manzanita

Arctostaphylos uva-ursi Kinnikinnick, bearberry
Ardisia japonica Ardisia
Calycanthus occidentalis Spicebush
Castanopsis orthacantha Castanopsis
Ceanothus thyrsiflorus Blueblossom
Cinnamomum camphora Camphor tree
Clintonia andrewsiana Andrew's clintonia bead lily
Cornus kousa × Cornus capitata Cornus Norman Haddon
Corylus cornuta California hazelnut
Distylium myricoides Myrtle-leafed distylium
Drimys winteri Winter's bark
Dryopteris arguta California wood fern
Eucalyptus haemastoma Scribbly gum
Euonymus kiautschovicus Spreading euonymus
Fraxinus latifolia Oregon ash
Gaultheria shallon Salal, Oregon wintergreen
Hamamelis mollis Chinese witch-hazel
Hamamelis × intermedia (*H. mollis & H. japonica*) Hybrid witchhazel
Ilex cornuta Buford holly, Chinese holly
Ilex purpurea Oriental holly
Illicium parviflorum Yellow anise
Larix kaempferi Japanese larch
Leucothoe axillaris Fetter-bush, dog hobble
Leucothoe fontanesiana Drooping leucothoe
Loropetalum chinense Lorapetalum
Magnolia denudata Lily tree
Magnolia grandiflora Southern magnolia
Magnolia stellata Star magnolia
Magnolia × loebneri Loebner magnolia
Magnolia × soulangeana Saucer magnolia
Mahonia nervosa Creeping Oregon grape
Manglietia insignis Red lotus tree
Michelia maudiae Michelia
Michelia wilsonii Michelia
Molinadendron sinaloense
Nerium oleander Oleander
Nothofagus obliqua Roble beech
Osmanthus decorus (≡*Phillyrea decora;* ≡*P. vilmoriniana*) Osmanthus
Osmanthus delavayi Delavay Osmanthus, Delavay tea olive
Osmanthus fragrans Sweet olive
Osmanthus heterophyllus Holly olive
Osmorhiza berteroi Sweet Cicely
Parakmeria lotungensis Eastern joy lotus tree
Pittosporum undulatum Victorian box
Prunus laurocerasus English laurel, cherry laurel
Prunus lusitanica Portuguese laurel cherry
Pyracantha koidzumii Formosa firethorn
Quercus acuta Japanese evergreen oak
Quercus petraea Sessile oak
Quercus rubra Northern red oak
Rosa (specific cultivars)
  Royal Bonica (tagged: "MEImodac")
  Pink Meidiland (tagged: "MEIpoque")
  Pink Sevillana (tagged: "MEIgeroka")
Rosa rugosa Rugosa rose
Rubus spectabilis Salmonberry
Schima wallichii Chinese guger tree
Taxus brevifolia Pacific yew
Taxus × media Yew

Torreya californica California nutmeg
Toxicodendron diversilobum Poison oak
Trachelospermum jasminoides Star jasmine, confederate jasmine
Vancouveria planipetala Redwood ivy
Veronica spicata Syn. *Pseudolysimachion spicatum* Spiked speedwell

[72 FR 8597, Feb. 27, 2007, as amended at 84 FR 16192, Apr. 18, 2019]

§ 301.92–3 Quarantined areas and regulated establishments.

(a) *Quarantined areas.* (1) Except as otherwise provided in paragraph (a)(2) of this section, the Administrator will designate as a quarantined area in paragraph (a)(3) of this section each State, or each portion of a State, in which *Phytophthora ramorum* has been confirmed by an inspector to be established in the natural environment, in which the Administrator has reason to believe that *Phytophthora ramorum* is present in the natural environment, or that the Administrator considers it necessary to quarantine because of its inseparability for quarantine enforcement purposes from localities in which *Phytophthora ramorum* has been found in the natural environment. Less than an entire area will be designated as a quarantined area only if the Administrator determines that:

(i) The State has adopted and is enforcing restrictions on the intrastate movement of regulated, restricted, and associated articles that are substantially the same as those imposed by this subpart on the interstate movement of regulated, restricted, and associated articles; and

(ii) The designation of less than the entire State as a quarantined area will prevent the interstate spread of *Phytophthora ramorum*.

(2) The Administrator or an inspector may temporarily designate any non-quarantined area as a quarantined area in accordance with paragraph (a)(1) of this section. The Administrator will give a copy of this regulation along with a written notice for the temporary designation to the owner or person in possession of the nonquarantined area. Thereafter, the interstate movement of any regulated, restricted, or associated article from an area temporarily designated as a quarantined area will be subject to this subpart. As soon as practicable, this area will be added to the list in paragraph (a)(3) of this section or the designation will be terminated by the Administrator or an inspector. The owner or person in possession of an area for which designation is terminated will be given notice of the termination as soon as practicable.

(3) The following areas are designated as quarantined areas:

California

Alameda County. The entire county.
Contra Costa County. The entire county.
Humboldt County. The entire county.
Lake County. The entire county.
Marin County. The entire county.
Mendocino County. The entire county.
Monterey County. The entire county.
Napa County. The entire county.
San Francisco County. The entire county.
San Mateo County. The entire county.
Santa Clara County. The entire county.
Santa Cruz County. The entire county.
Solano County. The entire county.
Sonoma County. The entire county.
Trinity County. The entire county.

Oregon

Curry County. The following portion of Curry County that lies inside the area starting at the point where the mouth of the Rogue River meets the Pacific Ocean and continuing east along the Rogue River to the northeast corner of T35S R12W section 31; then south to the northeast corner of T38S R12W section 18; then east to the northeast corner of T38S R12W section 13; then south to northeast corner of T38S R12W section 25; then east to the northeast corner of T38S R11W section 29; then south to the northeast corner of T40S R11W section 8; then east to the northeast corner of T40S R11W section 10; then south to the State border with California; then west to the intersection of the State border and U.S. Highway 101; then northwest along U.S. Highway 101 to the intersection with West Benham Lane; then west along West Benham Lane to the Pacific Coastline; then following the Pacific Coastline northwest to the point of beginning.

(b) *Regulated establishments*—(1) *Designation.* The Administrator will designate a nursery that is not located in a quarantined area for *Phytophthora ramorum* as a regulated establishment for *Phytophthora ramorum* if the nursery ships regulated, restricted, or associated articles interstate and sources of *Phytophthora ramorum* are detected on nursery stock, or in soil, growing media, pots used for nursery stock, standing water, drainage water, water used for irrigation, or any other regulated, restricted, or associated articles at the nursery.

(2) *Deregulation.* The Administrator will withdraw regulation of a regulated establishment if, for 3 consecutive years, each time the nursery is inspected by an inspector, it is found free of sources of *Phytophthora ramorum* inoculum.

(Approved by the Office of Management and Budget under control number 0579–0310)

[84 FR 16193, Apr. 18, 2019]

§ 301.92–4 **Conditions governing the interstate movement of regulated, restricted, and associated articles, and non-host nursery stock from quarantined and regulated establishments.**

(a) *Interstate movement of regulated and associated articles from quarantined areas.* Regulated and associated articles may be moved interstate from a quarantined area[5] only in accordance with this subpart.

(1) *With a certificate.* Any regulated or associated article may be moved interstate from a quarantined area if accompanied by a certificate issued and attached in accordance with §§ 301.92–5 and 301.92–8, and provided that the regulated or associated article is moved through the quarantined area without stopping except for refueling, rest stops, emergency repairs, and for traffic conditions, such as traffic lights or stop signs.

(2) *Without a certificate.* (i)(A) The regulated article or associated article originated outside the quarantined area and the point of origin of the arti-

cle is indicated on the waybill of the vehicle transporting the article; and

(B) The regulated or associated article is moved from outside of the quarantined area through the quarantined area without stopping except for refueling or for traffic conditions, such as traffic lights or stop signs, and the article is not unpacked or unloaded in the quarantined area.

(ii) Soil samples may be moved from a quarantined area for *Phytophthora ramorum* for chemical or physical (compositional) analysis provided that they are moved to a laboratory; and that laboratory:

(A) Has entered into and is operating under a compliance agreement with APHIS in accordance with § 301.92–6;

(B) Is abiding by all terms and conditions of that compliance agreement; and

(C) Is approved by APHIS to test and/or analyze such samples.

(b) *Interstate movement of restricted articles from quarantined areas.* Restricted articles may be moved interstate from a quarantined area[6] only in accordance with this section.

(1) *With a permit.* Any restricted article may be moved interstate from a quarantined area only if the article is moved pursuant to a permit issued by the Administrator in accordance with part 330 of this chapter.

(2) *Without a permit.* (i) The restricted article originated outside the quarantined area and the point of origin of the article is indicated on the waybill of the vehicle transporting the article; and

(ii) The restricted article is moved from outside the quarantined area through the quarantined area without stopping except for refueling or for traffic conditions, such as traffic lights or stop signs, and the article is not unpacked or unloaded in the quarantined area.

(c) *Interstate movement of nursery stock from nurseries in quarantined areas*—(1) *Regulated articles of nursery stock and associated articles.* Regulated articles of nursery stock and associated articles may only be moved interstate from

---

[5] Requirements under all other applicable Federal domestic plant quarantines and regulations must also be met.

[6] See footnote 4 of this subpart.

nurseries in quarantined areas in accordance with paragraph (a) of this section.

(2) *Non-host nursery stock.* Any nursery stock of a taxon not listed in § 301.92–2 as a regulated or associated article may only be moved interstate from nurseries in quarantined areas as follows:

(i) *With a certificate.* If the non-host nursery stock originates from a nursery in a quarantined area that contains regulated or associated articles, the nursery stock must be accompanied by a certificate issued and attached in accordance with §§ 301.92–5 and 301.92–8, and be moved through the quarantined area without stopping except for refueling, rest stops, emergency repairs, and for traffic conditions, such as traffic lights or stop signs.

(ii) *Without a certificate.* If the non-host nursery stock originates from a nursery in a quarantined area that does not contain regulated or associated articles, the nursery stock may be moved interstate without a certificate, provided that:

(A) The nursery from which plants originate has been inspected and found free of evidence of *Phytophthora ramorum* in accordance with § 301.92–11(b)(3), and

(B) The nursery stock is not rooted in soil or growing media. [7]

(d) *Interstate movement of regulated, restricted, and associated articles from regulated establishments.* Regulated, restricted, and associated articles may be moved interstate from a regulated establishment if the regulated establishment has entered into a compliance agreement with APHIS in accordance with § 301.92–6, and the articles are accompanied by a certificate issued in accordance with § 301.92–5.

[72 FR 8597, Feb. 27, 2007, as amended at 84 FR 16193, Apr. 18, 2019]

---

[7] To be eligible for interstate movement, non-host nursery stock that is rooted in soil or growing media requires certification that the soil or growing media meets the requirements of § 301.92–5(a)(1)(iii).

§ **301.92–5 Issuance and cancellation of certificates.**

(a) *Movements from quarantined areas.* (1) An inspector [8] may issue a certificate for the interstate movement of regulated articles, associated articles, or non-host nursery stock [9] from a quarantined area if the inspector determines that:

(i) The regulated articles have been treated under the direction of an inspector in accordance with part 305 of this chapter; or

(ii) The regulated articles are wood products such as firewood, logs, or lumber that are free of bark; [10] or

(iii) The regulated article is soil or growing media that has not been in direct physical contact with any article infected with *Phytophthora ramorum*, and from which all duff has been removed; or

(iv) The articles are nursery stock or regulated articles of decorative trees without roots, wreaths, garlands, or greenery that:

(A)(*1*) Are shipped from a nursery that has been inspected in accordance with the inspection and sampling protocol described in § 301.92–11(a)(1), and the nursery is free of evidence of *Phytophthora ramorum* infestation; or

(*2*) Are shipped from a nursery that has been inspected in accordance with the inspection and sampling protocol described in § 301.92–11(a)(2), and the nursery is free of evidence of *Phytophthora ramorum* infestation; or

(*3*) Are shipped from a nursery that has been inspected in accordance with the inspection and sampling protocol

---

[8] Services of an inspector may be requested by contacting local offices of Plant Protection and Quarantine, which are listed in telephone directories. The addresses and telephone numbers of local offices may also be obtained from the Animal and Plant Health Inspection Service, Plant Protection and Quarantine, Invasive Species and Pest Management, 4700 River Road Unit 160, Riverdale, MD 20737, or the APHIS Web site at *http://www.aphis.usda.gov/ppq/sphd/*.

[9] Paragraph (d)(2)(ii) of § 301.92–4 allows the interstate movement of non-host nursery stock without a certificate under certain conditions.

[10] Firewood, logs, lumber of species listed in 301.92–2(d) and marked with an asterisk are not regulated articles, as noted in § 301.92–2(b)(1).

described in § 301.92–11(a)(2), is not free of evidence of *Phytophthora ramorum* infestation, but has entered into and is operating under a compliance agreement with APHIS, and is determined by an inspector to be abiding by all terms and conditions of that agreement; and

(B) Are part of a shipment of nursery stock, decorative trees without roots, wreaths, garlands, or greenery that has been inspected prior to interstate movement in accordance with § 301.92–11(a)(2), and the regulated articles in the shipment are free of evidence of *Phytophthora ramorum* infection; and

(C) Have been kept separate from regulated and associated articles and non-host nursery stock not inspected between the time of the inspection and the time of interstate movement; and

(D) Have not been grown in, or moved from, other areas within a quarantined area except nurseries that are annually inspected for *Phytophthora ramorum* in accordance with § 301.92–11 and that have been found free of evidence of *Phytophthora ramorum* infestation, *except that* certified nurseries which receive articles from a non-certified nursery in a quarantined or regulated area may continue to ship other plants interstate, provided that the uncertified plants are safeguarded, segregated, and withheld from interstate movement until the plants are inspected and tested and found free of evidence of *Phytophthora ramorum*.

(v) The regulated or associated article or non-host nursery stock is to be moved in compliance with any additional emergency conditions the Administrator may impose under section 414 of the Plant Protection Act (7 U.S.C. 7714)[11] to prevent the spread of *Phytophthora ramorum*; and

---

[11] Sections 414, 421, and 434 of the Plant Protection Act (7 U.S.C. 7714, 7731, and 7754) provide that the Secretary of Agriculture may, under certain conditions, hold, seize, quarantine, treat, apply other remedial measures to destroy or otherwise dispose of any plant, plant pest, plant product, article, or means of conveyance that is moving, or has moved into or through the United States or interstate if the Secretary has reason to believe the article is a plant pest or is infested with a plant pest at the time of movement.

(vi) The regulated or associated article or non-host nursery stock is eligible for unrestricted movement under all other Federal domestic plant quarantines and regulations applicable to the regulated or associated article.

(2) [Reserved]

(b) *Movements from regulated establishments.* An inspector may issue a certificate for the movement of regulated, restricted, and/or associated articles from a regulated establishment if the inspector determines that:

(1) The nursery has entered into a compliance agreement in accordance with § 301.92–6 and is abiding by all terms and conditions of that agreement; and

(2) The nursery has been inspected in accordance with § 301.92–11(c); and

(3) The articles to be shipped interstate are free from *Phytophthora ramorum* inoculum; and

(4) The movement of the articles is not subject to additional restriction under section 414 of the Plant Protection Act (7 U.S.C. 7714) or other Federal domestic plant quarantines and regulations.

(c) Certificates issued under paragraphs (a) and (b) of this section may be issued by any person engaged in the business of growing, processing, handling, or moving regulated or associated articles or nursery stock provided such person has entered into and is operating under a compliance agreement in accordance with § 301.92–6. Any such person may execute and issue a certificate for the interstate movement of regulated or associated articles or nursery stock if an inspector has previously made the determination that the article is eligible for a certificate in accordance with any applicable section of this subpart.

(d) Any certificate that has been issued may be withdrawn, either orally or in writing, by an inspector if he or she determines that the holder of the certificate has not complied with all conditions in this subpart for the use of the certificate. If the withdrawal is oral, the withdrawal and the reasons for the withdrawal will be confirmed in writing as promptly as circumstances allow. Any person whose certificate has been withdrawn may appeal the decision in writing to the Administrator within 10 days after receiving the written notification of the withdrawal. The appeal must state all of the facts and reasons upon which the person relies to show that the certificate was wrongfully withdrawn. As promptly as circumstances allow, the Administrator will grant or deny the appeal, in writing, stating the reasons for the decision. A hearing will be held to resolve any conflict as to any material fact. Rules of practice concerning a hearing will be adopted by the Administrator.

(Approved by the Office of Management and Budget under control numbers 0579–0310 and 0579–0088)

[72 FR 8597, Feb. 27, 2007, as amended at 75 FR 4241, Jan. 26, 2010; 84 FR 16193, Apr. 18, 2019]

§ 301.92–6 Compliance agreements and cancellation.

(a) Any person engaged in growing, processing, handling, or moving regulated articles, associated articles, or non-host nursery stock may enter into a compliance agreement when an inspector determines that the person understands this subpart, agrees to comply with its provisions, and agrees to comply with all the provisions contained in the compliance agreement. [12]

(b) Any compliance agreement may be canceled, either orally or in writing, by an inspector whenever the inspector finds that the person who has entered into the compliance agreement has failed to comply with this subpart. If the cancellation is oral, the cancellation and the reasons for the cancellation will be confirmed in writing as promptly as circumstances allow. Any person whose compliance agreement has been canceled may appeal the decision, in writing, within 10 days after receiving written notification of the cancellation. The appeal must state all of the facts and reasons upon which the

---

[12] Compliance agreement forms are available without charge from the Animal and Plant Health Inspection Service, Plant Protection and Quarantine, Invasive Species and Pest Management, 4700 River Road Unit 160, Riverdale, MD 20737–1236, and from local offices of the Plant Protection and Quarantine, which are listed in telephone directories. Forms are also available on the Internet at *http://www.aphis.usda.gov/ppq/ispm/pramorum/resources.html.*

person relies to show that the compliance agreement was wrongfully canceled. As promptly as circumstances allow, the Administrator will grant or deny the appeal, in writing, stating the reasons for the decision. A hearing will be held to resolve any conflict as to any material fact. Rules of practice concerning a hearing will be adopted by the Administrator.

(Approved by the Office of Management and Budget under control numbers 0579–0088 and 0579–0310)

[72 FR 8597, Feb. 27, 2007, as amended at 84 FR 16194, Apr. 18, 2019]

§ 301.92–7 Availability of inspectors; assembly for inspection.

(a) Any person (other than a person authorized to issue certificates under §301.92–5(c)) who desires to move a regulated or associated article or non-host nursery stock interstate accompanied by a certificate must notify an inspector [13] as far in advance of the desired interstate movement as possible, but no less than 48 hours before the desired time of inspection.

(b) The regulated or associated article or non-host nursery stock must be assembled at the place and in the manner the inspector designates as necessary to comply with this subpart.

[72 FR 8597, Feb. 27, 2007, as amended at 84 FR 16194, Apr. 18, 2019]

§ 301.92–8 Attachment and disposition of certificates and recordkeeping.

(a) A certificate required for the interstate movement of a regulated article, associated article, or non-host nursery stock must, at all times during the interstate movement, be:

(1) Attached to the outside of the container containing the regulated article, associated article, or non-host nursery stock; or

(2) Attached to the regulated article, associated article, or non-host nursery stock itself if not in a container; or

(3) Attached to the consignee's copy of the accompanying waybill. If the certificate is attached to the consignee's copy of the waybill, the regulated article, associated article, or

---

[13] See footnote 7 of this subpart.

non-host nursery stock must be sufficiently described on the certificate and on the waybill to identify the regulated article, associated article, or non-host nursery stock.

(b) The certificate for the interstate movement of a regulated article, associated article, or non-host nursery stock must be furnished by the carrier to the consignee listed on the certificate upon arrival at the location provided on the certificate.

(c) All nurseries that are operating under compliance agreements must maintain records of all incoming shipments of plants for a minimum of 24 months and must make them available to inspectors upon request. In addition, all nurseries that are operating under compliance agreements, except retail dealers, must maintain records of outgoing shipments for a minimum of 24 months and must make them available to inspectors upon request.

(Approved by the Office of Management and Budget under control numbers 0579–0088 and 0579–0310)

§ 301.92–9　Costs and charges.

The services of the inspector during normal business hours (8 a.m. to 4:30 p.m., Monday through Friday, except holidays) will be furnished without cost. The user will be responsible for all costs and charges arising from inspection and other services provided outside normal business hours.

§ 301.92–10　[Reserved]

§ 301.92–11　Inspection and sampling protocols.

(a) *Nurseries in quarantine areas shipping regulated articles of nursery stock and associated articles interstate*—(1) *Nurseries in which Phytophthora ramorum has not been detected since March 31, 2011.* To meet the requirements of § 301.92–5(a)(1)(iv), nurseries that are located in quarantined areas, that move regulated articles of nursery stock, decorative trees without roots, wreaths, garlands, or greenery, associated articles, or non-host nursery stock interstate, and in which *Phytophthora ramorum* has not been detected since March 31, 2011, must meet the following requirements. Any such nurseries in quarantined areas that do

not meet the following requirements are prohibited from moving regulated articles and associated articles interstate. Any such nurseries in quarantined areas that do not meet the following requirements or those in paragraph (b) of this section are prohibited from moving non-host nursery stock interstate.

(i) *Annual inspection, sampling, and testing*—(A) *Inspection.* The nursery must be inspected annually for symptoms of *Phytophthora ramorum* by an inspector. Inspectors will visually inspect for symptomatic plants throughout the nursery, and inspection will focus on, but not be limited to, regulated articles and associated articles.

(B) *Sampling.* A minimum of 40 plant samples must be tested per nursery location. Samples must be taken from all symptomatic plants if symptomatic plants are present. If fewer than 40 symptomatic plants are present, each symptomatic plant must be sampled and the remainder of the 40 sample minimum must be taken from asymptomatic plants. If no symptomatic plants are present, 40 asymptomatic plants must be sampled; biased toward proven hosts. Each sample may contain more than one leaf, and may come from more than one plant, but all plants in the sample must be from the same lot. Asymptomatic samples, if collected, must be taken from regulated and associated articles and nearby plants. Inspectors must conduct inspections at times when the best expression of symptoms is anticipated and must take nursery fungicide programs into consideration. Nursery owners must keep records of fungicide applications for 2 years and must make them available to inspectors upon request.

(C) *Testing.* Samples must be labeled and sent for testing to a laboratory approved by APHIS and must be tested using a test method approved by APHIS, in accordance with § 301.92–12.

(D) *Annual certification.* If all plant samples tested in accordance with this section and § 301.92–12 return negative results for *Phytophthora ramorum*, an inspector may certify that the nursery is free of evidence of *Phytophthora ramorum* infestation at the time of the inspection, and the nursery is eligible to enter into or maintain its compliance agreement in accordance with § 301.92–6.

(ii) *Pre-shipment inspection, sampling, and testing*—(A) *Inspection.* During the 30 days prior to interstate movement from a nursery in a quarantine area, regulated articles or associated articles intended for interstate movement must be inspected for symptoms of *Phytophthora ramorum* by an inspector. Inspection will focus on, but not be limited to, regulated articles and associated articles. No inspections of shipments will be conducted unless the nursery from which the shipment originates has a current and valid annual certification in accordance with this section.

(1) If no symptomatic plants are found upon inspection, the shipment may be considered free from evidence of *Phytophthora ramorum* and is eligible for interstate movement, provided that the nursery is operating under a compliance agreement with APHIS in accordance with § 301.92–6.

(2) If symptomatic plants are found upon inspection, the inspector will collect at least one sample per symptomatic plant, and one sample per regulated article or associated article that is in close proximity to, or that has had physical contact with, a symptomatic plant.

(B) *Testing and withholding from interstate movement.* Samples taken in accordance with this paragraph (a)(1) must be labeled and sent for testing to a laboratory approved by APHIS and must be tested using a test method approved by APHIS, in accordance with § 301.92–12. The interstate movement of plants in the shipment is prohibited until the plants in the shipment are determined to be free of evidence of *Phytophthora ramorum* infection in accordance with § 301.92–12.

(2) *Nurseries in which Phytophthora ramorum has been detected since March 31, 2011.* To meet the requirements of § 301.92–5(a)(1)(iv), nurseries that are located in quarantined areas, that move regulated articles of nursery stock, decorative trees without roots, wreaths, garlands, or greenery, associated articles, or non-host nursery stock interstate, and in which *Phytophthora ramorum* has been de-

tected since March 31, 2011, must meet the following requirements. Any such nurseries in quarantined areas that do not meet the following requirements are prohibited from moving regulated articles and associated articles interstate. Any such nurseries in quarantined areas that do not meet the following requirements or those in paragraph (b) of this section are prohibited from moving non-host nursery stock interstate.

(i) *Inspections.* The nursery must be inspected at least twice annually for symptoms of *Phytophthora ramorum* infestation by an inspector. The inspection will focus on regulated plants and other potential sources of *Phytophthora ramorum* inoculum.

(ii) *Sampling.* Samples must be taken from host plants, soil, standing water, drainage water, water for irrigation, and any other articles determined by the inspector to be possible sources of *Phytophthora ramorum* inoculum. The number of samples taken may vary depending on the possible sources of inoculum identified at the nursery, as well as the number of host articles in the nursery.

(iii) *Testing.* Samples must be labeled and sent for testing to a laboratory approved by APHIS and must be tested using a test method approved by APHIS in accordance with § 301.92–12.

(iv) *Negative results; certification.* If all samples tested in accordance with this section and § 301.92–12 return negative results for *Phytophthora ramorum*, an inspector may certify that the nursery is free of *Phytophthora ramorum* at the time of the inspection. If the nursery is inspected and determined by an inspector to be free of *Phytophthora ramorum* inoculum each time it is inspected for 3 consecutive years, the nursery will thereafter be inspected in accordance with paragraph (a)(1) of this section.

(v) *Positive results.* If any samples tested in accordance with this section and § 301.92–12 return positive results for *Phytophthora ramorum*, the nursery may ship lots of regulated, restricted, and associated articles interstate pursuant to § 301.92–5(b) only if the lot is determined to be free from *Phytophthora ramorum* inoculum. The method for this determination will be

155

specified in the nursery's compliance agreement with APHIS.

(b) *Nurseries in quarantined areas shipping non-host nursery stock interstate.* Nurseries located in quarantined areas and that move non-host nursery stock interstate must meet the requirements of this paragraph or the requirements of paragraph (a) of this section. If such nurseries contain any regulated or restricted articles, the nursery must meet the requirements of paragraph (a) of this section. This paragraph (b) only applies if there are no regulated or associated articles or nursery stock at the nursery. Nurseries that do not meet the requirements of paragraph (a) of this section or this paragraph (b) are prohibited from moving non-host nursery stock interstate.

(1) *Annual visual inspection.* The nursery must be visually inspected annually for symptoms of *Phytophthora ramorum*. Inspections and determinations of freedom from evidence of *Phytophthora ramorum* infestation must occur at the time when the best expression of symptoms is anticipated.

(2) *Sampling.* All plants showing symptoms of infection with *Phytophthora ramorum* upon inspection will be sampled and tested in accordance with § 301.92–12. If symptomatic plants are found upon inspection, the following plants must be withheld from interstate shipment until testing is completed and the nursery is found free of evidence of *Phytophthora ramorum* in accordance with this paragraph (b) and § 301.92–12: All symptomatic plants, any plants located in the same lot as the suspect plant, and any plants located within 2 meters of this lot of plants.

(3) *Certification.* If all plant samples tested in accordance with this section and § 301.92–12 return negative results for *Phytophthora ramorum*, or if an inspector at the nursery determines that plants in a nursery exhibit no signs of infection with *Phytophthora ramorum*, the inspector may certify that the nursery free of evidence of *Phytophthora ramorum* infestation at the time of inspection. Certification is valid for 1 year and must be renewed each year to continue shipping plants interstate.

(c) *Regulated establishments shipping regulated, restricted, or associated articles*

of interstate—(1) *Inspections.* To meet the conditions of § 301.92–5(b), the regulated establishment must be inspected at least twice annually for symptoms of *Phytophthora ramorum* infestation by an inspector. The inspection will focus on regulated plants and other potential sources of *Phytophthora ramorum* inoculum.

(2) *Sampling.* Samples must be taken from host plants, soil, standing water, drainage water, water for irrigation, growing media, and any other articles determined by the inspector to be possible sources of *Phytophthora ramorum* inoculum. The number of samples taken may vary depending on the possible sources of inoculum identified at the nursery, as well as the number of host articles in the nursery.

(3) *Testing.* Samples must be labeled and sent for testing to a laboratory approved by APHIS and must be tested using a test method approved by APHIS in accordance with § 301.92–12.

(4) *Negative results; certification.* If all samples tested in accordance with this section and § 301.92–12 return negative results for *Phytophthora ramorum*, an inspector may certify that the nursery is free of *Phytophthora ramorum* at the time of the inspection. For purposes of § 301.92–5(b), regulated, restricted, and associated articles at a certified nursery are considered free from *Phytophthora ramorum* until the time of the next inspection.

(5) *Positive results.* If any samples tested in accordance with this section and § 301.92–12 return positive results for *Phytophthora ramorum*, the nursery may ship lots of regulated, restricted, and associated articles interstate pursuant to § 301.92–5(b) only if the lot is determined to be free from *Phytophthora ramorum* inoculum. The method for this determination will be specified in the nursery's compliance agreement with APHIS.

(Approved by the Office of Management and Budget under control number 0579–0310)

[84 FR 16194, Apr. 18, 2019]

§ 301.92–12   Testing protocols.

Samples must be analyzed using a methodology approved by APHIS at a laboratory approved by APHIS. The following methodology is approved by APHIS.

(a) *Optional ELISA Prescreening.* An APHIS-approved ELISA may be used to prescreen samples to determine the presence of *Phytophthora* spp.

(1) *Negative prescreening results.* If all samples from a single nursery are found to be negative through APHIS-approved ELISA prescreening, no further testing is required. The nursery may be considered free of evidence of *Phytophthora ramorum,* and plants in the nursery are eligible for interstate movement under certificate in accordance with § 301.92–5.

(2) *Positive prescreening results.* If ELISA prescreening reveals the presence of *Phytophthora* spp. in any plants, each sample that returns positive ELISA results must be tested as provided in paragraph (b) of this section.

(b) *Mandatory testing procedures.* If ELISA prescreening is not performed, or if results of ELISA prescreening are positive for *Phytophthora* spp. in any sample, the sample must be analyzed using an APHIS-approved test. Samples will be considered positive for *Phytophthora ramorum* based on positive results of any approved test. Positive PCR or other molecular tests do not require confirmatory culture tests, nor do positive culture tests require confirmatory PCR or other molecular tests; however, if culture tests return other than positive results, an APHIS-approved PCR or other molecular test must be conducted, as provided in paragraph (b)(1) of this section.

(1) *PCR or other molecular tests*—(i) *Negative results.* If the results of PCR or other molecular tests are negative for all samples in a nursery, no further testing is required. The nursery may be considered free of evidence of *Phytophthora ramorum* and plants in the nursery are eligible for interstate movement under certificate in accordance with § 301.92–5.

(ii) *Positive results.* If any samples tested using PCR or other molecular tests return positive results for *Phytophthora ramorum,* the nursery from which they originate is prohibited from moving plants interstate. The nursery will be eligible to ship certain plants interstate when an inspector de-

termines that those plants are free of evidence of *Phytophthora ramorum.*

(2) *Culture Test*—(i) *Negative results.* If the results of culture tests are other than positive for any samples taken from a single nursery, plants in the nursery must continue to be withheld from shipment in accordance with § 301.92–11 and each plant sample must be tested again using a PCR or other molecular test, as described in this section.

(ii) *Positive results.* If any culture tests return positive results for *Phytophthora ramorum,* the nursery from which they originate is prohibited from moving plants interstate as directed by an inspector. The nursery will be eligible to ship certain plants interstate when an inspector determines that those plants are free of evidence of *Phytophthora ramorum.*

(c) *Other test methods.* Other test methods may be acceptable if approved by APHIS.

[72 FR 8597, Feb. 27, 2007, as amended at 84 FR 16195, Apr. 18, 2019]

PART 302—DISTRICT OF COLUMBIA; MOVEMENT OF PLANTS AND PLANT PRODUCTS

Sec.
302.1 Definitions.
302.2 Movement of plants and plant products.

AUTHORITY: 7 U.S.C. 7701–7772 and 7781–7786; 7 CFR 2.22, 2.80, and 371.3.

SOURCE: 66 FR 1016, Jan. 5, 2001, unless otherwise noted.

§ 302.1 Definitions.

Inspector. Any employee of the Animal and Plant Health Inspection Service or other person authorized by the Administrator to inspect and certify the plant health status of plants and products under this part.

Interstate. From any State into or through any other State.

State. The District of Columbia, Puerto Rico, the Northern Mariana Islands, or any State, territory, or possession of the United States.

§ 302.2 Movement of plants and plant products.

Inspection or documentation of the plant health status of plants or plant products to be moved interstate from the District of Columbia may be obtained by contacting the State Plant Health Director, Plant Protection and Quarantine, APHIS, Wayne A. Cawley, Jr. Building, Room 350, 50 Harry S. Truman Parkway, Annapolis, MD 21401–7080; phone: (410) 224–3452; fax: (410) 224–1142.

[66 FR 54641, Oct. 30, 2001]

PART 305—PHYTOSANITARY TREATMENTS

AUTHORITY: 7 U.S.C. 7701–7772 and 7781–7786; 21 U.S.C. 136 and 136a; 7 CFR 2.22, 2.80, and 371.3.

SOURCE: 75 FR 4241, Jan. 26, 2010, unless otherwise noted.

§ 305.1 Definitions.

Administrator. The Administrator, Animal and Plant Health Inspection Service, United States Department of Agriculture, or any person delegated to act for the Administrator in matters affecting this part.

APHIS. The Animal and Plant Health Inspection Service, United States Department of Agriculture.

Cold treatment. Exposure of a commodity to a specified cold temperature that is sustained for a specific time period to kill targeted pests, especially fruit flies.

Dose mapping. Measurement of absorbed dose within a process load using dosimeters placed at specified locations to produce a one-, two-, or three-dimensional distribution of absorbed dose, thus rendering a map of absorbed-dose values.

Dosimeter. A device that, when irradiated, exhibits a quantifiable change in some property of the device that can be related to absorbed dose in a given material using appropriate analytical instrumentation and techniques.

Dosimetry system. A system used for determining absorbed dose, consisting of dosimeters, measurement instruments and their associated reference standards, and procedures for the system's use.

Fumigant. A gaseous chemical that easily diffuses and disperses in air and is toxic to the target organism.

Fumigation. Releasing and dispersing a toxic chemical in the air so that it reaches the target organism in a gaseous state.

Inspector. Any individual authorized by the Administrator of APHIS or the Commissioner of Customs and Border Protection, Department of Homeland Security, to enforce the regulations in this part.

Irradiation. Treatment with any type of ionizing radiation.

Methyl bromide. A colorless, odorless biocide used to fumigate a wide range of commodities.

Neutralize. To prevent the establishment of a plant pest by killing it, sterilizing it, preventing its development from an immature stage, or preventing its emergence from its host.

Plant Protection and Quarantine (PPQ). The Plant Protection and Quarantine program of APHIS.

PPQ Treatment Manual. A document that contains treatment schedules that are approved by the Administrator for use under this part. The Treatment Manual is available on the Internet at (*http://www.aphis.usda.gov/import_export/plants/manuals/index.shtml*) or by contacting the Animal and Plant Health Inspection Service, Plant Protection and Quarantine, Manuals Unit, 92 Thomas Johnson Drive, Suite 200, Frederick, MD 21702.

Quick freeze. A commercially acceptable method of quick freezing at subzero temperatures with subsequent storage and transportation at not higher than 20 °F. Methods that accomplish this are known as quick freezing, sharp freezing, cold pack, or frozen pack, but may be any equivalent commercially acceptable freezing method.

Section 18 of Federal Insecticide, Fungicide, and Rodenticide Act (FIFRA). An emergency exemption granted by the U.S. Environmental Protection Agency to Federal or State agencies authorizing an unregistered use of a pesticide for a limited time.

Treatment facility. Any APHIS-certified place, warehouse, or approved enclosure where a treatment is conducted to mitigate a plant pest.

Vacuum fumigation. Fumigation performed in a gas-tight enclosure. Most air in the enclosure is removed and replaced with a small amount of fumigant. The reduction in pressure reduces the required duration of the treatment.

[75 FR 4241, Jan. 26, 2010, as amended at 76 FR 60360, Sept. 29, 2011; 83 FR 5876, Feb. 12, 2018]

§ 305.2 **Approved treatments.**

(a) Certain commodities or articles require treatment, or are subject to treatment, prior to interstate movement within the United States or importation or entry into the United States. Treatment is required as indicated in parts 301, 318, and 319 of this chapter, on a permit, or by an inspector.

(b) Treatments may only be administered in accordance with the requirements of this part and in accordance with treatment schedules approved by the Administrator as effective at neutralizing quarantine pests. The treatment schedules found in the PPQ Treatment Manual have been approved by the Administrator. Treatment schedules may be added to the PPQ Treatment Manual in accordance with § 305.3. Treatment schedules may also be approved by the Administrator in accordance with paragraph (c) of this section.

(c) Persons who wish to have a treatment schedule approved by the Administrator as effective at neutralizing a quarantine pest or pests may apply for approval by submitting the treatment schedule, along with any supporting information and data, to the Animal and Plant Health Inspection Service, Plant Protection and Quarantine, Center for Plant Health Science and Technology, 1730 Varsity Drive, Suite 400, Raleigh, NC 27606–5202. Upon receipt of such an application, the Administrator will review the schedule and the supporting information and data and respond with approval or denial of the treatment schedule. If the Administrator determines the treatment schedule to be of potential general use, the Administrator may add the new treatment schedule to the PPQ Treatment Manual or revise an existing schedule, as appropriate, in accordance with § 305.3.

(d) APHIS is not responsible for losses or damages incurred during treatment and recommends that a sample be treated first before deciding whether to treat the entire shipment.

[75 FR 4241, Jan. 26, 2010, as amended at 76 FR 60360, Sept. 29, 2011]

§ 305.3 **Processes for adding, revising, or removing treatment schedules in the PPQ Treatment Manual.**

(a) *Normal process for adding, revising, or removing treatment schedules.* Unless there is a need to immediately add, revise, or remove a treatment schedule, as provided in paragraph (b)(1) of this section, a treatment schedule may be added to the PPQ Treatment Manual, revised, or removed from the PPQ Treatment Manual as follows:

(1) *Notice of change to treatment schedule.* APHIS will publish in the FEDERAL REGISTER a notice describing the reasons we have determined that it is necessary to add, revise, or remove a treatment schedule and, if necessary, making available the new or revised treatment schedule as it would be added to the PPQ Treatment Manual. In our notice, we will provide for a public comment period on the new or revised treatment schedule or on the removal of the treatment schedule from the PPQ Treatment Manual.

(2) *Response to comments.* (i) APHIS will issue a notice after the close of the public comment period indicating that the treatment schedule specified in the initial notice will be added to the PPQ Treatment Manual, revised as described in the notice, or removed from the PPQ Treatment Manual if:

(A) No comments were received on the notice;

(B) The comments on the notice supported our action; or

(C) The comments on the notice were evaluated but did not change our determination that it is necessary to add, revise, or remove the treatment schedule, as described in the notice.

(ii) If the notice issued after the close of the public comment period indicates that a change will be made to the PPQ Treatment Manual, APHIS will make available a new version of the PPQ Treatment Manual that reflects the addition, revision, or removal of the particular treatment schedule.

(iii) If comments present information that causes us to determine that the change described in the notice is not appropriate, APHIS will issue a notice informing the public of this determination after the close of the comment period.

(b) *Process for immediately adding, revising, or removing treatment schedules.* Treatment schedules may be immediately added to the PPQ Treatment Manual, revised, or removed from the PPQ Treatment Manual under the circumstances described in paragraph (b)(1) of this section and in accordance with the process described in paragraphs (b)(2) and (b)(3) of this section.

(1) *Circumstances in which the immediate process may be used.* Treatment schedules may be immediately added to the PPQ Treatment Manual, revised, or removed from the PPQ Treatment Manual if any of the following circumstances apply:

(i) PPQ has determined that an approved treatment schedule is ineffective at neutralizing the targeted plant pest(s);

(ii) PPQ has determined that, in order to neutralize the targeted plant pest(s), the treatment schedule must be administered using a different process than was previously used;

(iii) PPQ has determined that a new treatment schedule is effective, based on efficacy data, and that ongoing trade in an article or articles may be adversely impacted unless the new treatment schedule is approved for use; or

(iv) The use of a treatment schedule is no longer authorized by the U.S. Environmental Protection Agency or by any other Federal entity.

(2) *Process for immediate change to treatment schedules.* If PPQ determines that one or more of the circumstances in paragraph (b)(1) of this section applies and that it is necessary to take immediate action, APHIS will publish in the FEDERAL REGISTER a notice describing the reasons we have determined that it is necessary to immediately add, revise, or remove a treatment schedule and, if necessary, making available the new or revised treatment schedule as it has been added to the PPQ Treatment Manual. Treatment schedules that have been added to the PPQ Treatment Manual or revised under this process will be identified in the PPQ Treatment Manual as having been added or revised through the immediate process described in this paragraph (b). The PPQ Treatment Manual will indicate that these treatment schedules are subject to change or removal based on public comment. In our notice, we will provide for a public comment period on the new or revised treatment schedule or on the removal of the treatment schedule from the PPQ Treatment Manual.

(3) *Response to comments.* (i) APHIS will issue a notice after the close of the public comment period affirming the action described in the initial notice if:

(A) No comments were received on the notice;

(B) The comments on the notice supported our action; or

(C) The comments on the notice were evaluated but did not change our determination that it was necessary to add, revise, or remove the treatment schedule, as described in the notice.

(ii) If the notice issued after the close of the public comment period indicates that the initial change to the PPQ Treatment Manual is affirmed, APHIS will make available a new version of the PPQ Treatment Manual that will reflect the addition, revision, or removal of the particular treatment schedule in the main body of the PPQ Treatment Manual.

(iii) If comments present information that causes us to determine that it is necessary to change a treatment schedule added to the PPQ Treatment Manual under this process or to further revise a treatment schedule that was revised under this process, APHIS will publish a notice in the FEDERAL REGISTER informing the public of this determination after the close of the comment period and will revise the treatment schedule accordingly.

(iv) If comments present information that causes us to determine that the change described in the initial notice was not appropriate, APHIS will publish a notice in the FEDERAL REGISTER informing the public of this determination after the close of the comment period and will, if necessary, remove the new or revised treatment schedule from the separate section of the PPQ Treatment Manual.

§305.4 Monitoring and certification of treatments.

(a) All treatments approved under part 305 are subject to monitoring and verification by APHIS.

(b) Any treatment performed outside the United States must be monitored and certified by an inspector or an official authorized by APHIS. During the entire interval between treatment and export, the consignment must be stored and handled in a manner that prevents any infestation by pests and noxious weeds.

§305.5 Chemical treatment requirements.

(a) *Certified facility.* The fumigation treatment facility must be certified by APHIS. Facilities are required to be inspected and recertified annually, or as often as APHIS directs, depending upon treatments performed, commodities handled, and operations conducted at the facility. In order to be certified, a fumigation facility must:

(1) Be capable of administering the required dosage range for the required duration and at the appropriate temperature, as specified in the treatment schedules in the PPQ Treatment Manual or in another treatment schedule approved in accordance with §305.2.

(2) Be adequate to contain the fumigant and be constructed from material that is not reactive to the fumigant.

(3) For vacuum fumigation facilities, be constructed to withstand required negative pressure.

(b) *Monitoring.* Treatment must be monitored by an official authorized by APHIS to ensure proper administration of the treatment, including that the correct amount of gas reaches the target organism and that an adequate number and placement of blowers, fans, sampling tubes, or monitoring lines are used in the treatment enclosure. An official authorized by APHIS approves, adjusts, or rejects the treatment.

(c) *Compliance agreements.* Any person who conducts a fumigation in the United States or operates a facility where fumigation is conducted in the United States for phytosanitary purposes must sign a compliance agreement with APHIS.

(1) *Fumigation treatment facilities treating imported articles; compliance agreements with facility operators for fumigation in the United States.* If fumigation treatment of imported articles is conducted in the United States, the fumigation treatment facility operator or the person who conducts fumigation must sign a compliance agreement with APHIS. The fumigation facility operator or the person who conducts fumigation must agree to comply with the requirements of this section and any additional requirements found necessary by APHIS to prevent the escape of any pests of concern that may be associated with the articles to be treated.

(2) *Fumigation treatment facilities treating articles moved interstate from Hawaii and U.S. territories.* Fumigation treatment facilities treating articles moved interstate from Hawaii and U.S. territories must complete a compliance agreement with APHIS as provided in §318.13–3(d) of this chapter.

(3) *Fumigation treatment facilities treating articles moved interstate from areas quarantined for fruit flies.* Fumigation treatment facilities treating articles moved interstate from areas quarantined for fruit flies must complete a compliance agreement with APHIS as provided in §301.32–6 of this chapter.

(4) *Fumigation treatment facilities treating articles moved interstate from areas quarantined for Asian citrus psyllid.* Fumigation treatment facilities treating articles moved interstate from areas quarantined only for Asian citrus psyllid, and not for citrus greening, must complete a compliance agreement with APHIS as provided in §301.76–8 of this chapter.

(d) *Treatment procedures.* (1) To kill the pest, all chemical applications must be administered in accordance with an Environmental Protection Agency (EPA) approved pesticide label and the APHIS-approved treatment

161

schedule prescribed in the PPQ Treatment Manual or in another treatment schedule approved in accordance with § 305.2. If EPA cancels approval for the use of a pesticide on a commodity, then the treatment schedule prescribed in the PPQ Treatment Manual or approved in accordance with § 305.2 is no longer authorized for that commodity. If the commodity is not listed on the pesticide label and/or included in a Federal quarantine or crisis exemption in accordance with FIFRA section 18, then no chemical treatment is available.

(2) Temperature/concentration readings must be taken for items known to be sorptive or whose sorptive properties are unknown when treatment is administered in chambers at normal atmospheric pressure.

(3) Unless otherwise specified in the PPQ Treatment Manual or in another approved treatment schedule, the volume of the commodity stacked inside the treatment enclosure must not exceed ⅔ of the volume of the enclosure. Stacking must be approved by an official authorized by APHIS before treatment begins. All commodities undergoing treatment must be listed on the label or authorized under Section 18 of FIFRA.

(4) Recording and measuring equipment must be adequate to accurately monitor the gas concentration, to ensure the correct amount of gas reaches the pests, and to detect any leaks in the enclosure. At least three sampling tubes or monitoring lines must be used in the treatment enclosure.

(5) An adequate number of blowers or fans must be used inside of the treatment enclosure to uniformly distribute gas throughout the enclosure. The circulation system must be able to recirculate the entire volume of gas in the enclosure in 3 minutes or less.

(6) The exposure period begins after all gas has been introduced.

(7) For vacuum fumigation: The vacuum pump must be able to reduce pressure in the treatment enclosure to 1–2 inches of mercury in 15 minutes or less.

(Approved by the Office of Management and Budget under control number 0579–0450)

[75 FR 4241, Jan. 26, 2010, as amended at 76 FR 60361, Sept. 29, 2011; 83 FR 5876, Feb. 12, 2018]

§ 305.6 Cold treatment requirements.

(a) *Certification of treatment facilities.* All facilities or locations used for refrigerating fruits or vegetables in accordance with the cold treatment schedules in the PPQ Treatment Manual or in another treatment schedule approved in accordance with § 305.2 must be certified by APHIS. Recertification of the facility or carrier is required every 3 years, or as often as APHIS directs, depending on treatments performed, commodities handled, and operations conducted at the facility. A facility will only be certified or recertified if the Administrator determines that the location of the facility is such that those Federal agencies involved in its operation and oversight have adequate resources to conduct the necessary operations at the facility, that the pest risks can be managed at that location, and that the facility meets all criteria for approval. Other agencies that have regulatory oversight and requirements must concur in writing with the establishment of the facility prior to APHIS approval. In order to be certified, facilities and carriers must:

(1) Be capable of keeping treated and untreated fruits, vegetables, or other articles separate so as to prevent reinfestation of articles and spread of pests;

(2) Be capable of preventing the escape and spread of pests while regulated articles are at the facility; and

(3) Have equipment that is adequate to effectively perform cold treatment.

(b)(1) *Location of facilities.* Where certified cold treatment facilities are available, an approved cold treatment may be conducted for any imported regulated article either prior to shipment to the United States or in the United States. For any regulated article moved interstate from Hawaii or U.S. territories, cold treatment may be conducted either prior to movement to the mainland United States or in the mainland United States. Cold treatment facilities may be located in any State on the mainland United States. For cold treatment facilities located in the area south of 39° latitude and west of 104° longitude, the following additional conditions must be met:

(i) Prospective facility operators must submit a detailed layout of the facility site and its location to APHIS. APHIS will evaluate plant health risks based on the proposed location and layout of the facility site. APHIS will only approve a proposed facility if the Administrator determines that regulated articles can be safely transported to the facility from the port of entry or points of origin in the United States.

(ii) The government of the State in which the facility is to be located must concur in writing with the location of the facility or, if it does not concur, must provide a written explanation of concern based on pest risks. In instances where the State government does not concur with the proposed facility location, and provides a written explanation of concern based on pest risks, APHIS and the State must agree on a strategy to resolve the pest risk concerns prior to APHIS approval. If the State does not provide a written explanation of concern based on pest risks, then State concurrence will not be required before APHIS approves the facility location.

(iii) Untreated articles may not be removed from their packaging prior to treatment under any circumstances.

(iv) The facility must have contingency plans, approved by APHIS, for safely destroying or disposing of regulated articles if the facility is unable to properly treat a shipment.

(v) The facility may only treat articles approved by APHIS for treatment at the facility. Approved articles will be listed in the compliance agreement required in paragraph (f) of this section.

(vi) Arrangements for treatment must be made before the departure of a consignment from its port of entry or points of origin in the United States. APHIS and the facility must agree on all parameters, such as time, routing, and conveyance, by which the consignment will move from the port of entry or points of origin in the United States to the treatment facility. If APHIS and the facility cannot reach agreement in advance on these parameters then no consignments may be moved to that facility until an agreement has been reached.

(vii) Regulated articles must be conveyed to the facility in a refrigerated (via motorized refrigeration equipment) conveyance at a temperature that minimizes the mobility of the pests of concern for the article.

(viii) The facility must apply all post-treatment safeguards required for certification under paragraph (a) of this section before releasing the articles.

(ix) The facility must remain locked when not in operation.

(x) The facility must maintain and provide APHIS with an updated map identifying places where horticultural or other crops are grown within 4 square miles of the facility. Proximity of host material to the facility will necessitate trapping or other pest monitoring activities, funded by the facility, to help prevent establishment of any escaped pests of concern, as approved by APHIS; these activities will be listed in the compliance agreement required in paragraph (f) of this section. The treatment facility must have a pest management plan within the facility.

(xi) The facility must comply with any additional requirements including, but not limited to, the use of pest-proof packaging and container seals, that APHIS may require to prevent the escape of plant pests during transport to and from the cold treatment facility itself, for a particular facility based on local conditions, and for any other risk factors of concern. These activities will be listed in the compliance agreement required in paragraph (f) of this section.

(2) For articles that are moved interstate from areas quarantined for fruit flies, cold treatment facilities may be located either within or outside of the quarantined area. If the articles are treated outside the quarantined area, they must be accompanied to the facility by a limited permit issued in accordance with §301.32–5(b) of this chapter and must be moved in accordance with any safeguards determined to be appropriate by APHIS.

(c) *Cold treatment enclosures.* All enclosures, in which cold treatment is performed, including refrigerated containers, must:

(1) Be capable of maintaining the highest temperature of the treatment schedule under which the fruit will be treated specified in the PPQ Treatment Manual or in another approved treatment schedule before the treatment begins and holding fruit at or below the treatment temperature during the treatment.

(2) Maintain fruit pulp temperatures according to treatment schedules with no more than a 0.39 °C (0.7 °F) variation in temperature between two consecutive hourly readings.

(3) Be structurally sound and adequate to maintain required temperatures.

(d) *Treatment procedures.* (1) All material, labor, and equipment for cold treatment performed on a vessel must be provided by the vessel or vessel agent. An official authorized by APHIS monitors, manages, and advises in order to ensure that the treatment procedures are followed.

(2) Refrigeration must be completed in the container, compartment, or room in which it is begun.

(3) Fruit that may be cold treated must be safeguarded to prevent cross-contamination or mixing with other infested fruit.

(4) Fruit intended for in-transit cold treatment must be precooled to no more than the highest temperature of the treatment schedule under which the fruit will be treated prior to beginning treatment. The in-transit treatment enclosure may not be used for precooling unless an official authorized by APHIS approves the loading of the fruit in the treatment enclosure as adequate to allow for fruit pulp temperatures to be taken prior to beginning treatment. If the fruit is precooled outside the treatment enclosure, an official authorized by APHIS will take pulp temperatures manually from a sample of the fruit as the fruit is loaded for in-transit cold treatment to verify that precooling was completed. If the pulp temperatures for the sample are 0.28 °C (0.5 °F) or more above the highest temperature of the treatment schedule under which the fruit will be treated, the pallet from which the sample was taken will be rejected and returned for additional precooling until the fruit reaches the highest tempera-

ture of the treatment schedule under which the fruit will be treated. If fruit is precooled in the treatment enclosure, or if treatment is conducted at a cold treatment facility in the United States, the fruit must be precooled to the highest temperature of the treatment schedule under which the fruit will be treated, as verified by an official authorized by APHIS, prior to beginning treatment.

(5) Breaks, damage, etc., in the treatment enclosure that preclude maintaining correct temperatures must be repaired before the enclosure is used. An official authorized by APHIS must approve loading of compartment, number and placement of temperature probes or sensors, and initial fruit temperature readings before beginning the treatment. Hanging decks and hatch coamings within vessels may not be used as enclosures for in-transit cold treatment without prior written approval from APHIS. Double-stacking of pallets is not allowed.

(6) Only the same type of fruit in the same type of package may be treated together in a container; no mixture of fruits in containers may be treated. A numbered seal must be placed on the doors of the loaded container and may be removed only at the port of destination by an official authorized by APHIS.

(7) Temperature recording devices used during treatment must be secured using measures approved by APHIS as adequate to ensure the security and integrity of cold treatment data. The devices must be able to record the date, time, and sensor number and automatic and continuous records of the temperature during all calibrations and during treatment. Recording devices must be capable of generating temperature charts for verification by an inspector. If records of calibrations or treatments are found to have been manipulated, the vessel or container in which the treatment is performed may be suspended from conducting cold treatments until proper equipment is installed and an official authorized by APHIS has recertified it. APHIS' decision to recertify a vessel or container will take into account the severity of the infraction that led to suspension.

(8) A minimum of four temperature probes or sensors is required for vessel holds used as treatment enclosures. A minimum of three temperature probes or sensors is required for other treatment enclosures. An official authorized by APHIS will have the option to require that additional temperature probes or sensors be used, depending on the size of the treatment enclosure.

(9) Fruit pulp temperatures must be maintained at the temperature specified in the treatment schedule with no more than a 0.39 °C (0.7 °F) variation in temperature between two consecutive hourly readings. Failure to comply with this requirement will result in invalidation of the treatment unless an official authorized by APHIS can verify that the pulp temperature was maintained at or below the treatment temperature for the duration of the treatment.

(10) The time required to complete the treatment begins when all temperature probes reach the prescribed cold treatment schedule temperature. Refrigeration continues until the vessel arrives at the port of destination and the fruit is released for unloading by an inspector even though this may prolong the period required for the cold treatment.

(11) Temperatures must be recorded at intervals no longer than 1 hour apart. Gaps of longer than 1 hour will invalidate the treatment or indicate treatment failure unless an official authorized by APHIS can verify that the pulp temperature was maintained at or below the treatment temperature for the duration of the treatment.

(12) Cold treatment is not completed until so declared by an official authorized by APHIS or the certifying official of the foreign country; consignments of treated commodities may not be discharged until APHIS clearance has been fully completed, including review and approval of treatment record charts.

(13) Cold treatment of fruits in break bulk vessels or containers must be initiated by an official authorized by APHIS if there is not a treatment technician who has been trained to initiate cold treatments for either break bulk vessels or containers.

(14) An official authorized by APHIS may perform audits to ensure that the treatment procedures comply with the regulations in this section and that the treatment is administered in accordance with the treatment schedules in the PPQ Treatment Manual or in accordance with another approved treatment schedule. The official authorized by APHIS must be given the appropriate materials and access to the facility, container, or vessel necessary to perform the audits.

(15) An inspector will sample and cut fruit from each consignment after it has been cold treated to monitor treatment effectiveness. If a single live pest of concern in any stage of development is found, the consignment will be held until an investigation is completed and appropriate remedial actions have been implemented. If APHIS determines at any time that the safeguards contained in this section do not appear to be effective against the pests of concern, APHIS may suspend the importation of fruits from the originating country and conduct an investigation into the cause of the deficiency. APHIS may waive the sampling and cutting requirement of paragraph (d)(15) of this section, provided that the national plant protection organization (NPPO) of the exporting country has conducted such sampling and cutting in the exporting country as part of a biometric sampling protocol approved by APHIS.

(16) The cold treatments required for the entry of fruit are considered necessary for the elimination of plant pests, and no liability shall attach to the U.S. Department of Agriculture or to any officer or representative of that Department in the event injury results to fruit offered for entry in accordance with these instructions. In prescribing cold treatments of certain fruits, it should be emphasized that inexactness and carelessness in applying the treatments may result in injury to the fruit or its rejection for entry.

(e) *Monitoring.* Treatment must be monitored by an inspector to ensure proper administration of the treatment. An inspector must also approve the recording devices and sensors used to monitor temperatures and conduct an operational check of the equipment before each use and ensure sensors are

165

calibrated. An inspector may approve, adjust, or reject the treatment. Facilities must be located within the local commuting area for APHIS employees for inspection purposes. Facilities treating imported articles must also be located within an area over which the U.S. Department of Homeland Security is assigned authority to accept entries of merchandise, to collect duties, and to enforce the provisions of the customs and navigation laws in force.

(f) *Compliance agreements.* Any person who operates a facility where cold treatment is conducted for phytosanitary purposes must sign a compliance agreement with APHIS.

(1) *Compliance agreements with importers and facility operators for cold treatment in the United States.* If cold treatment of imported articles is conducted in the United States, both the importer and the operator of the cold treatment facility or the person who conducts the cold treatment must sign compliance agreements with APHIS. In the importer compliance agreement, the importer must agree to comply with any additional requirements found necessary by APHIS to ensure the shipment is not diverted to a destination other than an approved treatment facility and to prevent escape of plant pests from the articles to be treated during their transit from the port of first arrival to the cold treatment facility in the United States. In the facility compliance agreement, the facility operator or person conducting the cold treatment must agree to comply with the requirements of this section and any additional requirements found necessary by APHIS to prevent the escape of any pests of concern that may be associated with the articles to be treated.

(2) *Compliance agreements with cold treatment facilities outside the United States.* If cold treatment of imported articles is conducted outside the United States, the operator of the cold treatment facility must sign a compliance agreement or an equivalent agreement with APHIS and the NPPO of the country in which the facility is located. In this agreement, the facility operator must agree to comply with the requirements of this section, and the NPPO of the country in which the facility is located must agree to mon-

itor that compliance and inform the Administrator of any noncompliance.

(3) *Cold treatment facilities treating articles moved interstate from Hawaii and U.S. territories.* Cold treatment facilities treating articles moved interstate from Hawaii and the U.S. territories must complete a compliance agreement with APHIS as provided in § 318.13–3(d) of this chapter.

(Approved by the Office of Management and Budget under control number 0579–0450)

[75 FR 4241, Jan. 26, 2010, as amended at 75 FR 52217, Aug. 25, 2010; 76 FR 60361, Sept. 29, 2011; 78 FR 63374, Oct. 24, 2013; 83 FR 5876, Feb. 12, 2018]

§ 305.7 Quick freeze treatment requirements.

Quick freeze treatment for fruits and vegetables imported into the United States or moved interstate from Hawaii or Puerto Rico must be conducted in accordance with §§ 319.56-12 or 318.13-13, respectively, of this chapter. The PPQ Treatment Manual indicates fruits and vegetables for which quick freeze is an authorized treatment. Requests to authorize quick freeze as a treatment for other fruits and vegetables may be made in accordance with § 305.2(c).

[75 FR 4241, Jan. 26, 2010, as amended at 76 FR 60361, Sept. 29, 2011]

§ 305.8 Heat treatment requirements.

(a) *Certified facility.* The treatment facility must be certified by APHIS. Recertification is required annually, or as often as APHIS directs, depending upon treatments performed, commodities handled, and operations conducted at the facility. In order to be certified, a heat treatment facility must:

(1) Have equipment that is capable of adequately circulating air or water (as relevant to the treatment), changing the temperature, and maintaining the changed temperature sufficient to meet the treatment schedule parameters in the PPQ Treatment Manual or in another treatment schedule approved in accordance with § 305.2.

(2) Have equipment used to record, monitor, or sense temperature, maintained in proper working order.

(3) Keep treated and untreated fruits, vegetables, or articles separate so as to

prevent reinfestation and spread of pests.

(b) *Monitoring.* Treatment must be monitored by an official authorized by APHIS to ensure proper administration of the treatment. An official authorized by APHIS approves, adjusts, or rejects the treatment.

(c) *Compliance agreements.* Facilities located in the United States must operate under a compliance agreement with APHIS. The compliance agreement must be signed by a representative of the heat treatment facilities located in the United States and APHIS. The compliance agreement must contain requirements for equipment, temperature, water quality, circulation, and other measures for performing heat treatments to ensure that treatments are administered properly. Compliance agreements must allow officials of APHIS to inspect the facility to monitor compliance with the regulations.

(d) *Workplans.* Facilities located outside the United States must operate in accordance with a workplan. The workplan must be signed by a representative of the heat treatment facilities located outside the United States, the national plant protection organization of the country of origin (NPPO), and APHIS. The workplan must contain requirements for equipment, temperature, water quality, circulation, and other measures to ensure that heat treatments are administered properly. Workplans for facilities outside the United States must include trust fund agreement information regarding payment of the salaries and expenses of APHIS employees on site. Workplans must allow officials of the NPPO and APHIS to inspect the facility to monitor compliance with APHIS regulations.

(e) *Treatment procedures.* (1) Before each treatment can begin, an official authorized by APHIS must approve the loading of the commodity in the treatment container.

(2) Sensor equipment must be adequate to monitor the treatment, its type and placement must be approved by an official authorized by APHIS, and the equipment must be tested by an official authorized by APHIS prior to beginning the treatment. Sensor equipment must be locked before each treatment to prevent tampering.

(3) Fruits, vegetables, or articles of substantially different sizes must be treated separately; oversized fruit may be rejected by an official authorized by APHIS.

(4) The treatment period begins when the temperature specified by the treatment schedule has been reached. An official authorized by APHIS may abort the treatment if the facility requires an unreasonably long time to achieve the required temperature.

[75 FR 4241, Jan. 26, 2010, as amended at 76 FR 60361, Sept. 29, 2011]

§305.9 Irradiation treatment requirements.

Irradiation, carried out in accordance with the provisions of this section, is approved as a treatment for any imported regulated article (i.e., fruits, vegetables, cut flowers, and foliage); for any regulated article moved interstate from Hawaii, Puerto Rico, the U.S. Virgin Islands, Guam, and the Commonwealth of the Northern Marianas Islands (referred to collectively, in this section, as Hawaii and U.S. territories); for any berry, fruit, nut, or vegetable listed as a regulated article in §301.32–2(a) of this chapter; and for any regulated article listed in 301.76–2 of this chapter and intended for consumption, as apparel or as a similar personal accessory, or for decorative use.

(a) *Location of facilities.* (1) Where certified irradiation facilities are available, an approved irradiation treatment may be conducted for any imported regulated article either prior to shipment to the United States or in the United States. For any regulated article moved interstate from Hawaii or U.S. territories, irradiation treatment may be conducted either prior to movement to the mainland United States or in the mainland United States. Irradiation facilities may be located in any State on the mainland United States. For irradiation facilities located in the States of Alabama, Arizona, California, Florida, Georgia, Kentucky, Louisiana, Mississippi, Nevada, New Mexico,

North Carolina, South Carolina, Tennessee, Texas, and Virginia, the following additional conditions must be met:

(i) Prospective facility operators must submit a detailed layout of the facility site and its location to APHIS. APHIS will evaluate plant health risks based on the proposed location and layout of the facility site. APHIS will only approve a proposed facility if the Administrator determines that regulated articles can be safely transported to the facility from port of entry or points of origin in the United States.

(ii) The government of the State in which the facility is to be located must concur in writing with the location of the facility or, if it does not concur, must provide a written explanation of concern based on pest risks. In instances where the State government does not concur with the proposed facility location, and provides a written explanation of concern based on pest risks, APHIS and the State must agree on a strategy to resolve the pest risk concerns prior to APHIS approval. If the State does not provide a written explanation of concern based on pest risks, then State concurrence will not be required before APHIS approves the facility location.

(iii) Untreated articles may not be removed from their packaging prior to treatment under any circumstances.

(iv) The facility must have contingency plans, approved by APHIS, for safely destroying or disposing of regulated articles if the facility is unable to properly treat a shipment.

(v) The facility may only treat articles approved by APHIS for treatment at the facility. Approved articles will be listed in the compliance agreement required in paragraph (c)(1)(i) of this section.

(vi) Arrangements for treatment must be made before the departure of a consignment from its port of entry or points of origin in the United States. APHIS and the facility must agree on all parameters, such as time, routing, and conveyance, by which the consignment will move from the port of entry or points of origin in the United States to the treatment facility. If APHIS and the facility cannot reach agreement in advance on these parameters then no consignments may be moved to that facility until an agreement has been reached.

(vii) Regulated articles must be conveyed to the facility in a refrigerated (via motorized refrigeration equipment or other methods including ice or insulation) or air-conditioned conveyance at a temperature that minimizes the mobility of the pests of concern for the article.

(viii) The facility must maintain and provide APHIS with an updated map identifying places where horticultural or other crops are grown within 4 square miles of the facility. Proximity of host material to the facility will necessitate trapping or other pest monitoring activities to help prevent establishment of any escaped pests of concern, as approved by APHIS; these activities will be listed in the compliance agreement required in paragraph (c)(1)(i) of this section. The treatment facility must have a pest management plan within the facility.

(ix) The facility must comply with any additional requirements that APHIS may require to prevent the escape of plant pests during transport to and from the irradiation facility itself, for a particular facility based on local conditions, and for any other risk factors of concern. These activities will be listed in the compliance agreement required in paragraph (c)(1)(i) of this section.

(2) For articles that are moved interstate from areas quarantined for fruit flies, irradiation facilities may be located either within or outside of the quarantined area. If the articles are treated outside the quarantined area, they must be accompanied to the facility by a limited permit issued in accordance with §301.32-5(b) of this chapter and must be moved in accordance with any safeguards determined to be appropriate by APHIS.

(3) For articles that are moved interstate from areas quarantined only for Asian citrus psyllid, and not for citrus greening, irradiation facilities must be located within an area that is not quarantined for citrus greening.

(b) *Approved facilities.* The irradiation treatment facility must be approved by APHIS. Other agencies that have regulatory oversight and requirements must concur in writing with the establishment of the facility prior to APHIS approval. In order to be approved, a facility must fulfill the requirements in paragraphs (c) and (d) of this section.

(c) *Compliance agreements.* Compliance agreements for facilities located in States listed in paragraph (a)(1) of this section may also contain additional provisions as described in paragraphs (a)(1)(i) through (a)(1)(ix) of this section. (1) *Irradiation facilities treating imported articles.* (i) *Compliance agreements with importers and facility operators for irradiation in the United States.* If irradiation of imported articles is conducted in the United States, both the importer and the operator of the irradiation facility must sign compliance agreements with APHIS. In the facility compliance agreement, the facility operator must agree to comply with any additional requirements found necessary by APHIS to prevent the escape, prior to irradiation, of any pests of concern that may be associated with the articles to be irradiated. In the importer compliance agreement, the importer must agree to comply with any additional requirements found necessary by APHIS to ensure the shipment is not diverted to a destination other than an approved treatment facility and to prevent escape of plant pests from the articles to be irradiated during their transit from the port of first arrival to the irradiation facility in the United States.

(ii) *Compliance agreement with irradiation facilities outside the United States.* If irradiation of imported articles is conducted outside the United States, the operator of the irradiation facility must sign a compliance agreement with APHIS and the national plant protection organization (NPPO) of the country in which the facility is located. In this agreement, the facility operator must agree to comply with the requirements of this section, and the NPPO of the country in which the facility is located must agree to monitor that compliance and to inform the Administrator of any noncompliance.

(2) *Irradiation facilities treating articles moved interstate from Hawaii and U.S. territories.* Irradiation facilities treating articles moved interstate from Hawaii and U.S. territories must com-

plete a compliance agreement with APHIS as provided in § 318.13-3(d) of this chapter.

(3) *Irradiation facilities treating articles moved interstate from areas quarantined for fruit flies.* Irradiation facilities treating articles moved interstate from areas quarantined for fruit flies must complete a compliance agreement with APHIS as provided in § 301.32-6 of this chapter.

(4) Irradiation facilities treating articles moved interstate from areas quarantined only for Asian citrus psyllid, and not for citrus greening, must complete a compliance agreement with APHIS as provided in § 301.76-8 of this chapter.

(d) *Certified facility.* The irradiation treatment facility must be certified by APHIS. Recertification is required in the event of an increase in the amount of radioisotope, a decrease in the amount of radioisotope for a reason other than natural decay, a major modification to equipment that affects the delivered dose, or a change in the owner or managing entity of the facility. Recertification also may be required in cases where a significant variance in dose delivery has been measured by the dosimetry system. In order to be certified, a facility must:

(1) Be capable of administering the minimum absorbed ionizing radiation doses specified in the PPQ Treatment Manual or in another treatment schedule approved in accordance with § 305.2 to the regulated articles;[1]

(2) Be constructed so as to provide physically separate locations for treated and untreated articles, except that articles traveling by conveyor directly into the irradiation chamber may pass through an area that would otherwise be separated. The locations must be separated by a permanent physical barrier such as a wall or chain link fence 6 or more feet high to prevent transfer of cartons, or some other means approved during certification to prevent reinfestation of articles and spread of pests.

---

[1] The maximum absorbed ionizing radiation dose and the irradiation of food is regulated by the Food and Drug Administration under 21 CFR part 179.

(3) If the facility is to be used to treat imported articles and is located in the United States, the facility will only be certified if APHIS determines that regulated articles will be safely transported to the facility from the port of arrival without significant risk that plant pests will escape in transit or while the regulated articles are at the facility.

(e) *Monitoring and interagency agreements.* Treatment must be monitored by an inspector. This monitoring will include inspection of treatment records and unannounced inspections of the facility by an inspector, and may include inspection of articles prior to or after irradiation. Facilities must be located within the local commuting area for APHIS employees for inspection purposes.

(1) *Irradiation facilities treating imported articles; irradiation treatment framework equivalency workplan.* Facilities shall be located within an area over which the U.S. Department of Homeland Security is assigned authority to accept entries of merchandise, to collect duties, and to enforce the provisions of the customs and navigation laws in force. The NPPO of a country from which articles are to be imported into the United States in accordance with this section must sign a framework equivalency workplan with APHIS. In this plan, both the NPPO and APHIS will specify the following items for their respective countries:

(A) Citations for any requirements that apply to the importation of irradiated fruits and vegetables;

(B) The type and amount of inspection, monitoring, or other activities that will be required in connection with allowing the importation of irradiated fruits and vegetables into that country; and

(C) Any other conditions that must be met to allow the importation of irradiated fruits and vegetables into that country.

(2) *Irradiation facilities located in foreign countries.* Facilities in foreign countries that carry out irradiation operations must notify the Director of Preclearance, PPQ, APHIS, 4700 River Road Unit 140, Riverdale, MD 20737-1236, of scheduled operations at least 30 days before operations commence, except where otherwise provided in the facility preclearance workplan. To ensure the appropriate level of monitoring, before articles may be imported in accordance with this section, the following agreements must be signed, in addition to the irradiation treatment framework equivalency workplan required in paragraph (e)(1) of this section:

(i) *Facility preclearance workplan.* Prior to commencing importation into the United States of articles treated at a foreign irradiation facility, APHIS and the NPPO of the country from which articles are to be imported must jointly develop a preclearance workplan that details the activities that APHIS and the foreign NPPO will carry out in connection with each irradiation facility to verify the facility's compliance with the requirements of this section. Typical activities to be described in this workplan may include frequency of visits to the facility by APHIS and foreign plant protection inspectors, methods for reviewing facility records, and methods for verifying that facilities are in compliance with the requirements for separation of articles, packaging, labeling, and other requirements of this section. This facility preclearance workplan will be reviewed and renewed by APHIS and the foreign NPPO on an annual basis.

(ii) *Trust fund agreement.* Irradiated articles may be imported into the United States in accordance with this section only if the NPPO of the country in which the irradiation facility is located or a private export group has entered into a trust fund agreement with APHIS. That agreement requires the NPPO or the private export group to pay, in advance of each shipping season, all costs that APHIS estimates it will incur in providing inspection and treatment monitoring services at the irradiation facility during that shipping season. Those costs include administrative expenses and all salaries (including overtime and the Federal share of employee benefits), travel expenses (including per diem expenses), and other incidental expenses incurred by APHIS in performing these services. The agreement will describe the general nature and scope of APHIS services provided at irradiation facilities covered by the agreement, such as whether APHIS inspectors will monitor operations continuously or intermittently, and will generally describe the extent of inspections APHIS will perform on articles prior to and after irradiation. The agreement requires the NPPO or private export group to deposit a certified or cashier's check with APHIS for the amount of those costs, as estimated by APHIS. If the deposit is not sufficient to meet all costs incurred by APHIS, the agreement further requires the NPPO or the private export group to deposit with APHIS a certified or cashier's check for the amount of the remaining costs, as determined by APHIS, before any more articles irradiated in that country may be imported into the United States. After a final audit at the conclusion of each shipping season, any overpayment of funds would be returned to the NPPO or the private export group or held on account until needed, at the option of the NPPO or the private export group.

(3) *Irradiation facilities located within the United States.* Facilities located within the United States must notify an inspector at least 24 hours (excluding Saturday, Sunday, and Federal holidays) before scheduled operations.[2] If the facility will be used to treat imported articles, the NPPO of the country from which the articles are to be imported into the United States in accordance with this section must also sign the irradiation treatment framework equivalency workplan required in paragraph (e)(1) of this section.

(f) *Packaging.* Articles that are irradiated in accordance with this section must be packaged in cartons in the following manner:

(1) Irradiated articles may not be packaged for shipment in a carton with nonirradiated articles.

(2) For all imported articles irradiated prior to arrival in the United States, all articles moved interstate from Hawaii or U.S. territories and irradiated prior to arrival in the mainland United States, and all regulated articles to be moved interstate from an area quarantined for fruit flies or Asian citrus psyllid that are treated within the quarantined area:

(i) The fruits and vegetables must be packaged either:

(A) In insect-proof cartons that have no openings that will allow the entry of the pests of concern. The cartons must be sealed with seals that will visually indicate if the cartons have been opened. The cartons may be constructed of any material that prevents entry or oviposition (if applicable) by the pests of concern into the articles in the carton;[3] or

(B) In noninsect-proof cartons that are stored immediately after irradiation in a room completely enclosed by walls or screening that completely precludes access by the pests of concern. If stored in noninsect-proof cartons in a room that precludes access by the pests of concern, prior to leaving the room, each pallet of cartons must be completely enclosed in polyethylene shrink wrap, or another solid or netting covering that completely precludes access to the cartons by the pests of concern.

(ii) To preserve the integrity of treated lots, each pallet-load of cartons containing the fruits and vegetables must be secured before leaving the irradiation facility in one of the following ways:

(A) With polyethylene shrink wrap;

(B) With net wrapping; or

(C) With strapping.

(iii) Packaging must be labeled in a manner that allows an inspector to determine treatment lot numbers, packing and treatment facility identification and location, and dates of packing and treatment.

(A) For imported articles that are treated prior to arrival in the United States, pallets that remain intact as one unit until entry into the United States may have one such label per pallet. Pallets that are broken apart into smaller units prior to or during entry

---

[2] Inspectors are assigned to local offices of the Animal and Plant Health Inspection Service, which are listed in telephone directories.

[3] If there is a question as to the adequacy of a carton, send a request for approval of the carton, together with a sample carton, to the Animal and Plant Health Inspection Service, Plant Protection and Quarantine, Center for Plant Health Inspection and Technology, 1730 Varsity Drive, Suite 400, Raleigh, NC 27606-5202.

into the United States, or that will be broken apart into smaller units after entry into the United States, must have the required label information on each individual carton.

(B) For articles moved interstate from Hawaii or U.S. territories that are treated prior to arrival in the mainland United States, pallets that remain intact as one unit until entry into the mainland United States may have one such label per pallet. Pallets that are broken apart into smaller units prior to or during entry into the mainland United States, or that will be broken apart into smaller units after entry into the mainland United States, must have the required label information on each individual carton.

(3) For all articles imported to be irradiated upon arrival in the United States, moved interstate from Hawaii or U.S. territories to be irradiated upon arrival in the mainland United States, or moved interstate from areas quarantined for fruit flies or Asian citrus psyllid to be irradiated outside the quarantined area, the articles must be packed in cartons that have no openings that will allow the exit of the pests of concern and that are sealed with seals that will visually indicate if the cartons have been opened. They may be constructed of any material that prevents the pests of concern from exiting the carton. Cartons of untreated articles must be shipped in shipping containers sealed prior to their shipment with seals that will visually indicate if the shipping containers have been opened.

(g) *Containers or vans.* Containers or vans that will transport treated articles must be free of pests of concern prior to loading the treated articles.

(h) *Certification of treatment for articles treated outside the United States.* For each consignment treated in an irradiation facility outside the United States, a phytosanitary certificate, with the treatment section completed and issued by the NPPO, must accompany the consignment.

(i) *Dosage.* The regulated articles must receive the minimum absorbed ionizing radiation dose specified in the PPQ Treatment Manual or in another approved treatment schedule.

(j) *Dosimetry systems at the irradiation facility.* (1) Dosimetry must indicate the doses needed to ensure that all the articles will receive the minimum dose prescribed.

(2) The absorbed dose, as measured using an accurate dosimetry system, must meet or exceed the absorbed dose for the pest(s) of concern required by the PPQ Treatment Manual or by another approved treatment schedule.

(3) When designing the facility's dosimetry system and procedures for its operation, the facility operator must address guidance and principles from the International Standards Organization/American Society for Testing and Materials standard[4] or an equivalent standard recognized by APHIS.

(k) *Records.* An irradiation processor must maintain records of each treated lot for 1 year following the treatment date, and must make these records available for inspection by an inspector during normal business hours (8 a.m. to 4:30 p.m., Monday through Friday, except holidays). These records must include the lot identification, scheduled process, evidence of compliance with the scheduled process, ionizing energy source, source calibration, dosimetry, dose distribution in the product, and the date of irradiation.

(1) *Request for initial certification and inspection of facility.* Persons requesting initial certification of an irradiation treatment facility must submit the request for approval in writing to the Animal and Plant Health Inspection Service, Plant Protection and Quarantine, Center for Plant Health Inspection and Technology, 1730 Varsity Drive, Suite 400, Raleigh, NC 27606-5202. The initial request must identify the owner, location, and radiation source of the facility, and the applicant must supply additional information about the facility construction, treatment protocols, and operations upon request by APHIS if APHIS requires additional information to evaluate the request. Before the Administrator determines

---

[4] Designation ISO/ASTM 51261-2002(E), "Standard Guide for Selection and Calibration of Dosimetry Systems for Radiation Processing," American Society for Testing and Materials, *Annual Book of ASTM Standards.*

172

whether an irradiation facility is eligible for certification, an inspector will make a personal inspection of the facility to determine whether it complies with the standards of this section.

(m) *Denial and withdrawal of certification.* (1) The Administrator will withdraw the certification of any irradiation treatment facility upon written request from the irradiation processor.

(2) The Administrator will deny or withdraw certification of an irradiation treatment facility when any provision of this section is not met. Before withdrawing or denying certification, the Administrator will inform the irradiation processor in writing of the reasons for the proposed action and provide the irradiation processor with an opportunity to respond. The Administrator will give the irradiation processor an opportunity for a hearing regarding any dispute of a material fact, in accordance with rules of practice that will be adopted for the proceeding. However, the Administrator will suspend certification pending final determination in the proceeding if he or she determines that suspension is necessary to prevent the spread of any dangerous insect. The suspension will be effective upon oral or written notification, whichever is earlier, to the irradiation processor. In the event of oral notification, written confirmation will be given to the irradiation processor within 10 days of the oral notification. The suspension will continue in effect pending completion of the proceeding and any judicial review of the proceeding.

(n) *Department not responsible for damage.* This treatment is approved to assure quarantine security against the plant pests listed in the PPQ Treatment Manual or the plant pests for which another treatment schedule is approved in accordance with § 305.2. From the literature available, the articles authorized for treatment under this section are believed tolerant to the treatment; however, the facility operator and shipper are responsible for determination of tolerance. The Department of Agriculture and its inspectors assume no responsibility for any loss or damage resulting from any treatment prescribed or monitored. Additionally, the Nuclear Regulatory Commission is responsible for ensuring that irradiation facilities are constructed and operated in a safe manner. Further, the Food and Drug Administration is responsible for ensuring that irradiated foods are safe and wholesome for human consumption.

(o) *Substitution of irradiation for other treatments.* Treatment of fruits and vegetables that are from foreign localities, from Hawaii, Puerto Rico, and the U.S. Virgin Islands, or from domestic areas under quarantine with irradiation in accordance with this section may be substituted for other approved treatments if the target pests of the other approved treatments are approved for treatment with irradiation in the PPQ Treatment Manual or approved for treatment with irradiation in accordance with § 305.2.

(Approved by the Office of Management and Budget under control numbers 0579–0155, 0579–0215, and 0579–0198, 0579–0383)

[75 FR 4241, Jan. 26, 2010, as amended at 75 FR 34336, June 17, 2010; 76 FR 60361, Sept. 29, 2011; 77 FR 42624, July 20, 2012; 83 FR 5878, Feb. 12, 2018]

PART 318—STATE OF HAWAII AND TERRITORIES QUARANTINE NOTICES

Subpart A—Regulated Articles From Hawaii and the Territories

173

Subpart B—Territorial Cotton, Cottonseed, and Cottonseed Products

Subpart C—Sand, Soil, or Earth, with Plants From Territories and Districts

AUTHORITY: 7 U.S.C. 7701–7772 and 7781–7786; 7 CFR 2.22, 2.80, and 371.3.

SOURCE: 24 FR 10777, Dec. 29, 1959, unless otherwise noted.

Subpart A—Regulated Articles From Hawaii and the Territories

SOURCE: 74 FR 2775, Jan 16, 2009, unless otherwise noted. Redesignated at 84 FR 2428, Feb. 7, 2019.

§ 318.13–1   Notice of quarantine.

(a) Under the authority of section 412 of the Plant Protection Act, the Secretary of Agriculture may prohibit or restrict the movement in interstate commerce of any plant or plant product if the Secretary determines that the prohibition or restriction is necessary to prevent the introduction into the United States or the dissemination within the United States of a plant pest or noxious weed.

(b) The Secretary has determined that it is necessary to prohibit the interstate movement of cut flowers and fruits and vegetables and plants and portions of plants from Hawaii, Puerto Rico, the U.S. Virgin Islands, Guam, and the Commonwealth of the Northern Mariana Islands except as provided in this subpart or as provided in "Subpart B—Territorial Cotton, Cottonseed, and Cottonseed Products" and "Subpart C—Sand, Soil, or Earth, with

Plants from Territories and Districts" in this part.

[74 FR 2775, Jan. 16, 2009, as amended at 74 FR 15641, Apr. 7, 2009; 84 FR 2428, Feb. 7, 2019]

§ 318.13–2   Definitions.

Administrator. The Administrator of the Animal and Plant Health Inspection Service (APHIS), U.S. Department of Agriculture, or any other employee of APHIS to whom authority has been delegated to act in the Administrator's stead.

Animal and Plant Health Inspection Service. The Animal and Plant Health Inspection Service (APHIS) of the U.S. Department of Agriculture.

Certification (certified). A type of authorization, issued by an inspector, evidencing freedom from infestation, to allow the movement of certain regulated articles in accordance with the regulations in this subpart. "Certified" shall be construed accordingly.

Commercial consignment. A lot of fruits or vegetables that an inspector identifies as having been produced for sale or distribution in mass markets. Such identification will be based on a variety of indicators, including, but not limited to: Quantity of produce, type of packaging, identification of grower and packinghouse on the packaging, and documents consigning the fruits or vegetables to a wholesaler or retailer.

Compliance agreement. Any agreement to comply with stipulated conditions as prescribed under § 318.13–3 or § 318.13–4 or § 305.34 of this chapter, executed by any person to facilitate the interstate movement of regulated articles under this subpart.

Consignment. A quantity of plants, plant products, and/or other articles, including fruits or vegetables, being moved from one country to another and covered, when required, by a single certificate or limited permit (a consignment may be composed of one or more commodities or lots).

Continental United States. The 48 contiguous States, Alaska, and the District of Columbia.

Cut flower. Any cut blooms, fresh foliage, and dried decorative plant material customarily used in the florist trade and not for planting; and being the severed portion of a plant, including the inflorescence, and any parts of the plant attached thereto, in a fresh state.

Disinfection (disinfect and disinfected). The application to parts or all of a ship, vessel, other surface craft, or aircraft of a treatment that may be designated by the inspector as effective against such plant pests as may be present. ("Disinfect" ,and "disinfected" shall be construed accordingly.)

Fruits and vegetables. A commodity class for fresh parts of plants intended for consumption or processing and not planting.

Inspector. A State agricultural inspector or any individual authorized by the Administrator of APHIS or the Commissioner of Customs and Border Protection, Department of Homeland Security, to enforce the regulations in this subpart.

Interstate. From one State into or through any other State; or within the District of Columbia, Guam, the Virgin Islands of the United States, or any other territory or possession of the United States.

Limited permit. A document issued by an inspector or a person operating under a compliance agreement for the interstate movement of regulated articles to a specified destination for:

(1) Consumption, limited utilization or processing, or treatment; or

(2) Movement into or through the continental United States in conformity with a transit permit.

Lot. A number of units of a single commodity, identifiable by its homogeneity of composition and origin, forming all or part of a consignment.

Means of conveyance. A ship, truck, aircraft, or railcar.

Moved (move and movement). Shipped, offered for shipment to a common carrier, received for transportation or transported by a common carrier, or carried, transported, moved, or allowed to be moved, directly or indirectly, from Hawaii, Puerto Rico, Guam, the Commonwealth of the Northern Marina Islands, or the U.S. Virgin Islands into or through the continental United States or any other State or territory of the United States (or from or into or through other places as specified in this subpart). "Move" and "movement" shall be construed accordingly.

Packing materials. Any plant or plant product, soil, or other substance associated with or accompanying any commodity or consignment to serve for filling, wrapping, ties, lining, mats, moisture retention, protection, or any other auxiliary purpose. The word "packing," as used in the expression "packing materials," includes the presence of such materials within, in contact with, or accompanying a consignment.

Person. Any individual, partnership, corporation, association, joint venture, or other legal entity.

Plant debris. Detached leaves, twigs, or other portions of plants, or plant litter or rubbish as distinguished from approved parts of clean fruits and vegetables, or other commercial articles.

Plant pests. Any living stage of any of the following that can directly or indirectly injure, cause damage to, or cause disease in any plant or plant product: A protozoan, nonhuman animal, parasitic plant, bacterium, fungus, virus or viroid, infectious agent or other pathogen, or any article similar to or allied with any of those articles.

Plant Protection and Quarantine (PPQ). The Plant Protection and Quarantine program of APHIS.

Regulated articles. Fruits or vegetables in the raw or unprocessed state; cut flowers; seeds; and plants or plant products for nonpropagative or propagative use.

Sealed (sealable) container. A completely enclosed container designed for the storage and/or transportation of commercial air, sea, rail, or truck cargo, and constructed of metal or fiberglass, or other similarly sturdy and impenetrable material, providing an enclosure accessed through doors that are closed and secured with a lock or seal. Sealed (sealable) containers used for sea consignments are distinct and separable from the means of conveyance carrying them when arriving in and in transit through the continental United States. Sealed (sealable) containers used for air consigments are distinct and separable from the means of conveyance carrying them before any transloading in the continental

United States. Sealed (sealable) containers used for air consignments after transloading in the continental United States or for overland consignments in the continental United States may either be distinct and separable from the means of conveyance carrying them, or be the means of conveyance itself.

Soil. The loose surface material of the earth in which plants grow, in most cases consisting of disintegrated rock with an admixture of organic material and soluble salts.

State. Any of the several States of the United States, the Commonwealth of the Northern Mariana Islands, the Commonwealth of Puerto Rico, the District of Columbia, Guam, the Virgin Islands of the United States, or any other territory or possession of the United States.

Transit permit. A written authorization issued by the Administrator for the movement of fruits and vegetables en route to a foreign destination that are otherwise prohibited movement by this subpart into the continental United States. Transit permits authorize one or more consignments over a designated period of time.

Transloading. The transfer of cargo from one sealable container to another, from one means of conveyance to another, or from a sealable container directly into a means of conveyance.

United States. All of the States.

[74 FR 2775, Jan 16, 2009, as amended at 83 FR 46637, Sept. 14, 2018]

§318.13–3 General requirements for all regulated articles.

All regulated articles that are allowed movement under this subpart must be moved in accordance with the following requirements, except as specifically provided otherwise in this subpart.

(a) *Freedom from plant debris.* All regulated articles moved under this subpart must be free from plant debris.

(b) *Certification.* Certification may be issued for the movement of regulated articles under the following conditions:

(1) *Certification on basis of inspection or nature of lot involved.* Regulated articles may be certified when they have been inspected by an inspector and found apparently free from infestation and infection, or without such inspec-

tion when the inspector determines that the lot for consignment is of such a nature that no danger of infestation or infection is involved.

(i) Persons intending to move any articles that may be certified must contact the local Plant Protection and Quarantine office as far as possible in advance of the contemplated date of shipment in order to request an inspection.

(ii) Persons intending to move any articles that may be certified must prepare, handle, and safeguard such articles from infestation or reinfestation, and assemble them at such points as the inspector may designate, placing them so that inspection may be readily made.

(2) *Certification on basis of treatment.* (i) Regulated articles for which treatments are approved under part 305 of this chapter may be certified if such treatments have been applied in accordance with part 305 of this chapter and if the articles were handled after such treatment in accordance with a compliance agreement executed by the applicant for certification or under the supervision of an inspector.

(ii) Regulated articles certified after treatment in accordance with part 305 of this chapter that are taken aboard any ship, vessel, other surface craft, or aircraft must be segregated and protected in a manner as required by the inspector.

(c) *Limited permits.* (1) Limited permits [1] may be issued by an inspector for the movement of certain noncertified regulated articles to restricted destinations.

(2) Limited permits may be issued by an inspector for the movement of regulated articles that would otherwise be prohibited movement under this subpart, if the articles are to be moved in accordance with §318.13–6.

(3) Except when the regulations specify that an inspector must issue the limited permit, limited permits may be issued by a person operating under a compliance agreement.

---

[1] Limited permits can be obtained from each State or territory's local Plant Protection and Quarantine office.

(d) *Compliance agreements.* As a condition for the movement of regulated articles for which a compliance agreement is required, the person entering the compliance agreement must agree to the following:

(1) That he or she will use any permit or certification issued to him or her in accordance with the provisions in the permit, the requirements in this subpart, and the compliance agreement;

(2) That he or she will maintain at his or her establishment such safeguards against the establishment and spread of infestation and infection and comply with such conditions as to the maintenance of identity, handling (including post-treatment handling), and interstate movement of regulated articles and the cleaning and treatment of means of conveyance and containers used in such movement of the articles, as may be required by the inspector in each specific case to prevent the spread of infestation or infection; and

(3) That he or she will allow inspectors to inspect the establishment and its operations.

(e) *Attachment of limited permit or verification of certification.* Except as otherwise provided for certain air cargo and containerized cargo on ships moved in accordance with §318.13-10, each box, bale, crate, or other container of regulated articles moved under certification or limited permit shall have the limited permit attached to the outside of the container or bear a U.S. Department of Agriculture stamp or inspection sticker verifying that the consignment has been certified in accordance with paragraph (b) of this section: *Provided,* That if a limited permit or certification is issued for a consignment of more than one container or for bulk products, certification shall be stamped on or the limited permit shall be attached to the accompanying waybill, manifest, or bill of lading.

(f) *Withdrawal of certification, transit permits, limited permits, or compliance agreements.* Any certification, transit permit, limited permit, or compliance agreement which has been issued or authorized may be withdrawn by an inspector orally or in writing, if such inspector determines that the holder thereof has not complied with all conditions under the regulations for the use of such document. If the cancellation is oral, the decision and the reasons for the withdrawal shall be confirmed in writing as promptly as circumstances allow. Any person whose certification, transit permit, limited permit, or compliance agreement has been withdrawn may appeal the decision in writing to the Administrator within 10 days after receiving the written notification of the withdrawal. The appeal shall state all of the facts and reasons upon which the person relies to show that the certification, transit permit, limited permit, or compliance agreement was wrongfully withdrawn. The Administrator shall grant or deny the appeal, in writing, stating the reasons for such decision, as promptly as circumstances allow. If there is a conflict as to any material fact, a hearing shall be held to resolve such conflict. Rules of practice concerning such a hearing will be adopted by the Administrator.

(g) *Container marking and identity.* Except as provided in §318.13-6(c), consignments of regulated articles moved in accordance with this subpart must have the following information clearly marked on each container or on the waybill, manifest, or bill of lading accompanying the articles: Nature and quantity of contents; name and address of shipper, owner, or person shipping or forwarding the articles; name and address of consignee; shipper's identifying mark and number; and the certification stamp or number of the limited permit authorizing movement, if one was issued.

(h) *Refusal of movement.* An inspector may refuse to allow the interstate movement of a regulated article if the inspector finds that the regulated article is prohibited, is not accompanied by required documentation, is so infested with a plant pest or noxious weed that, in the judgment of the inspector, it cannot be cleaned or treated, or contains soil or other prohibited contaminants.

(i) *Costs and charges.* Services of the inspector during regularly assigned hours of duty at the usual places of duty shall be furnished without cost to the one requesting such services. APHIS will not assume responsibility for any costs or charges, other than those indicated in this section, in connection with the inspection, treatment, conditioning, storage, forwarding, or any other operation of any character incidental to the physical movement of regulated articles or plant pests.

(j) *APHIS not responsible for damage.* APHIS assumes no responsibility for any damage to regulated articles that results from the application of treatment or other measures required under this subpart (or under part 305 of this chapter) to protect against the dissemination of plant pests within the United States.

(Approved by the Office of Management and Budget under control number 0579–0346)

[74 FR 2775, Jan. 16, 2009, as amended at 75 FR 4249, Jan. 26, 2010]

§ 318.13–4 Authorization of certain fruits and vegetables for interstate movement.

(a) *Determination by the Administrator.* No fruit or vegetable is authorized for interstate movement from Hawaii or the territories unless the Administrator has determined that the risk posed by each quarantine pest associated with the fruit or vegetable can be reasonably mitigated by the application of one or more phytosanitary measures designated by the Administrator.

(b) *Designated phytosanitary measures.* (1) The fruits and vegetables are subject to phytosanitary treatments, which could include, but are not limited to, pest control treatments in the field or growing site, and post-harvest treatments.

(2) The fruits and vegetables are subject to growing area pest mitigations, which could include, but are not limited to, detection surveys, trapping requirements, pest exclusionary structures, and field inspections.

(3) The fruits and vegetables are subject to safeguarding and movement mitigations, which could include, but are not limited to, safeguarded transport, box labeling, limited distribution, insect-proof boxes, and importation as commercial consignments only.

(4) The fruits and vegetables are subject to administrative mitigations, which could include, but are not limited to, registered fields or orchards, registered growing sites, registered packinghouses, inspection in the State of origin by an inspector, and operational workplan monitoring.

(5) The fruits and vegetables are subject to any other measures deemed appropriate by the Administrator.

(c) *Authorized fruits and vegetables*—(1) *Comprehensive list.* The name and origin of all fruits and vegetables authorized for interstate movement under this section, as well as the applicable requirements for their movement, may be found on the internet at *https://www.aphis.usda.gov/aphis/ourfocus/planthealth/complete-list-of-electronic-manuals.*

(2) *Fruits and vegetables authorized for interstate movement prior to October 15, 2018.* Fruits and vegetables that were authorized for interstate movement under this subpart as of October 15, 2018 may continue to be moved interstate under the same requirements that applied before October 15, 2018, except as provided in paragraph (c)(4) of this section.

(3) *Other fruits and vegetables.* Fruits and vegetables not already authorized for interstate movement as described in paragraph (c)(2) of this section may be authorized for interstate movement only after:

(i) APHIS has analyzed the pest risk posed by the interstate movement of a fruit or vegetable and has determined that the risk posed by each quarantine pest associated with the fruit or vegetable can be reasonably mitigated by the application of one or more phytosanitary measures;

(ii) APHIS has made its pest risk analysis and determination available for public comment for at least 60 days through a notice published in the FEDERAL REGISTER; and

(iii) The Administrator has announced his or her decision in a subsequent FEDERAL REGISTER notice to begin allowing interstate movement of the fruit or vegetable subject to the phytosanitary measures specified in the notice.

(4) *Changes to phytosanitary measures.* (i) If the Administrator determines that the phytosanitary measures required for a fruit or vegetable that has been authorized interstate movement

under this subpart are no longer sufficient to reasonably mitigate the pest risk posed by the fruit or vegetable, APHIS will prohibit or further restrict interstate movement of the fruit or vegetable. APHIS will also publish a notice in the FEDERAL REGISTER advising the public of its finding. The notice will specify the amended interstate movement requirements, provide an effective date for the change, and invite public comment on the subject.

(ii) If the Administrator determines that any of the phytosanitary measures required for a fruit or vegetable that has been authorized interstate movement under this subpart are no longer necessary to reasonably mitigate the pest risk posed by the fruit or vegetable, APHIS will make new pest risk documentation available for public comment, in accordance with paragraph (c)(3) of this section, prior to allowing interstate movement of the fruit or vegetable subject to the phytosanitary measures specified in the notice.

(Approved by the Office of Management and Budget under control number 0579–0346)

[83 FR 46637, Sept. 14, 2018]

§318.13–5 Pest-free areas.

Certain fruits or vegetables may be moved interstate provided that the fruits or vegetables originate from an area that is free of a specific pest or pests. In some cases, fruits or vegetables may only be moved interstate if the area of origin is free of all plant pests that attack the fruits or vegetables. In other cases, fruits or vegetables may be moved interstate if the area of origin is free of one or more plant pests that attack the fruit or vegetable and the risk posed by the remaining plant pests that attack the fruit or vegetable is mitigated by other specific phytosanitary measures contained in the regulations in this subpart.

(a) *Application of standards for pest-free areas.* APHIS will make a determination of an area's pest-free status based on information provided by the State. The information used to make this determination will include trapping and surveillance data, survey pro-

tocols, and protocols for actions to be performed upon detection of a pest.

(b) *Survey protocols.* APHIS must approve the survey protocol used to determine and maintain pest-free status, as well as protocols for actions to be performed upon detection of a pest. Pest-free areas are subject to audit by APHIS to verify their status.

(c) *Determination of pest freedom.* (1) For an area to be considered free of a specified pest for the purposes of this subpart, the Administrator must determine, and announce in a notice published in the FEDERAL REGISTER for a public comment period of 60 days, that the area meets the criteria of paragraphs (a) and (b) of this section.

(2) The Administrator will announce his or her decision in a subsequent FEDERAL REGISTER notice. If appropriate, APHIS will allow movement of the regulated article from a pest-free area because:

(i) No comments were received on the notice or

(ii) The comments on the notice did not affect the overall conclusions of the notice and the Administrator's determination of risk.

(d) *Decertification of pest-free areas; reinstatement.* If a pest is detected in an area that is designated as free of that pest, APHIS will publish in the FEDERAL REGISTER a notice announcing that the pest-free status of the area in question has been withdrawn and that interstate movement of host crops for the pest in question is subject to application of an approved treatment for the pest. If a treatment for the pest is not available, interstate movement of the host crops would be prohibited. In order for a decertified pest-free area to be reinstated, it would have to meet the criteria of paragraphs (a) through (c) of this section.

(e) *General requirements for the interstate movement of regulated articles from pest-free areas*—(1) *Labeling.* Each box of fruits or vegetables that is moved interstate from a pest-free area under this subpart must be clearly labeled with:

(i) The name of the orchard or grove of origin, or the name of the grower; and

(ii) The name of the municipality and State or territory in which the fruits or vegetables were produced; and

(iii) The type and amount of fruits or vegetables the box contains.

(2) *Compliance agreement.* Persons wishing to move fruits or vegetables from a pest-free area in Hawaii, Puerto Rico, Guam, the Commonwealth of the Northern Mariana Islands, or the U.S. Virgin Islands must enter into a compliance agreement with APHIS in accordance with § 318.13–3(d).

(3) *Safeguarding.* If fruits or vegetables are moved from a pest-free area into or through an area that is not free of that pest, the fruits or vegetables must be safeguarded during the time they are present in a non-pest-free area by being covered with insect-proof mesh screens or plastic tarpaulins, including while in transit to the packinghouse and while awaiting packaging. If fruits or vegetables are moved through an area that is not free of that pest during transit to a port, they must be packed in insect-proof cartons or containers or be covered by insect-proof mesh or plastic tarpaulins during transit to the port and subsequent movement into or through the United States. These safeguards described in this section must remain intact until the fruits or vegetables reach their final destination.

(Approved by the Office of Management and Budget under control number 0579–0346)

§ 318.13–6 **Transit of fruits and vegetables from Hawaii or the territories into or through the continental United States.**

Fruits and vegetables from Hawaii, Puerto Rico, Guam, the Commonwealth of the Northern Mariana Islands, or the U.S. Virgin Islands that are otherwise prohibited interstate movement into the continental United States by this subpart may transit the continental United States en route to a foreign destination when moved in accordance with this section.

(a) *Transit permit.* (1) A transit permit is required for the arrival, unloading, and movement through the continental United States of fruits and vegetables otherwise prohibited by this subpart from being moved through the continental United States from Hawaii,

Puerto Rico, Guam, the Commonwealth of the Northern Mariana Islands, or the U.S. Virgin Islands. Application for a transit permit may be made in writing or with PPQ Form 586.[2] The transit permit application must include the following information:

(i) The specific types of fruits and vegetables to be shipped (only scientific or English common names are acceptable);

(ii) The means of conveyance to be used to transport the fruit or vegetable through the continental United States;

(iii) The port of arrival in the continental United States, and the location of any subsequent stop;

(iv) The location of, and the time needed for, any storage in the continental United States;

(v) Any location in the continental United States where the fruits or vegetables are to be transloaded;

(vi) The means of conveyance to be used for transporting the fruits or vegetables from the port of arrival in the continental United States to the port of export;

(vii) The estimated time necessary to accomplish exportation, from arrival at the port of arrival in the continental United States to exit at the port of export;

(viii) The port of export; and

(ix) The name and address of the applicant and, if the applicant's address is not within the territorial limits of the continental United States, the name and address in the continental United States of an agent whom the applicant names for acceptance of service of process.

(2) A transit permit will be issued only if the following conditions are met:

(i) APHIS inspectors are available at the port of arrival, port of export, and any locations at which transloading of cargo will take place and, in the case of

---

[2] PPQ Form 586 can be obtained from PPQ Permit Services or at *http://www.aphis.usda.gov/plant_health/permits/transit.shtml* . Applications for transit permits should be submitted to USDA, APHIS, PPQ Permit Services, 4700 River Road Unit 136, Riverdale, MD 20737 or through e-permits *http://www.aphis.usda.gov/permits/learn_epermits.shtml.*

air consignments, at any interim stop in the continental United States, as indicated on the application for the transit permit;

(ii) The application indicates that the proposed movement would comply with the provisions in this section applicable to the transit permit; and

(iii) During the 12 months prior to receipt of the application by APHIS, the applicant has not had a transit permit withdrawn under §318.13–3(f), unless the transit permit has been reinstated upon appeal.

(b) *Limited permit.* Fruits or vegetables shipped from Hawaii, Puerto Rico, Guam, the Commonwealth of the Northern Mariana Islands, or the U.S. Virgin Islands through the continental United States under this section must be accompanied by a limited permit, a copy of which must be presented to an inspector at the port of arrival and the port of export in the continental United States, and at any other location in the continental United States where an air consignment is authorized to stop or where overland consignments change means of conveyance. An inspector will issue a limited permit if the following conditions are met:

(1) The inspector determines that the specific type and quantity of the fruits or vegetables being shipped are accurately described by accompanying documentation, such as the accompanying manifest, waybill, and bill of lading. (Only scientific or English common names are acceptable.) The fruits or vegetables shall be assembled at whatever point and in whatever manner the inspector designates as necessary to comply with the requirements of this section; and

(2) The inspector establishes that the consignment of fruits or vegetables has been prepared in compliance with the provisions of this section.

(c) *Marking requirements.* Each of the smallest units, including each of the smallest bags, crates, or cartons, containing regulated articles for transit through the continental United States under this section must be conspicuously marked, prior to the locking and sealing of the container in the State of origin, with a printed label that includes a description of the specific type and quantity of the fruits or vegetables

(only scientific or English common names are acceptable), the transit permit number under which the regulated articles are to be shipped, and, in English, the State in which they were grown and the statement "Distribution in the United States is Prohibited."

(d) *Handling of fruits and vegetables.* Fruits or vegetables shipped through the United States from Hawaii, Puerto Rico, Guam, the Commonwealth of the Northern Mariana Islands, or the U.S. Virgin Islands in accordance with this section may not be commingled in the same sealed container with fruits or vegetables that are intended for entry and distribution in the United States. The fruits or vegetables must be kept in sealed containers from the time the limited permit required by paragraph (b) of this section is issued, until the fruits or vegetables exit the United States, except as otherwise provided in the regulations in this section. Transloading must be carried out in accordance with the requirements of paragraphs (a), (h), and (i) of this section.

(e) *Area of movement.* The port of arrival, the port of export, ports for air stops, and overland movement within the continental United States of fruits or vegetables shipped under this section is limited to a corridor that includes all States of the continental United States except Alabama, Arizona, California, Florida, Georgia, Kentucky, Louisiana, Mississippi, Nevada, New Mexico, North Carolina, South Carolina, Tennessee, Texas, and Virginia, except that movement is allowed through Dallas/Fort Worth, TX, as an authorized stop for air cargo, or as a transloading location for consignments that arrive by air but that are subsequently transloaded into trucks for overland movement from Dallas/Fort Worth, TX, into the designated corridor by the shortest route. Movement through the United States must begin and end at locations staffed by APHIS inspectors.

(f) *Movement of regulated articles.* Transportation through the continental United States shall be by the most direct route to the final destination of the consignment in the country to which it is exported, as determined by APHIS based on commercial shipping routes and timetables and set forth in the transit permit. No change in the quantity of the original consignment from that described in the limited permit is allowed. No remarking is allowed. No diversion or delay of the consignment from the itinerary described in the transit permit and limited permit is allowed unless authorized by an APHIS inspector upon determination by the inspector that the change will not significantly increase the risk of plant pests or diseases in the United States, and unless each port to which the consignment is diverted is staffed by APHIS inspectors.

(g) *Notification in case of emergency.* In the case of an emergency such as an accident, a mechanical breakdown of the means of conveyance, or an unavoidable deviation from the prescribed route, the person in charge of the means of conveyance must, as soon as practicable, notify the APHIS office at the port where the cargo arrived in the United States.

(h) *Consignments by sea.* Except as authorized by this paragraph, consignments arriving in the United States by sea from Hawaii, Puerto Rico, Guam, the Commonwealth of the Northern Mariana Islands, or the U.S. Virgin Islands may be transloaded once from a ship to another ship or, alternatively, once to a truck or railcar at the port of arrival and once from a truck or railcar to a ship at the port of export, and must remain in the original sealed container, except under extenuating circumstances and when authorized by an inspector upon determination by the inspector that the transloading would not significantly increase the risk of the introduction of plant pests or diseases into the United States, and provided that APHIS inspectors are available to provide supervision. No other transloading of the consignment is allowed, except under extenuating circumstances (*e.g.* , equipment breakdown) and when authorized by an inspector upon determination by the inspector that the transloading would not significantly increase the risk of the introduction of plant pests or diseases into the continental United States, and provided that APHIS inspectors are available to provide supervision.

(i) *Consignments by air.* (1) Consignments arriving in the United States by air from Hawaii, Puerto Rico, Guam, the Commonwealth of the Northern Mariana Islands, or the U.S. Virgin Islands may be transloaded only once in the United States. Transloading of air consignments must be carried out in the presence of an APHIS inspector. Consignments arriving by air that are transloaded may be transloaded either into another aircraft or into a truck trailer for export by the most direct route to the final destination of the consignment through the designated corridor set forth in paragraph (e) of this section. This may be done at either the port of arrival in the United States or at the second air stop within the designated corridor, as authorized in the transit permit and as provided in paragraph (i)(2) of this section. No other transloading of the consignment is allowed, except under extenuating circumstances (*e.g.*, equipment breakdown) and when authorized by an APHIS inspector upon determination by the inspector that the transloading would not significantly increase the risk of the introduction of plant pests or diseases into the United States, and provided that APHIS inspectors are available to provide supervision. Transloading of air consignments will be authorized only if the following conditions are met:

(i) The transloading is done into sealable containers;

(ii) The transloading is carried out within the secure area of the airport (*i.e.*, that area of the airport that is open only to personnel authorized by the airport security authorities);

(iii) The area used for any storage is within the secure area of the airport; and

(iv) APHIS inspectors are available to provide the supervision required by paragraph (i)(1) of this section.

(2) Except as authorized by paragraph (f) of this section, consignments that continue by air from the port of arrival in the continental United States may be authorized by APHIS for only one additional stop in the continental United States, provided the second stop is within the designated corridor set

forth in paragraph (e) of this section and is staffed by APHIS inspectors. As an alternative to transloading a consignment arriving in the United States into another aircraft, consignments that arrive by air may be transloaded into a truck trailer for export by the most direct route to the final destination of the consignment through the designated corridor set forth in paragraph (e) of this section. This may be done at either the port of arrival in the United States or at the second authorized air stop within the designated corridor. No other transloading of the consignment is allowed, except under extenuating circumstances (*e.g.*, equipment breakdown) and when authorized by an APHIS inspector upon determination by the inspector that the transloading would not significantly increase the risk of the introduction of plant pests or diseases into the United States, and provided that APHIS inspectors are available to provide supervision.

(j) *Duration and location of storage.* Any storage in the United States of fruits or vegetables shipped under this section must be for a duration and in a location authorized in the transit permit required by paragraph (a) of this section. Areas where such fruits or vegetables are stored must be either locked or guarded at all times the fruits and vegetables are present. Cargo shipped under this section must be kept in a sealed container while stored in the continental United States.

(k) *Temperature requirement.* Except for time spent on aircraft and except during storage and transloading of air consignments, the temperature in the sealed containers containing fruits and vegetables moved under this section must be 60 °F or lower from the time the regulated articles leave Hawaii, Puerto Rico, Guam, the Commonwealth of the Northern Mariana Islands, the U.S. Virgin Islands, or any other territory or possession of the United States until they exit the United States.

(l) *Prohibited materials.* (1) The person in charge of or in possession of a sealed container used for movement into or through the United States under this section must ensure that the sealed

container is carrying only those fruits or vegetables authorized by the transit permit required under paragraph (a) of this section; and

(2) The person in charge of or in possession of any means of conveyance or container returned to the United States without being reloaded after being used to export fruits or vegetables from the United States under this section must ensure that the means of conveyance or container is free of materials prohibited importation into the United States under this chapter.

(m) Authorization by APHIS of the movement of fruits or vegetables through the United States under this section does not imply that such fruits or vegetables are enterable into the destination country. Consignments returned to the United States from the destination country shall be subject to all applicable regulations, including "Subpart L—Fruits and Vegetables" of part 319 and "Plant Quarantine Safeguard Regulations" of part 352 of this chapter.

(n) Any restrictions and requirements with respect to the arrival, temporary stay, unloading, transloading, transiting, exportation, or other movement or possession in the United States of any fruits or vegetables under this section shall apply to any person who brings into, maintains, unloads, transloads, transports, exports, or otherwise moves or possesses in the United States such fruits or vegetables, whether or not that person is the one who was required to have a transit permit or limited permit for the fruits or vegetables or is a subsequent custodian of the fruits or vegetables. Failure to comply with all applicable restrictions and requirements under this section by such a person shall be deemed to be a violation of this section.

(Approved by the Office of Management and Budget under control number 0579–0346)

[74 FR 2775, Jan 16, 2009, as amended at 84 FR 2428, Feb. 7, 2019]

§318.13–7 Products as ships' stores or in the possession of passengers or crew.

(a) *In the possession of passengers or crew members.* Small quantities of fruits, vegetables, or cut flowers subject to the quarantine and regulations in this subpart, when loose and free of packing materials, may be taken aboard any ship, vessel, or other surface craft by passengers or members of the crew without inspection and certification in the State of origin. However, if such articles are not eligible for certification under § 318.13-3, they must be entirely consumed or disposed of before arrival within the territorial waters of the continental United States, Hawaii, Puerto Rico, Guam, the Commonwealth of the Northern Mariana Islands, or the U.S. Virgin Islands.

(b) *As ships' stores or decorations.* Fruits, vegetables, or cut flowers subject to the quarantine and regulations in this subpart may be taken aboard a ship, vessel, or other surface craft in Hawaii, Puerto Rico, Guam, the Commonwealth of the Northern Mariana Islands, or the U.S. Virgin Islands without inspection or certification. Fruits, vegetables, and cut flowers that are so taken aboard such a carrier must be either:

(1) Entirely consumed or removed from the ship, vessel, or other surface craft before arrival within the territorial waters of the continental United States, Hawaii, Puerto Rico, Guam, the Commonwealth of the Northern Mariana Islands, the U.S. Virgin Islands, or any other territory or possession of the United States; or

(2) In the case of a surface carrier, retained aboard such carrier under seal or otherwise disposed of subject to safeguards equivalent to those imposed on other prohibited or restricted products by paragraphs (b) and (c) of § 352.10 of this chapter.

§ 318.13-8 Articles and persons subject to inspection.

In addition to the inspection requirements in §§ 318.13-9 and 318.13-10, persons, means of conveyance (including ships, other oceangoing craft, and aircraft), baggage, cargo, and any other articles, that are destined for movement, are moving, or have been moved from Hawaii, Puerto Rico, Guam, the Commonwealth of the Northern Mariana Islands, or the U.S. Virgin Islands to a destination elsewhere in the United States are subject to agricultural inspection at the port of departure, the port of arrival, or any other

authorized port. If an inspector finds any article prohibited movement by the quarantine and regulations of this subpart, he or she, taking the least drastic action, shall order the return of the article to the place of origin, or the exportation of the article, under safeguards satisfactory to him or her, or otherwise dispose of it, in whole or part, to comply with the quarantine and regulations of this subpart.

§ 318.13-9 Inspection and disinfection of means of conveyance.

(a) *Inspection of aircraft prior to departure.* No person shall move any aircraft from Hawaii, Puerto Rico, Guam, the Commonwealth of the Northern Mariana Islands, or the U.S. Virgin Islands to any other State unless the person moving the aircraft has contacted an inspector and offered the inspector the opportunity to inspect the aircraft prior to departure and the inspector has informed the person proposing to move the aircraft that the aircraft may depart.

(b) *Inspection of aircraft moving to Guam.* Any person who has moved an aircraft from Hawaii, Puerto Rico, the Commonwealth of the Northern Mariana Islands, or the U.S. Virgin Islands to Guam shall contact an inspector and offer the inspector the opportunity to inspect the aircraft upon the aircraft's arrival in Guam.

(c) *Inspection of ships upon arrival.* Any person who has moved a ship or other oceangoing craft from Hawaii, Puerto Rico, Guam, the Commonwealth of the Northern Mariana Islands, or the U.S. Virgin Islands to any other State shall contact an inspector and offer the inspector the opportunity to inspect the ship or other oceangoing craft upon its arrival.

(d) *Disinfection of means of conveyance.* If an inspector finds that a means of conveyance is infested with or contains plant pests, and the inspector orders disinfection of the means of conveyance, then the person in charge or in possession of the means of conveyance shall disinfect the means of conveyance and its cargo in accordance with an approved method contained in part 305 of this chapter under the supervision of an inspector and in a manner prescribed by the inspector, prior to any movement of the means of conveyance or its cargo.

§318.13–10 Inspection of baggage, other personal effects, and cargo.

(a) *Offer for inspection by aircraft passengers.* Passengers destined for movement by aircraft from Hawaii, Puerto Rico, Guam, the Commonwealth of the Northern Mariana Islands, or the U.S. Virgin Islands to any other State shall offer their carry-on baggage and other personal effects for inspection at the place marked for agricultural inspections, which will be located at the airport security checkpoint or the aircraft boarding gate, at the time they pass through the checkpoint or the gate. Passengers shall offer their check-in baggage for inspection at agricultural inspection stations prior to submitting their baggage to the check-in baggage facility. When an inspector has inspected and passed such baggage or personal effects, he or she shall apply a U.S. Department of Agriculture stamp, inspection sticker, or other identification to such baggage or personal effects to indicate that such baggage or personal effects have been inspected and passed as required. Passengers shall disclose any fruits, vegetables, plants, plant products, or other articles that are requested to be disclosed by the inspector. When an inspection of a passenger's baggage or personal effects discloses an article in violation of the regulations in this part, the inspector shall seize the article. The passenger shall state his or her name and address to the inspector, and provide the inspector with corroborative identification. The inspector shall record the name and address of the passenger, the nature of the identification presented for corroboration, the nature of the violation, the types of articles involved, and the date, time, and place of the violation.

(b) *Offer for inspection by aircraft crew.* Aircraft crew members destined for movement by aircraft from Hawaii, Puerto Rico, Guam, the Commonwealth of the Northern Mariana Islands, or the U.S. Virgin Islands to any other State, shall offer their baggage and personal effects for inspection at the inspection station designated for the employing airline not less than 20

minutes prior to the scheduled departure time of the aircraft or the rescheduled departure time as posted in the public areas of the airport. When an inspector has inspected and passed such baggage or personal effects, he or she shall apply a U.S. Department of Agriculture stamp, inspection sticker, or other identification to the baggage or personal effects to indicate that such baggage or personal effects have been inspected and passed as required. Aircraft crew members shall disclose any fruits, vegetables, plants, plant products, or other articles that are requested to be disclosed by the inspector. When an inspection of a crew member's baggage or personal effects discloses an article in violation of the regulations in this part, the inspector shall seize the article. The crew member shall state his or her name and address to the inspector, and provide the inspector with corroborative identification. The inspector shall record the name and address of the crew member, the nature of the identification presented for corroboration, the nature of the violation, the types of articles involved, and the date, time, and place of the violation.

(c) *Baggage inspection for persons traveling to Guam on aircraft.* No person who has moved from Hawaii, Puerto Rico, or the U.S. Virgin Islands to Guam on an aircraft shall remove or attempt to remove any baggage or other personal effects from the area secured for customs inspections before the person has offered to an inspector, and has had passed by the inspector, his or her baggage and other personal effects. Persons shall disclose any fruits, vegetables, plants, plant products, or other articles that are requested to be disclosed by the inspector. When an inspection of a person's baggage or personal effects discloses an article in violation of the regulations in this part, the inspector shall seize the article. The person shall state his or her name and address to the inspector, and provide the inspector with corroborative identification. The inspector shall record the name and address of the person, the nature of the identification presented for corroboration, the nature of the violation, the types of articles

§ 318.13–11   7 CFR Ch. III (1–1–22 Edition)

involved, and the date, time, and place of the violation.

(d) *Baggage acceptance and loading on aircraft.* No person shall accept or load any check-in aircraft baggage destined for movement from Hawaii, Puerto Rico, Guam, the Commonwealth of the Northern Mariana Islands, or the U.S. Virgin Islands to any other State unless the baggage bears a U.S. Department of Agriculture stamp, inspection sticker, or other indication applied by an inspector representing that the baggage has been inspected and certified.

(e) *Offer for inspection by persons moving by ship.* No person who has moved on any ship or other oceangoing craft from Hawaii, Puerto Rico, Guam, the Commonwealth of the Northern Mariana Islands, or the U.S. Virgin Islands to any other territory, State, or District of the United States, shall remove or attempt to remove any baggage or other personal effects from the designated inspection area as provided in paragraph (h) of this section on or off the ship or other oceangoing craft unless the person has offered to an inspector for inspection, and has had passed by the inspector, the baggage and other personal effects. Persons shall disclose any fruits, vegetables, plants, plant products, or other articles that are requested to be disclosed by the inspector. When an inspection of a person's baggage or personal effects discloses an article in violation of the regulations in this part, the inspector shall seize the article. The person shall state his or her name and address to the inspector, and provide the inspector with corroborative identification. The inspector shall record the name and address of the person, the nature of the identification presented for corroboration, the nature of the violation, the types of articles involved, and the date, time, and place of the violation.

(f) *Loading of certain cargoes.* (1) Except as otherwise provided in paragraph (f)(2) of this section, no person shall present to any common carrier or contract carrier for movement, and no common carrier or contract carrier shall load, any cargo containing fruits, vegetables, or other articles regulated under this subpart that are destined for movement from Hawaii, Puerto Rico, Guam, or the U.S. Virgin Islands to

any other State unless the cargo has been offered for inspection, passed by an inspector, and bears a U.S. Department of Agriculture stamp or inspection sticker, or unless a limited permit is attached to the cargo as specified in § 318.13–3(e).

(2) Cargo designated may be loaded without a U.S. Department of Agriculture stamp or inspection sticker attached to the cargo or a limited permit attached to the cargo if the cargo is moved:

(i) As containerized cargo on ships or other oceangoing craft or as air cargo;

(ii) The carrier has on file documentary evidence that a valid limited permit was issued for the movement or that the cargo was certified; and

(iii) A notation of the existence of these documents is made by the carrier on the waybill, manifest, or bill of lading that accompanies the consignment.

(3) Cargo moved in accordance with § 318.13–6(b) that does not have a limited permit attached to the cargo must have a limited permit attached to the waybill, manifest, or bill of lading accompanying the consignment.

(g) *Removal of certain cargoes in Guam.* No person shall remove or attempt to remove from a designated inspection area as provided in paragraph (h) of this section, on or off the means of conveyance, any cargo moved from Hawaii, Puerto Rico, the Commonwealth of the Northern Mariana Islands, or the U.S. Virgin Islands to Guam containing fruits, vegetables, or other articles regulated under this subpart, unless the cargo has been inspected and passed by an inspector in Guam.

(h) *Space and facilities for baggage and cargo inspection.* Baggage and cargo inspection will not be performed until the person in charge or possession of the ship, other oceangoing craft, or aircraft provides space and facilities on the means of conveyance, pier, or airport that are adequate, in the inspector's judgment, for the performance of inspection.

§ 318.13–11 **Posting of warning notice and distribution of baggage declarations.**

(a) Before any aircraft or any ship, vessel, or other surface craft moving to Guam, the Commonwealth of Northern

Mariana Islands, or American Samoa from Hawaii or any other territory or possession of the United States arrives in Guam, the Commonwealth of Northern Mariana Islands, or American Samoa, a baggage declaration, to be furnished by the U.S. Department of Agriculture, calling attention to the provisions of the Plant Protection Act and the quarantine and regulations in this subpart, must be distributed to each adult passenger. These baggage declarations shall be executed and signed by the passengers and shall be collected and delivered by the master or other responsible officer of the aircraft, ship, vessel, or other surface craft to the inspector on arrival at the quarantine or inspection area.

(b) Every person owning or controlling any dock, harbor, or landing field in Hawaii, Puerto Rico, Guam, the Commonwealth of Northern Mariana Islands, or the U.S. Virgin Islands from which ships, vessels, other surface craft, or aircraft leave for ports in any other State shall post, and keep posted at all times, in one or more conspicuous places in passenger waiting rooms on or in said dock, harbor, or landing field a warning notice directing attention to the quarantine and regulations in this subpart. Every master, or other responsible officer of any ship, vessel, other surface craft, or aircraft leaving Hawaii, Puerto Rico, Guam, the Commonwealth of the Northern Mariana Islands, or the U.S. Virgin Islands destined to a port in any other State, shall similarly post, and keep posted at all times, such a warning notice in the ship, vessel, other surface craft, or aircraft under his charge.

§318.13-12 Movement by the U.S. Department of Agriculture.

Notwithstanding any other restrictions of this subpart, regulated articles may be moved if they are moved by the U.S. Department of Agriculture for experimental or scientific purposes and are moved under conditions found by the Administrator to be adequate to prevent the spread of plant pests and diseases.

§318.13-13 Movement of frozen fruits and vegetables.

Frozen fruits and vegetables may be certified for movement from Hawaii, Puerto Rico, Guam, the Commonwealth of the Northern Mariana Islands, or the U.S. Virgin Islands, into or through any other territory, State, or District of the United States in accordance with §318.13-3. Such fruits and vegetables must be held at a temperature not higher than 20 °F during shipping and upon arrival in the continental United States, and in accordance with the requirements for the interstate movement of frozen fruits and vegetables in part 305 of this chapter.

[74 FR 2775, Jan 16, 2009, as amended at 83 FR 46638, Sept. 14, 2018]

§318.13-14 Movement of processed fruits, vegetables, and other products.

(a) Fruits, vegetables, and other products that are processed sufficiently as to preclude the survival of any live pests can be moved interstate from Hawaii, Puerto Rico, the U.S. Virgin Islands, Guam, and the Commonwealth of the Northern Mariana Islands. Those processed products which are approved for interstate movement from those States can be found in the fruits and vegetables manuals for those States. These manuals are available on the Internet at *http://www.aphis.usda.gov/import_export/plants/manuals/ports/downloads/hawaii.pdf* and *http://www.aphis.usda.gov/import_export/plants/manuals/ports/downloads/puerto_rico.pdf.*

(b) Consignments of processed fruits, vegetables, or other products that have not been processed sufficiently as to be incapable of harboring fruit flies are subject to the interstate movement requirements which apply to the fruit, vegetable, or other product in its unprocessed state.

§318.13-15 Parcel post inspection.

Inspectors are authorized to inspect, with the cooperation of the U.S. Postal Service, parcel post packages placed in the mails in Hawaii, Puerto Rico, Guam, the Commonwealth of the Northern Mariana Islands, or the U.S. Virgin Islands to determine whether such packages contain products whose movement is not authorized under this subpart, to examine any such products that are found for insect infestation, and to notify the postmaster in writing of any violations of this subpart that are found as a result of an inspection.

§ 318.13-16 Regulated articles from Guam.

(a)(1) Regulated articles, other than soil, may be moved from Guam into or through any other State only if they meet the strictest plant quarantine requirements in accordance with part 319 of this chapter for similar articles offered for entry into such States from the countries of East and Southeast Asia, including Cambodia, India, Japan, Korea, Laos, the northeastern provinces of Manchuria, the Philippines, Taiwan, and Vietnam, or the islands of the Central and South Pacific, including Micronesia, Melanesia, and Polynesia, as well as Australia, New Zealand, and the Malay Archipelago, except requirements for permits, phytosanitary certificates, notices of arrival, and notices of consignment from port of arrival. Soil must meet the requirements of § 330.300 of this chapter.

(2) Regulated articles that do not meet the requirements of paragraph (a)(1) of this section are prohibited movement from Guam into or through any other State.

(b)(1) Regulated articles moved from Guam into or through any other State shall be subject to inspection at the port of first arrival in another part of the United States to determine whether they are free of plant pests and otherwise meet the requirements applicable to them under this subpart, and shall be subject to release, in accordance with § 330.105(a) of this chapter as if they were foreign arrivals. Such articles shall be released only if they meet all applicable requirements under this subpart.

(2) A release shall be issued in writing unless the inspection involves small quantities of regulated articles, in which case a release may be issued orally by the inspector.

[74 FR 2775, Jan 16, 2009. Redesignated and amended at 83 FR 46638, Sept. 14, 2018]

§ 318.13-17 Cut flowers from Hawaii.

(a) Except for cut blooms and leis of mauna loa and jade vine and except for cut blooms of gardenia not grown in accordance with paragraph (b) of this section, cut flowers may be moved interstate from Hawaii under limited permit, to a destination specified in the permit, directly from an establishment operated in accordance with the terms of a compliance agreement executed by the operator of the establishment, if the articles have not been exposed to infestation and they are not accompanied by any articles prohibited interstate movement under this subpart.

(b) Cut blooms of gardenia may be moved interstate from Hawaii if grown and inspected in accordance with the provisions of this section. [4]

(1) The grower's production area must be inspected annually by an inspector and found free of green scale. If green scale is found during an inspection, a 2-month ban will be placed on the interstate movement of cut blooms of gardenia from that production area. Near the end of the 2 months, an inspector will reinspect the grower's production area to determine whether green scale is present. If reinspection determines that the production area is free of green scale, shipping may resume. If reinspection determines that green scale is still present in the production area, another 2-month ban on shipping will be placed on the interstate movement of gardenia from that production area. Each ban will be followed by reinspection in the manner specified, and the production area must be found free of green scale prior to interstate movement.

(2) The grower must establish a buffer area surrounding gardenia production areas. The buffer area must extend 20 feet from the edge of the production area. Within the buffer area, the growing of gardenias and the following green scale host plants is prohibited: Ixora, ginger (*Alpinia purpurata*), plumeria, coffee, rambutan, litchi, guava, citrus, anthurium, avocado, banana, cocoa, macadamia, celery,

---

[4] Cut blooms of gardenia are also eligible for interstate movement with treatment in accordance with part 305 of this chapter.

Pluchea indica, mango, orchids, and annona.

(3) An inspector must visually inspect the cut blooms of gardenias in each consignment prior to interstate movement from Hawaii to the mainland United States. If the inspector does not detect green scale in the consignment, the inspector will certify the consignment in accordance with §318.13–3(b). If the inspector finds green scale in a consignment, that consignment will be ineligible for interstate movement from Hawaii.

(Approved by the Office of Management and Budget under control number 0579–0198)

[74 FR 2775, Jan 16, 2009. Redesignated at 83 FR 46638, Sept. 14, 2018]

Subpart B—Territorial Cotton, Cottonseed, and Cottonseed Products

SOURCE: Redesignated at 84 FR 2428, Feb, 7, 2019.

QUARANTINE

§318.47 Notice of quarantine.

(a) The Secretary of Agriculture having previously quarantined Hawaii and Puerto Rico on account of the pink bollworm of cotton (Pectinophora gossypiella Saunders) and the cotton blister mite (Eriophyes gossypii Banks), insect pests new to and not widely prevalent or distributed within and throughout the United States, now determines that it is necessary to extend the quarantine to prevent the spread of these insects from the Virgin Islands of the United States, where they are known to occur.

(b) Under the authority of sections 411, 412, 414, and 434 of the Plant Protection Act (7 U.S.C. 7711, 7712, 7714, and 7754), Hawaii, Puerto Rico, and the Virgin Islands of the United States are quarantined to prevent the spread of the aforementioned insect pests.

(c) All parts and products of plants of the genus Gossypium, such as seeds including seed cotton; cottonseed; cotton lint, linters, and other forms of cotton fiber; cottonseed hulls, cake, meal, and other cottonseed products, except oil; cotton waste; and all other unmanufactured parts of cotton plants; and all

second-hand burlap and other fabric which have been used, or are of the kinds ordinarily used, for wrapping or containing cotton, are hereby prohibited movement from the Hawaii, Puerto Rico, and the Virgin Islands of the United States into or through any other State, Territory or District of the United States, in manner or method or under conditions other than those prescribed in the regulations hereinafter made or amendments thereto: *Provided,* That whenever the Deputy Administrator of the Plant Protection and Quarantine Programs shall find that existing conditions as to the pest risk involved in the movement of the articles to which the regulations supplemental hereto apply, make it safe to modify, by making less stringent, the restrictions contained in any such regulations, he shall set forth and publish such findings in administrative instructions, specifying the manner in which the regulations should be made less stringent, whereupon such modification shall become effective.

(d) As used in this subpart, unless the context otherwise requires, the term *State, Territory, or District of the United States* means State, the District of Columbia, Alaska, Guam, Hawaii, Puerto Rico, or the Virgin Islands of the United States.

[24 FR 10777, Dec. 29, 1959, as amended at 66 FR 21054, Apr. 27, 2001]

§318.47a Administrative instructions relating to Guam.

The plants, products and articles specified in §318.47(c) may be moved from Hawaii into or through Guam without restriction under this subpart.

RULES AND REGULATIONS

CROSS REFERENCE: For rules and regulations governing the importation of cotton and cottonseed products into the United States, see §§319.8 to 319.8–27 of this chapter.

§318.47–1 Definitions.

For the purpose of the regulations in this subpart the following words, names, and terms shall be construed, respectively, to mean:

(a) *Cotton.* Parts and products of plants of the genus Gossypium, including seed cotton; cottonseed; cotton

189

lint, linters and other forms of cotton fiber; cottonseed hulls, cake, meal, and other cottonseed products, except oil; cotton waste; and all other unmanufactured parts of cotton plants; and second-hand burlap and other fabric which have been used, or are of the kinds ordinarily used, for wrapping or containing cotton.

(b) *Seed cotton.* The unginned lint and seed admixture, just as it is picked from the cotton boll.

(c) *Cottonseed.* The seed of the cotton plant, either separated from the lint or as a component part of seed cotton.

(d) *Lint.* All forms of raw or unmanufactured ginned cotton, either baled or unbaled, including all cotton fiber, except linters, which has not been woven or spun, or otherwise manufactured.

(e) *Linters.* All forms of unmanufactured cotton fiber separated from cottonseed after the lint has been removed, including that form referred to as "hull fiber."

(f) *Waste.* All forms of cotton waste derived from the manufacture of cotton lint, in any form or under any trade designation, including gin waste; and waste products derived from the milling of cottonseed.

(g) *Seedy waste.* Picker waste, gin waste, and oil mill waste, and any other cotton by-products capable of carrying a high percentage of cottonseed.

(h) *Clean waste.* Wastes derived from the processing of lint in machines after the card machine, including card strips but not card fly.

(i) *Bale covers.* Second-hand burlap and other second-hand fabric by whatever trade designation, which have been used, or are of the kinds ordinarily used, for wrapping or otherwise containing cotton. Burlap and other fabric of the kinds ordinarily used for wrapping cotton, when new or unused, are excluded from this definition.

(j) *Certificate (certification, certified).* A type of authorization, evidencing freedom from infestation, issued by the Deputy Administrator of the Plant Protection and Quarantine Programs to allow the movement of lint, linters, waste, seed cotton, cottonseed, cottonseed hulls, cake, and meal, and bale covers in accordance with the regulations in this subpart. "Certification"

and "certified" shall be construed accordingly.

(k) *Permit.* A type of general authorization issued by the Deputy Administrator of the Plant Protection and Quarantine Programs to allow the movement of lint, linters, waste other than seedy waste, cottonseed cake and meal, and bale covers in accordance with the regulations in this subpart.

(l) *Fumigated.* Fumigated under the supervision of an inspector of the Plant Protection and Quarantine Programs in a fumigation plant approved by the Deputy Administrator of said Programs and in accordance with methods approved by him.

(m) *Moved (movement, move).* Shipped, offered for shipment to a common carrier, received for transportation or transported by a common carrier, or carried, transported, moved, or allowed to be moved, directly or indirectly, from Hawaii, Puerto Rico, or the Virgin Islands of the United States, into or through any other State, Territory, or District of the United States. "Movement" and "move" shall be construed accordingly.

§318.47–2 Articles the movement of which is prohibited or regulated.

(a) *Articles prohibited movement.* The movement of seed cotton, cottonseed, and seedy waste, when unfumigated, is prohibited except as provided in §318.47–3(b)(2).

(b) *Articles the movement of which is regulated.* Lint; linters; waste; seed cotton; cottonseed; cottonseed hulls, cake, and meal; and bale covers may be moved upon compliance with the conditions prescribed in §318.47–3.

§318.47–3 Conditions governing the issuance of certificates and permits.

(a) *Fumigated lint; linters; waste; seed cotton; cottonseed; cottonseed hulls, cake, and meal; and bale covers.* Lint; linters; waste; seed cotton; cottonseed; cottonseed hulls, cake, and meal; and bale covers, fumigated in the Territory or District of origin in accordance with part 305 of this chapter and so certified, are allowed unrestricted movement to any port.

(b) *Unfumigated lint, linters, waste, and bale covers.* (1) Unfumigated Hawaiian, Puerto Rican, or Virgin Islands of the United States lint, linters, waste other than seedy waste, and bale covers will be allowed to move under permit, by all-water route, for entry only at the ports of Norfolk, Baltimore, New York, Boston, San Francisco, and Seattle, or other port of arrival designated in the permit, and at such designated port of arrival shall become subject to the regulations governing the handling of cotton imported from foreign countries.

(2) Fumigation may be waived and certificates issued for lint, linters, and waste which have been determined by an inspector of the Plant Protection and Quarantine Programs to have been so manufactured or processed by bleaching, dyeing, or other means, as to have removed all seeds, or to have destroyed all insect life therein.

(c) *Cottonseed cake and meal.* (1) Cottonseed cake and meal which have been inspected in the Territory or District of origin and certified by an inspector of the Plant Protection and Quarantine Programs as being free from contamination with whole, uncrushed cottonseed, will be allowed unrestricted movement to any port.

(2) Hawaiian, Puerto Rican, and Virgin Islands of the United States cottonseed cake and meal, when neither fumigated nor inspected in accordance with the provisions of this section, will be allowed entry under permit through any port at which the services of an inspector are available, subject to examination by an inspector for freedom from contamination with uncrushed cottonseed. If found to be free from such contamination, the cottonseed cake or meal may be released from further entry restrictions. Cottonseed cake or meal found to be contaminated shall be refused entry or subjected as a condition of entry and release to such safeguards as may be prescribed by the inspector from such administratively approved methods as will, in his judgment, be necessary to eliminate infestations of the pink bollworm or cotton blister mite.

[24 FR 10777, Dec. 29, 1959, as amended at 75 FR 4251, Jan. 26, 2010]

§ 318.47–4　Shipments by the Department of Agriculture.

Cotton may be moved by the Department of Agriculture for experimental or scientific purposes under such conditions as may be prescribed by the Deputy Administrator of the Plant Protection and Quarantine Programs, which conditions may include clearance through the New Crops Research Branch of the Plant Science Research Division, Agricultural Research Service.

Subpart C—Sand, Soil, or Earth, with Plants from Territories and Districts

SOURCE: Redesignated at 84 FR 2428, Feb. 7, 2019.

§ 318.60　Notice of quarantine.

(a) The Secretary of Agriculture, having previously quarantined Hawaii and Puerto Rico to prevent the spread to other parts of the United States, by means of sand, soil, or earth about the roots of plants, of immature stages of certain dangerous insects, including Phyllophaga spp. (White grubs), Phytalus sp., and Adoretus sp., and of several species of termites or white ants, new to and not heretofore widely prevalent or distributed within and throughout the United States, now determines that it is necessary also to quarantine the Virgin Islands of the United States to prevent the spread of such dangerous insects from said Virgin Islands.

(b) Under the authority of sections 411, 412, 414, and 434 of the Plant Protection Act (7 U.S.C. 7711, 7712, 7714, and 7754), Hawaii, Puerto Rico, and the Virgin Islands of the United States are quarantined to prevent the spread of the aforementioned dangerous insects.

(c) Sand (other than clean ocean sand), soil, or earth around the roots of plants must not be shipped, offered for shipment to a common carrier, received for transportation or transported by a common carrier, or carried, transported, moved, or allowed to be moved by any person from Hawaii, Puerto Rico, or the Virgin Islands of the United States into or through any other State, Territory, or District of the United States: *Provided,* That the prohibitions in this paragraph (c) do not apply to the movement of soil from Hawaii, Puerto Rico, and the Virgin Islands other than that soil around the roots of plants; movement of soil that is not around the roots of plants is regulated under part 330 of this chapter: *Provided further,* That the prohibitions of this section shall not apply to the movement of such products in either direction between Puerto Rico and the Virgin Islands of the United States: *Provided further,* That such prohibitions shall not prohibit the movement of such products by the United States Department of Agriculture for scientific or experimental purposes, nor prohibit the movement of sand, soil, or earth around the roots of plants which are carried, for ornamental purposes, on vessels into mainland ports of the United States and which are not intended to be landed thereat, when evidence is presented satisfactory to the inspector of the Plant Protection and Quarantine Programs of the Department of Agriculture that such sand, soil, or earth has been so processed or is of such nature that no pest risk is involved, or that the plants with sand, soil, or earth around them are maintained on board under such safeguards as will preclude pest escape: *And provided further,* That such prohibitions shall not prohibit the movement of plant cuttings or plants that have been—

(1) Freed from sand, soil, and earth;

(2) Subsequently potted and established in sphagnum moss or other packing material approved under §319.37–11 of this chapter that had been stored under shelter and had not been previously used for growing or packing plants;

(3) Grown thereafter in a manner satisfactory to an inspector of the Plant Protection and Quarantine Programs to prevent infestation through contact with sand, soil, or earth; and

(4) Certified by an inspector of the Plant Protection and Quarantine Programs as meeting the requirements of paragraphs (c)(1) through (3) of this section.

(d) As used in this section, the term *State, Territory, or District of the United States* means "Guam, Hawaii, Puerto Rico, the Virgin Islands of the United States, or the continental United States."

[24 FR 10777, Dec. 29, 1959, as amended at 66 FR 21054, Apr. 27, 2001; 83 FR 11855, Mar. 19, 2018; 84 FR 29957, June 25, 2019]

PART 319—FOREIGN QUARANTINE NOTICES

Subpart A—Preemption

Sec.
319.1   Preemption of State and local laws.

Subpart B—Requests To Amend The Regulations

319.5   Requirements for submitting requests to change the regulations in 7 CFR part 319.

Subpart C—Controlled Import Permits

319.6   Controlled import permits.

Subpart D—Permits: Allocation, Issuance, Denial, and Revocation

319.7   Definitions.
319.7–1   Applying for a permit.
319.7–2   Issuance of permits and labels.
319.7–3   Denial of permits.
319.7–4   Withdrawal, cancellation, and revocation of permits.
319.7–5   Appeal of denial or revocation.

Subpart E—Foreign Cotton and Covers

QUARANTINE

319.8   Notice of quarantine.
319.8a   Administrative instructions relating to the entry of cotton and covers into Guam.

REGULATIONS; GENERAL

319.8–1   Definitions.

CONDITIONS OF IMPORTATION AND ENTRY OF COTTON AND COVERS

319.8–2   Permit procedure.
319.8–3   Refusal and cancellation of permits.
319.8–4   Notice of arrival.
319.8–5   Marking of containers.
319.8–6   Cottonseed cake and cottonseed meal.
319.8–7   Processed lint, linters, and waste.
319.8–8   Lint, linters, and waste.
319.8–9   Hull fiber and gin trash.
319.8–10   Covers.

SPECIAL CONDITIONS FOR THE ENTRY OF COTTON AND COVERS FROM MEXICO

319.8–11   From approved areas of Mexico.
319.8–12   From the West Coast of Mexico.

AUTHORITY: 7 U.S.C. 1633, 7701–7772, and 7781–7786; 21 U.S.C. 136 and 136a; 7 CFR 2.22, 2.80, and 371.3.

SOURCE: 24 FR 10788, Dec. 29, 1959, unless otherwise noted.

Subpart A—Preemption

SOURCE: 75 FR 17292, Apr. 6, 2010, unless otherwise noted. Redesignated at 84 FR 2428, Feb. 7, 2019.

§ 319.1 Preemption of State and local laws.

(a) Under section 436 of the Plant Protection Act (7 U.S.C. 7756), a State or political subdivision of a State may not regulate in foreign commerce any plant or plant product in order to control, eradicate, or prevent the introduction or dissemination of a biological control organism, plant pest, or noxious weed within the United States.

(b) Therefore, in accordance with section 436 of the Plant Protection Act, the regulations in this part preempt all State and local laws that are inconsistent with or exceed the regulations in this part.

Subpart B—Requests To Amend The Regulations

SOURCE: Redesignated at 84 FR 2428, Feb. 7, 2019.

§319.5 Requirements for submitting requests to change the regulations in 7 CFR part 319.

(a) *Definitions.*

Commodity. A plant, plant product, or other agricultural product being moved for trade or other purpose.

(b) *Procedures for submitting requests and supporting information.* Persons who request changes to the import regulations contained in this part and who wish to import plants, plant parts, or plant products that are not allowed importation under the conditions of this part must file a request with the Animal and Plant Health Inspection Service (APHIS) in order for APHIS to consider whether the new commodity can be safely imported into the United States. The initial request can be formal (e.g., a letter) or informal (e.g., made during a bilateral discussion between the United States and another country), and can be made by any person. Upon APHIS confirmation that granting a person's request would require amendments to the regulations in this part, the national plant protection organization of the country from which the commodity would be exported must provide APHIS with the information listed in paragraph (d) of this section before APHIS can proceed with its consideration of the request; requests that are not supported with this information in a timely manner will be considered incomplete and APHIS may not take further action on such requests until all required information is submitted.

(c) *Addresses.* The national plant protection organization of the country from which commodities would be exported must submit the information listed in paragraph (d) of this section to: Commodity Import Analysis and Operations, PPQ, APHIS, 4700 River Road Unit 140, Riverdale, MD 20737.

(d) *Information.* The following information must be provided to APHIS in order for APHIS to consider a request to change the regulations in part 319:

(1) *Information about the party submitting the request.* The address, telephone and fax numbers, and e-mail addresses of the national plant protection organization of the country from which commodities would be exported; or, for requests that address a multi-country region, the address, telephone and fax numbers, and e-mail addresses of the exporting countries' national and regional plant protection plant protection organizations.

(2) *Information about the commodity proposed for importation into the United States.* (i) A description and/or map of the specific location(s) of the areas in the exporting country where the plants, plant parts, or plant products are produced;

(ii) The scientific name (including genus, species, and author names), synonyms, and taxonomic classification of the commodity;

(iii) Identification of the particular plant or plant part (*i.e.,* fruit, leaf, root, entire plant, etc.) and any associated plant part proposed for importation into the United States;

(iv) The proposed end use of the imported commodity (e.g., propagation, consumption, milling, decorative, processing, etc.); and

(v) The months of the year when the commodity would be produced, harvested, and exported.

(3) *Shipping information:* (i) Detailed information as to the projected quantity and weight/Volume of the proposed importation, broken down according to varieties, where applicable, and;

(ii) Method of shipping in international commerce and under what conditions, including type of conveyance, and type, size, and capacity of packing boxes and/or shipping containers.

(4) *Description of pests and diseases associated with the commodity*[1] (i) Scientific name (including genus, species,

---

[1] When a change is being sought to the conditions governing the importation of a commodity that is already authorized for importation into the United States, an update to or confirmation of previously submitted pest and disease information, rather than a new, complete submission of that information, may be appropriate. Persons seeking such a

Continued

and author names) and taxonomic classification of arthropods, fungi, bacteria, nematodes, virus, viroids, mollusks, phytoplasmas, spiroplasmas, etc., attacking the crop;

(ii) Plant part attacked by each pest, pest life stages associated with each plant part attacked, and location of pest (in, on, or with commodity); and

(iii) References.

(5) *Current strategies for risk mitigation or management.* (i) Overview of agronomic or horticultural management practices used in production of the commodity, including methods of pest risk mitigation or control programs; and

(ii) Identification of parties responsible for pest management and control.

(e) *Additional information.* None of the additional information listed in this paragraph need be provided at the same time as information required under paragraphs (a) through (d) of this section; it is required only upon request by APHIS. If APHIS determines that additional information is required in order to complete a pest risk analysis in accordance with international standards for pest risk analysis, we will notify the party submitting the request in writing what specific additional information is required. If this information is not provided, and is not available to APHIS from other sources, a request may be considered incomplete and APHIS may be unable to take further action on the request until the necessary additional information is submitted. The additional information may include one or more of the following types of information:

(1) *Contact information:* Address, phone and fax numbers, and/or e-mail address for local experts (e.g., academicians, researchers, extension agents) most familiar with crop production, entomology, plant pathology, and other relevant characteristics of the commodity proposed for importation.

(2) *Additional information about the commodity:* (i) Common name(s) in English and the language(s) of the exporting country;

(ii) Cultivar, variety, or group description of the commodity;

(iii) Stage of maturity at which the crop is harvested and the method of harvest;

(iv) Indication of whether the crop is grown from certified seed or nursery stock, if applicable;

(v) If grown from certified seed or stock, indication of the origin of the stock or seed (country, State); and

(vi) Color photographs of plant, plant part, or plant product itself.

(3) *Information about the area where the commodity is grown:* (i) Unique characteristics of the production area in terms of pests or diseases;

(ii) Maps of the production regions, pest-free areas, etc.;

(iii) Length of time the commodity has been grown in the production area;

(iv) Status of growth of production area (*i.e.,* acreage expanding or stable); and

(v) Physical and climatological description of the growing area.

(4) *Information about post-harvest transit and processing:* (i) Complete description of the post-harvest processing methods used; and

(ii) Description of the movement of the commodity from the field to processing to exporting port (e.g., method of conveyance, shipping containers, transit routes, especially through different pest risk areas).

(5) *Shipping methods:* (i) Photographs of the boxes and containers used to transport the commodity; and

(ii) Identification of port(s) of export and import and expected months (seasons) of shipment, including intermediate ports-of-call and time at intermediate ports-of-call, if applicable.

(6) *Additional description of all pests and diseases associated with the commodity to be imported:* (i) Common name(s) of the pest in English and local language(s);

(ii) Geographic distribution of the pest in the country, if it is a quarantine pest and it follows the pathway;

(iii) Period of attack (e.g., attacks young fruit beginning immediately after blooming) and records of pest incidence (e.g., percentage of infested plants or infested fruit) over time (e.g., during the different phenological

---

change may contact APHIS for a determination as to whether an update will be appropriate in a particular case.

stages of the crops and/or times of the year);

(iv) Economic losses associated with pests of concern in the country;

(v) Pest biology or disease etiology or epidemiology; and

(vi) Photocopies of literature cited in support of the information above.

(7) *Current strategies for risk mitigation or management:* (i) Description of pre-harvest pest management practices (including target pests, treatments [e.g., pesticides], or other control methods) as well as evidence of efficacy of pest management treatments and other control methods;

(ii) Efficacy of post-harvest processing treatments in pest control;

(iii) Culling percentage and efficacy of culling in removing pests from the commodity; and

(iv) Description of quality assurance activities, efficacy, and efficiency of monitoring implementation.

(8) *Existing documentation:* Relevant pest risk analyses, environmental assessment(s), biological assessment(s), and economic information and analyses.

(f) *Availability of additional guidance.* Information related to the processing of requests to change the import regulations contained in this part may be found on the APHIS Web site at *http://www.aphis.usda.gov/ppq/pra/.*

(Approved by the Office of Management and Budget under control number 0579–0261)

[71 FR 30567, May 30, 2006]

Subpart C—Controlled Import Permits

SOURCE: 78 FR 25568, May 2, 2013, unless otherwise noted. Redesignated at 84 FR 2428, Feb. 7, 2019.

§319.6 Controlled import permits.

(a) *Definitions.*

Administrator. The Administrator of the Animal and Plant Health Inspection Service, United States Department of Agriculture, or any employee of the United States Department of Agriculture delegated to act in his or her stead.

Developmental purposes. The evaluation, monitoring, or verification of plant material for plant health risks and/or the adaptability of the material for certain uses or environments.

Experimental purposes. Scientific testing which utilizes collected data and employs analytical processes under controlled conditions to create qualitative or quantitative results.

Therapeutic purposes. The application of specific scientific processes designed to eliminate, isolate, or remove potential plant pests or diseases.

(b) *Purpose and scope.* The regulations in this part prohibit or restrict the importation into the United States of certain plants, plant products, and other articles to prevent the introduction and dissemination of plant pests and noxious weeds within and throughout the United States. The regulations in this subpart provide a process under which a controlled import permit (CIP) may be issued to authorize the importation, for experimental, therapeutic, or developmental purposes, of an article whose importation is prohibited under this part. A CIP may also be issued to authorize, for those same purposes, the importation of an article under conditions that differ from those prescribed in the relevant regulations in this part.

(c) *Application process.* Applications for a CIP are available without charge from the Animal and Plant Health Inspection Service (APHIS), Plant Protection and Quarantine (PPQ), Permit Unit, 4700 River Road Unit 136, Riverdale, MD 20737–1236, or from local PPQ offices. Applications may be submitted by mail, by fax, or electronically and must be submitted at least 60 days prior to arrival of the article at the port of entry. Mailed applications must be submitted to the address above, faxed applications may be submitted to 301–734–4300, and electronic applications may be submitted through the ePermits Web site at *https://epermits.aphis.usda.gov/epermits.*

(1) The completed application for a CIP must provide the following information:

(i) Name, address in the United States, and contact information of the applicant;

(ii) Identity (common and botanical [genus and species] names) of the plant material to be imported, quantity of importation, country of origin, and country shipped from;

(iii) Intended experimental, therapeutic, or developmental purpose for the importation; and

(iv) Intended ports of export and entry, means of conveyance, and estimated date of arrival.

(2) APHIS may issue a CIP if the Administrator determines that the plant pest risks associated with the plant material and its intended experimental, therapeutic, or developmental use can be effectively mitigated. The CIP will contain the applicable conditions for importation and subsequent handling of the plant material if it is deemed eligible to be imported into the United States, including the specifications for the facility where the plant will be held. The plant material may be imported only if all applicable requirements are met.

(d) *Shipping conditions.* Consignments of plant material to be offered for importation under a CIP must meet the following requirements, unless otherwise specified under the conditions of the CIP:

(1) The plant material must be selected from apparently disease-free and pest-free sources.

(2) The plant material must be free of soil, other foreign matter or debris, other prohibited plants, noxious weed seeds, and living organisms such as parasitic plants, pathogens, insects, snails, and mites.

(3) Fungicides, insecticides, and other treatments such as coatings, dips, or sprayings must not be applied before shipment, unless otherwise specified. Plant materials may be refused entry if they are difficult or hazardous to inspect because of the presence of such treatments. Plant materials must not be wrapped or otherwise packaged in a manner that impedes or prevents adequate inspection or treatment.

(4) The plant material must be moved in an enclosed container or one completely enclosed by a covering adequate to prevent the possible escape or introduction of plant pests during shipment. Any packing material used in the consignment of the plant material must meet the requirements of § 319.37–11, and wood packing material used in

the consignment must meet the requirements of § 319.40–3(b) and (c).

(5) Consignments may be shipped as cargo, by mail or air freight, or hand-carried, as specified in the conditions of the CIP.

(6) The plant material must be offered for importation at the port of entry or plant inspection station as specified in the conditions of the CIP.

(7) A copy of the CIP must accompany each consignment, and all consignments must be labeled in accordance with instructions in the CIP.

(8) Each consignment must be accompanied by an invoice or packing list indicating its contents.

(e) *Post-importation conditions.* (1) At the approved facility where the plant material will be maintained following its importation, plant material imported under a CIP must be identified and labeled as quarantine material to be used only in accordance with a valid CIP.

(2) Plant material must be stored in a secure place or in the manner indicated in the CIP and be under the supervision and control of the permit holder. During regular business hours, properly identified officials, either Federal or State, must be allowed to inspect the plant material and the facilities in which the plant material is maintained.

(3) The permit holder must keep the permit valid for the duration of the authorized experimental, therapeutic, or developmental purpose. The PPQ Permit Unit must be informed of a change in contact information for the permit holder within 10 business days of such change.

(4) Plant material imported under a CIP must not be moved or distributed to another person without prior authorization from the PPQ Permit Unit.

(5) Should the permit holder leave the institution in which the plant material imported under a CIP is kept, the plant material must be destroyed unless, prior to the departure of the original permit holder, another person assumes responsibility for the continued maintenance of the plant material and such person obtains a new CIP for the plant material. Should the permit holder be otherwise unavailable to maintain the plant material for which

the CIP was issued, the plant material must be destroyed unless another person assumes responsibility for the continued maintenance of the plant material and such person obtains a new CIP for the plant material. Permission to move or distribute plant material that was authorized for importation under a CIP to another person must be obtained by contacting the PPQ Permit Unit.

(6) CIPs issued by APHIS are valid for a period of 1 year. The permittee may request the existing permit be renewed for up to an additional 2 years prior to the expiration of the CIP and if no adverse indications exist from the previous year.

(f) Failure to comply with all of the conditions specified in the CIP or any applicable regulations or administrative instructions, or forging, counterfeiting, or defacing permits or shipping labels, may result in immediate revocation of the permit, denial of future permits, and civil or criminal penalties for the permit holder.

(g) *Denial, withdrawal, cancellation, or revocation of permit.* The Administrator may deny a permit application in accordance with §319.7-3, and a permit may be withdrawn, canceled, or revoked in accordance with §319.7-4.

(1) *Action upon cancellation or revocation of permit.* Upon cancellation or revocation of a permit, the permittee must surrender, destroy, or remove all regulated plant material covered by the permit in accordance with §319.7-4(e).

(2) *Appeal of denial or revocation.* Any person whose application for a permit has been denied or whose permit has been revoked may appeal the denial or revocation in accordance with §319.7-5.

(Approved by the Office of Management and Budget under control number 0579-0384)

[78 FR 25568, May 2, 2013, as amended at 79 FR 19807, Apr. 10, 2014; 81 FR 40150, June 21, 2016; 83 FR 11855, Mar. 19, 2018]

Subpart D—Permits: Allocation, Issuance, Denial, and Revocation

SOURCE: 78 FR 19807, Apr. 10, 2014, unless otherwise noted. Redesignated at 84 FR 2428, Feb. 7, 2019.

§319.7 Definitions.

The following definitions apply to this subpart:

Administrative instructions. Published documents related to the enforcement of this part and issued under authority of the Plant Protection Act, as amended (7 U.S.C. 7701 *et seq.*), by the Administrator.

Administrator. The Administrator of the Animal and Plant Health Inspection Service or any employee of the United States Department of Agriculture delegated to act in his or her stead.

Animal and Plant Health Inspection Service (APHIS). The Animal and Plant Health Inspection Service of the United States Department of Agriculture.

Applicant. A person at least 18 years of age who, on behalf of him- or herself or another person, submits an application for a permit to import into the United States or move interstate a regulated article in accordance with this part.

Approved. Approved by the Administrator of the Animal and Plant Health Inspection Service.

Article. Any material or tangible objects that could harbor or be a vector of plant pests or noxious weeds.

Consignment. A quantity of plants, plant products, and/or other articles being moved from one country to another authorized when required, by a single permit. A consignment may be composed of one or more commodities or lots.

Country of origin. The country where the plants, or plants from which the plant products are derived, were grown or where the non-plant articles were produced.

Enter, entry. To move into, or the act of movement into, the commerce of the United States.

Import, importation. To move into, or the act of movement into, the territorial limits of the United States.

Inspector. Any individual authorized by the Administrator of the Animal and Plant Health Inspection Service or the Commissioner of the Bureau of Customs and Border Protection, Department of Homeland Security, to enforce the regulations in this part.

Intended use. The purpose for the importation of the regulated article, including, but not limited to, consumption, propagation, or research purposes.

Lot. All the regulated. articles on a single means of conveyance that are derived from the same species of plant or are the same type of non-plant article, were subjected to the same treatments prior to importation, and are consigned to the same person.

Means of conveyance. Any personal property used for or intended for use for the movement of any other .personal property.

Move. To carry,. enter, import, mail, ship, or transport; to aid, abet, cause, or induce the carrying, entering, importing, mailing, shipping, or transporting; to offer to carry, enter, import, mail, ship, or transport; to receive to carry, enter, import, mail, ship, or transport; to release into the environment; or to allow any of the activities described in this definition.

Oral authorization. Verbal permission to import that may be granted by an inspector at the port of entry.

Permit. A written authorization, including by electronic methods, to move plants, plant products, biological control organisms, plant pests, noxious weeds, or articles under conditions prescribed by the Administrator.

Permittee. The person who, on behalf of self or another person, is legally the importer of an article, meets the requirements of §319.7–2(f), and·is responsible for compliance with the conditions for the importation that is the subject of a permit issued in accordance with this.part.

Person. Any individual, partnership, corporation, association, joint venture, or other legal entity.

Plant. Any plant (including any plant part) for or capable of propagation, including a tree, a tissue culture, a plantlet culture, pollen, a shrub, a vine, a cutting, a graft, a scion, a bud, a bulb, a root, and a.seed.

Plant pest. Any living stage of any of the following that can directly or indirectly injure, cause damage to, or cause disease in any plant or plant product: A protozoan; a nonhuman animal; a parasitic plant; a bacterium; a fungus; a virus or viroid; an infectious agent or other pathogen; or any article similar to or allied with any of the foregoing enumerated articles.

Plant product. Any flower, fruit, vegetable, root, bulb, seed, or other plant part that is not included in the definition of *plant,* or any manufactured or processed plant or plant part.

Port of entry. A port at which a specified shipment or means of conveyance is accepted for entry or admitted without entry into the United States for transit purposes.

Port· of first arrival. The area (such as a seaport, airport, or .land border) where a person or means of conveyance first arrives in the United States, and where inspection of regulated articles may be carried out by inspectors.

PPQ. The Plant Protection and Quarantine Program, Animal and Plant Health Inspection Service of the United States Department of Agriculture, delegated responsibility for enforcing provisions of the Plant Protection Act and related legislation, quarantines and regulations.

Regulated article. Any material or tangible object regulated by this part for entry into the United States or interstate movement.

Soil. The unconsolidated material from the earth's surface that consists of rock and mineral particles mixed with organic material and that supports or is capable of supporting biotic communities.

State. Any of the several States of the United States, the Commonwealth of the Northern Mariana Islands, the Commonwealth of Puerto Rico, the District of Columbia, Guam, the Virgin Islands of the United States, or any other territory or possession of the United States.

Treatment. A procedure approved by the Administrator for neutralizing infestations or infections of plant pests or diseases, such as fumigation, application of chemicals or dry or moist heat, or processing, utilization, or storage.

United States. All of the States.

§ 319.7–1   Applying for a permit.

(a) Persons who wish to import regulated articles into the United States must apply for a permit, unless the regulated articles are not subject to a requirement under this part that a permit be issued prior to a consignment's arrival. An applicant for a permit to import regulated articles into the United States in accordance with this part must be:

(1) Capable of acting in the capacity of the permittee in accordance with §319.7–2(e), or must designate a permittee who is so capable should the permit be issued;

(2) Applying for a permit on behalf of self or on behalf of another person as permittee; and

(3) At least 18 years of age.

(b) Permit applications must be submitted by the applicant in writing or electronically through one of the means listed at *http:// www.aphis.usda.gov/plant_health/permits/index.shtml* in advance of the action(s) proposed on the permit application.

(c) The application for a permit must contain the following information:

(1) Legal name, address, and contact information of the applicant, and affirmation by the applicant that the applicant is at least 18 years of age;

(2) The same information of the permittee if different from the applicant, and, if the permittee is an individual, affirmation by the permittee that the permittee is at least 18 years of age;

(3) Specific type of regulated article (common and scientific names, if applicable);

(4) Country of origin;

(5) Intended use of the regulated article;

(6) Intended port(s) of first arrival; and

(7) A description of any processing, treatment, or handling of the regulated article to be performed prior to or following importation, including the location where any processing or treatment was or will be performed and the names and dosage of any chemical employed in treatments of the regulated article.

(d) The application for a permit may also require the following information:

(1) Means of conveyance;

(2) Quantity of the regulated article;

(3) Estimated date of arrival;

(4) Name, address, and contact information of any broker or subsequent custodian of the regulated article;

(5) Exporting country from which the article is to be moved, when not the country of origin; and

(6) Any other information determined to be necessary by APHIS to inform the decision to issue the permit.

(e) Application for a permit to import regulated articles into the United States must be submitted at least 30 days prior to arrival of the article at the port of entry.

(1) If, through no fault of the importer, a consignment of regulated articles subject to a requirement under this part that a permit be issued prior to a consignment's arrival arrives at a U.S. port before a permit is received, the consignment may be held, under suitable safeguards prescribed by the inspector, in custody at the risk and expense of the importer pending issuance of a permit or authorization from APHIS.

(2) An oral authorization may be granted by an inspector at the port of entry for a consignment, provided that:

(i) All applicable entry requirements are met;

(ii) Proof of application for a written permit is provided to the inspector; and

(iii) PPQ verifies that the application for a written permit has been received and that PPQ intends to issue the permit.

§319.7–2 Issuance of permits and labels.

(a) Upon receipt of an application, APHIS will issue a permit if, after review of the application, APHIS determines that the regulated articles are eligible to be imported into the United States under any applicable conditions. The permit will specify the applicable conditions of entry and the port of entry, and a copy will be provided to the permittee. The permit will only be valid for the time period indicated on the permit.

(b) The applicant for a permit for the importation of regulated articles into the United States must designate the person who will be named as the permittee upon the permit's issuance. The applicant and the permittee may be the same person or different persons.

(c) The act, omission, or failure of the permittee as an officer, agent, or person acting for or employed by any other person within the scope of his or her employment or office will be deemed also to be the act, omission, or failure of the other person.

(d) Failure to comply with all of the conditions specified in the permit or any applicable regulations or administrative instructions, or forging, counterfeiting, or defacing permits or shipping labels, may result in immediate revocation of the permit, denial of any future permits, and civil or criminal penalties for the permittee.

(e) The permittee will remain responsible for the consignment regardless of any delegation to a subsequent custodian of the importation.

(f) A permittee must:

(1) If an individual, be at least 18 years of age and have and maintain an address in the United States that is specified on the permit and be physically present during normal business hours at that address during any periods when articles are being imported or moved interstate under the permit; or

(2) If another legal entity, maintain an address or business office in the United States with a designated individual for service of process; and

(3) Serve as the contact for the purpose of communications associated with the movement of the regulated article for the duration of the permit. The PPQ Permit Unit must be informed of a change in contact information for the permittee within 10 business days of such change;

(4) Ensure compliance with the applicable regulatory requirements and permit conditions associated with the movement of the regulated article for the duration of the permit;

(5) Provide written or electronic acknowledgment and acceptance of permit conditions when APHIS requests such acknowledgment;

(6) Serve as the primary contact for communication with APHIS regarding the permit; and

(7) Maintain all conditions of the permit for the entirety of its prescribed duration.

(g) The regulated article may be imported only if all applicable requirements of the permit issued for the importation of the regulated article or any other documents or instructions issued by APHIS are met and complied with as determined by APHIS.

(h) In accordance with the regulations in this part, labels may be issued to the permittee for the importation of regulated articles. Such labels may contain information about the shipment's nature, origin, movement conditions, or other matters relevant to the permit and will indicate that the importation is authorized under the conditions specified in the permit.

(1) If issued, the quantity of labels will be sufficient for the permittee to attach one to each parcel. Labels must be affixed to the outer packaging of the parcel.

(2) Importations without such required labels will be refused entry into the United States, unless a label is not required and not issued for the importation.

(i) Even if a permit has been issued for the importation of a regulated article, the regulated article may be imported only if an inspector at the port of entry determines that no remedial measures pursuant to the Plant Protection Act are necessary to mitigate or address any plant pest or noxious weed risks.[1]

(j) A permit application may be withdrawn at the request of the applicant prior to the issuance of the permit.

(k) A permit may be canceled after issuance at the request of the permittee.

(l) A permit may be amended if APHIS finds that the permit is incomplete or contains factual errors.

(m) In accordance with Section 7734 of the Plant Protection Act, as amended (7 U.S.C. 7701 *et seq.*), the actions, omissions, or failures of any agent of the permittee may be deemed the actions, omissions, or failures of a permittee as well; and that failure to comply with all of the conditions specified in the permit or any applicable regulations or administrative instructions, or forging, counterfeiting, or defacing permits or shipping labels, may result in immediate revocation of the permit,

---

[1] An inspector may hold, seize, quarantine, treat, apply other remedial measures to, destroy, or otherwise dispose of plants, plant pests, and other articles in accordance with sections 414, 421, and 434 of the Plant Protection Act (7 U.S.C. 7714, 7731, and 7754).

denial of any future permits, and civil or criminal penalties for the permittee.

§319.7–3 Denial of permits.

(a) APHIS may deny an application for a permit to import a regulated article into the United States. A denial, including the reason for the denial, will be provided in writing, including by electronic methods, to the applicant as promptly as circumstances permit. The denial of a permit may be appealed in accordance with §319.7–5.

(b) APHIS may deny an application for a permit to import a regulated article:

(1) If APHIS determines that the applicant is not likely to abide by permit conditions. Factors that may lead to such a determination include, but are not limited to, the following:

(i) The applicant, or a partnership, firm, corporation, or other legal entity in which the applicant has a substantial interest, financial or otherwise, has not complied with any permit that was previously issued by APHIS;

(ii) APHIS determines that issuing the permit would circumvent any order revoking or denying a permit under the Plant Protection Act;

(iii) APHIS determines that the applicant has previously failed to comply with any APHIS regulation;

(iv) APHIS determines that the applicant has previously failed to comply with any Federal, State, or local law, regulation, or instruction concerning the importation of prohibited or restricted foreign agricultural products;

(v) APHIS determines that the applicant has failed to comply with the laws or regulations of a national plant protection organization or equivalent body, as these pertain to plant health;

(vi) APHIS determines that the applicant has made false or fraudulent statements or provided false or fraudulent records to APHIS; or

(vii) The applicant has been convicted or has pled nolo contendere to any crime involving fraud, bribery, extortion, or any other crime involving a lack of integrity.

(2) If the application for a permit contains information that is found to be materially false, fraudulent, deceptive, or misrepresentative;

(3) If APHIS concludes that the actions proposed under the permit would present an unacceptable risk to plants and plant products because of the potential for introduction or dissemination of a plant pest or noxious weed within the United States;

(4) If the importation is adverse to the conduct of an eradication, suppression, control, or phytosanitary program of APHIS or a program recognized by APHIS;

(5) If the importation is not in compliance with any applicable import regulations or any administrative instructions or measures, including, but not limited to, all the requirements of this part; or

(6) If a State executive official, or a State plant protection official authorized to do so, objects to the movement in writing and provides specific, detailed information that there is a risk the movement will result in the dissemination of a plant pest or noxious weed into the State, and APHIS determines that such plant pest risk cannot be adequately addressed or mitigated.

§319.7–4 Withdrawal, cancellation, and revocation of permits.

(a) *Withdrawal of an application.* If the applicant wishes to withdraw a permit application before issuance of a permit, he or she must provide the request in writing to APHIS. APHIS will provide written notification to the applicant as promptly as circumstances allow regarding reception of the request and withdrawal of the application.

(b) *Cancellation of permit by permittee.* If a permittee wishes to cancel a permit after its issuance, he or she must provide the request in writing to APHIS. APHIS will provide written notification to the applicant as promptly as circumstances allow regarding reception of the request and cancellation of the permit.

(c) *Revocation of permit by APHIS.* APHIS may revoke any outstanding permit to import regulated articles into the United States. A revocation, including the reason for the revocation, will be provided in writing, including by electronic methods, to the permittee as promptly as circumstances permit. The revocation of a permit may be appealed in accordance with §319.7–5.

(d) APHIS may revoke a permit to import a regulated article if:

(1) Information is received subsequent to the issuance of the permit of circumstances that APHIS determines would constitute cause for the denial of an application under §319.7–3; or

(2) APHIS determines that the permittee has failed to maintain the safeguards or otherwise observe the conditions specified in the permit or in any applicable regulations or administrative instructions, including, but not limited to, all of the requirements of this part.

(e) Upon revocation of a permit, the permittee must, without cost to the Federal Government and in the manner and method APHIS considers appropriate, either:

(1) Surrender all regulated articles covered by the revoked permit and any other affected plant material to an inspector;

(2) Destroy, under the supervision of an inspector, all regulated articles covered by the revoked permit and any other affected plant material; or

(3) Remove all regulated articles covered by the revoked permit and any other affected plant material from the United States.

[78 FR 19807, Apr. 10, 2014, as amended at 81 FR 5888, Feb. 4, 2016]

§319.7–5 Appeal of denial or revocation.

(a) All denials of an application for a permit, or revocations of an existing permit, will be provided in writing, including by electronic methods, as promptly as circumstances permit and will include the reasons for the denial or revocation.

(b) Any person whose application for a permit has been denied or whose permit has been revoked may appeal the decision in writing to APHIS within 10 business days from the date the communication of notification of the denial or revocation of the permit was received. The appeal must state all facts and reasons upon which the person is relying to show that the denial or revocation was incorrect.

(c) APHIS will grant or deny the appeal in writing and will state in writing the reason for the decision. The denial or revocation will remain in effect during the resolution of the appeal.

Subpart E—Foreign Cotton and Covers

Source: Redesignated at 84 FR 2428, Feb. 7, 2019.

QUARANTINE

§319.8 Notice of quarantine.

(a) Pursuant to sections 411–414 and 434 of the Plant Protection Act (7 U.S.C. 7711–7714 and 7754), the Administrator of the Animal and Plant Health Inspection Service has determined that the unrestricted importation into the United States from all foreign countries and localities of any parts or products of plants of the genus *Gossypium,* including seed cotton; cottonseed; cotton lint, linters, and other forms of cotton fiber (not including yarn, thread, and cloth); cottonseed hulls, cake, meal, and other cottonseed products, except oil; cotton waste, including gin waste and thread waste; any other unmanufactured parts of cotton plants; second-hand burlap and other fabrics, shredded or otherwise, that have been used or are of the kinds ordinarily used, for containing cotton, grains (including grain products), field seeds, agricultural roots, rhizomes, tubers, or other underground crops, may result in the entry into the United States of the pink bollworm (*Pectinophora gossypiella* (Saund.)), the golden nematode of potatoes (*Heterodera rostochiensis* Wr.), the flag smut disease (*Urocystis tritici* Koern.), and other injurious plant diseases and insect pests. Accordingly, to prevent the introduction into the United States of plant pests, the importation of those articles into the United States is prohibited unless they are imported in accordance with the regulations in this subpart or their importation has been authorized for experimental, therapeutic, or developmental purposes by a controlled import permit issued in accordance with §319.6.

(b) The importation of cotton plants (including any plant parts) that are for planting or capable of being planted is restricted in Subpart H—Plants for Planting of this part.

[78 FR 25569, May 2, 2013, as amended at 83 FR 11855, Mar. 19, 2018; 84 FR 2429, Feb. 7, 2019]

§319.8a Administrative instructions relating to the entry of cotton and covers into Guam.

The plants and products specified in §319.8(a) may be imported into Guam without further permit, other than the authorization contained in this paragraph. Sections 319.8–2 and 319.8–3 shall not be applicable to such importations. In addition, such importations need not comply with the requirements of §319.8–4 relating to notice of arrival inasmuch as there is available to the inspector the essential information normally supplied by the importer at the time of importation. Sections 319.8–5 through 319.8–27 shall not be applicable to importations into Guam. Inspection of such importations may be made under the general authority of §330.105(a) of this chapter. If an importation is found infected, infested, or contaminated with any plant pest and is not subject to disposal under this part, disposition may be made in accordance with §330.106 of this chapter.

REGULATIONS; GENERAL

§319.8–1 Definitions.

For the purposes of the regulations in this subpart, the following words shall be construed, respectively, to mean:

Administrator. The Administrator of the Animal and Plant Health Inspection Service, United States Department of Agriculture, or any employee of the United States Department of Agriculture delegated to act in his or her stead.

Approved. Approved by the Administrator.

Approved areas of Mexico. Any areas of Mexico, other than Northwest Mexico and the west coast of Mexico, which are designated by the Administrator as areas in which cotton and cotton products are produced and handled under conditions comparable to those under which like cotton and cotton products are produced and handled in the generally infested pink bollworm regulated area in the United States.

Approved fumigation facilities. Approved vacuum fumigation plant at a port where an inspector is available to supervise the fumigation.

Approved mill or plant. A mill or plant operating under a signed agreement with the Plant Protection and Quarantine Programs required for approval of a mill or plant as specified in §319.8–8(a)(2).

Authorized. Authorized by the Administrator.

Compressed. Compressed or pressed and baled or packaged to a density greater than approximately 20 pounds and less than approximately 28 pounds per cubic foot.

Compressed to high density. Compressed or pressed and baled or packaged to a density of approximately 28 or more pounds per cubic foot.

Contamination (contaminate). Containing or bearing whole cottonseed or seed cotton or other material which may carry the pink bollworm, the golden nematode of potatoes, the flag smut disease, or other injurious plant diseases or insect pests. (The verb contaminate shall be construed accordingly.)

Cotton. Parts and products of plants of the genus Gossypium, including seed cotton; cottonseed; cotton lint, linters and other forms of cotton fiber, not including yarn, thread and cloth; cottonseed hulls, cake, meal, and other cottonseed products, except oil; waste; and all other unmanufactured parts of cotton plants.

Cottonseed. Cottonseed from which the lint has been removed and that is intended for processing or consumption.

Covers. Second-hand burlap and other fabrics, shredded or otherwise, including any whole bag, any bag that has been slit open, and any part of a bag, which have been used, or are of the kinds ordinarily used, for containing cotton, grains (including grain products), field seeds, agricultural roots, rhizomes, tubers, or other underground crops. Burlap and other fabrics, when new or unused are excluded from this definition.

Gin trash. All of the material produced during the cleaning and ginning of seed cotton, bollies or snapped cotton except the lint, cottonseed, and gin waste.

Inspector. A properly identified employee of the U.S. Department of Agriculture or other person authorized to enforce the provisions of the Plant Protection Act.

Lint. All forms of raw ginned cotton, either baled or unbaled, except linters and waste.

Linters. All forms of cotton fiber separated from cottonseed after the lint has been removed, excluding so-called hull fiber.

North, northern. When used to designate ports of arrival, these terms mean the port of Norfolk, VA, and all Atlantic Coast ports north thereof, ports along the Canadian border, and Pacific Coast ports in the States of Washington and Oregon. When used in a geographic sense to designate areas or locations, these terms mean any State in which cotton is not grown commercially. However, when cotton is grown commercially in certain portions of a State, as is the case in Illinois, Kansas, and Missouri, these terms include those portions of such State as may be determined by the Administrator as remote from the main area of cotton production.

Northwest Mexico. All of the State of Baja California, Mexico, and that part of the State of Sonora, Mexico, lying between San Luis Mesa and the Colorado River.

Permit. A form of authorization to allow the importation of cotton or covers in accordance with the regulations in this subpart and in §§ 319.7 through 319.7–5.

Person. Any individual, firm, corporation, company, society, or association, or any organized group of any of the foregoing.

Pink bollworm regulated area; generally infested pink bollworm regulated area. The pink bollworm regulated area consists of those States or parts thereof designated as regulated area in Administrative Instructions issued under § 301.52–2 of this chapter. The generally infested pink bollworm regulated area is that part of the regulated area designated as generally infested in the said Administrative Instructions.

Plant Protection and Quarantine Programs. The Plant Protection and Quarantine Programs, Animal and Plant Health Inspection Service, of the United States Department of Agriculture.

Root crop. The underground crop portions of any plants.

Samples. Samples of lint, linters, waste, cottonseed cake, and cottonseed meal, of the amount and character usually required for trade purposes.

Seed cotton. Cotton as it comes from the field.

Treatment. Procedures administratively approved by the Administrator for destroying infestations or infections of insect pests or plant diseases, such as fumigation, application of chemicals or dry or moist heat, or processing, utilization, or storage.

Uncompressed. Baled or packaged to a density not exceeding approximately 20 pounds per cubic foot.

United States. Any of the States, the District of Columbia, Guam, Puerto Rico, or the Virgin Islands of the United States.

Utilization. Processing or manufacture, in lieu of fumigation at time of entry, at a mill or plant authorized by APHIS through a compliance agreement for foreign cotton processing or manufacturing.

Waste. All forms of cotton waste derived from the manufacture of cotton lint, in any form or under any trade designation, including gin waste and thread waste; and waste products derived from the milling of cottonseed. Gin trash is not within the definition of waste.

West Coast of Mexico. The State of Sinaloa, the State of Sonora (except that part of the Imperial Valley lying between San Luis Mesa and the Colorado River), and the Southern Territory of Baja California, in Mexico.

[24 FR 10788, Dec. 29, 1959, as amended at 27 FR 5389, June 7, 1962; 36 FR 24917, Dec. 24, 1971; 37 FR 10554, May 25, 1972; 66 FR 21055, Apr. 27, 2001; 78 FR 25569, May 2, 2013; 79 FR 19870, Apr. 10, 2014; 83 FR 11855, Mar. 19, 2018]

CONDITIONS OF IMPORTATION AND ENTRY OF COTTON AND COVERS

§319.8–2 Permit procedure.

(a) Except as otherwise provided for in §§319.8–10 and 319.8–18, permits shall be obtained for importations into the United States of all cotton and covers. Permits will be issued only for cotton and covers authorized entry under §§319.8–6 through 319.8–20. Persons desiring to import cotton or covers under §§319.8–6 through 319.8–20 shall, in advance of departure of such material from a foreign port, submit to the Plant Protection and Quarantine Programs an application for a permit in accordance with §§319.7 through 319.7–5 . Applications to import cottonseed shall state the approximate quantity and the proposed United States port of entry. Applications to import lint, linters, or waste shall state whether such materials are compressed.

(b) Applications to import lint, linters, or waste at a port[1] other than one in the North, in California, or on the Mexican Border shall also specify whether the commodity is compressed to high density.

(c) Upon receipt of an application to import lint, linters, waste, or covers, without treatment, for utilization under agreement as defined in §319.8–8(a)(2), an investigation will be made by an inspector to determine that the receiving mill or plant is satisfactorily located geographically, is equipped with all necessary safeguards, and is apparently in a position to fulfill all precautionary conditions to which it may agree. Upon determination by the inspector that these qualifications are fulfilled, the owner or operator of the mill or plant may sign an agreement specifying that the required precautionary conditions will be maintained. Such signed agreement will be a necessary requisite to the release at the port of entry of any imported lint, linters, waste, or covers for forwarding to and utilization at such mill or plant in lieu of vacuum fumigation or other treatment otherwise required by this subpart. Permits for the importation of

---

[1] Including ports in Guam, Hawaii, Puerto Rico, and the Virgin Islands of the United States.

such materials will be issued in accordance with paragraph (a) of this section.

(d) Permits for importation of any cotton or covers are conditioned upon compliance with all of the conditions specified in the permit and any applicable regulations or administrative instructions of this part.

(e) Pending development of adequate treating facilities in Guam, any cotton or covers that are subject to treatment as a condition of entry therein must first be entered and treated in accordance with the requirements of this subpart at a U.S. port of arrival where such treating facilities are available.

(Approved by the Office of Management and Budget under control number 0579–0049)

[24 FR 10788, Dec. 29, 1959, as amended at 48 FR 57466, Dec. 30, 1983; 78 FR 25570, May 2, 2013; 79 FR 19811, Apr. 10, 2014]

§319.8–3 Refusal and cancellation of permits.

(a) Permits for entry from the West Coast of Mexico, as authorized in §319.8–12 of lint, linters, waste, cottonseed, and cottonseed hulls may be refused and existing permits cancelled by the Administrator if he or she has determined that the pink bollworm is present in the West Coast of Mexico or in Northwest Mexico, or that other conditions exist therein that would increase the hazard of pest introduction into the United States.

(b) Permits for entry from Northwest Mexico as authorized in §319.8–13 of lint, linters, waste, cottonseed, cottonseed hulls, and covers that have been used for cotton, may be refused and existing permits cancelled by the Administrator if he or she has determined that the pink bollworm is present in Northwest Mexico or in the West Coast of Mexico, or that other conditions exist therein that would increase the hazard of pest introduction into the United States.

[27 FR 5389, June 7, 1962, as amended at 36 FR 24917, Dec. 24, 1971; 70 FR 33324, June 7, 2005; 78 FR 25570, May 2, 2013]

§319.8–4 Notice of arrival.

Immediately upon arrival at a port of entry of any shipment of cotton or covers, the importer shall submit to an inspector or, in the case of Guam, through the Customs officer of the Government of Guam, notice of such arrival using a form provided for that purpose (Form PPQ–368). Forms will be submitted using a U.S. Government electronic information exchange system or other authorized method.

(Approved by the Office of Management and Budget under control number 0579–0049)

[81 FR 40150, June 21, 2016]

§ 319.8–5 Marking of containers.

Every bale or other container of cotton lint, linters, waste, or covers imported or offered for entry shall be plainly marked or tagged with a bale number or other mark to distinguish it from other bales or containers of similar material. Bales of lint, linters, and waste from approved areas of Mexico, the West Coast of Mexico, or Northwest Mexico shall be tagged or otherwise marked to show the gin or mill of origin unless they are immediately exported.

(Approved by the Office of Management and Budget under control number 0579–0049)

[27 FR 5389, June 7, 1962, as amended at 48 FR 57466, Dec. 30, 1983]

§ 319.8–6 Cottonseed cake and cottonseed meal.

Entry of cottonseed cake and cottonseed meal will be authorized through any port at which the services of an inspector are available, subject to examination by an inspector for freedom from contamination. If found to be free of contamination, importations of such cottonseed cake and cottonseed meal will be released from further plant quarantine entry restrictions. If found to be contaminated such importations will be refused entry or subjected as a condition of entry to such safeguards as the inspector may prescribe, according to a method selected by the inspector from administratively authorized procedures known to be effective under the conditions under which the safeguards are applied.

[24 FR 10788, Dec. 29, 1959, as amended at 70 FR 33324, June 7, 2005]

§ 319.8–7 Processed lint, linters, and waste.

Entry of lint, linters, and waste will be authorized without treatment but upon compliance with other applicable requirements of this subpart when the inspector can determine that such lint, linters, and waste have been so processed by bleaching, dyeing, or other means, as to have removed all cottonseed or to have destroyed all insect life.

§ 319.8–8 Lint, linters, and waste.

(a) *Compressed to high density.* (1)(i) Entry of lint, linters, and waste, compressed to high density, will be authorized subject to vacuum fumigation by approved methods at any port where approved fumigation facilities are available.

(ii) Importations of such lint, linters, and waste, arriving at a northern port where there are no approved fumigation facilities may be entered for transportation in bond to another northern port where such facilities are available, for the required vacuum fumigation.

(iii) Such lint, linters, and waste compressed to high density arriving at a port in the State of California where there are no approved fumigation facilities may be entered for immediate transportation in bond via an all-water route if available, otherwise by overland transportation in van-type trucks or box cars after approved surface treatment, or under such other conditions as may be deemed necessary and are prescribed by the inspector to (*a*) any port where approved fumigation facilities are available, there to receive the required vacuum fumigation before release, or (*b*) to an approved mill or plant for utilization.

(2) Entry of lint, linters, and waste compressed to high density, will be authorized without vacuum fumigation at any northern port, subject to movement to an approved mill or plant, the owner or operator of which has executed an agreement with the Plant Protection and Quarantine Programs to the effect that, in consideration of the waiving, of vacuum fumigation as a condition of entry and the substitution of approved utilization therefor:

(i) The lint, linters, and waste so entered will be processed or manufactured at the mill or plant and until so used will be retained thereat, unless written authority is granted by the

Plant Protection and Quarantine Programs to move the material to another mill or plant;

(ii) Sanitary measures satisfactory to the Plant Protection and Quarantine Programs will be taken with respect to the collection and disposal of any waste, residues, and covers, including the collection and disposal of refuse from railroad cars, trucks, or other carriers used in transporting the material to the mill or plant;

(iii) Inspectors of the Plant Protection and Quarantine Programs will have access to the mill or plant at any reasonable time to observe the methods of handling the material, the disposal of refuse, residues, waste, and covers, and otherwise to check compliance with the terms of the agreement;

(iv) Such reports of the receipt and utilization of the material, and disposal of waste therefrom as may be required by the inspector will be submitted to him promptly;

(v) Such other requirements as may be necessary in the opinion of the Administrator to assure retention of the material, including all wastes and residues, at the mill or plant and its processing, utilization or disposal in a manner that will eliminate all pest risk, will be complied with.

(3) Failure to comply with any of the conditions of an agreement specified in paragraph (a)(2) of this section may be cause for immediate cancellation of the agreement by the inspector and refusal to release, without vacuum fumigation, lint, linters, and waste for transportation to the mill or plant.

(4) Agreements specified in paragraph (a)(2) of this section may be executed only with owners or operators of mills or plants located in States in which cotton is not grown commercially and at locations in such other States as may be administratively designated by the Administrator after due consideration of possible pest risk involved and the proximity of growing cotton.

(b) *Uncompressed or compressed.* (1)(i) Entry of uncompressed or compressed lint, linters, and waste will be authorized, subject to vacuum fumigation by approved methods, through any northern port, through any port in the State of California, and through any port on the Mexican Border, where approved fumigation facilities are available.

(ii) Importations of such lint, linters, and waste arriving at a northern port where there are no approved fumigation facilities may be entered for immediate transportation in bond to another northern port where such facilities are available, for the required vacuum fumigation.

(iii) Compressed lint, linters, and waste arriving at a port in the State of California where there are no approved fumigation facilities may be entered for immediate transportation in bond by an all-water route if available, otherwise by overland transportation in van-type trucks or box cars after approved surface treatment, or under such other conditions as may be deemed necessary and are prescribed by the inspector, to any port in California or any northern port where approved fumigation facilities are available, there to receive the required vacuum fumigation before release, or to any northern port for movement to an approved mill or plant for utilization.

(iv) Uncompressed lint, linters, and waste arriving at a port in the State of California where there are no approved fumigation facilities may be entered for immediate transportation in bond by an all-water route to any port in California or any northern port where approved fumigation facilities are available, there to receive the required vacuum fumigation before release, or to a northern port for movement to an approved mill or plant for utilization.

(2) Entry without vacuum fumigation will be authorized for compressed lint, linters, and waste, and for uncompressed waste derived from cotton milled in countries that do not produce cotton,[2] arriving at a northern

---

[2] For the purposes of this subpart the following countries are considered to be those in which cotton is not produced: Austria, Belgium, Canada, Denmark, Republic of Ireland (Eire), Finland, France, Germany (both East and West), Great Britain and Northern Ireland (United Kingdom), Iceland, Liechtenstein, Luxembourg, Netherlands, Norway, Portugal, Sweden, and Switzerland.

port, subject to movement to an approved mill or plant.

[24 FR 10788, Dec. 29, 1959, as amended at 27 FR 5389, June 7, 1962; 36 FR 24917, Dec. 24, 1971; 78 FR 25570, May 2, 2013; 79 FR 19810, Apr. 10, 2014]

§ 319.8–9 Hull fiber and gin trash.

(a) Entry of hull fiber will be authorized under the same conditions as are applicable to waste under this subpart.

(b) Gin trash may be imported only under the provisions of § 319.8–20.

[24 FR 10788, Dec. 29, 1959, as amended at 27 FR 5390, June 7, 1962]

§ 319.8–10 Covers.

(a) Entry of covers (including bags, slit bags, and parts of bags) which have been used as containers for cotton grown or processed in countries other than the United States may be authorized either (1) through a Mexican border port named in the permit for vacuum fumigation by an approved method in that part of the United States within the generally infested pink bollworm regulated area; or (2) through a northern port or a port in the State of California subject to vacuum fumigation by an approved method or without vacuum fumigation when the covers are to be moved to an approved mill or plant for utilization. When such covers are forwarded from a northern port to a mill or plant in California for utilization, or from a California port to another California or northern port for vacuum fumigation thereat or for movement to a mill or plant for utilization such movement shall be made by an all-water route unless the bales are compressed to a density of 20 pounds or more per cubic foot in which case the bales may be moved overland in van-type trucks or box cars if all-water transportation is not available. Such overland movement may be made only after approved surface treatment or under such other conditions as may be deemed necessary and are prescribed by the inspector. When such covers arrive at a port other than a northern, California, or Mexican border port they will be required to be transported therefrom immediately in bond by an all-water route to a northern or California port where approved vacuum fumigation facilities are available for

vacuum fumigation thereat by an approved method or for forwarding therefrom to an approved mill or plant for utilization.

(b) American cotton bagging, commonly known as coarse gunny, which has been used to cover only cotton grown or processed in the United States, may be authorized entry at any port under permit and upon compliance with §§ 319.8–4 and 319.8–5, without fumigation or other treatment. Marking patches of the finer burlaps or other fabrics when attached to bales of such bagging may be disregarded if, in the judgment of the inspector, they do not present a risk of carrying live pink bollworms, golden nematode cysts or flag smut spores.

(c) Bags, slit bags, parts of bags, and other covers which have been used as containers for root crops or are of a kind ordinarily used as containers for root crops may be authorized entry subject to immediate treatment in such manner and according to such method as the inspector may select from administratively authorized procedures known to be effective under the conditions under which the treatment is applied, and subject to any additional safeguard measures that may be prescribed by the inspector pursuant to § 319.8–24, or that he may prescribe in regard to the manner of discharge from the carrier and conveyance to the place of treatment: *Provided,* That such covers may be authorized entry from Canada without treatment as prescribed in this paragraph unless the covers are found to be contaminated.

(d) Bags, slit bags, parts of bags, and other covers that have been used as containers for wheat or wheat products that have not been so processed as to have destroyed all flag smut disease spores, or that have been used as containers for field seeds separated from wheat during the process of screening, and which arrive from a country named in § 319.59–2(a)(2) of this part, if intended for reuse in this country as grain containers may be authorized entry, subject to immediate treatment at the port of arrival. If such covers are not intended to be reused in this country as grain containers their entry may be authorized subject to movement for utilization to an approved mill or plant the owner or operator of which has executed an appropriate agreement with the Plant Protection and Quarantine Programs similar to that described in § 319.8–8(a)(2). Covers coming within this paragraph only, may be entered without permit other than the authorization provided in this paragraph and without other restriction under this subpart upon presentation to an inspector of satisfactory evidence that they have been used only for grains exported from the United States and are being returned empty without use abroad and that while abroad they have been handled in a manner to prevent their contamination.

(e) When upon arrival at a port of entry any shipment of bags, slit bags, parts of bags, or other covers, is found to include one or more bales containing material the importation of which is regulated by paragraph (a), (c), or (d) of this section, the entire shipment, or any portion thereof, may be required by the inspector to be treated as specified in the applicable paragraph.

(f) If upon their arrival at a port of entry covers are classified by the inspector as coming within more than one paragraph of this section, they will be authorized entry only upon compliance with such requirements of the applicable paragraphs as the inspector may deem necessary to prevent the introduction of plant diseases and insect pests.

(g) Notwithstanding the provisions of any other paragraph of this section the entry from any country of bags, slit bags, parts of bags, and other covers will be authorized without treatment but upon compliance with other applicable sections of this subpart if the inspector finds that they have obviously not been used in a manner that would contaminate them or when in the inspector's opinion there is otherwise no plant pest risk associated with their entry.

[24 FR 10788, Dec. 29, 1959, as amended at 27 FR 5390, June 7, 1962; 36 FR 24917, Dec. 24, 1971; 63 FR 31101, June 8, 1998]

SPECIAL CONDITIONS FOR THE ENTRY OF COTTON AND COVERS FROM MEXICO

SOURCE: Sections 319.8–11 through 319.8–14 appear at 27 FR 5309, June 7, 1962, unless otherwise noted.

§ 319.8–11 From approved areas of Mexico.

(a) Entry of lint, linters, and waste (including gin and oil mill wastes) which were derived from cotton grown in, and which were produced and handled only in approved areas of Mexico [3] may be authorized through Mexican Border ports in Texas named in the permits

(1) For movement into the generally infested pink bollworm regulated area such products becoming subject immediately upon release by the inspector to the requirements, in § 301.52 of this chapter, applicable to like products originating in the pink bollworm regulated area, or

(2) For movement to an approved mill or plant for utilization, or

(3) For movement to New Orleans for immediate vacuum fumigation.

(b) Entry of cottonseed or cottonseed hulls in bulk, or in covers that are new or which have not been used previously to contain cotton or unmanufactured cotton products, may be authorized through Mexican Border ports in Texas named in the permits, for movement into the generally infested pink bollworm regulated area when certified by an inspector as having been produced in an approved area and handled subsequently in a manner satisfactory to the inspector. Upon arrival in the generally infested pink bollworm regulated area such cottonseed or cottonseed hulls will be released from further plant quarantine entry requirements and shall become subject immediately to the requirements in § 301.52 of this chapter.

[27 FR 5309, June 7, 1962, as amended at 63 FR 31101, June 8, 1998; 78 FR 25570, May 2, 2013; 79 FR 19810, Apr. 10, 2014]

---

[3] See § 319.8–1(p) for definition of "Approved areas of Mexico." These are within that part of Mexico not included in the "West Coast of Mexico" (§ 319.8–1(q)) or "Northwest Mexico" (§ 319.8–1(r)).

§ 319.8–12 From the West Coast of Mexico.

Contingent upon continued freedom of the West Coast of Mexico and of Northwest Mexico from infestations of the pink bollworm, entry of the following products may be authorized under permit subject to inspection to determine freedom from hazardous plant pest conditions:

(a) Compressed lint and linters.

(b) Uncompressed lint and linters for movement into the generally infested pink bollworm regulated area, movement thereafter to be in accordance with § 301.52 of this chapter.

(c) Compressed or uncompressed cotton waste for movement under bond to Fabens, Texas, for vacuum fumigation after which it will be released from further plant quarantine entry requirements.

(d) Cottonseed when certified by an inspector as having been treated, stored, and transported in a manner satisfactory to the Administrator.

(e) Untreated, non-certified cottonseed contained in new bags for movement by special manifest to any destination in the generally infested pink bollworm regulated area, movement thereafter to be in accordance with § 301.52 of this chapter.

(f) Cottonseed hulls when certified by an inspector as having been treated, stored, and transported in a manner satisfactory to the Administrator.

(g) Any cotton products for movement through Mexican border ports in Texas directly into the generally infested pink bollworm regulated area, movement thereafter to be in accordance with § 301.52 of this chapter.

[27 FR 5309, June 7, 1962, as amended at 36 FR 24917, Dec. 24, 1971; 78 FR 25570, May 2, 2013]

§ 319.8–13 From Northwest Mexico.

Contingent upon continued freedom of Northwest Mexico and of the West Coast of Mexico from infestations of the pink bollworm and other plant pest conditions that would increase risk of pest introduction into the United States with importations authorized under this section, entry of the following products may be authorized under permit subject to inspection upon arrival to determine freedom from hazardous plant pest conditions:

(a) Lint, linters, and waste.

(b) Cottonseed.

(c) Cottonseed hulls.

(d) Covers that have been used for cotton only.

§ 319.8–14 Mexican cotton and covers not otherwise enterable.

Mexican cotton and covers not enterable under § 319.8–11, § 319.8–12, or § 319.8–13 may be entered in accordance with §§ 319.8–6 through 319.8–10 and §§ 319.8–16 through 319.8–20 insofar as said sections are applicable.

Miscellaneous Provisions

§ 319.8–16 Importation into United States of cotton and covers exported therefrom.

(a) Cotton and covers grown, produced, or handled in the United States and exported therefrom, and in the original bales or other containers in which such material was exported therefrom, may be imported into the United States at any port under permit, without vacuum fumigation or other treatment or restriction as to utilization, upon compliance with §§ 319.8–2, 319.8–4, and § 319.8–5, and upon the submission of evidence satisfactory to the inspector that such material was grown, produced, or handled in the United States and does not constitute a risk of introducing the pink bollworm into the United States.

(b) Cotton and covers of foreign origin imported into the United States in accordance with this subpart and exported therefrom, when in the original bales or other original containers, may be reimported into the United States under the conditions specified in paragraph (a) of this section.

§ 319.8–17 Importation for exportation, and importation for transportation and exportation; storage.

(a) Importation of cotton and covers for exportation, or for transportation and exportation, in accordance with this subpart shall also be subject to §§ 352.1 through 352.8 of this chapter, as amended.

(b) Importation at northern ports of unfumigated lint, linters, waste, cottonseed cake, cottonseed meal and covers used only for cotton, for exportation or for transportation and exportation through another northern port, may be authorized by the inspector under permit if, in his judgment, such procedures can be authorized without risk of introducing the pink bollworm.

(c) Entry under permit of lint, linters, or waste compressed to high density will be authorized for purposes of storage in the north pending exportation, fumigation, or utilization in an approved mill or plant provided the owner or operator of such proposed storage place has executed an agreement with the Plant Protection and Quarantine Programs similar to those required for mills or plants to utilize lint, linters, and waste as specified in § 319.8–8(a)(2), and provided further that

(1) Inspectors are available to supervise the storage,

(2) The bales of material to be stored are free from surface contamination,

(3) The material is kept segregated from other cotton and covers in a manner satisfactory to the inspector, and

(4) The waste is collected and disposed of in a manner satisfactory to the inspector.

(d) Except as provided in § 319.8–23(a)(4), compressed lint, linters, and waste, uncompressed waste derived from cotton milled in a non-cotton-producing country,[4] and covers, arriving at a port in the north for entry for exportation, vacuum fumigation, or utilization in accordance with the requirements in this subpart, may be allowed movement in Customs custody for storage at a point in the north pending such exportation, or movement to an approved mill or plant for vacuum fumigation or utilization, when there are inspectors available to supervise such storage, if the bales are free of surface contamination, if they are kept segregated from other cotton

and covers in a manner satisfactory to the inspector, and if waste is collected and disposed of in a manner satisfactory to the inspector. Such lint, linters, waste, and covers shall remain under Customs custody until released by the inspector.

(e) Importation of lint, linters, and waste from Mexico for transportation and exportation will be authorized under permit if such material is compressed before, or immediately upon entering into the United States, or is compressed while en route to the port of export at a compress specifically authorized in the permit. The ports of export which may be named in the permit shall be limited to those that have been administratively approved for such exportation. Storage of such compressed cotton may be authorized, in approved bonded warehouses in Texas.

(f) Entry of uncompressed lint, linters, and waste from Mexico may be authorized at ports named in the permit for exportation at ports within the generally infested pink bollworm regulated area or for transportation and exportation via rail to Canada under such conditions and over such routes as may be specified in the permit.

[24 FR 10788, Dec. 29, 1959, as amended at 27 FR 5390, June 7, 1962; 36 FR 24917, Dec. 24, 1971; 63 FR 31101, June 8, 1998; 78 FR 25570, May 2, 2013; 79 FR 19810, Apr. 10, 2014]

§ 319.8–18 Samples.

(a) Samples of lint, linters, waste, cottonseed cake, and cottonseed meal may be entered without further permit other than the authorization contained in this section, but subject to inspection and such treatment as the inspector may deem necessary. Samples which represent either such products of United States origin or such products imported into the United States in accordance with the requirements of this subpart, and which were exported from the United States, may be entered into the United States without inspection when the inspector is satisfied as to the identity of the samples.

(b) Samples of cottonseed or seed cotton may be entered subject to the conditions and requirements provided in §§ 319.8–2, 319.8–4, and 319.8–19.

(c) Bales or other containers of cotton shall not be broken or opened for

[4] For the purposes of this subpart the following countries are considered as non-cotton-producing countries: Austria, Belgium, Canada, Denmark, Eire, Finland, France, Germany, Great Britain (United Kingdom), Iceland, Liechtenstein, Luxembourg, Netherlands, Norway, Portugal, Sweden and Switzerland.

sampling and samples shall not be drawn until the inspector has so authorized and has prescribed the conditions and safeguards under which such samples shall be obtained.

§§ 319.8–19—319.8–20 [Reserved]

§ 319.8–21 Release of cotton and covers after 18 months' storage.

Cotton and covers, the entry of which has been authorized subject to vacuum fumigation or other treatment because of the pink bollworm only, and which have not received such treatment but have been stored for a period of 18 months or more will be released from further plant quarantine entry restrictions.

§ 319.8–22 Ports of entry or export.

When ports of entry or export are not specifically designated in this subpart but are left to the judgment of the inspector, the inspector shall designate only such ports as have been administratively approved for such entry or export.

§ 319.8–23 Treatment.

(a)(1) Vacuum fumigation as required in this subpart must be conducted in accordance with part 305 of this chapter.

(2) After cotton and covers have been vacuum fumigated they shall be so marked under the supervision of an inspector. Such material may thereafter be distributed, forwarded, or shipped without further plant quarantine entry restriction.

(3) Cotton and covers held by an importer for vacuum fumigation must be stored under conditions satisfactory to the inspector.

(4) Prompt vacuum fumigation of cotton and covers (other than high density cotton free of surface contamination) will be required at non-northern ports. Similar prompt vacuum fumigation will be required at Norfolk, Virginia, during the period June 15 to October 15 of each year, except for covers which have been used to contain only lint, linters, or waste, and the bales of which are compressed to a density of 28 or more pounds per cubic foot and are free of surface contamination.

(b) An inspector may authorize the substitution of processing, utilization, or other form of treatment for vacuum fumigation when in his opinion such other treatment, selected by him from administratively authorized procedures, will be effective in eliminating infestation of the pink bollworm.

[24 FR 10788, Dec. 29, 1959, as amended at 75 FR 4251, Jan. 26, 2010]

§ 319.8–24 Collection and disposal of waste.

(a) Importers shall handle imported, unfumigated cotton and covers in a manner to avoid waste. If waste does occur, the importer or his or her agent shall collect and dispose of such waste in a manner satisfactory to the inspector.

(b) If, in the judgment of an inspector, it is necessary as a safeguard against risk of pest dispersal to clean railway cars, lighters, trucks, and other vehicles and vessels used for transporting such cotton or covers, or to clean piers, warehouses, fumigation plants, mills, or other premises used in connection with importation of such cotton or covers, the importer or his or her agent shall perform such cleaning, in a manner satisfactory to the inspector.

(c) All costs incident to such collection, disposal, and cleaning other than the services of the inspector during his or her regular tour of duty and at the usual place of duty, shall be borne by the importer or his or her agent.

[24 FR 10788, Dec. 29, 1959, as amended at 70 FR 33324, June 7, 2005]

§ 319.8–25 Costs and charges.

The services of the inspector during regularly assigned hours of duty and at the usual places of duty shall be furnished without cost to the importer. The Plant Protection and Quarantine Programs will not assume responsibility for any costs or charges, other than those indicated in this section, in connection with the entry, inspection, treatment, conditioning, storage, forwarding, or any other operation of any character incidental to the physical entry of an importation of a restricted material.

§ 319.8–26 Material refused entry.

Any material refused entry for noncompliance with the requirements of this subpart shall be promptly removed from the United States or abandoned by the importer for destruction, and pending such action shall be subject to the immediate application of such safeguards against escape of plant pests as the inspector may prescribe. If such material is not promptly safeguarded by the importer, removed from the United States, or abandoned for destruction to the satisfaction of the inspector it may be seized, destroyed, or otherwise disposed of in accordance with sections 414 and 421 of the Plant Protection Act (7 U.S.C. 7714 and 7731). Neither the Department of Agriculture nor the inspector will be responsible for any costs accruing for demurrage, shipping charges, cartage, labor, chemicals, or other expenses incidental to the safeguarding or disposal of material refused entry by the inspector, nor will the Department of Agriculture or the inspector assume responsibility for the value of material destroyed.

[24 FR 10788, Dec. 29, 1959, as amended at 66 FR 21055, Apr. 27, 2001]

Subpart F—Sugarcane

Source: Redesignated at 84 FR 2429, Feb. 7, 2019.

§ 319.15 Notice of quarantine.

(a) The importation into the United States of sugarcane and its related products, including cuttings, canes, leaves and bagasse, from all foreign countries and localities is prohibited, except for importations for experimental, therapeutic, or developmental purposes under the conditions specified in a controlled import permit issued in accordance with § 319.6.

(b) The importation of sugarcane plants (including any plant parts) that are for planting or capable of being planted is restricted under Subpart H—Plants for Planting of this part.

(c) As used in this subpart, unless the context otherwise requires, the term "United States" means the States, the District of Columbia, Guam, Puerto Rico, and the Virgin Islands of the United States.

[24 FR 10788, Dec. 29, 1959, as amended at 66 FR 21055, Apr. 27, 2001; 78 FR 25570, May 2, 2013; 83 FR 11855, Mar. 19, 2018; 84 FR 2429, Feb. 7, 2019]

§ 319.15a Administrative instructions and interpretation relating to entry into Guam of bagasse and related sugarcane products.

Bagasse and related sugarcane products have been so processed that, in the judgment of the Department, their importation into Guam will involve no pest risk, and they may be imported into Guam without further permit, other than the authorization contained in this paragraph. Such importations may be made without the submission of a notice of arrival inasmuch as there is available to the inspector the essential information normally supplied by the importer at the time of importation. Inspection of such importations may be made under the general authority of § 330.105(a) of this chapter. If an importation is found infected, infested, or contaminated with any plant pest and is not subject to disposal under this part, disposition may be made in accordance with § 330.106 of this chapter.

Subpart G—Corn Diseases

Source: Redesignated at 84 FR 2429, Feb. 7, 2019.

Quarantine

§ 319.24 Notice of quarantine.

(a) The fact has been determined by the Secretary of Agriculture, and notice is hereby given, that maize or Indian corn (Zea mays L.) and closely related plants are subject to certain injurious diseases, especially Peronospora maydis Raciborski, Sclerospora sacchari Miyake and other downy mildews; also the Physoderma diseases of maize, Physoderma zeae-maydis Shaw, and Physoderma maydis Miyake, new to and not heretofore widely prevalent or distributed within and throughout the United States, and that these diseases occur in southeastern Asia (including India, Siam, Indo-China and

China), Malayan Archipelago, Australia, Oceania, Philippine Islands, Formosa, Japan, and adjacent islands.

(b) The importation of corn plants (including any plant parts) that are for planting or capable of being planted is restricted in Subpart H—Plants for Planting of this part.

(c) Except as otherwise provided in this subpart, the importation into the United States of raw or unmanufactured corn seed and all other portions of Indian corn or maize and related plants, including all species of teosinte (*Euchlaena*), jobs-tears (*Coix*), *Polytoca*, *Chionachne*, and *Sclerachne*, from southeastern Asia (including India, Indochina, and the People's Republic of China), Malayan Archipelago, Australia, New Zealand, Oceania, Philippine Islands, Manchuria, Japan, and adjacent islands is prohibited. However, this prohibition does not apply to importations of such items for experimental, therapeutic, or developmental purposes under the conditions specified in a controlled import permit issued in accordance with § 319.6.

(d) As used in this subpart, unless the context otherwise requires, the term "United States" means the States, the District of Columbia, Guam, Puerto Rico, and the Virgin Islands of the United States.

(e) Seed of Indian corn or maize (*Zea mays L.*) that is free from the cob and from all other parts of corn may be imported into the United States from New Zealand without further restriction.

[24 FR 10788, Dec. 29, 1959, as amended at 58 FR 44745, Aug. 25, 1993; 66 FR 21055, Apr. 27, 2001; 78 FR 25570, May 2, 2013; 83 FR 11855, Mar. 19, 2018; 84 FR 2429, Feb. 7, 2019]

§ 319.24a Administrative instructions relating to entry of corn into Guam.

Corn may be imported into Guam without further permit, other than the authorization contained in this section but subject to compliance with § 319.24–3. Such imports need not comply with the notice of arrival requirements of § 319.24–4 inasmuch as information equivalent to that in a notice of arrival is available to the inspector from another source. Section 319.24–5 shall not be applicable to importations of corn into Guam. Such importations shall be subject to inspection at the port of entry. Corn found upon inspection to contain disease infection will be subject to sterilization in accordance with methods selected by the inspector from administratively authorized procedures known to be effective under the conditions in which applied.

REGULATIONS GOVERNING ENTRY OF INDIAN CORN OR MAIZE

§ 319.24–1 Application for permits for importation of corn.

Persons contemplating the importation of corn into the United States shall obtain a permit in accordance with §§ 319.7 through 319.7–5.

(Approved by the Office of Management and Budget under control number 0579–0049)

[79 FR 19810, Apr. 10, 2014]

§ 319.24–2 [Reserved]

§ 319.24–3 Marking as condition of entry.

Every bag or other container of corn offered for entry shall be plainly marked with such numbers or marks as will make it easily possible to associate the bags or containers with a particular importation.

(Approved by the Office of Management and Budget under control number 0579–0049)

[24 FR 10788, Dec. 29, 1959, as amended at 48 FR 57466, Dec. 30, 1983]

§ 319.24–4 [Reserved]

§ 319.24–5 Condition of entry.

The corn shall not be removed from the port of entry, nor shall any bag or other container thereof be broken or opened, except for the purpose of sterilization, until a written notice is given to the United States Collector of Customs, or, in the case of Guam, the Customs officer of the Government of Guam, by an inspector of the Plant Protection and Quarantine Programs, that the corn has been properly sterilized and released for entry without further restrictions so far as the jurisdiction of the Department of Agriculture extends thereto. All apparatus and methods for accomplishing such sterilization must be satisfactory to the Plant Protection and Quarantine Programs. Corn will be delivered to the permittee for sterilization, upon the filing with the appropriate customs official of a bond in the amount of $5,000, or in an amount equal to the invoice value of the corn if such value is less than $5,000, with approved sureties, and conditioned upon sterilization of the corn under the supervision and the satisfaction of an inspector of the Plant Protection and Quarantine Programs; and upon the redelivery of the corn to said customs official within 40 days from the arrival of the corn at the port of entry.

Subpart H—Plants for Planting

Source: 83 FR 11856, Mar. 19, 2018, unless otherwise noted. Redesignated at 84 FR 2429, Feb. 7, 2019.

§319.37–1 Notice of quarantine.

(a) Under section 412(a) of the Plant Protection Act, the Secretary of Agriculture may prohibit or restrict the importation and entry of any plant or plant product if the Secretary determines that the prohibition or restriction is necessary to prevent the introduction into the United States or the dissemination within the United States of a plant pest or noxious weed.

(b) The Secretary has determined that it is necessary to designate the importation of certain taxa of plants for planting as not authorized pending pest risk analysis, as provided in §319.37–4. The Secretary has determined that it is necessary to restrict the importation into the United States of all other plants for planting and to impose additional restrictions on the importation of specific types of plants for planting, in accordance with this subpart and as described in the Plants for Planting Manual.

(c) The importation of plants that are imported for processing or consumption, as determined by an inspector based on documentation accompanying the articles, is not subject to this subpart but may be subject to restrictions elsewhere in this part.

(d) The importation of taxa of plants for planting that are listed in parts 360 and 361 of this chapter is subject to the restrictions in those parts.

(e) The Plant Protection and Quarantine Programs also enforces regula-tions promulgated under the Endangered Species Act of 1973 (16 U.S.C. 1531–1544) which contain additional prohibitions and restrictions on importation into the United States of plants for planting subject to this subpart (see 50 CFR parts 17 and 23).

(f) Within the Plants for Planting Manual, one or more common names of plants for planting may be given in parentheses after most scientific names (when common names are known) for the purpose of helping to identify the plants for planting represented by such scientific names; however, unless otherwise specified, a reference to a scientific name includes all plants for planting within the taxon represented by the scientific name regardless of whether the common name or names are as comprehensive in scope as the scientific name. When restrictions apply to the importation of a taxon of plants for planting for which there are taxonomic synonyms, those restrictions apply to the importation of all the synonyms of that taxon as well.

§319.37–2 Definitions.

The following definitions apply to this subpart:

Administrator. The Administrator of the Animal and Plant Health Inspection Service, United States Department of Agriculture, or any other employee of the United States Department of Agriculture authorized to act in his or her stead.

Animal and Plant Health Inspection Service (APHIS). The Animal and Plant Health Inspection Service, United States Department of Agriculture.

Bulb. The portion of a plant commonly known as a bulb, bulbil, bulblet, corm, cormel, rhizome, tuber, or pip, and including fleshy roots or other underground fleshy growths, a unit of which produces an individual plant.

Consignment. A quantity of plants for planting being moved from one country to another and covered, when required, by a single phytosanitary certificate (a consignment may be composed of one or more lots or taxa).

217

Controlled import permit. A written or electronically transmitted authorization issued by APHIS for the importation into the United States of otherwise prohibited or restricted plant material for experimental, therapeutic, or developmental purposes, under controlled conditions as prescribed by the Administrator in accordance with § 319.6.

Earth. The softer matter composing part of the surface of the globe, in distinction from the firm rock, and including the soil and subsoil, as well as finely divided rock and other soil formation materials down to the rock layer.

From. Plants for planting are considered to be "from" any country or locality in which they are grown. *Provided,* That plants for planting imported into Canada from another country or locality shall be considered as being solely from Canada if they meet the following conditions:

(1) They are imported into the United States directly from Canada after having been grown for at least 1 year in Canada;

(2) They have never been grown in a country from which their importation would not be authorized pending pest risk analysis under § 319.37–4;

(3) They have never been grown in a country, other than Canada, from which it would be subject to certain restrictions on the importation of specific types of plants for planting under § 319.37–20, which are listed in the Plants for Planting Manual; *Provided,* that plants for planting that would be subject to postentry quarantine if imported into the United States may be imported from Canada after growth in another country if they were grown in Canada in postentry quarantine under conditions equivalent to those specified in the Plants for Planting Manual; and

(4) They were not imported into Canada in growing media.

Inspector. Any individual authorized by the Administrator or the Commissioner of Customs and Border Protection, Department of Homeland Security, to enforce the regulations in this part.

Lot. A number of units of a single commodity, identifiable by its homogeneity of composition and origin, forming all or part of a consignment.

Mother stock. A group of plants from which plant parts are taken to produce new plants.

National plant protection organization (NPPO). The official service established by a government to discharge the functions specified by the International Plant Protection Convention.

Noxious weed. Any plant or plant product that can directly or indirectly injure or cause damage to crops (including plants for planting or plant products), livestock, poultry, or other interests of agriculture, irrigation, navigation, the natural resources of the United States, the public health, or the environment.

Official control. The active enforcement of mandatory phytosanitary regulations and the application of mandatory phytosanitary procedures with the objective of eradication or containment of quarantine pests.

Person. Any individual, partnership, corporation, association, joint venture, or other legal entity.

Phytosanitary certificate. A document, including electronic versions, that is related to a restricted article and is issued not more than 15 days prior to shipment of the restricted article from the country in which it was grown and that:

(1) Is patterned after the model certificate of the International Plant Protection Convention, a multilateral convention on plant protection under the authority of the Food and Agriculture Organization of the United Nations (FAO);

(2) Is issued by an official of a foreign national plant protection organization in one of the five official languages of the FAO;

(3) Is addressed to the national plant protection organization of the United States (Animal and Plant Health Inspection Service);

(4) Describes the shipment;

(5) Certifies the place of origin for all contents of the shipment;

(6) Certifies that the shipment has been inspected and/or tested according to appropriate official procedures and is considered free from quarantine pests of the United States;

(7) Contains any additional declarations required in the Plants for Planting Manual; and

(8) Certifies that the shipment conforms with the phytosanitary requirements of the United States and is considered eligible for importation pursuant to the laws and regulations of the United States.

Place of production. Any premises or collection of fields operated as a single production or farming unit. This may include production sites that are separately managed for phytosanitary purposes.

Plant. Any plant (including any plant part) for or capable of propagation, including a tree, a tissue culture, a plantlet culture, pollen, a shrub, a vine, a cutting, a graft, a scion, a bud, a bulb, a root, and a seed.

Plant broker. An entity that purchases or takes possession of plants for planting from an approved place of production for the purpose of exporting those plants without further growing beyond maintaining the plants until export.

Plant pest. Any living stage of any of the following that can directly or indirectly injure, cause damage to, or cause disease in any plant or plant product: A protozoan, a nonhuman animal, a parasitic plant, a bacterium, a fungus, a virus or viroid, an infectious agent or other pathogen, or any article similar to or allied with any of these articles.

Plant Protection and Quarantine Programs. The organizational unit within APHIS that is delegated responsibility for enforcing provisions of the Plant Protection Act (7 U.S.C. 7701 *et seq.*) and related legislation, quarantines, and regulations.

Planting. Any operation for the placing of plants in a growing medium, or by grafting or similar operations, to ensure their subsequent growth, reproduction, or propagation.

Plants for planting. Plants intended to remain planted, to be planted, or replanted.

Plants for Planting Manual. The document that contains restrictions on the importation of specific types of plants for planting, as provided in § 319.37–20, and other information about the importation of plants for planting as provided in this subpart. The Plants for Planting Manual is available on the internet at *https://www.aphis.usda.gov/ import_export/plants/manuals/ports/ downloads/plants_for_planting.pdf.* or by contacting the Animal and Plant Health Inspection Service, Plant Protection and Quarantine, 4700 River Road, Unit 133, Riverdale, MD 20737–1236.

Port of first arrival. The land area (such as a seaport, airport, or land border station) where a person, or a land, water, or air vehicle, first arrives after entering the territory of the United States, and where inspection of plants for planting is carried out by inspectors.

Preclearance. Phytosanitary inspection and/or clearance in the country in which the plants for planting were grown, performed by or under the regular supervision of APHIS.

Production site. A defined portion of a place of production utilized for the production of a commodity that is managed separately for phytosanitary purposes. This may include the entire place of production or portions of it. Examples of portions of places of production are a defined orchard, grove, field, greenhouse, screenhouse, or premises.

Quarantine pest. A plant pest or noxious weed that is of potential economic importance to the United States and not yet present in the United States, or present but not widely distributed and being officially controlled.

Regulated plant. A vascular or nonvascular plant. Vascular plants include gymnosperms, angiosperms, ferns, and fern allies. Gymnosperms include cycads, conifers, and gingko. Angiosperms include any flowering plant. Fern allies include club mosses, horsetails, whisk ferns, spike mosses, and quillworts. Nonvascular plants include mosses, liverworts, hornworts, and green algae.

Secretary. The Secretary of Agriculture, or any other officer or employee of the Department of Agriculture to whom authority to act in his/her stead has been or may hereafter be delegated.

Soil. The loose surface material of the earth in which plants, trees, and shrubs grow, in most cases consisting of disintegrated rock with an admixture of organic material and soluble salts.

Species (spp.). All species, clones, cultivars, strains, varieties, and hybrids of a genus.

State. Any of the several States of the United States, the Commonwealth of the Northern Mariana Islands, the Commonwealth of Puerto Rico, the District of Columbia, Guam, the Virgin Islands of the United States, or any other territory or possession of the United States.

State Plant Regulatory Official. The official authorized by the State to sign agreements with Federal agencies involving operations of the State plant protection agency.

Taxon (taxa). Any grouping within botanical nomenclature, such as family, genus, species, or cultivar.

Type of plants for planting. A grouping of plants for planting based on shared characteristics such as biological traits, morphology, botanical nomenclature, or risk factors.

United States. All of the States.

§319.37-3 General restrictions on the importation of plants for planting.

(a) The importation of certain taxa of plants for planting is not authorized pending pest risk analysis in accordance with §319.37-4.

(b) General restrictions that apply to the importation of all plants for planting other than those whose importation is not authorized pending pest risk analysis are found in §§319.37-5 through 319.37-11.

(c) In accordance with §319.37-20, the Administrator may impose restrictions on the importation of specific types of plants for planting. These restrictions are listed in the Plants for Planting Manual. Additional information on restrictions applicable to the importation of specific types of plants for planting can be found in §§319.37-20 through 319.37-23.

§319.37-4 Taxa of plants for planting whose importation is not authorized pending pest risk analysis.

(a) *Determination by the Administrator.* The importation of certain taxa of plants for planting poses a risk of introducing quarantine pests into the United States. Therefore, the importation of these taxa is not authorized pending the completion of a pest risk analysis, except as provided in paragraph (f) of this section. These taxa are listed in the Plants for Planting Manual. There are two categories of taxa whose importation is not authorized pending pest risk analysis: Taxa of plants for planting that are quarantine pests, and taxa of plants for planting that are hosts of quarantine pests. For taxa of plants for planting that have been determined to be quarantine pests, the list includes the names of the taxa. For taxa of plants for planting that are hosts of quarantine pests, the list includes the names of the taxa, the foreign places from which the taxa's importation is not authorized, and the quarantine pests of concern.

(b) *Addition of taxa.* A taxon of plants for planting may be added to one of the lists of taxa not authorized for importation pending pest risk analysis under this section as follows:

(1) *Data sheet.* APHIS will publish in the FEDERAL REGISTER a document that announces our determination that a taxon of plants for planting is either a quarantine pest or a host of a quarantine pest. This notice will make available a data sheet that details the scientific evidence APHIS evaluated in making the determination that the taxon is a quarantine pest or a host of a quarantine pest. The data sheet will include references to the scientific evidence that APHIS used in making the determination. In our notice, we will provide for a public comment period of a minimum of 60 days on our additions to the list.

(2) *Response to comments.* (i) APHIS will issue a notice after the close of the public comment period indicating that the taxon will be added to the list of taxa not authorized for importation pending pest risk analysis if:

(A) No comments were received on the data sheet;

(B) The comments on the data sheet revealed that no changes to the data sheet were necessary; or

(C) Changes to the data sheet were made in response to public comments, but the changes did not affect APHIS' determination that the taxon poses a risk of introducing a quarantine pest into the United States.

(ii) If comments present information that leads us to determine that the importation of the taxon does not pose a risk of introducing a quarantine pest into the United States, APHIS will not add the taxon to the list of plants for planting whose importation is not authorized pending pest risk analysis. APHIS will issue a notice giving public notice of this determination after the close of the comment period.

(c) *Criterion for listing a taxon of plants for planting as a quarantine pest.* A taxon will be added to the list of taxa whose importation is not authorized pending pest risk analysis if scientific evidence causes APHIS to determine that the taxon is a quarantine pest.

(d) *Criteria for listing a taxon of plants for planting as a host of a quarantine pest.* A taxon will be added to the list of taxa whose importation is not authorized pending pest risk analysis if scientific evidence causes APHIS to determine that the taxon is a host of a quarantine pest. The following criteria must be fulfilled in order to make this determination:

(1) The plant pest in question must be determined to be a quarantine pest; and

(2) The taxon of plants for planting must be determined to be a host of that quarantine pest.

(e) *Removing a taxon from the list of taxa not authorized pending pest risk analysis.* (1) Requests to remove a taxon from the list of taxa whose importation is not authorized pending pest risk analysis (NAPPRA) must be made in accordance with §319.5. APHIS will conduct a pest risk analysis in response to such a request. The pest risk analysis will examine the risk associated with the importation of that taxon as well as measures available to mitigate that risk. The pest risk analysis may analyze importation of the taxon from a specific area, country, or countries, or from all areas of the world. The conclusions of the pest risk analysis will apply accordingly.

(2) If the pest risk analysis indicates that the taxon is a quarantine pest or a host of a quarantine pest and the Administrator determines that there are no measures available that adequately mitigate the risk of introducing a quarantine pest into the United States through the taxon's importation, we will continue to list the taxon as not authorized for importation pending pest risk analysis. We will publish a notice making the pest risk analysis available for comment. If comments cause us to change our determination, we will publish another notice in accordance with either paragraph (e)(3) or (4) of this section, as appropriate. If comments do not cause us to change our determination, we will publish a second notice responding to the comments and affirming our determination that the taxon should continue to be listed as NAPPRA.

(3) If the pest risk analysis supports a determination that importation of the taxon be allowed subject to taxon-specific restrictions, APHIS will publish a notice making the pest risk analysis available to the public for comment in accordance with the process in §319.37–20(c).

(4) If the pest risk analysis supports a determination that importation of the taxon be allowed subject to the general restrictions of this subpart, APHIS will publish a notice announcing our intent to remove the taxon from the list of taxa whose importation is not authorized pending pest risk analysis and making the pest risk analysis supporting the taxon's removal available for public comment.

(i) APHIS will issue a notice after the close of the public comment period indicating that the importation of the taxon will be subject only to the general restrictions of this subpart if:

(A) No comments were received on the pest risk analysis;

(B) The comments on the pest risk analysis revealed that no changes to the pest risk analysis were necessary; or

(C) Changes to the pest risk analysis were made in response to public comments, but the changes did not affect the overall conclusions of the analysis and the Administrator's determination that the importation of the taxon does not pose a risk of introducing a quarantine pest into the United States.

(ii) If information presented by commenters indicates that the pest risk

221

analysis needs to be revised, APHIS will issue a notice after the close of the public comment period indicating that the importation of the taxon will continue to be listed as not authorized pending pest risk analysis while the information presented by commenters is analyzed and incorporated into the pest risk analysis. APHIS will subsequently publish a new notice announcing the availability of the revised pest risk analysis.

(5) APHIS may also remove a taxon from the list of taxa whose importation is not authorized pending pest risk analysis when APHIS determines that the evidence used to add the taxon to the list was erroneous (for example, involving a taxonomic misidentification).

(f) *Controlled import permits.* Any plants for planting whose importation is not authorized pending pest risk analysis in accordance with this section may be imported or offered for entry into the United States if:

(1) Imported for experimental, therapeutic, or developmental purposes under the conditions specified in a controlled import permit issued in accordance with § 319.6;

(2) Imported at the National Plant Germplasm Inspection Station, Building 580, Beltsville Agricultural Research Center East, Beltsville, MD 20705 or through any USDA plant inspection station listed in the Plants for Planting Manual;

(3) Imported pursuant to a controlled import permit issued for such plants for planting and kept on file at the port of entry;

(4) Imported under conditions specified on the controlled import permit and found by the Administrator to be adequate to prevent the introduction into the United States of quarantine pests, *i.e.,* conditions of treatment, processing, growing, shipment, disposal; and

(5) Imported with a controlled import tag or label securely attached to the outside of the container containing the plants for planting or securely attached to the plant itself if not in a container, and with such tag or label bearing a controlled import permit number corresponding to the number of

the controlled import permit issued for such plants for planting.

(Approved by the Office of Management and Budget under control number 0579–0380)

§ 319.37–5 Permits.

(a)(1) Plants for planting may be imported or offered for importation into the United States only after issuance of a written permit by the Plant Protection and Quarantine Programs, except as provided in the Plants for Planting Manual. Exceptions from the requirement for a written permit will be added, changed, or removed in accordance with § 319.37–20.

(2) Plants for planting whose importation is subject to postentry quarantine, as listed in the Plants for Planting Manual, must also be imported under an importer postentry quarantine growing agreement in accordance with § 319.37–23(c).

(b) An application for a written permit should be submitted to the Plant Protection and Quarantine Programs (Animal and Plant Health Inspection Service, Plant Protection and Quarantine, Permits, Permit Unit, 4700 River Road, Unit 133, Riverdale, MD 20737–1236) at least 30 days prior to arrival of the plants for planting at the port of entry. Application forms are available without charge from that address or on the internet at *http://www.aphis.usda.gov/permits/ppq_epermits.shtml.* The completed application shall include the following information:

(1) Name, address, and telephone number of the importer;

(2) The taxon or taxa and the approximate quantity of plants for planting intended to be imported;

(3) Country(ies) or locality(ies) where grown;

(4) Intended United States port of entry;

(5) Means of transportation, *e.g.,* mail, airmail, express, air express, freight, airfreight, or baggage; and

(6) Expected date of arrival.

(c) A permit indicating the applicable conditions for importation under this subpart will be issued by Plant Protection and Quarantine Programs if, after review of the application, the plants for planting are deemed eligible to be imported into the United States under the conditions specified in the permit. However, even if such a permit is issued, the plants for planting may be imported only if all applicable requirements of this subpart are met and only if an inspector at the port of entry determines that no remedial measures pursuant to the Plant Protection Act are necessary with respect to the plants for planting.[1]

(d) Any permit that has been issued may be revoked by an inspector or APHIS in accordance with § 319.7-4.

(e) Any plants for planting not required to be imported with a permit in accordance with paragraph (a) of this section may be imported or offered for importation into the United States only after issuance of an oral authorization for importation issued by an inspector at the port of entry.

(f) An oral authorization for importation of plants for planting shall be issued at a port of entry by an inspector only if all applicable requirements of this subpart are met, such plants for planting are eligible to be imported under an oral authorization, and an inspector at the port of entry determines that no measures pursuant to section 414 of the Plant Protection Act (7 U.S.C. 7714) are necessary with respect to such plants for planting.

(g) Persons wishing to import plants for planting into the United States for experimental, therapeutic, or developmental purposes must apply for a controlled import permit in accordance with §§ 319.6 and 319.37-3.

(Approved by the Office of Management and Budget under control numbers 0579-0190, 0579-0285, and 0579-0319)

§ 319.37-6 Phytosanitary certificates.

(a) *Phytosanitary certificates.* Any plants for planting offered for importation into the United States must be accompanied by a phytosanitary certificate, except as described in paragraphs (b) and (c) of this section. The phytosanitary certificate must identify the genus of the plants for planting it

accompanies. When the importation of individual species or cultivars within a genus is restricted in accordance with § 319.37-20, the phytosanitary certificate must also identify the species or cultivar of the plants for planting it accompanies. Otherwise, identification of the species is strongly preferred, but not required. Intergeneric and interspecific hybrids must be designated by placing the multiplication sign "×" between the names of the parent taxa. If the hybrid is named, the multiplication sign may instead be placed before the name of an intergeneric hybrid or before the epithet in the name of an interspecific hybrid.

(b) *Small lots of seed.* Lots of seed may be imported without a phytosanitary certificate required by paragraph (a) of this section under the following conditions:

(1) The importation of the seed is authorized by a written permit issued in accordance with § 319.37-5.

(2) The seed is not listed as not authorized pending pest risk analysis, as provided in § 319.37-4; is not of any noxious weed species listed in part 360 of this chapter; is not subject to restrictions on specific types of plants for planting as provided in § 319.37-20; is not restricted under the regulations in parts 330 and 340 of this chapter; and meets the requirements of part 361 of this chapter.

(3) The seed meets the following packaging and shipping requirements:

(i) Each seed packet is clearly labeled with the name of the collector/shippor, the country of origin, and the scientific name at least to the genus, and preferably to the species, level;

(ii) There are a maximum of 50 seeds of 1 taxon (taxonomic category such as genus, species, cultivar, etc.) per packet; or a maximum weight not to exceed 10 grams of seed of 1 taxon per packet;

(iii) There are a maximum of 50 seed packets per shipment;

(iv) The seeds are free from pesticides;

(v) The seeds are securely packaged in packets or envelopes and sealed to prevent spillage;

(vi) The shipment is free from soil, plant material other than seed, other foreign matter or debris, seeds in the fruit or seed pod, and living organisms

---

[1] An inspector may hold, seize, quarantine, treat, apply other remedial measures to, destroy, or otherwise dispose of plants, plant pests, or other articles in accordance with sections 414, 421, and 434 of the Plant Protection Act (7 U.S.C. 7714, 7731, and 7754).

such as parasitic plants, pathogens, insects, snails, mites; and

(vii) At the time of importation, the shipment is sent to either the Plant Germplasm Quarantine Center in Beltsville, MD, or a USDA plant inspection station.

(c) *Importation of other plants for planting without phytosanitary certificates.* (1) The Administrator may authorize the importation of types of plants for planting without a phytosanitary certificate if the plants for planting are accompanied by equivalent documentation agreed upon by the Administrator and the NPPO of the exporting country as sufficient to establish the eligibility of the plants for importation into the United States. The documentation must be provided by the NPPO or refer to documentation provided by the NPPO. The documentation must be agreed upon before the plants for planting are exported from the exporting country to the United States.

(2) The Administrator may impose additional restrictions on the importation of plants for planting that are not accompanied by a phytosanitary certificate to ensure that the plants are appropriately identified and free of quarantine pests.

(3) The Plants for Planting Manual lists types of plants for planting that are not required to be accompanied by a phytosanitary certificate; the countries from which their importation without a phytosanitary certificate is authorized; the approved documentation of eligibility for importation; and any additional conditions on their importation.

(4) Types of plants for planting may be added to or removed from the list of plants for planting that are not required to be accompanied by a phytosanitary certificate in accordance with § 319.37–20. The requirements for importing types of plants for planting without a phytosanitary certificate may also be changed by a notice issued in accordance with § 319.37–20. The notice published for comment will describe the documentation agreed upon by the Administrator and the NPPO of the exporting country and any additional restrictions to be imposed on the importation of the type of plants for planting.

(Approved by the Office of Management and Budget under control numbers 0579–0142, 0579–0190, 0579–0285, and 0579–0319)

§ 319.37–7 Marking and identity.

(a) Any consignment of plants for planting for importation, other than by mail at the time of importation, or offer for importation into the United States shall plainly and correctly bear on the outer container (if in a container) or the plants for planting (if not in a container) the following information:

(1) General nature and quantity of the contents;

(2) Country and locality where grown;

(3) Name and address of shipper, owner, or person shipping or forwarding the plants for planting;

(4) Name and address of consignee;

(5) Identifying shipper's mark and number; and

(6) Number of written permit authorizing the importation, if one was required under § 319.37–5.

(b) Any consignment of plants for planting for importation by mail shall be plainly and correctly addressed and mailed to the Plant Protection and Quarantine Programs at a port of entry listed in the Plants for Planting Manual as approved to receive imported plants for planting, shall be accompanied by a separate sheet of paper within the package plainly and correctly bearing the name, address, and telephone number of the intended recipient, and shall plainly and correctly bear on the outer container the following information:

(1) General nature and quantity of the contents;

(2) Country and locality where grown;

(3) Name and address of shipper, owner, or person shipping or forwarding the plants for planting; and

(4) Number of written permit authorizing the importation, if one was required under § 319.37–5.

(c) Any consignment of plants for planting for importation (by mail or otherwise), at the time of importation or offer for importation into the United

States shall be accompanied by an invoice or packing list indicating the contents of the consignment.

(Approved by the Office of Management and Budget under control numbers 0579–0190 and 0579–0319)

§319.37-8 Ports of entry: Approved ports, notification of arrival, inspection, and refusal of entry.

(a) *Approved ports of entry.* Any plants for planting required to be imported under a written permit in accordance with §319.37–5(a), if not precleared, must be imported or offered for importation only at a USDA plant inspection station, unless the Plants for Planting Manual indicates otherwise. Ports of entry through which plants for planting must pass through before arriving at these USDA plant inspection stations are listed in the Plants for Planting Manual. All other plants for planting may be imported or offered for importation at any Customs designated port of entry indicated in 19 CFR 101.3(b)(1). Exceptions may be listed in §330.104 of this chapter. Plants for planting that are required to be imported under a written permit that are also precleared in the country of export are not required to enter at an inspection station and may enter through any Customs port of entry. Exceptions may be listed in §330.104 of this chapter.

(b) *Notification upon arrival at the port of entry.* Promptly upon arrival of any plants for planting at a port of entry, the importer shall notify the Plant Protection and Quarantine Programs of the arrival by such means as a manifest, Customs entry document, commercial invoice, waybill, a broker's document, or a notice form provided for that purpose.

(c) *Inspection and treatment.* Any plants for planting may be sampled and inspected by an inspector at the port of first arrival and/or under preclearance inspection arrangements in the country in which the plants for planting were grown, and must undergo treatment in accordance with part 305 of this chapter if treatment is ordered by the inspector. Any plants for planting found upon inspection to contain or be contaminated with quarantine pests that cannot be eliminated by treat-

ment will be denied entry at the first United States port of arrival and must be destroyed or shipped to a point outside the United States.

(d) *Disposition of plants for planting not in compliance with this subpart.* The importer of any plants for planting denied entry for noncompliance with this subpart must, at the importer's expense and within the time specified in an emergency action notification (PPQ Form 523), destroy, ship to a point outside the United States, treat in accordance with part 305 of this chapter, or apply other safeguards to the plants for planting, as prescribed by an inspector, to prevent the introduction into the United States of quarantine pests. In choosing which action to order and in setting the time limit for the action, the inspector shall consider the degree of pest risk presented by the plant pest associated with the plants for planting, whether the plants for planting are a host of the pest, the types of other host materials for the pest in or near the port, the climate and season at the port in relation to the pest's survival range, and the availability of treatment facilities for the plants for planting.

(e) *Removal of plants for planting from port of first arrival.* No person shall remove any plants for planting from the port of first arrival unless and until notice is given to the collector of customs by the inspector that the plants for planting has satisfied all requirements under this subpart.

(Approved by the Office of Management and Budget under control numbers 0579–0190, 0579–0310, and 0579–0319)

§319.37-9 Treatment of plants for planting; costs and charges for inspection and treatment; treatments applied outside the United States.

(a) The services of a Plant Protection and Quarantine inspector during regularly assigned hours of duty and at the usual places of duty shall be furnished without cost to the importer.[2] No charge will be made to the importer for Government-owned or -controlled special inspection facilities and equipment

[2] Provisions relating to costs for other services of an inspector are contained in part 354 of this chapter.

225

used in treatment, but the inspector may require the importer to furnish any special labor, chemicals, packing materials, or other supplies required in handling an importation under the regulations in this subpart. The Plant Protection and Quarantine Programs will not be responsible for any costs or charges, other than those indicated in this section.

(b) Any treatment performed in the United States on plants for planting must be performed at the time of importation into the United States. Treatment shall be performed by an inspector or under an inspector's supervision at a Government-operated special inspection facility, except that an importer may have such treatment performed at a nongovernmental facility if the treatment is performed at nongovernment expense under the supervision of an inspector and in accordance with part 305 of this chapter and in accordance with any treatment required by an inspector as an emergency measure in order to prevent the dissemination of any quarantine pests. However, treatment may be performed at a nongovernmental facility only in cases of unavailability of government facilities and only if, in the judgment of an inspector, the plants for planting can be transported to such nongovernmental facility without the risk of introduction into the United States of quarantine pests.

(c) Any treatment performed outside the United States must be monitored and certified by an APHIS inspector or an official from the NPPO of the exporting country. If monitored and certified by an official of the NPPO of the exporting country, then a phytosanitary certificate must be issued with the following declaration: "The consignment of (fill in taxon) has been treated in accordance with 7 CFR part 305." During the entire interval between treatment and export, the consignment must be stored and handled in a manner that prevents any infestation by quarantine pests.

(Approved by the Office of Management and Budget under control number 0579–0190)

§ 319.37-10   Growing media.

(a) Any plants for planting at the time of importation or offer for impor-

tation into the United States shall be free of sand, soil, earth, and other growing media, except as provided in paragraph (b), (c), or (d) of this section.

(b)(1) Plants for planting from Canada may be imported in any growing medium, except as restricted in the Plants for Planting Manual. Restrictions on growing media for specific types of plants for planting imported from Canada will be added, changed, or removed in accordance with § 319.37-20.

(2) Plants for planting from an area of Canada regulated by the national plant protection organization of Canada for a soil-borne plant pest may only be imported if the phytosanitary certificate accompanying it contains an additional declaration that the plant was grown in a manner to prevent infestation by that soil-borne plant pest.

(c) Certain types of plants for planting growing solely in certain growing media listed in the Plants for Planting Manual may be imported established in such growing media. The Administrator has determined that the importation of the specified types of plants for planting in these growing media does not pose a risk of introducing quarantine pests into the United States. If the Administrator determines that a new growing medium may be added to the list of growing media in which imported plants for planting may be established, or that a growing medium currently listed for such purposes is no longer suitable for establishment of imported plants for planting, APHIS will publish in the FEDERAL REGISTER a notice that announces our proposed determination and requests comment on the change. After the close of the comment period, APHIS will publish another notice informing the public of the Administrator's decision on the change to the list of growing media in which imported types of plants for planting may be established.

(d) Certain types of plants for planting, as listed in the Plants for Planting Manual, may be imported when they are established in a growing medium approved by the Administrator and they are produced in accordance with additional requirements specified in the Plants for Planting Manual.

Changes to the list of plants for planting that may be imported in growing media, and to the requirements for the importation of those types of plants for planting, will be made in accordance with §319.37–20.

(Approved by the Office of Management and Budget under control numbers 0579–0190, 0579–0439, 0579–0454, 0579–0458, and 0579–0463)

[83 FR 11856, Mar. 19, 2018, as amended at 84 FR 29958, June 25, 2019]

§319.37–11 Packing and approved packing material.

(a) Plants for planting for importation into the United States must not be packed in the same container as plants for planting whose importation into the United States is not authorized pending pest risk analysis in accordance with §319.37–4.

(b) Any plants for planting at the time of importation or offer for importation into the United States shall not be packed in a packing material unless the plants were packed in the packing material immediately prior to shipment; such packing material is free from sand, soil, or earth (except as designated in the Plants for Planting Manual); has not been used previously as packing material or otherwise; and is approved by the Administrator as not posing a risk of introducing quarantine pests. Approved packing materials are listed in the Plants for Planting Manual.

(c) If the Administrator determines that a new packing material may be added to the list of packing materials, or that a packing material currently listed should no longer be approved, APHIS will publish in the FEDERAL REGISTER a notice that announces our proposed determination and requests comment on the change. After the close of the comment period, APHIS will publish another notice informing the public of the Administrator's decision on the change to the list of approved packing materials.

(Approved by the Office of Management and Budget under control number 0579–0190)

§§319.37–12—319.37–19 [Reserved]

§319.37–20 Restrictions on the importation of specific types of plants for planting.

(a) *Plant type-specific restrictions.* In addition to the general restrictions in this subpart, the Administrator may impose additional restrictions on the importation of specific types of plants for planting necessary to effectively mitigate the risk of introducing quarantine pests into the United States through the importation of specific plants for planting. Additional restrictions may be placed on the importation of the entire plant or on certain plant parts. A list of the types of plants for planting whose importation is subject to additional restrictions, and the specific restrictions that apply to the importation of each type of plants for planting, may be found in the Plants for Planting Manual.

(b) *Basis for changing restrictions.* The Administrator may determine that it is necessary to add, change, or remove restrictions on the importation of a specific type of plants for planting, based on the risk of introducing a quarantine pest through the importation of that type of plants for planting. The Administrator will make this determination based on the findings of a pest risk analysis or on other scientific evidence.

(c) *Process for adding, changing, or removing restrictions.* Restrictions on the importation of a specific type of plants for planting beyond the general restrictions in §§319.37–5 through 319.37–11 will be changed through the following process:

(1) *Document describing restrictions.* APHIS will publish in the FEDERAL REGISTER a notice that announces our proposed determination that it is necessary to add, change, or remove restrictions on the importation of a specific type of plants for planting. This notice will make available for public comment a document describing the restrictions that the Administrator has determined are necessary and how these restrictions will mitigate the risk of introducing quarantine pests into the United States.

(2) *Response to comments.* APHIS will issue a second notice after the close of

the public comment period on the notice described in paragraph (c)(1) of this section. This notice will inform the public of the specific restrictions, if any, that the Administrator has determined to be necessary in order to mitigate the risk of introducing quarantine pests into the United States through the importation of the type of plants for planting. In response to the public comments submitted, the Administrator may implement the restrictions described in the document made available by the initial notice, amend the restrictions in response to public comment, or determine that changes to the restrictions on the importation of the type of plants for planting are unnecessary.

(d) *Previously imposed restrictions on specific types of plants for planting.* Types of plants for planting whose importation was subject to specific restrictions by specific regulation as of April 18, 2018, will continue to be subject to those restrictions, except as changed in accordance with the process specified in paragraph (c) of this section. The restrictions are found in the Plants for Planting Manual.

§ 319.37-21 Integrated pest risk management measures.

If a type of plants for planting is a host of a quarantine pest or pests, APHIS may require the type of plants for planting to be produced in accordance with integrated pest risk management measures as a condition of importation. This section sets out a general framework for integrated pest risk management measures. When APHIS determines that integrated measures are necessary to mitigate risk, APHIS will use this framework to develop integrated pest risk management measures that mitigate the quarantine pest risks associated with that type of plants for planting through the process described in § 319.37-20.

(a) *Responsibilities of the place of production.* The place of production is responsible for identifying, developing, and implementing procedures that meet the requirements of both the NPPO of the exporting country and APHIS. Participants in the export program must be approved by the NPPO or its designee and APHIS. Approval will be conferred by the NPPO or its designee and APHIS after the participant meets the conditions required for integrated pest risk management. Approval will be withdrawn if the participant fails to meet the conditions at any time. All documentation required under paragraphs (a)(5) and (6) of this section will be maintained by the exporting place of production and made available to official representatives of the NPPO of the exporting country and APHIS upon request. The place of production must be open to necessary and reasonable audit, monitoring, and evaluation of compliance by the NPPO of the exporting country and APHIS. The management of the place of production will be responsible for complying with the integrated pest risk management measures. Management must specify the roles and responsibilities of its personnel to perform program activities. The place of production must notify the NPPO of the exporting country of deficiencies detected during internal audits. The NPPO of the exporting country will be responsible for ensuring that the place of production is in compliance with the integrated pest risk management measures.

(1) *Pest management program.* The place of production must develop and implement an approved pest management program that contains ongoing pest monitoring and procedures for the exclusion and control of plant pests. The place of production must obtain material used to produce plants for planting from sources that are free of quarantine pests and that are approved by the NPPO of the exporting country and APHIS. All sources of plants for planting and the phytosanitary status of those plants must be well-documented and the program for producing plants for planting carefully monitored.

(2) *Training.* A training program approved by the NPPO of the exporting country and APHIS must be established, documented, and regularly conducted at the place of production. The training program must ensure that all those involved in the export program possess specific knowledge related to the relevant components of the program and a general understanding of its requirements.

(3) *Internal audits.* The place of production must perform, or designate parties to perform internal audits that ensure that a plan approved and documented by APHIS and the NPPO of the exporting country is being followed and is achieving the appropriate level of pest management.

(4) *Traceability.* The place of production must implement a procedure approved by APHIS and the NPPO of the exporting country or its designee that documents and identifies plants from propagation through harvest and sale to ensure that plants can be traced forward and back from the place of production. Depending on the nature of the quarantine pests, the system may need to account for:

(i) The origin and pest status of mother stock;

(ii) The year of propagation and the place of production of all plant parts that make up the plants for planting intended for export;

(iii) Geographic location of the place of production;

(iv) Location of plants for planting within the place of production;

(v) The plant taxon; and

(vi) The purchaser's identity.

(5) *Documentation of program procedures.* The place of production must develop a manual approved by the NPPO of the exporting country and APHIS that guides the place of production's operation and that includes the following components:

(i) Administrative procedures (including roles and responsibilities and training procedures);

(ii) Pest management plan;

(iii) Place of production internal audit procedures;

(iv) Management of noncompliant product or procedures;

(v) Traceability procedures; and

(vi) Recordkeeping systems.

(6) *Records.* A place of production must maintain records on its premises as specified by APHIS and the NPPO of the exporting country. These records must be made available to APHIS and the NPPO of the exporting country upon request. These documents include all the elements described in this paragraph (a) and copies of all internal and external audit documents and reports.

(b) *Responsibilities of APHIS and the NPPO of the exporting country.* APHIS and the NPPO of the exporting country are responsible for collaborating to establish program requirements, including workplans and compliance agreements as necessary, for recognizing and implementing particular import programs. Technically justified modifications to the program may be negotiated. The administration of program requirements must include such elements as clarification of terminology, testing and retesting requirements, eligibility, the nomenclature of certification levels, horticultural management, isolation and sanitation requirements, inspection, documentation, identification and labeling, quality assurance, noncompliance and remedial measures, and postentry quarantine requirements. The criteria for approving, suspending, removing, and reinstating approval for a particular program should be jointly developed and agreed upon by APHIS and the NPPO of the exporting country. Information should be exchanged between APHIS and the NPPO of the exporting country through officially designated points of contact.

(c) *Responsibilities of the NPPO of the exporting country.* (1) The NPPO of the exporting country must provide sufficient information to APHIS to support the evaluation and acceptance of export programs. This may include:

(i) Specific identification of the commodity, place of production, and expected volume and frequency of consignments;

(ii) Relevant production, harvest, packing, handling, and transport details;

(iii) Pests associated with the plant including prevalence, distribution, and damage potential;

(iv) Risk management measures proposed for a pest management program; and

(v) Relevant efficacy data.

(2) A phytosanitary certificate should be issued by the NPPO of the exporting country unless APHIS and the NPPO of the exporting country agree to use other documentation in accordance with §319.37-6(c).

(3) Other responsibilities of the NPPO of the exporting country include:

(i) Establishing and maintaining compliance agreements as necessary;

(ii) Oversight and enforcement of program provisions;

(iii) Arrangements for monitoring and audit; and

(iv) Maintaining appropriate records.

(d) *Responsibilities of plant brokers trading in plants for planting produced in accordance with integrated pest risk management measures.* Plant brokers trading in plants for planting produced in accordance with integrated pest risk management measures must be approved by the NPPO of the exporting country or its designee. The list of plant brokers must be provided to APHIS upon request. Approval may only be conferred by the NPPO or its designee after the participant demonstrates that it can meet the requirements of this paragraph (d). Approval must be withdrawn if the participant fails to meet the conditions at any time. Plant brokers must ensure the traceability of export consignments to an approved place of production or production site. Brokers must maintain the phytosanitary status of the plants equivalent to an approved place of production from purchase, storage, and transportation to the export destination. Plant brokers must document these processes for verifying status and maintaining traceability.

(e) *External audits.* APHIS and the NPPO of the exporting country will agree to the requirements for external audits.

(1) *APHIS audits.* APHIS will evaluate the integrated pest risk management measures of the NPPO of the exporting country before acceptance. This could consist of documentation review, site visits, and inspection and testing of plants produced under the system. Following approval, APHIS or its designee will monitor and periodically audit the system to ensure that it continues to meet the stated objectives. Audits will include inspection of imported plants for planting, site visits, and review of the integrated pest risk management measures and internal audit processes of the place of production and the NPPO of the exporting country.

(2) *Audits by the NPPO of the exporting country.* The NPPO must arrange for audits of the exporting system. Audits may be conducted by the NPPO or its designee and may consist of inspection and testing of plants for planting and the documentation and management practices as they relate to the program. Audits should verify that:

(i) The places of production in the program are free of quarantine pests;

(ii) Program participants are complying with the specified standards;

(iii) The integrated pest management measures continue to meet APHIS requirements; and

(iv) Arrangements with designees are complied with.

(f) *Noncompliance.* (1) The exporting NPPO must notify APHIS of noncompliance within the integrity of the system or noncompliance by a place of production that affects the phytosanitary integrity of the commodity. The requirements for notification will be determined between the NPPO of the exporting country and APHIS.

(2) Regulatory responses to program failures will be based on existing bilateral agreements. Contingency plans may be established in advance to ensure that alternative measures are available in the event that all or part of a program fails. APHIS will specify the consequences of noncompliance to the NPPO of the exporting country. The NPPO must specify the consequences of noncompliance to the participants in the program. These may vary depending on the nature and severity of the infraction. In addition, remedial measures should be specified to enable a suspended or decertified place of production or plant broker to become eligible for reinstatement or recertification.

(3) Places of production or plant brokers that do not meet the conditions of the program must be suspended. Plants for planting must not be exported from a place of production or a plant broker that has failed to meet the program requirements.

(4) The effectiveness of remedial measures taken must be verified before reinstatement to the program by the exporting NPPO and, where appropriate, by APHIS.

(Approved by the Office of Management and Budget under control number 0579–0190)

§319.37–22 Trust fund agreements.

If APHIS personnel need to be physically present in an exporting country or region to facilitate the exportation of plants for planting and APHIS services are to be funded by the NPPO of the exporting country or a private export group, then the NPPO or the private export group must enter into a trust fund agreement with APHIS that is in effect at the time APHIS' services are needed. Under the agreement, the NPPO of the exporting country or the private export group must pay in advance all estimated costs that APHIS expects to incur in providing inspection services in the exporting country. These costs will include administrative expenses incurred in conducting the services and all salaries (including overtime and the Federal share of employee benefits), travel expenses (including per diem expenses), and other incidental expenses incurred by the inspectors in performing services. The agreement must require the NPPO of the exporting country or region or a private export group to deposit a certified or cashier's check with APHIS for the amount of those costs, as estimated by APHIS. The agreement must further specify that, if the deposit is not sufficient to meet all costs incurred by APHIS, the NPPO of the exporting country or a private export group must deposit with APHIS, before the services will be completed, a certified or cashier's check for the amount of the remaining costs, as determined by APHIS. After a final audit at the conclusion of each shipping season, any overpayment of funds would be returned to the NPPO of the exporting country or region or a private export group, or held on account.

(Approved by the Office of Management and Budget under control number 0579–0190)

§319.37–23 Postentry quarantine.

(a) *Postentry quarantine.* One specific restriction that may be placed upon the importation of a type of plants for planting in accordance with §319.37–20 is that it be grown in postentry quarantine. The Plants for Planting Manual lists the taxa required to be imported into postentry quarantine. Plants for planting grown in postentry quarantine must be grown under postentry quarantine conditions specified in paragraphs (c) and (d) of this section, and may be imported or offered for importation into the United States only:

(1) If destined for a State that has completed a State postentry quarantine agreement with APHIS in accordance with paragraph (b) of this section;

(2) If an importer postentry quarantine growing agreement has been completed and submitted to Plant Protection and Quarantine in accordance with paragraph (c) of this section. The agreement must be signed by the person (the importer) applying for the importation of the plants for planting in accordance with §319.6; and,

(3) If Plant Protection and Quarantine has determined that the completed postentry quarantine growing agreement fulfills the applicable requirements of this section and that services by State inspectors are available to monitor and enforce the postentry quarantine.

(b) *State postentry quarantine agreement.* Plants for planting required to undergo postentry quarantine in accordance with §319.37–20 may only be imported if destined for postentry quarantine growing in a State which has entered into a written agreement with the Animal and Plant Health Inspection Service, signed by the Administrator or his or her designee and by the State Plant Regulatory Official. In accordance with the laws of individual States, inspection and other postentry quarantine services provided by a State may be subject to charges imposed by the State. A list of States that have entered into a postentry quarantine agreement in accordance with this paragraph can be found in the Plants for Planting Manual.

(c) *Importer postentry quarantine growing agreements.* Any plants for planting required to be grown under postentry quarantine conditions, as well as any increase therefrom, shall be grown in accordance with an importer postentry quarantine growing agreement signed by the person (the importer) applying for a written permit in accordance with §319.37–5 for importation of the plants for planting and submitted to Plant Protection and Quarantine. On each importer postentry quarantine growing agreement, the person shall also obtain the signature of the State Plant Regulatory Official for the State in which plants for planting covered by the agreement will be grown. The importer postentry quarantine growing agreement shall specify the kind, number, and origin of plants to be imported; the conditions specified in the Plants for Planting Manual under which the plants for planting will be grown, maintained, and labeled; and the reporting requirements in the case of abnormal or dead plants for planting. The agreement shall certify to APHIS and to the State in which the plants for planting are grown that the signer of the agreement will comply with the conditions of the agreement for the postentry quarantine growing period prescribed for the type of plants for planting in the Plants for Planting Manual.

(d) *Applications for permits.* A completed importer postentry quarantine agreement shall accompany the application for a written permit for plants for planting required to be grown under postentry quarantine conditions. Importer postentry quarantine agreement forms are available without charge from the Animal and Plant Health Inspection Service, Plant Protection and Quarantine, Permit Unit, 4700 River Road, Unit 136, Riverdale, MD 20737-1236 or on the internet at *http:// www.aphis.usda.gov/permits/ ppq_epermits.shtml.*

(e) *Inspector-ordered disposal, movement, or safeguarding of plants for planting; costs and charges, civil and criminal liabilities*—(1) *Growing at unauthorized sites.* If an inspector determines that any plants for planting subject to the postentry quarantine growing requirements of this section, or any increase therefrom, is being grown at an unauthorized site, the inspector may file an emergency action notification (PPQ Form 523) with the owner of the plants for planting or the person who owns or is in possession of the site on which the plants for planting is being grown. The person named in the PPQ Form 523 must, within the time specified in PPQ Form 523, sign a postentry quarantine growing agreement, destroy, ship to a point outside the United States, move

to an authorized postentry quarantine site, and/or apply treatments or other safeguards to the plants for planting, the increase therefrom, or any portion of the plants for planting or the increase therefrom, as prescribed by an inspector to prevent the introduction of quarantine pests into the United States. In choosing which action to order and in setting the time limit for the action, the inspector shall consider the degree of pest risk presented by the quarantine pests associated with the type of plants for planting (including increase therefrom), the types of other host materials for the pest in or near the growing site, the climate and season at the site in relation to the pest's survival, and the availability of treatment facilities.

(2) *Growing at authorized sites.* If an inspector determines that any plants for planting, or any increase therefrom, grown at a site specified in an authorized postentry quarantine growing agreement is being grown contrary to the provisions of this section, including in numbers greater than the number approved by the postentry quarantine growing agreement, or in a manner that otherwise presents a risk of introducing quarantine pests into the United States, the inspector shall issue an emergency action notification (PPQ Form 523) to the person who signed the postentry quarantine growing agreement. That person shall be responsible for carrying out all actions specified in the emergency action notification. The emergency action notification may extend the time for which the plants for planting and the increase therefrom must be grown under the postentry quarantine conditions specified in the authorized postentry quarantine growing agreement, or may require that the person named in the notification must destroy, ship to a point outside the United States, or apply treatments or other safeguards to the plants for planting, the increase therefrom, or any portion of the plants for planting or the increase therefrom, within the time specified in the emergency action notification. In choosing which action to order and in setting the time limit for the action, the inspector shall consider the degree of pest risk presented by the quarantine

pests associated with the type of plants for planting (including increase therefrom), the types of other host materials for the pest in or near the growing site, the climate and season at the site in relation to the pest's survival, and the availability of treatment facilities.

(3) *Costs and charges.* All costs pursuant to any action ordered by an inspector in accordance with this section shall be borne by the person who signed the postentry quarantine growing agreement covering the site where the plants for planting were grown, or if no such agreement was signed, by the owner of the plants for planting at the growing site.

(Approved by the Office of Management and Budget under control number 0579–0190)

Subpart I—Logs, Lumber, and Other Wood Articles

SOURCE: 60 FR 27674, May 25, 1995, unless otherwise noted. Redesignated at 84 FR 2429, Feb. 7, 2019.

§319.40–1 Definitions.

Administrator. The Administrator of the Animal and Plant Health Inspection Service, United States Department of Agriculture, or any employee of the United States Department of Agriculture delegated to act in his or her stead.

APHIS. The Animal and Plant Health Inspection Service, United States Department of Agriculture.

Bark chips. Bark fragments broken or shredded from log or branch surfaces.

Certificate. A certificate of inspection relating to a regulated article, which is issued by an official authorized by the national government of the country in which the regulated article was produced or grown, which is addressed to the plant protection service of the United States (Plant Protection and Quarantine Programs), which contains a description of the regulated article, which certifies that the regulated article has been inspected, is believed to be free of plant pests, and is believed to be eligible for importation pursuant to the laws and regulations of the United States, and which may contain any specific additional declarations required under this subpart.

Compliance agreement. A written agreement between APHIS and a person engaged in processing, handling, or moving regulated articles, in which the person agrees to comply with requirements contained in the agreement.

Controlled import permit. A written or electronically transmitted authorization issued by APHIS for the importation into the United States of otherwise prohibited or restricted plant material for experimental, therapeutic, or developmental purposes, under controlled conditions as prescribed by the Administrator in accordance with §319.6.

Fines. Small particles or fragments of wood, slightly larger than sawdust, that result from chipping, sawing, or processing wood.

Free from rot. No more than two percent by weight of the regulated articles in a lot show visual evidence of fructification of fungi or growth of other microorganisms that cause decay and the breakdown of cell walls in the regulated articles.

General permit. A written authorization contained in §319.40–3 for any person to import the articles named by the general permit, in accordance with the requirements specified by the general permit, without being issued a specific permit.

Humus, compost, and litter. Partially or wholly decayed plant matter.

Import (imported, importation). To bring or move into the territorial limits of the United States.

Importer document. A written declaration signed by the importer of regulated articles, which must accompany the regulated articles at the time of importation, in which the importer accurately declares information about the regulated articles required to be disclosed by §319.40–2(b).

Inspector. Any individual authorized by the Administrator to enforce this subpart.

Log. The bole of a tree; trimmed timber that has not been sawn further than to form cants.

Loose wood packing material. Excelsior (wood wool), sawdust, and wood shavings, produced as a result of sawing or shaving wood into small, slender, and curved pieces.

Lot. All the regulated articles on a single means of conveyance that are derived from the same species of tree and were subjected to the same treatments prior to importation, and that are consigned to the same person.

Lumber. Logs that have been sawn into boards, planks, or structural members such as beams.

Permit. A specific permit to import a regulated article issued in accordance with § 319.40-4, or a general permit promulgated in § 319.40-3.

Plant pest. Any living stage of any insects, mites, nematodes, slugs, snails, protozoa, or other invertebrate animals, bacteria, fungi, other parasitic plants or reproductive parts of parasitic plants, noxious weeds, viruses, or any organism similar to or allied with any of the foregoing, or any infectious substances, which can injure or cause disease or damage in any plants, parts of plants, or any products of plants.

Port of first arrival. The area (such as a seaport, airport, or land border station) where a person or a means of conveyance first arrives in the United States, and where inspection of regulated articles is carried out by inspectors.

Primary processing. Any of the following processes: cleaning (removal of soil, limbs, and foliage), debarking, rough sawing (bucking or squaring), rough shaping, spraying with fungicide or insecticide sprays, and fumigation.

Regulated article. The following articles, if they are unprocessed, have received only primary processing, or contain parts that are either unprocessed or have received only primary processing and are not feasibly separable from the other parts of the article: Logs; lumber; any whole tree; any cut tree or any portion of a tree, not solely consisting of leaves, flowers, fruits, buds, or seeds; bark; cork; laths; hog fuel; sawdust; painted raw wood products; excelsior (wood wool); wood chips; wood mulch; wood shavings; pickets; stakes; shingles; solid wood packing materials; humus; compost; litter; and wooden handicrafts.

Regulated wood packaging material. Wood packaging material other than manufactured wood materials, loose wood packing materials, and wood pieces less than 6 mm thick in any dimension, that are used or for use with cargo to prevent damage, including, but not limited to, dunnage, crating, pallets, packing blocks, drums, cases, and skids.

Sealed container; sealable container. A completely enclosed container designed for the storage or transportation of cargo, and constructed of metal or fiberglass, or other rigid material, providing an enclosure which prevents the entrance or exit of plant pests and is accessed through doors that can be closed and secured with a lock or seal. Sealed (sealable) containers are distinct and separable from the means of conveyance carrying them.

Specific permit. A written document issued by APHIS to the applicant in accordance with § 319.40-4 that authorizes importation of articles in accordance with this subpart and specifies or refers to the regulations applicable to the particular importation.

Statement of origin and movement. A signed, accurate statement certifying the area or areas where the regulated articles originated and, if applicable, the area or areas they were moved through prior to importation. The statement may be printed directly on the documentation accompanying the shipment of regulated articles, or it may be provided on a separate document. The statement does not require the signature of a public officer of a national plant protection organization; exporters may sign the document.

Tropical hardwoods. Hardwood timber species which grow only in tropical climates.

United States. All of the States of the United States, the District of Columbia, Guam, the Northern Mariana Islands, Puerto Rico, the Virgin Islands of the United States, and all other territories and possessions of the United States.

Wood chips. Wood fragments broken or shredded from any wood.

Wood mulch. Bark chips, wood chips, wood shavings, or sawdust intended for use as a protective or decorative ground cover.

Wood packaging material. Wood or wood products (excluding paper products) used in supporting, protecting or carrying a commodity (includes dunnage).

Wooden handicraft. A commodity class of articles derived or made from natural components of wood, twigs, and vines, and including bamboo poles and garden stakes. Handicrafts include the following products where wood is present: Carvings, baskets, boxes, bird houses, garden and lawn/patio furniture (rustic), potpourri, artificial trees (typically artificial ficus trees), trellis towers, garden fencing and edging, and other items composed of wood.

[60 FR 27674, May 25, 1995, as amended at 63 FR 50110, Sept. 18, 1998; 63 FR 69542, Dec. 17, 1998; 65 FR 21127, Apr. 20, 2000; 69 FR 55732, Sept. 16, 2004; 69 FR 61587, Oct. 20, 2004; 70 FR 33324, June 7, 2005; 72 FR 30467, June 1, 2007; 77 FR 12443, Mar. 1, 2012; 78 FR 25571, May 2, 2013]

§319.40-2 General prohibitions and restrictions; relation to other regulations.

(a) *Permit required.* Except for regulated articles exempted from this requirement by paragraph (c) of this section or §319.40-3, no regulated article may be imported unless a specific permit has been issued for importation of the regulated article in accordance with §319.40-4, and unless the regulated article meets all other applicable requirements of this subpart and any requirements specified by APHIS in the specific permit.

(b) *Importer document; documentation of type, quantity, and origin of regulated articles.* Except for regulated articles exempted from this requirement by paragraph (c) of this section or §319.40-3, no regulated article may be imported unless it is accompanied by an importer document stating the following information. A certificate that contains this information may be used in lieu of an importer document at the option of the importer:

(1) The genus and species of the tree from which the regulated article was derived;

(2) The country, and locality if known, where the tree from which the regulated article was derived was harvested;

(3) The quantity of the regulated article to be imported;

(4) The use for which the regulated article is imported; and

(5) Any treatments or handling of the regulated article required by this subpart that were performed prior to arrival at the port of first arrival.

(c) *Regulation of articles imported for propagation or human consumption.* The requirements of this subpart do not apply to regulated articles that are allowed importation in accordance with Subpart H—Plants for Planting of this part or to regulated articles imported for human consumption that are allowed importation in accordance with Subpart L—Fruits and Vegetables of this part.

(d) *Regulated articles imported for experimental, therapeutic, or developmental purposes.* Any regulated article may be imported without further restriction under this subpart if:

(1) Imported for experimental, therapeutic, or developmental purposes under the conditions specified in a controlled import permit issued in accordance with §319.6;

(2) Imported pursuant to a controlled import permit issued by APHIS for the regulated article prior to its importation and kept on file at the port of first arrival; and

(3) Imported under conditions specified on the controlled import permit and found by the Administrator to be adequate to prevent the introduction into the United States of plant pests.

(e) *Designation of additional regulated articles.* An inspector may designate any article as a regulated article by giving written notice of the designation to the owner or person in possession or control of the article. APHIS will implement rulemaking to add articles designated as regulated articles to the definition of regulated article in §319.40-1 if importation of the article appears to present a recurring significant risk of introducing plant pests. Inspectors may designate an article as a regulated article after determining that:

(1) The article was imported in the same container or hold as a regulated article;

(2) Other articles of the same type imported from the same country have been found to carry plant pests; or

(3) The article appears to be contaminated with regulated articles or soil.

(f) In addition to meeting the requirements of this subpart, bark and bark products and logs and pulpwood with bark attached, as well as cut trees (*e.g.*, Christmas trees), imported from Canada are subject to the inspection and certification requirements for gypsy moth in § 319.77–4 of this part.

(Approved by the Office of Management and Budget under control number 0579–0049)

[60 FR 27674, May 25, 1995, as amended at 63 FR 13485, Mar. 20, 1998; 64 FR 45866, Aug. 23, 1999; 69 FR 52418, Aug. 26, 2004; 69 FR 61587, Oct. 20, 2004; 71 FR 40878, July 19, 2006; 72 FR 39501, July 18, 2007; 78 FR 25571, May 2, 2013; 83 FR 11865, Mar. 19, 2018; 84 FR 2429, Feb. 7, 2019]

§ 319.40–3 **General permits; articles that may be imported without a specific permit; articles that may be imported without either a specific permit or an importer document.**

(a) *Canada and Mexico.* (1) The following articles may be imported into the United States under general permit:

(i) From Canada: Regulated articles, other than the following:

(A) Regulated articles of the subfamilies Aurantioideae, Rutoideae, and Toddalioideae of the botanical family Rutaceae; and

(B) Regulated articles of *Fraxinus* spp. (ash), which are subject to the requirements in § 319.40–5(n).

(ii) From States in Mexico adjacent to the United States: Commercial and noncommercial shipments of mesquite wood for cooking; commercial and noncommercial shipments of unmanufactured wood for firewood; and small, noncommercial packages of unmanufactured wood for personal cooking or personal medicinal purposes.

(2) Commercial shipments allowed in paragraph (a)(1) of this section are subject to the inspection and other requirements in § 319.40–9 and must be accompanied by an importer document stating that they are derived from trees harvested in Canada or States in Mexico adjacent to the United States border.

(3) Noncommercial shipments allowed in paragraph (a)(1) of this section are subject to inspection and other requirements of § 319.40–9 and must be accompanied by an importer document or oral declaration stating that they are derived from trees harvested in Canada or States in Mexico adjacent to the United States border.

(b) *Regulated wood packaging material.* Regulated wood packaging material, whether in actual use as packing for regulated or nonregulated articles or imported as cargo, may be imported into the United States under a general permit in accordance with the following conditions:

(1) The wood packaging material must have been treated in accordance with part 305 of this chapter.

(2) *Marking.* The wood packaging material must be marked in a visible location on each article, preferably on at least two opposite sides of the article, with a legible and permanent mark that indicates that the article meets the requirements of this paragraph. The mark must be approved by the International Plant Protection Convention in its International Standards for Phytosanitary Measures to certify that wood packaging material has been subjected to an approved measure, and must include a unique graphic symbol, the ISO two-letter country code for the country that produced the wood packaging material, a unique number assigned by the national plant protection agency of that country to the producer of the wood packaging material, and an abbreviation disclosing the type of treatment (*e.g.*, HT for heat treatment or MB for methyl bromide fumigation). The currently approved format for the mark is as follows, where XX would be replaced by the country code, 000 by the producer number, and YY by the treatment type (HT or MB):

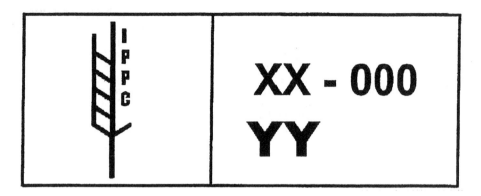

(3) *Immediate reexport of regulated wood packaging material without required mark.* An inspector at the port of first arrival may order the immediate reexport of regulated wood packaging material that is imported without the mark required by paragraph (b)(2) of this section, in addition to or in lieu of any port of first arrival procedures required by §319.40–9 of this part.

(4) *Exception for Department of Defense.* Regulated wood packaging material used by the Department of Defense (DOD) of the U.S. Government to package nonregulated articles, including commercial shipments pursuant to a DOD contract, may be imported into the United States without the mark required by paragraph (b)(2) of this section.

(Approved by the Office of Management and Budget under control numbers 0579–0049 and 0579–0225)

(c) *Loose wood packing materials.* APHIS hereby issues a general permit to import regulated articles authorized by this paragraph. Loose wood packing materials (whether in use as packing or imported as cargo) that are dry may be imported subject to the inspection and other requirements in §319.40–9 and without further restriction under this subpart.

(d) *Bamboo timber.* APHIS hereby issues a general permit to import regulated articles authorized by this paragraph. Bamboo timber which is free of leaves and seeds and has been sawn or split lengthwise and dried may be imported subject to the inspection and other requirements in §319.40–9 and

without further restriction under this subpart.

(e) *Regulated articles the permit process has determined to present no plant pest risk.* Regulated articles for which a specific permit has been issued in accordance with §319.40–4(b)(2)(i) may be imported without other restriction under this subpart, except that they are subject to the inspection and other requirements in §319.40–9.

(Approved by the Office of Management and Budget under control numbers 0579–0049 and 0579–0257)

[60 FR 27674, May 25, 1995, as amended at 63 FR 50110, Sept. 18, 1998; 63 FR 69542, Dec. 17, 1998; 69 FR 52418, Aug. 26, 2004; 69 FR 55732, Sept. 16, 2004; 69 FR 61587, Oct. 20, 2004; 71 FR 57386, Sept. 29, 2006; 72 FR 30462, 30467, June 1, 2007; 75 FR 4251, Jan. 26, 2010; 85 FR 61809, Oct. 1, 2020]

§319.40–4 Application for a permit to import regulated articles; issuance and withdrawal of permits.

(a) *Application procedure.* An application for a permit must be obtained and submitted in accordance with §§319.7 through 319.7–5.

(b) *Review of application and issuance of permit.* After receipt and review of the application, APHIS shall determine whether it appears that the regulated article at the time of importation will meet either the specific importation requirements in §319.40–5 or the universal importation requirements in §319.40–6.

(1) If it appears that the regulated article proposed for importation will meet the requirements of either §319.40–5 or §319.40–6, a permit stating

the applicable conditions for importation under this subpart shall be issued for the importation of the regulated article identified in the application.

(2) If it appears that the regulated article proposed for importation will not meet the requirements of either § 319.40–5 or § 319.40–6 because these sections do not address the particular regulated article identified in the application, APHIS shall review the application by applying the plant pest risk assessment standards specified in § 319.40–11.

(i) If this review reveals that importation of the regulated article under a permit and subject to the inspection and other requirements in § 319.40–9, but without any further conditions, will not result in the introduction of plant pests into the United States, a permit for importation of the regulated article shall be issued. The permit may only be issued in unique and unforeseen circumstances when the importation of the regulated article is not expected to recur.

(ii) If this review reveals that the regulated article may be imported under conditions that would reduce the plant pest risk to an insignificant level, APHIS may implement rulemaking to add the additional conditions to this subpart, and after the regulations are effective, may issue a permit for importation of the regulated article.

(Approved by the Office of Management and Budget under control number 0579–0049)

[60 FR 27674, May 25, 1995, as amended at 66 FR 21056, Apr. 27, 2001; 69 FR 52418, Aug. 26, 2004; 79 FR 19810, Apr. 10, 2014; 81 FR 40150, June 21, 2016]

§ 319.40–5 Importation and entry requirements for specified articles.

(a) *Bamboo timber.* Bamboo timber consisting of whole culms or canes may be imported into Guam or the Northern Mariana Islands subject to inspection and other requirements of § 319.40–9. Bamboo timber consisting of whole culms or canes that are completely dry as evidenced by lack of moisture in node tissue may be imported into any part of the United States subject to inspection and other requirements of § 319.40–9.

(b) *Monterey pine logs and lumber from Chile and New Zealand; Douglas-fir logs and lumber from New Zealand*—(1) *Logs*—(i) *Requirements prior to importation.* Monterey or Radiata pine (*Pinus radiata*) logs from Chile or New Zealand and Douglas-fir (*Pseudotsuga menziesii*) logs from New Zealand that are accompanied by a certificate stating that the logs meet the requirements of paragraph (b)(1)(i) (A) through (D) of this section, and that are consigned to a facility in the United States that operates in accordance with § 319.40–8, may be imported in accordance with paragraphs (b)(1)(i)(A) through (b)(1)(iii) of this section.

(A) The logs must be from live healthy trees which are apparently free of plant pests, plant pest damage, and decay organisms.

(B) The logs must be debarked in accordance with § 319.40–7(b) prior to fumigation.

(C) The logs and any regulated wood packaging material to be used with the logs during shipment to the United States must be fumigated in accordance with part 305 of this chapter within 45 days following the date the trees are felled and prior to arrival of the logs in the United States, in the holds or in sealable containers. Fumigation must be conducted in the same sealable container or hold in which the logs and regulated wood packaging material are exported to the United States.

(D) During shipment to the United States, no other regulated article is permitted on the means of conveyance with the logs, unless the logs and the other regulated articles are in separate holds or separate sealed containers, or, if the logs and other regulated articles are mixed in a hold or sealed container, the other regulated articles either have been heat treated with moisture reduction in accordance with part 305 of this chapter, or have been fumigated in the hold or sealable container in accordance with paragraph (b)(1)(i)(C) of this section.

(ii) *Requirements upon arrival in the United States.* The following requirements apply upon arrival of the logs in the United States.

(A) The logs must be kept segregated from other regulated articles from the time of discharge from the means of conveyance until the logs are completely processed at a facility in the United States that operates under a compliance agreement in accordance with § 319.40–8.

(B) The logs must be moved from the port of first arrival to the facility that operates under a compliance agreement in accordance with § 319.40–8 by as direct a route as reasonably possible.

(iii) *Requirements at the processing facility.* The logs must be consigned to a facility operating under a compliance agreement in accordance with § 319.40–8 that includes the following requirements:

(A) Logs or any products generated from logs, including lumber, must be heat treated in accordance with part 305 of this chapter, or heat treated with moisture reduction in accordance with part 305 of this chapter.

(B) The logs, including sawdust, wood chips, or other products generated from the logs in the United States, must be processed in accordance with paragraph (b)(1)(iii) of this section within 60 days from the time the logs are released from the port of first arrival.

(C) Sawdust, wood chips, and waste generated by sawing or processing the logs must be disposed of by burning, heat treatment in accordance with part 305 of this chapter , heat treatment with moisture reduction in accordance with part 305 of this chapter , or other processing that will destroy any plant pests associated with the sawdust, wood chips, and waste. Composting and use of the sawdust, wood chips, and waste as mulch are prohibited unless composting and use as mulch are preceded by fumigation in accordance with part 305 of this chapter , heat treatment in accordance with part 305 of this chapter , or heat treatment with moisture reduction in accordance with part 305 of this chapter . Wood chips, sawdust, and waste may be moved in enclosed trucks for processing at another facility operating under a compliance agreement in accordance with § 319.40–8.

(2) *Raw lumber.* Raw lumber, including regulated wood packaging material imported as cargo, from Chile or New Zealand derived from Monterey or Radiata pine (*Pinus radiata*) logs and raw lumber from New Zealand derived

from Douglas-fir (*Pseudotsuga menziesii*) logs may be imported in accordance with paragraphs (b)(2) (i) and (ii) of this section.

(i) During shipment to the United States, no other regulated article (other than regulated wood packaging material) is permitted on the means of conveyance with the raw lumber, unless the raw lumber and the other regulated articles are in separate holds or separate sealed containers; *Except for* mixed shipments of logs and raw lumber fumigated in accordance with part 305 of this chapter and moved in accordance with paragraph (b)(1)(i)(D) of this section. Raw lumber on the vessel's deck must be in a sealed container.

(ii) The raw lumber must be consigned to a facility operating under a compliance agreement in accordance with § 319.40–8 that requires the raw lumber to be heat treated in accordance with part 305 of this chapter or heat treated with moisture reduction in accordance with part 305 of this chapter before any cutting, planing, or sawing of the raw lumber, and within 30 days from the time the lumber is released from the port of first arrival.

(c) *Tropical hardwoods—*(1) *Debarked.* Tropical hardwood logs and lumber that have been debarked in accordance with § 319.40–7(b) may be imported subject to the inspection and other requirements of § 319.40–9.

(2) *Not debarked.* Tropical hardwood logs that have not been debarked may be imported if fumigated in accordance with part 305 of this chapter prior to arrival in the United States.

(3) *Not debarked; small lots.* Tropical hardwood logs that have not been debarked may be imported into the United States, other than into Hawaii, Puerto Rico, or the Virgin Islands of the United States, if imported in a lot of 15 or fewer logs and subject to the inspection and other requirements of § 319.40–9.

(d) *Temperate hardwoods.* Temperate hardwood logs and lumber (with or without bark) from all places except places in Asia that are east of 60° East Longitude and north of the Tropic of Cancer may be imported if fumigated in accordance with part 305 of this chapter prior to arrival in the United

States and subject to the inspection and other requirements of §319.40-9.

(e) *Regulated articles associated with exclusively tropical climate pests.* Regulated articles that have been identified by a plant pest risk assessment as associated solely with plant pests that can successfully become established only in tropical or subtropical climates may be imported if:

(1) The regulated article is imported only to a destination in the continental United States; and,

(2) the regulated article is not imported into any tropical or subtropical areas of the United States specified in the permit.

(f) Cross-ties (railroad ties) from all places, except places in Asia that are east of 60° East Longitude and north of the Tropic of Cancer, may be imported if completely free of bark and accompanied by an importer document stating that the cross-ties will be pressure treated with a preservative within 30 days following the date of importation at a U.S. facility under compliance agreement. Cross-ties (railroad ties) may also be imported if heat treated in accordance with part 305 of this chapter.

(g) through (k) [Reserved]

(l) *Cross-ties (railroad ties) and pine and fir lumber from Mexican States adjacent to the United States/Mexico border.*[1] Cross-ties (railroad ties) 8 inches or less at maximum thickness and lumber derived from pine and fir may be imported from Mexican States adjacent to the United States/Mexico border into the United States if they:

(1) Originate from Mexican States adjacent to the United States/Mexico border;

(2) Are 100 percent free of bark; and

(3) Are fumigated in accordance with part 305 of this chapter prior to arrival in the United States.

(m) [Reserved]

(n) *Regulated articles of the genus Fraxinus from Canada.* Except for articles prohibited under paragraph (n)(4) of this section, regulated articles of the genus *Fraxinus* (ash) from Canada may be imported in accordance with this

paragraph (n) and subject to the certification requirements in §319.40-2(a) and the inspection and other requirements in §319.40-9. Articles being moved from counties or municipal regional counties in Canada not regulated for the emerald ash borer (EAB) may not transit an EAB-regulated area in Canada en route to the United States unless they are moving directly through the EAB-regulated area without stopping (except for refueling or for traffic conditions, such as traffic lights or stop signs). If these articles are being moved through the regulated area between May 1 and August 31 or when the ambient air temperature is 40 °F or higher, they must be in an enclosed vehicle or completely covered to prevent access by the emerald ash borer.

(1) Firewood of all hardwood (non-coniferous) species, and ash logs and wood, including cants and stumps, that originate in a county or municipal regional county regulated for the emerald ash borer within a Province or Territory regulated by the Canadian Government for the emerald ash borer require a permit issued under §319.40-2(a) and must be accompanied by a certificate bearing an additional declaration that the articles in the shipment were:

(i) Debarked, and vascular cambium removed to a depth of 1.27 cm (½ inch) during the debarking process; or

(ii) Heat treated in accordance with part 305 of this chapter. The phytosanitary certificate accompanying such articles must describe the treatment method employed.

(2) Firewood of all hardwood (non-coniferous) species, and ash logs and wood, including cants and stumps, that originate in a county or municipal regional county not regulated for the emerald ash borer within a Province or Territory regulated for the emerald ash borer require a permit issued under §319.40-2(a) and must be accompanied by a certificate with an additional declaration stating that the articles in the shipment were produced/harvested in a county or municipal regional county where the emerald ash borer does not occur, based on official surveys.

(3) Firewood of all hardwood (non-coniferous) species, and ash logs and wood, including cants and stumps, that

---

[1] Cross-ties (railroad ties) may also be imported in accordance with paragraph (f) of this section, or may be imported if heat treated in accordance with §319.40-7(c).

originate in a Province or Territory that is not regulated for the emerald ash borer must be accompanied by an importer document that certifies that the article originated in a county or municipal regional county free of the emerald ash borer.

(4) The importation of ash wood chips or bark chips larger than 1 inch diameter in any two dimensions that originate in a county or municipal regional county regulated for the emerald ash borer within a Province or Territory regulated for the emerald ash borer is prohibited.

(5) Ash wood chips or bark 1 inch or less in diameter that originate in an area regulated for the emerald ash borer within a Province or Territory regulated for the emerald ash borer must be accompanied by a permit issued under §319.40-2(a) and a phytosanitary certificate with an additional declaration stating that the wood or bark chips in the shipment were ground to 1 inch (2.54 cm) or less in diameter in any two dimensions.

(6) Ash wood chips or bark chips that originate in a county or municipal regional county not regulated for the emerald ash borer within a Province or Territory regulated for the emerald ash borer must be accompanied by a permit issued under §319.40-2(a), and a valid certificate with an additional declaration stating that the articles in the shipment were produced/harvested in a county or municipal regional county where the emerald ash borer does not occur, based on official surveys.

(7) Ash wood chips or bark chips that originate in a Province or Territory that is not regulated for the emerald ash borer must be accompanied by an importer document that certifies that the article originates in a Province or Territory free of the emerald ash borer.

(o) *Wooden handicrafts from China.* Wooden handicrafts more than 1 centimeter in diameter may be imported into the United States from China only in accordance with this paragraph and all other applicable provisions of this title. Wooden handicrafts less than 1 centimeter in diameter are exempt from the requirements of this paragraph, but are still subject to all other applicable provisions of this chapter.

(1) *Treatment.* Wooden handicrafts must be treated in accordance with part 305 of this chapter.

(2) *Identification tag.* All packages in which wooden handicrafts are shipped must be labeled with a merchandise tag containing the identity of the product manufacturer. The identification tag must be applied to each shipping package in China prior to exportation and remain attached to the shipping package until it reaches the location at which the wooden handicraft will be sold in the United States.

(Approved by the Office of Management and Budget under control numbers 0579-0049, 0579-0257, 0579-0319, and 0579-0367)

[60 FR 27674, May 25, 1995, as amended at 63 FR 69542, Dec. 17, 1998; 64 FR 59604, Nov. 3, 1999; 69 FR 52418, Aug. 26, 2004; 69 FR 55733, Sept. 16, 2004; 69 FR 61587, Oct. 20, 2004; 70 FR 33325, June 7, 2005; 72 FR 30467, June 1, 2007; 75 FR 4251, Jan. 26, 2010; 77 FR 12443, Mar. 1, 2012; 79 FR 19810, Apr. 10, 2014; 85 FR 61809, Oct. 1, 2020]

§319.40-6 Universal importation options.

(a) *Logs.* Logs may be imported if prior to importation the logs have been debarked in accordance with §319.40-7(b) and heat treated in accordance with part 305 of this chapter. During the entire interval between treatment and export, the logs must be stored and handled in a manner which excludes any access to the logs by plant pests.

(b) *Lumber*—(1) *Heat treated or heat treated with moisture reduction.* Lumber that prior to importation has been heat treated in accordance with part 305 of this chapter , or heat treated with moisture reduction in accordance with part 305 of this chapter , may be imported in accordance with paragraphs (b)(1) (i) and (ii) of this section.

(i) During shipment to the United States, no other regulated article (other than solid wood packing materials) is permitted on the means of conveyance with the lumber, unless the lumber and the other regulated articles are in separate holds or separate sealed containers, or, if the lumber and other regulated articles are mixed in a hold or sealed container, all the regulated articles have been heat treated in accordance with part 305 of this chapter , or heat treated with moisture reduction in accordance with part 305 of this chapter . Lumber on the vessel's deck must be in a sealed container, unless it has been heat treated with moisture reduction in accordance with part 305 of this chapter

(ii) If lumber has been heat treated in accordance with part 305 of this chapter , that fact must be stated on the importer document, or by a permanent marking on each piece of lumber in the form of the letters "HT" or the words "Heat Treated." If lumber has been heat treated with moisture reduction in accordance with part 305 of this chapter, that fact must be stated on the importer document, or by a permanent marking, on each piece of lumber or on the cover of bundles of lumber, in the form of the letters "KD" or the words "Kiln Dried."

(2) *Raw lumber*. Raw lumber, including solid wood packing materials imported as cargo, from all places except places in Asia that are east of 60° East Longitude and north of the Tropic of Cancer may be imported in accordance with paragraphs (b)(2) (i) and (ii) of this section.

(i) During shipment to the United States, no other regulated article (other than solid wood packing materials) is permitted on the means of conveyance with the raw lumber, unless the raw lumber and the other regulated articles are in separate holds or separate sealed containers. Raw lumber on the vessel's deck must be in a sealed container.

(ii) The raw lumber must be consigned to a facility operating under a compliance agreement in accordance with § 319.40–8 that requires the raw lumber to be heat treated in accordance with part 305 of this chapter or heat treated with moisture reduction in accordance with part 305 of this chapter , within 30 days from the time the lumber is released from the port of first arrival. Heat treatment must be completed before any cutting, planing, or sawing of the raw lumber.

(c) *Wood chips and bark chips*—(1) *From Chile (pine) and South America (eucalyptus)*. Wood chips from Chile that are derived from Monterey or Radiata pine (*Pinus radiata*) logs and wood chips from South America that are derived

from temperate species of *Eucalyptus* may be imported in accordance with paragraph (c)(2) of this section or in accordance with the following requirements:

(i) The wood chips must be accompanied by a certificate stating that the wood chips meet the requirements in paragraphs (c)(1)(i)(A) through (c)(1)(i)(C) of this section.

(A) The wood chips were treated with a surface pesticide treatment in accordance with part 305 of this chapter within 24 hours after the log was chipped and were retreated with a surface pesticide treatment in accordance with part 305 of this chapter if more than 30 days elapsed between the date of the first treatment and the date of export to the United States.

(B) The wood chips were derived from logs from live, healthy, plantation-grown trees that were apparently free of plant pests, plant pest damage, and decay organisms, and the logs used to make the wood chips were debarked in accordance with § 319.40–7(b) before being chipped.

(C) No more than 45 days elapsed from the time the trees used to make the wood chips were felled to the time the wood chips were exported.

(ii) During shipment to the United States, no other regulated articles (other than solid wood packing materials) are permitted in the holds or sealed containers carrying the wood chips. Wood chips on the vessel's deck must be in a sealed container.

(iii) The wood chips must be consigned to a facility in the United States that operates under a compliance agreement in accordance with § 319.40–8. The following requirements apply upon arrival of the wood chips in the United States:

(A) Upon arrival in the United States, the wood chips must be unloaded by a conveyor that is covered to prevent the chips from being blown by the wind and from accidental spillage. The facility receiving the wood chips must have a procedure in place to retrieve any chips that fall during unloading.

(B) If the wood chips must be transported after arrival, the chips must be covered or safeguarded in a manner that prevents the chips from spilling or falling off the means of conveyance or from being blown off the means of conveyance by wind.

(C) The wood chips must be stored at the facility on a paved surface and must be kept segregated from other regulated articles from the time of discharge from the means of conveyance until the chips are processed. The storage area must not be adjacent to wooded areas.

(D) The wood chips must be processed within 45 days of arrival at the facility. Any fines or unusable wood chips must be disposed of by burning within 45 days of arrival at the facility.

(2) *From locations other than certain places in Asia.* Wood chips and bark chips from any place except places in Asia that are east of 60° east longitude and north of the Tropic of Cancer may be imported in accordance with this paragraph.

(i) The wood chips or bark chips must be accompanied by an importer document stating that the wood chips or bark chips were either:

(A) Derived from live, healthy, tropical species of plantation-grown trees grown in tropical areas; or

(B) Fumigated with methyl bromide in accordance with part 305 of this chapter, heat treated in accordance with part 305 of this chapter , or heat treated with moisture reduction in accordance with part 305 of this chapter.

(ii) During shipment to the United States, no other regulated articles (other than solid wood packing materials) are permitted in the holds or sealed containers carrying the wood chips or bark chips. Wood chips or bark chips on the vessel's deck must be in a sealed container; *Except that:* If the wood chips or bark chips are derived from live, healthy, plantation-grown trees in tropical areas, they may be shipped on deck if no other regulated articles are present on the vessel and the wood chips or bark chips are completely covered by a tarpaulin during the entire journey directly to the United States.

(iii) The wood chips or bark chips must be free from rot at the time of importation, unless accompanied by an importer document stating that the entire lot was fumigated with methyl bromide in accordance with part 305 of

this chapter, heat treated in accordance with part 305 of this chapter, or heat treated with moisture reduction in accordance with part 305 of this chapter.

(iv) Wood chips or bark chips imported in accordance with this paragraph must be consigned to a facility operating under a compliance agreement in accordance with §319.40-8. The wood chips or bark chips must be burned, heat treated in accordance with part 305 of this chapter, heat treated with moisture reduction in accordance with part 305 of this chapter, or otherwise processed in a manner that will destroy any plant pests associated with the wood chips or bark chips within 30 days of arrival at the facility. If the wood chips or bark chips are to be used for mulching or composting, they must first be fumigated in accordance with part 305 of this chapter , heat treated in accordance with part 305 of this chapter, or heat treated with moisture reduction in accordance with part 305 of this chapter.

(d) *Wood mulch, humus, compost, and litter.* Wood mulch, humus, compost, and litter may be imported if accompanied by an importer document stating that the wood mulch, humus, compost, or litter was fumigated in accordance with part 305 of this chapter, heat treated in accordance with part 305 of this chapter, or heat treated with moisture reduction in accordance with §319.40-7(d).

(e) *Cork and bark.* Cork and cork bark, cinnamon bark, and other bark to be used for food, manufacture of medicine, or chemical extraction may be imported if free from rot at the time of importation and subject to the inspection and other requirements of §319.40-9.

(Approved by the Office of Management and Budget under control number 0579-0049)

[60 FR 27679, May 25, 1995; 60 FR 30157, June 7, 1995, as amended at 65 FR 21127, Apr. 20, 2000; 69 FR 2295, Jan. 15, 2004; 69 FR 52418, Aug. 26, 2004; 75 FR 4252, Jan. 26, 2010]

§319.40-7 Treatments and safeguards.

(a) *Certification of treatments or safeguards.* If APHIS determines that a document required for the importation of regulated articles is inaccurate, the regulated articles which are the subject of the certificate or other document shall be refused entry into the United States. In addition, APHIS may determine not to accept any further certificates for the importation of regulated articles in accordance with this subpart from a country in which an inaccurate certificate is issued, and APHIS may determine not to allow the importation of any or all regulated articles from any such country, until corrective action acceptable to APHIS establishes that certificates issued in that country will be accurate.

(b) *Debarking.* Except for raw lumber, no more than 2 percent of the surface of all regulated articles in a lot may retain bark, with no single regulated article retaining bark on more than 5 percent of its surface. For raw lumber, debarking must remove 100 percent of the bark.

(c) *Treatments.* Treatment of regulated articles under this subpart must be conducted in accordance with part 305 of this chapter.

(d) *Preservatives.* All preservative treatments that use a preservative product that is registered by the United States Environmental Protection Agency are authorized for treatment of regulated articles imported in accordance with this subpart. Preservative treatments must be performed in accordance with label directions approved by the United States Environmental Protection Agency.

(Approved by the Office of Management and Budget under control number 0579–0049)

[60 FR 27674, May 25, 1999, as amended at 64 FR 59604, Nov. 3, 1999; 65 FR 21128, Apr. 20, 2000; 67 FR 8465, Feb. 25, 2002; 69 FR 2295, Jan. 15, 2004; 69 FR 52418, Aug. 26, 2004; 70 FR 33325, June 7, 2005; 75 FR 4252, Jan. 26, 2010]

§ 319.40–8 Processing at facilities operating under compliance agreements.

(a) Any person who operates a facility in which imported regulated articles are processed may enter into a compliance agreement to facilitate the importation of regulated articles under this subpart. The compliance agreement shall specify the requirements necessary to prevent spread of plant pests from the facility, requirements to ensure the processing method effec-

tively destroys plant pests, and the requirements for the application of chemical materials in accordance with part 305 of this chapter. The compliance agreement shall also state that inspectors must be allowed access to the facility to monitor compliance with the requirements of the compliance agreement and of this subpart. Compliance agreement forms may be obtained from the Administrator or an inspector.

(b) Any compliance agreement may be canceled by the inspector who is supervising its enforcement, orally or in writing, whenever the inspector finds that the person who entered into the compliance agreement has failed to comply with the conditions of the compliance agreement. If the cancellation is oral, the decision to cancel the compliance agreement and the reasons for cancellation of the compliance agreement shall be confirmed in writing, as promptly as circumstances permit. Any person whose compliance agreement has been canceled may appeal the decision in writing to the Administrator within 10 days after receiving written notification of the cancellation. The appeal shall state all of the facts and reasons upon which the person relies to show that the compliance agreement was wrongfully canceled. The Administrator shall grant or deny the appeal, in writing, stating the reasons for granting or denying the appeal, as promptly as circumstances permit. If there is a conflict as to any material fact and the person whose compliance agreement has been canceled requests a hearing, a hearing shall be held to resolve the conflict. Rules of practice concerning the hearing will be adopted by the Administrator.

(Approved by the Office of Management and Budget under control number 0579–0049)

[60 FR 27674, May 25, 1995, as amended at 69 FR 52418, Aug. 26, 2004; 70 FR 33325, June 7, 2005]

§ 319.40–9 Inspection and other requirements at port of first arrival.

(a) *Procedures for all regulated articles.* (1) All imported regulated articles shall be inspected at the port of first arrival. If the inspector finds signs of plant pests on or in the regulated article, or finds that the regulated article may have been associated with other articles infested with plant pests, the regulated article shall be cleaned or treated as required by an inspector, and the regulated article and any products of the regulated article shall also be subject to reinspection, cleaning, and treatment at the option of an inspector at any time and place before all applicable requirements of this subpart have been accomplished.

(2) Regulated articles shall be assembled for inspection at the port of first arrival, or at any other place prescribed by an inspector, at a place and time and in a manner designated by an inspector.

(3) If an inspector finds that an imported regulated article is so infested with a plant pest that, in the judgment of the inspector, the regulated article cannot be cleaned or treated, or contains soil or other prohibited contaminants, the entire lot may be refused entry into the United States.

(4) No person shall move any imported regulated article from the port of first arrival unless and until an inspector notifies the person, in writing or through an electronic database, that the regulated article:

(i) Is in compliance with all applicable regulations and has been inspected and found to be apparently free of plant pests;[2] or,

(ii) Has been inspected and the inspector requires reinspection, cleaning, or treatment of the regulated article at a place other than the port of first arrival.

(b) *Notice of arrival; visual examination of regulated articles at port of first arrival.* (1) At least 7 days prior to the expected date of arrival in the United States of a shipment of regulated articles imported in accordance with this subpart, the permittee or his or her agent must notify the APHIS Officer in Charge at the port of arrival of the date of expected arrival. The address and telephone number of the APHIS Officer in Charge will be specified in any specific permit issued by APHIS[3]. This notice may be by any authorized method. The notice must include the number of any specific permit issued for the regulated articles; the name, if any, of the means of conveyance carrying the regulated articles; the type and quantity of the regulated articles; the expected date of arrival; the country of origin of the regulated articles; the name and the number, if any, of the dock or area where the regulated articles are to be unloaded; and the name of the importer or broker at the port of arrival.

(2) Imported regulated articles which have been debarked in accordance with §319.40-7(b) and can be safely and practically inspected will be visually examined for plant pests by an inspector at the port of first arrival. If plant pests are found on or in the regulated articles or if the regulated article cannot be safely and practically inspected, the regulated articles must be treated in accordance with part 305 of this chapter.

(c) *Marking and identity of regulated articles.* Any regulated article, at the time of importation shall bear on the outer container (if in a container), on the regulated article (if not in a container), or on a document accompanying the regulated article the following information:

(1) General nature and quantity of the regulated articles;

(2) Country and locality, if known, where the tree from which the regulated article was derived was harvested;

(3) Name and address of the person importing the regulated article;

(4) Name and address of consignee of the regulated article;

(5) Identifying shipper's mark and number; and

(6) Number of the permit (if one was issued) authorizing the importation of the regulated article into the United States.

---

[2] Certain regulated articles may also be subject to "Subpart L—Fruits and Vegetables," or to the noxious weed regulations under part 360 of this chapter, or to Endangered Species Act regulations under parts 355 and 356 of this chapter and 50 CFR parts 17 and 23.

[3] A list of APHIS Officers in Charge may be obtained from the Administrator, c/o Port Operations, Plant Protection and Quarantine, Animal and Plant Health Inspection Service, 4700 River Road, Riverdale, MD 20737.

(d) *Sampling for plant pests at port of first arrival.* Any imported regulated article may be sampled for plant pests at the port of first arrival. If an inspector finds it necessary to order treatment of a regulated article at the port of first arrival, any sampling will be done prior to treatment.

(Approved by the Office of Management and Budget under control number 0579–0049)

[60 FR 27674, May 25, 1995, as amended at 66 FR 21056, Apr. 27, 2001; 69 FR 52418, Aug. 26, 2004; 70 FR 33325, June 7, 2005; 72 FR 39501, July 18, 2007; 79 FR 19810, Apr. 10, 2014; 81 FR 40150, June 21, 2016; 84 FR 2429, Feb. 7, 2019]

§ 319.40–10 Costs and charges.

The services of an inspector during regularly assigned hours of duty and at the usual places of duty shall be furnished without cost to the importer.[4] The inspector may require the importer to furnish any labor, chemicals, packing materials, or other supplies required in handling regulated articles under this subpart. APHIS will not be responsible for any costs or charges, other than those identified in this section.

[60 FR 27674, May 25, 1995, as amended at 63 FR 50111, Sept. 18, 1998; 69 FR 52418, Aug. 26, 2004; 69 FR 55733, Sept. 16, 2004; 79 FR 19810, Apr. 10, 2014]

§ 319.40–11 Plant pest risk assessment standards.

When evaluating a request to import a regulated article not allowed importation under this subpart, or a request to import a regulated article under conditions other than those prescribed by this subpart, APHIS will conduct the following analysis to determine the plant pest risks associated with each requested importation in order to determine whether or not to issue a permit under this subpart or to propose regulations establishing conditions for the importation into the United States of the regulated article.

---

[4] Provisions relating to costs for other services of an inspector, including services related to extra inspection and separation of cargo from packing material for shipments that arrive without meeting the requirements of this subpart as required, are contained in part 354 of this chapter.

(a) *Collecting commodity information.* (1) APHIS will evaluate the application for information describing the regulated article and the origin, processing, treatment, and handling of the regulated article; and

(2) APHIS will evaluate history of past plant pest interceptions or introductions (including data from foreign countries) associated with the regulated article.

(b) *Cataloging quarantine pests.* For the regulated article specified in an application, APHIS will determine what plant pests or potential plant pests are associated with the type of tree from which the regulated article was derived, in the country and locality from which the regulated article is to be exported. A plant pest that meets one of the following criteria is a quarantine pest and will be further evaluated in accordance with paragraph (c) of this section:

(1) Non-indigenous plant pest not present in the United States;

(2) Non-indigenous plant pest, present in the United States and capable of further dissemination in the United States;

(3) Non-indigenous plant pest that is present in the United States and has reached probable limits of its ecological range, but differs genetically from the plant pest in the United States in a way that demonstrates a potential for greater damage potential in the United States;

(4) Native species of the United States that has reached probable limits of its ecological range, but differs genetically from the plant pest in the United States in a way that demonstrates a potential for greater damage potential in the United States; or

(5) Non-indigenous or native plant pest that may be able to vector another plant pest that meets one of the criteria in paragraphs (b)(1) through (4) of this section.

(c) *Determining which quarantine pests to assess.* (1) APHIS will divide quarantine pests identified in paragraph (b) of this section into groups depending upon where the plant pest is most likely to be found. The plant pests would be grouped as follows:

(i) Plant pests found on the bark;

(ii) Plant pests found under the bark; and

(iii) Plant pests found in the wood.

(2) APHIS will subdivide each of the groups in paragraph (c)(1) of this section into associated taxa.

(3) APHIS will rank the plant pests in each group in paragraph (c)(2) of this section according to plant pest risk, based on the available biological information and demonstrated plant pest importance.

(4) APHIS will identify any plant pests ranked in paragraph (c)(3) of this section for which plant pest risk assessments have previously been performed in accordance with this section. APHIS will conduct individual plant pest risk assessments for the remaining plant pests, starting with the highest ranked plant pest(s) in each group.

(5) The number of plant pests in each group to be evaluated through individual plant pest risk assessment will be based on biological similarities of members of the group as they relate to measures taken in connection with the importation of the regulated article to mitigate the plant pest risk associated with the regulated article. For example, if the plant pest risk assessment for the highest ranked plant pest indicates a need for a mitigation measure that would result in the same reduction of risk for other plant pests ranked in the group, the other members need not be subjected to individual plant pest risk assessment.

(d) *Conducting individual plant pest risk assessments.* APHIS will evaluate each of the plant pests identified in paragraph (c)(4) of this section by:

(1) Estimation of the probability of the plant pest being on, with, or in the regulated article at the time of importation;

(2) Estimation of the probability of the plant pest surviving in transit on the regulated article and entering the United States undetected;

(3) Estimation of the probability of the plant pest colonizing once it has entered into the United States;

(4) Estimation of the probability of the plant pest spreading beyond any colonized area; and

(5) Estimation of the damage to plants that could be expected upon introduction and dissemination within the United States of the plant pest.

(e) *Estimating unmitigated overall plant pest risk.* APHIS will develop an estimation of the overall plant pest risk associated with importing the regulated article based on compilation of individual plant pest risk assessments performed in accordance with paragraph (d) of this section.

(f) *Evaluating available requirements to determine whether they would allow safe importation of the regulated article.* The requirements of this subpart, and any other requirements relevant to the regulated article and plant pests involved, will be compared with the individual plant pest risk assessments in order to determine whether particular conditions on the importation of the regulated article would reduce the plant pest risk to an insignificant level. If APHIS determines that the imposition of particular conditions on the importation of the regulated article could reduce the plant pest risk to an insignificant level, and determines that sufficient APHIS resources are available to implement or ensure implementation of the conditions, APHIS will implement rulemaking to allow importation of the requested regulated article under the conditions identified by the plant pest risk assessment process.

Subpart J—Indian Corn or Maize, Broomcorn, and Related Plants

SOURCE: Redesignated at 84 FR 2429, Feb. 7, 2019.

QUARANTINE

§319.41 Notice of quarantine.

(a) The fact has been determined by the Secretary of Agriculture, and notice given, that dangerous plant pests, including the so-called European corn borer (Ostrinia nubilalis Hubn.), and also other dangerous insects, as well as plant diseases not heretofore widely prevalent or distributed within and throughout the United States, exist, as to one or more of such pests, in Europe, Asia, Africa, Dominion of Canada, Mexico, Central and South America, and other foreign countries and localities, and may be introduced into this country through importations of the stalks or other parts of Indian corn or maize, broomcorn, and related plants.

(b) To prevent the introduction of these plant pests, the following articles may not be imported into the United States except in accordance with this subpart: The raw or unmanufactured stalk and all other parts of Indian corn or maize (*Zea mays* L.), broomcorn (*Andropogon sorghum* var. *technicus*), sweet sorghums (*Andropogon sorghum*), grain sorghums (*Andropogon sorghum*), Sudan grass (*Andropogon sorghum* sudanensis), Johnson grass (*Andropogon halepensis*), sugarcane (*Saccharum officinarum*), including Japanese varieties, pearl millet (*Pennisetum glaucum*), napier grass (*Pennisetum purpureum*), teosinte (*Euchlaena luxurians*), and jobs-tears (*Coix lachryma-Jobi*).

(c) The Administrator may authorize the importation of articles otherwise prohibited under paragraph (b) of this section under conditions specified in a controlled import permit issued in accordance with § 319.6.

(d) The importation of plants (including any plant parts) of any of the taxa listed in paragraph (b) of this section that are for planting or capable of being planted is restricted under Subpart H—Plants for Planting of this part.

(e) As used in this subpart, unless the context otherwise requires, the term "United States" means the States, the District of Columbia, Guam, Puerto Rico, and the Virgin Islands of the United States.

[24 FR 10788, Dec. 29, 1959, as amended at 66 FR 21056, Apr. 27, 2001; 78 FR 25571, May 2, 2013; 83 FR 11865, Mar. 19, 2018; 84 FR 2429, Feb. 7, 2019]

§ 319.41a Administrative instructions relating to entry into Guam of broomcorn, brooms, and similar articles.

(a) Broomcorn for manufacturing purposes, and brooms and similar articles made of broomcorn may be imported into Guam without further permit, other than the authorization contained in this section, and without other restriction under this subpart. Notice of arrival for such importations is not necessary inasmuch as there is available to the inspector the essential information normally supplied by the importer at time of importation. Inspection of such importations may be made under the general authority of § 330.105(a) of this chapter. If an importation is found infected, infested, or contaminated with any plant pest and is not subject to disposal under this part 319, disposition may be made in accordance with § 330.106 of this chapter.

(b) Shelled corn and seeds of other plants listed in § 319.41, and mature corn on the cob, may be imported into Guam without further permit, other than the authorization contained in this section and without other restriction under this subpart, but such importations are subject to the requirements of § 319.37–6(a).

(c) Green corn on the cob may be imported into Guam without restriction under this subpart, but such importations are subject to the requirements of § 319.56–3.

[24 FR 10788, Dec. 29, 1959, as amended at 72 FR 39501, July 18, 2007; 83 FR 11865, Mar. 19, 2018]

§ 319.41b Administrative instructions prescribing conditions for entry of broomstraw without treatment.

Broomstraw, sometimes referred to as "combed stalkless", when consisting of individual straws entirely free from stems, stalks, stubs of stalks, and leaves, may be imported from all countries without seasonal limitation through ports of entry designated in the permit, provided it is bundled and baled to prevent breakage and scattering and to facilitate inspection, in the following manner:

(a) The broomstraw shall be assembled into bundles with the base of the individual straws at the same end, no alternating of layers being permitted.

(b) Each bundle shall be securely tied to prevent breakage.

(c) Individual bundles shall be compacted, grouped into bales, and so arranged that the butt of each bundle is exposed on the outside of the bale.

(d) Each bale shall be securely bound to prevent shifting or loosening of the bundles in transit.

(e) Broomstraw found upon inspection at the port of entry to contain stems, stalks, stubs of stalks, or leaves shall be sterilized under the supervision of an inspector. Broomstraw contaminated in the aforesaid manner, from countries other than those on the North or South American Continents or the West Indies, shall be considered as broomcorn and shall be subject to compliance with §319.41-3(b).

[25 FR 12809, Dec. 14, 1960]

RULES AND REGULATIONS

§319.41-1 Plant products permitted entry. [1]

Except as restricted from certain countries and localities by special quarantines and other orders now in force, [2] and by such as may hereafter be promulgated, the following articles may be imported:

(a) Subject only to the requirements of paragraphs (a), (b), and (c) of §319.41-5:

(1) Green corn on the cob, in small lots for local use only, from adjacent areas of Canada.

(2) Articles made of the stalks, leaves, or cobs of corn, when prepared, manufactured, or processed in such manner that in the judgment of the inspector no pest risk is involved in their entry.

---

[1] Except as provided in §319.41-6 the regulations in this subpart do not authorize importations through the mails.

[2] The entry of the following plants and plant products is prohibited or restricted by specific quarantines and other restrictive orders now in force.

(a) Living canes of sugarcane, or cuttings or parts thereof, from all foreign countries. (§319.15.)

(b) Except as provided for in paragraph (c) for corn seed from New Zealand, seed and all other portions in the raw or unmanufactured state of Indian corn or maize (Zea mays L.), and the closely related plants, including all species of Teosinte (Euchlaena), jobs-tears (Coix), Polytoca, Chionachne, Sclerachne, and Trilobachne, from Australia, Burma, Cambodia, China, Formosa, India, Indonesia, Japan and adjacent islands, Laos, Malaya, Manchuria, New Guinea, New Zealand, North Viet-Nam, Oceania, Pakistan, Philippines, Ryukyu Islands, Thailand, and Viet-Nam. (§319.24.)

(c) Seed of Indian corn or maize (Zea mays L.) that is free from the cob and from all other parts of corn may be imported into the United States from New Zealand without further restriction. (§319.24.).

(3) Corn silk.

(b) Upon compliance with the regulations in this subpart:

(1) Broomcorn for manufacturing purposes, brooms or similar articles made of broomcorn, clean shelled corn, and clean seed of the other plants covered by §319.41.

(2) Corn on the cob, green or mature, from the provinces of Canada west of and including Manitoba, [3] and from Mexico, Central America, South America, the West Indies, the Bahamas, and Bermuda.

(c) Seed of Indian corn or maize (Zea mays L.) that is free from the cob and from all other parts of corn may be imported into the United States from New Zealand without further restriction.

(d) Immature, dehusked "baby" sweet corn may be imported from Zambia in accordance with §319.56-2f(a).

[24 FR 10788, Dec. 29, 1959, as amended at 58 FR 44745, Aug. 25, 1993; 71 FR 29769, May 24, 2006]

§319.41-2 Application for permits.

Persons contemplating the importation of any of the articles specified in §319.41-1(b) shall first make application to the Plant Protection and Quarantine Program for a permit in accordance with §§319.7 through 319.7-5.

(Approved by the Office of Management and Budget under control number 0579-0049)

[79 FR 19810, Apr. 10, 2014]

§319.41-3 Issuance of permits.

(a) On approval by the Administrator of the application mentioned in §319.41-2, a permit will be issued.

(b) For broomcorn and brooms and similar articles made of broomcorn, permits will be issued by the Administrator for such ports as may be designated therein, except that permits will be issued for the entry of broomcorn originating in countries other than those in the North or South American Continents or the West Indies only through the ports of Baltimore, Boston, New York, and Norfolk,

---

[3] A quarantine is maintained by Canada to prevent spread of the European corn borer from the infested eastern areas to the still uninfested Provinces west of Ontario.

or through other northeastern ports which may from time to time be designated in the permit, and at which facilities for treatment of infested material may be available, such entry to be limited to those shipments accompanied by on-board bills of lading dated within the period September 15 through February 15 of the succeeding year, both dates inclusive. Permits will not be issued for the entry of broomcorn from any source through ports on the Pacific Coast.

(c) For shelled corn and for seeds of other plants listed in § 319.41, and for corn on the cob, green or mature, from the land areas designated in § 319.41(b)(2), permits will be issued for ports where the Plant Protection and Quarantine Programs maintains an inspection service and for such other ports as may be designated in the permit.

(d) Pending development of adequate treating facilities in Guam, any of the articles specified in § 319.41–1 that are subject to treatment as a condition of entry therein must first be entered and treated in accordance with the requirements of this subpart at a U.S. port of arrival where such treating facilities are available.

[24 FR 10788, Dec. 29, 1959, as amended at 33 FR 11811, Aug. 21, 1968; 36 FR 24917, Dec. 24, 1971; 78 FR 25571, May 2, 2013]

§ 319.41–4 Notice of arrival by permittee.

Immediately upon arrival of the importation at the port of arrival the permittee shall submit, in duplicate, notice to the Plant Protection and Quarantine Programs, through the U.S. Collector of Customs, or, in the case of Guam, through the Customs officer of the Government of Guam, on forms provided for that purpose, stating the number of the permit, the date of entry, the name of ship or vessel, railroad, or other carrier, the country and locality where the articles were grown, the name of the foreign shipper, the quantity or number of bales or containers, and the marks and numbers on the bales or containers, the port of ar-

rival, and the name of the importer or broker at the port of arrival.

(Approved by the Office of Management and Budget under control number 0579–0049)

[24 FR 10788, Dec. 29, 1959, as amended at 48 FR 57466, Dec. 30, 1983]

§ 319.41–5 Condition of entry.

(a) The entry of the articles covered by § 319.41–1 is conditioned on their freedom from the European corn borer and other injurious insects and plant diseases, and upon their freedom from contamination with plant materials prohibited entry under other quarantines. All shipments of these articles shall be subject to inspection at the port of arrival by an inspector of the Plant Protection and Quarantine Programs, in order to determine their freedom from such insects and diseases and from contaminating materials, and to such sterilization, grinding, or treatment in accordance with part 305 of this chapter, as the inspector may prescribe. Should an importation be found on inspection to be so infested or infected or contaminated that, in the judgment of the inspector, it can not be made safe by sterilization or other treatment in accordance with part 305 of this chapter, the entire shipment may be refused entry.

(b) When entry under sterilization or other treatment in accordance with part 305 of this chapter is permitted, the importation will be released to the permittee for such treatment, upon the filing with the appropriate customs official of a bond in the amount of $5,000, or in an amount equal to the invoice value, if such value be less than $5,000, with approved sureties, and conditioned that the importation shall be sterilized or otherwise treated under the supervision of the inspector; that no bale or container shall be broken, opened, or removed from the port of arrival unless and until a written notice is given to said customs official by an inspector that the importation has been properly sterilized or treated; and that the importation shall be redelivered to said customs official within 30 days after its arrival.

(c) Should a shipment requiring sterilization or other treatment in accordance with part 305 of this chapter under the provisions of the regulation in this subpart arrive at a port where facilities for such sterilization or other treatment in accordance with part 305 of this chapter are not maintained, such shipment shall either be promptly shipped under safeguards and by routing prescribed by the inspector to an approved port where facilities for sterilization or other treatment in accordance with part 305 of this chapter are available, or it shall be refused entry.

(d) Other conditions of entry as applying to the certain classes of articles enumerated in § 319.41–1 are:

(1) *Broomcorn.* All importations of broomcorn shall be so baled as to prevent breakage and scattering in connection with the necessary handling and sterilization; if in the judgment of the inspector they are not so baled, entry may be refused. All importations of broomcorn shall be subject to such sterilization or other treatment in accordance with part 305 of this chapter as the inspector may require.

(2) *Articles made of broomcorn.* Brooms or similar articles made of broomcorn shall be subject to sterilization unless their manufacture involves the substantial elimination of stems or such treatment of the included stems as in the judgment of the inspector shall preclude such articles from being the means of carriage of the European corn borer and of other injurious insects and plant diseases.

(3) *Shelled corn and other seeds.* If shipments of shelled corn and seeds of the other plants from countries other than those named in § 319.41–1 (b)(2) are found upon inspection at the port of arrival to be appreciably fouled with cobs or other portions of the plants the inspector may require sterilization or other treatment in accordance with part 305 of this chapter or may refuse entry.

[24 FR 10788, Dec. 29, 1959, as amended at 75 FR 4252, Jan. 26, 2010]

§ 319.41–6 Importations by mail.

In addition to entries by freight or express provided for in § 319.41–5, importations are permitted by mail of mature corn on the cob from the countries specified in § 319.41–1(b)(2), and clean shelled corn and clean seed of the other plants covered by § 319.41, provided that a permit has been issued for the impor-

tation in accordance with §§ 319.7 through 319.7–5 and all conditions of the permit are met.

(Approved by the Office of Management and Budget under control number 0579–0049)

[79 FR 19810, Apr. 10, 2014]

Subpart K—Rice

SOURCE: Redesignated at 84 FR 2429, Feb. 7, 2019.

QUARANTINE

§ 319.55 Notice of quarantine.

(a) The fact has been determined by the Secretary of Agriculture, and notice is hereby given:

(1) That injurious fungal diseases of rice, including downy mildew (*Sclerospora macrospora*), leaf smut (*Entyloma oryzae*), blight (*Oospora oryzetorum*), and glume blotch (*Melanomma glumarum*), as well as dangerous insect pests, new to and not heretofore widely prevalent or distributed within and throughout the United States, exist, as to one or more of such diseases and pests, in Europe, Asia, Africa, Central America, South America, and other foreign countries and localities, and may be introduced into this country through importations of rice straw and rice hulls; and

(2) That the unrestricted importation of rice straw and rice hulls may result in the entry into the United States of the injurious plant diseases heretofore enumerated, as well as insect pests.

(b) To prevent the introduction into the United States of the plant pests and diseases indicated above, the Secretary has determined that it is necessary to restrict the importation of rice straw and rice hulls from all foreign locations, except as otherwise provided in this subpart.

(c) The Administrator may authorize the importation of articles otherwise prohibited by this subpart under conditions specified in a controlled import permit issued in accordance with § 319.6.

(d) The importation of seed or paddy rice is restricted under Subpart H—Plants for Planting of this part.

(e) As used in this subpart, unless the context otherwise requires, the term

"United States" means the States, the District of Columbia, Guam, Puerto Rico, and the Virgin Islands of the United States.

[24 FR 10788, Dec. 29, 1959, as amended at 66 FR 21056, Apr. 27, 2001; 78 FR 25571, May 2, 2013; 83 FR 11865, Mar. 19, 2018; 84 FR 2429, Feb. 7, 2019]

§ 319.55a Administrative instructions relating to entry of rice straw and rice hulls into Guam.

Rice straw and rice hulls may be imported into Guam without further permit, other than the authorization contained in this paragraph. The port of entry shall be Agana or such other port as may be satisfactory to the inspector. Such importations may be made without the submission of a notice of arrival inasmuch as there is available to the inspector the essential information normally supplied by an importer at the time of importation. The requirements of §§ 319.55–6 and 319.55–7 shall not apply. Inspections of such importations may be made under the general authority of § 330.105(a) of this chapter. If an importation is found infected, infested, or contaminated by any plant pest and is not subject to disposal under this part, disposition may be made in accordance with § 330.106 of this chapter.

RULES AND REGULATIONS

§ 319.55–1 Definitions.

(a) *Seed or paddy rice.* Unhusked rice in the form commonly used for seed purposes; the regulations in this subpart do not apply to husked or polished rice imported for food purposes.

(b) *Port of first arrival.* The first port within the United States where the shipment is (1) offered for consumption entry or (2) offered for entry for immediate transportation in bond.

(c) *Inspector.* An Inspector of the Plant Protection and Quarantine Programs of the United States Department of Agriculture.

§ 319.55–2 Application for permit.

Application for a permit to import from any country rice straw or rice hulls may be made to the Plant Protec-

tion and Quarantine Programs in accordance with §§ 319.7 through 319.7–5.

(Approved by the Office of Management and Budget under control number 0579–0049)

[79 FR 19811, Apr. 10, 2014, as amended at 83 FR 11865, Mar. 19, 2018]

§ 319.55–3 Ports of entry.

(a) For importations of rice straw and rice hulls, permits will be issued for entry at New York and Boston and at such other ports as may later be approved by the Plant Protection and Quarantine Programs.

(b) Pending development of adequate treating facilities in Guam, rice straw and rice hulls that are subject to treatment as a condition of entry therein must first be entered and treated in accordance with the requirements of this subpart at a United States port of arrival where such treating facilities are available.

(c) Should a shipment requiring treatment arrive at a port where facilities for such treatment are not maintained, such shipment shall either be promptly shipped under safeguards and by routing prescribed by the inspector to an approved port where facilities for treatment are available, or it shall be refused entry.

[79 FR 19811, Apr. 10, 2014, as amended at 83 FR 11865, Mar. 19, 2018]

§ 319.55–4 [Reserved]

§ 319.55–5 Notice of arrival by permittee.

Immediately upon the arrival of a shipment at the port of first arrival, the permittee or his agent shall submit a notice, in duplicate, to the Plant Protection and Quarantine Programs, through the United States Collector of Customs, or, in the case of Guam, through the Customs officer of the Government of Guam, on a form provided for that purpose, stating the number of the permit, the quantity in the shipment, the locality where grown, the date of arrival, and, if by rail, the name of the railroad company, the car numbers, and the terminal where the shipment is to be unloaded, or, if by vessel, the name of the vessel and the designation of the dock where the shipment is to be landed.

(Approved by the Office of Management and Budget under control number 0579–0049)

[24 FR 10788, Dec. 29, 1959, as amended at 48 FR 57466, Dec. 30, 1983]

§319.55–6 Inspection and disinfection at port of arrival.

(a) [Reserved]

(b) *Rice straw and rice hulls.* (1) As a condition of entry, rice straw and rice hulls shall be subject to inspection and to treatment in accordance with part 305 of this chapter at the port of arrival, under the supervision of the inspector, by methods and at plants approved by the Plant Protection and Quarantine Programs and, as a further condition of entry, in order to permit effective treatment in accordance with part 305 of this chapter, the contents of packages or bales shall not be compressed to a density of more than 30 pounds per cubic foot. Rice straw and rice hulls will be admitted only at ports where adequate facilities are available for such treatment. The required treatment must be given within 20 days after arrival, but if any shipment of rice straw or rice hulls shall be found upon arrival to be dangerously infested or infected the inspector may direct immediate treatment under adequate safeguards; and, if the treatment and safeguards are not put into effect as directed, the shipment shall be removed from the country immediately or destroyed.

(2) Unless, within 20 days after the date of arrival of a shipment at the port at which the formal entry was filed, the importation has received the required treatment, due notice of which shall be given to the collector of customs by the inspector, demand will be made by the collector for redelivery of the shipment into customs custody under the terms of the entry bond, and, if such redelivery is not made, the shipment shall be removed from the country or destroyed.

(3) All charges for storage, cartage, and labor incident to inspection and disinfection, other than the services of the inspector, shall be paid by the importer.

(4) All shipments shall be so baled, bagged, or wrapped as to prevent scattering or wastage. If, in the judgment of the inspector, a shipment is not so bagged, baled, or wrapped, it shall be reconditioned at the expense of the permittee or entry may be refused.

[24 FR 10788, Dec. 29, 1959, as amended at 75 FR 4252, Jan. 26, 2010; 83 FR 11865, Mar. 19, 2018]

§319.55–7 Importations by mail.

Importations of rice straw and rice hulls may be made by mail or cargo, provided that a permit has been issued for the importation in accordance with §§319.7 through 319.7–5 and all conditions of the permit are met.

(Approved by the Office of Management and Budget under control number 0579–0049)

[83 FR 11865, Mar. 19, 2018]

Subpart L—Fruits and Vegetables

SOURCE: 72 FR 39501, July 18, 2007, unless otherwise noted. Redesignated at 84 FR 2429, Feb. 7, 2019.

§319.56–1 Notice of quarantine.

(a) Under section 412(a) of the Plant Protection Act, the Secretary of Agriculture may prohibit or restrict the importation and entry of any plant or plant product if the Secretary determines that the prohibition or restriction is necessary to prevent the introduction into the United States or the dissemination within the United States of a plant pest or noxious weed.

(b) The Secretary has determined that it is necessary to prohibit the importation into the United States of fruits and vegetables and associated plants and portions of plants except as provided in this part.

§319.56–2 Definitions.

Administrator. The Administrator of the Animal and Plant Health Inspection Service, United States Department of Agriculture, or any other employee of the United States Department of Agriculture delegated to act in his or her stead.

APHIS. The Animal and Plant Health Inspection Service, United States Department of Agriculture.

Commercial consignment. A lot of fruits or vegetables that an inspector identifies as having been imported for sale and distribution. Such identification will be based on a variety of indicators, including, but not limited to: Quantity of produce, type of packaging, identification of grower or packinghouse on the packaging, and documents consigning the fruits or vegetables to a wholesaler or retailer.

Commodity. A type of plant, plant product, or other regulated article being moved for trade or other purpose.

Consignment. A quantity of plants, plant products, and/or other articles, including fruits or vegetables, being moved from one country to another and covered, when required, by a single phytosanitary certificate (a consignment may be composed of one or more commodities or lots).

Continental United States. The 48 contiguous States, Alaska, and the District of Columbia.

Country of origin. Country where the plants from which the plant products are derived were grown.

Frozen fruit or vegetable. Any variety of raw fruit or vegetable preserved by commercially acceptable freezing methods in such a way that the commodity remains at −6.7 °C (20 °F) or below for at least 48 hours prior to release.

Fruits and vegetables. A commodity class for fresh parts of plants intended for consumption or processing and not for planting.

Import and importation. To move into, or the act of movement into, the territorial limits of the United States.

Inspector. Any individual authorized by the Administrator of APHIS or the Commissioner of the Bureau of Customs and Border Protection, Department of Homeland Security, to enforce the regulations in this subpart.

Lot. A number of units of a single commodity, identifiable by its homogeneity of composition and origin, forming all or part of a consignment.

National plant protection organization (NPPO). Official service established by a government to discharge the functions specified by the International Plant Protection Convention.

Noncommercial consignment. A lot of fruits or vegetables that an inspector identifies as having been imported for personal use and not for sale.

Permit. A written, oral, or electronically transmitted authorization to import fruits or vegetables in accordance with this subpart.

Phytosanitary certificate. A document, including electronic versions, that is related to a consignment and that:

(1) Is patterned after the model certificate of the International Plant Protection Convention (IPPC), a multilateral convention on plant protection under the authority of the Food and Agriculture Organization of the United Nations (FAO);

(2) Is issued by an official of a foreign national plant protection organization in one of the five official languages of the FAO;

(3) Is addressed to the plant protection service of the United States (Animal and Plant Health Inspection Service);

(4) Describes the consignment;

(5) Certifies the place of origin for all contents of the consignment;

(6) Certifies that the consignment has been inspected and/or tested according to appropriate official procedures and is considered to be free from quarantine pests of the United States;

(7) Contains any additional declarations required by this subpart; and

(8) Certifies that the consignment conforms with the phytosanitary requirements of the United States and is considered eligible for importation pursuant to the laws and regulations of the United States.

Phytosanitary measure. Any legislation, regulation, or official procedure having the purpose to prevent the introduction and/or spread of quarantine pests, or to limit the economic impact of regulated non-quarantine pests.

Plant litter and debris. Discarded or decaying organic matter; detached leaves, twigs, or stems that do not add commercial value to the product.

Port of first arrival. The first port within the United States where a consignment is offered for consumption entry or offered for entry for immediate transportation in bond.

Portions of plants. Stalks or stems, including the pediculus, pedicel, peduncle, raceme, or panicle, that are normally attached to fruits or vegetables.

Quarantine pest. A pest of potential economic importance to the area endangered by it and not yet present there, or present but not widely distributed there and being officially controlled.

United States. All of the States of the United States, the Commonwealth of Northern Mariana Islands, the Commonwealth of Puerto Rico, the District of Columbia, Guam, the Virgin Islands of the United States, and any other territory or possession of the United States.

[72 FR 39501, July 18, 2007, as amended at 73 FR 10972, Feb. 29, 2008; 80 FR 55018, Sept. 14, 2015; 83 FR 46638, Sept. 14, 2018]

§319.56–3 General requirements for all imported fruits and vegetables.

All fruits and vegetables that are allowed importation under this subpart must be imported in accordance with the following requirements, except as specifically provided otherwise in this subpart.

(a) *Freedom from unauthorized plant parts.* All fruits and vegetables imported under this subpart, whether in commercial or noncommercial consignments, must be free from plant litter or debris and free of any portions of plants that are specifically prohibited in the regulations in this subpart.

(b) *Permit.* (1) All fruits and vegetables imported under this subpart, whether commercial or noncommercial consignments, must be imported under permit issued by APHIS, must be imported under the conditions specified in the permit, and must be imported in accordance with all applicable regulations in this part; *except* for:

(i) Dried, cured, or processed fruits and vegetables (except frozen fruits and vegetables), including cured figs and dates, raisins, nuts, and dried beans and peas, except certain acorns and chestnuts subject to §319.56–11 of this subpart;

(ii) Fruits and vegetables grown in Canada (except potatoes from Newfoundland and that portion of the Municipality of Central Saanich in the Province of British Columbia east of the West Saanich Road, which are prohibited importation into the United States); and

(iii) Fruits and vegetables, except mangoes, grown in the British Virgin Islands that are imported into the U.S. Virgin Islands.

(2) Persons contemplating the importation of any fruits or vegetables under this subpart must apply for a permit in accordance with §§319.7 through 319.7–5.

(c) *Ports of entry.* (1) Fruits and vegetables must be imported into specific ports if so required by this subpart or by part 305 of this chapter, or if so required by a permit issued in accordance with this section and with §§319.7 through 319.7–5 for the importation of the particular fruit or vegetable. If a permit issued for the importation of fruits or vegetables names specific port(s) where the fruits or vegetables must be imported, the fruits and vegetables may only be imported into the port(s) named in the permit. If a permit issued for the importation of fruits or vegetables does not name specific port(s) where the fruits or vegetables must be imported, the fruits and vegetables may be imported into any port referenced in paragraph (c)(2) of this section.

(2) Fruits and vegetables imported under this subpart may be imported into any port listed in 19 CFR 101.3(b)(1), except as otherwise provided by part 319 or by a permit issued in accordance with part 319, and except as provided in §330.104 of this chapter. Fruits and vegetables that are to be cold treated at ports in the United States may only be imported into specific ports as provided in part 305 of this chapter.

(d) *Inspection, treatment, and other requirements.* All imported fruits or vegetables are subject to inspection, are subject to such disinfection at the port of first arrival as may be required by an inspector, and are subject to reinspection at other locations at the option of an inspector. If an inspector finds plants or portions of plants, or a plant pest or noxious weed, or evidence of a plant pest or noxious weed on or in any fruit or vegetable or its container, or finds that the fruit or vegetable may have been associated with other articles infested with plant pests or noxious weeds, the owner or agent of the owner of the fruit or vegetable must clean or treat the fruit or vegetable and its container as required by an inspector, and the fruit or vegetable is also subject to reinspection, cleaning, and treatment at the option of an inspector at any time and place until all applicable requirements of this subpart have been accomplished.

(1) *Notice of arrival; assembly for inspection.* Any person importing fruits and vegetables into the United States must offer those agricultural products for inspection and entry at the port of first arrival. The owner or agent must assemble the fruits and vegetables for inspection at the port of first arrival, or at any other place designated by an inspector, and in a manner designated by the inspector. All fruits and vegetables must be accurately disclosed and made available to an inspector for examination. The owner or the agent must provide an inspector with the name and address of the consignee and must make full disclosure of the type, quantity, and country and locality of origin of all fruits and vegetables in the consignment, either orally for noncommercial consignments or on an invoice or similar document for commercial consignments.

(2) *Refusal of entry.* If an inspector finds that an imported fruit or vegetable is prohibited, or is not accompanied by required documentation, or is so infested with a plant pest or noxious weed that, in the judgment of the inspector, it cannot be cleaned or treated, or contains soil or other prohibited contaminants, the entire lot or consignment may be refused entry into the United States.

(3) *Release for movement.* No person may move a fruit or vegetable from the port of first arrival unless an inspector has either:

(i) Released it;

(ii) Ordered treatment at the port of first arrival and, after treatment, released the fruit or vegetable;

(iii) Authorized movement of the fruit or vegetable to another location for treatment, further inspection, or destruction; or

(iv) Ordered the fruit or vegetable to be reexported.

(4) *Notice to owner of actions ordered by inspector.* If an inspector orders any disinfection, cleaning, treatment, reexportation, recall, destruction, or

other action with regard to imported fruits or vegetables while the consignment is in foreign commerce, the inspector will issue an emergency action notification (PPQ Form 523) to the owner of the fruits or vegetables or to the owner's agent. The owner must, within the time and in the manner specified in the PPQ Form 523, destroy the fruits and vegetables, ship them to a point outside the United States, move them to an authorized site, and/or apply treatments or other safeguards to the fruits and vegetables as prescribed to prevent the introduction of plant pests or noxious weeds into the United States.

(e) *Costs and charges.* APHIS will be responsible only for the costs of providing the services of an inspector during regularly assigned hours of duty and at the usual places of duty.[1] The owner of imported fruits or vegetables is responsible for all additional costs of inspection, treatment, movement, storage, destruction, or other measures ordered by an inspector under this subpart, including any labor, chemicals, packing materials, or other supplies required. APHIS will not be responsible for any costs or charges, other than those identified in this section.

(f) *APHIS not responsible for damage.* APHIS assumes no responsibility for any damage to fruits or vegetables that results from the application of treatments or other measures required under this subpart (or under part 305 of this chapter) to protect against the introduction of plant pests into the United States.

(Approved by the Office of Management and Budget under control number 0579-0049)

[72 FR 39501, July 18, 2007, as amended at 73 FR 10972, Feb. 29, 2008; 75 FR 4252, Jan. 26, 2010; 79 FR 19811, Apr. 10, 2014]

§ 319.56-4 Authorization of certain fruits and vegetables for importation.

(a) *Determination by the Administrator.* No fruit or vegetable is authorized importation into the United States unless the Administrator has determined that the risk posed by each quarantine pest

---

[1] Provisions relating to costs for other services of an inspector are contained in part 354 of this chapter.

associated with the fruit or vegetable can be reasonably mitigated by the application of one or more phytosanitary measures designated by the Administrator and the fruit or vegetable is imported into the United States in accordance with, and as stipulated in, the permit issued by the Administrator.

(b) *Designated phytosanitary measures.* (1) The fruits and vegetables are subject to phytosanitary treatments, which could include, but are not limited to, pest control treatments in the field or growing site, and post-harvest treatments.

(2) The fruits and vegetables are subject to growing area pest mitigations, which could include, but are not limited to detection surveys, trapping requirements, pest exclusionary structures, and field inspections.

(3) The fruits and vegetables are subject to safeguarding and movement mitigations, which could include, but are not limited to, safeguarded transport, box labeling, limited distribution, insect-proof boxes, and importation as commercial consignments only.

(4) The fruits and vegetables are subject to administrative mitigations, which could include, but are not limited to, registered fields or orchards, registered growing sites, registered packinghouses, inspection in the country of origin by an inspector or an official of the national plant protection organization of the exporting country, and operational workplan monitoring.

(5) The fruits and vegetables are subject to any other measures deemed appropriate by the Administrator.

(c) *Authorized fruits and vegetables*—(1) *Comprehensive list.* The name and origin of all fruits and vegetables authorized importation under this section, as well as the applicable requirements for their importation, may be found on the internet at *https://epermits.aphis.usda.gov/manual.*

(2) *Fruits and vegetables authorized importation prior to October 15, 2018.* Fruits and vegetables that were authorized importation under this subpart either directly by permit or by specific regulation as of October 15, 2018 may continue to be imported into the United States under the same requirements that applied before October 15, 2018, ex-

cept as provided in paragraph (c)(4) of this section.

(3) *Other fruits and vegetables.* Fruits and vegetables not already authorized for importation as described in paragraph (c)(2) of this section may be authorized importation only after:

(i) APHIS has analyzed the pest risk posed by the importation of a fruit or vegetable from a specified foreign region and has determined that the risk posed by each quarantine pest associated with the fruit or vegetable can be reasonably mitigated by the application of one or more phytosanitary measures;

(ii) APHIS has made its pest risk analysis and determination available for public comment for at least 60 days through a notice published in the FEDERAL REGISTER; and

(iii) The Administrator has announced his or her decision in a subsequent FEDERAL REGISTER notice to authorize the importation of the fruit or vegetable subject to the phytosanitary measures specified in the notice.

(4) *Changes to phytosanitary measures.* (i) If the Administrator determines that the phytosanitary measures required for a fruit or vegetable that has been authorized importation under this subpart are no longer sufficient to reasonably mitigate the pest risk posed by the fruit or vegetable, APHIS will prohibit or further restrict importation of the fruit or vegetable. APHIS will also publish a notice in the FEDERAL REGISTER advising the public of its finding. The notice will specify the amended importation requirements, provide an effective date for the change, and will invite public comment on the subject.

(ii) If the Administrator determines that any of the phytosanitary measures required for a fruit or vegetable that has been authorized importation under this subpart are no longer necessary to reasonably mitigate the pest risk posed by the fruit or vegetable, APHIS will make new pest risk documentation available for public comment, in accordance with paragraph (c)(3) of this section, prior to allowing importation of the fruit or vegetable subject to the phytosanitary measures specified in the notice.

(Approved by the Office of Management and Budget under control number 0579–0049)

[83 FR 46638, Sept. 14, 2018]

§ 319.56–5 Pest-free areas.

As provided elsewhere in this subpart, certain fruits and vegetables may be imported into the United States provided that the fruits or vegetables originate from an area that is free of a specific pest or pests. In some cases, fruits or vegetables may only be imported if the area of export is free of all quarantine pests that attack the fruit or vegetable. In other cases, fruits and vegetables may be imported if the area of export is free of one or more quarantine pests that attack the fruit or vegetable, and provided that the risk posed by the remaining quarantine pests that attack the fruit or vegetable is mitigated by other specific phytosanitary measures contained in the regulations in this subpart.

(a) *Application of international standard for pest free areas.* APHIS requires that determinations of pest-free areas be made in accordance with the criteria for establishing freedom from pests found in International Standard for Phytosanitary Measures No. 4, "Requirements for the establishment of pest free areas." The international standard was established by the International Plant Protection Convention of the United Nations' Food and Agriculture Organization and is incorporated by reference in § 300.5 of this chapter.

(b) *Survey protocols.* APHIS must approve the survey protocol used to determine and maintain pest-free status, as well as protocols for actions to be performed upon detection of a pest. Pest-free areas are subject to audit by APHIS to verify their status.

(c) *Determination of pest freedom.* (1) For an area to be considered free of a specified pest for the purposes of this subpart, the Administrator must determine, and announce in a notice or rule published in the FEDERAL REGISTER for 60 days public comment, that the area meets the criteria of paragraphs (a) and (b) of this section.

(2) The Administrator will announce his or her decision in a subsequent FEDERAL REGISTER notice. If appropriate, APHIS would begin issuing permits for importation of the fruit or vegetable from a pest-free area because:

(i) No comments were received on the notice or

(ii) The comments on the notice did not affect the overall conclusions of the notice and the Administrator's determination of risk.

(d) *Decertification of pest-free areas; reinstatement.* If a pest is detected in an area that is designated as free of that pest, APHIS would publish in the FEDERAL REGISTER a notice announcing that the pest-free status of the area in question has been withdrawn, and that imports of host crops for the pest in question are subject to application of an approved treatment for the pest. If a treatment for the pest is not available, importation of the host crops would be prohibited. In order for a decertified pest-free area to be reinstated, it would have to meet the criteria of paragraphs (a) and (b) of this section.

(e) *General requirements for fruits and vegetables imported from pest-free areas.* (1) *Labeling.* Each box of fruits or vegetables that is imported into the United States from a pest-free area under this subpart must be clearly labeled with:

(i) The name of the orchard or grove of origin, or the name of the grower; and

(ii) The name of the municipality and State in which the fruits or vegetables were produced; and

(iii) The type and amount of fruit the box contains.

(2) *Phytosanitary certificate.* A phytosanitary certificate must accompany the imported fruits or vegetables, and must contain an additional declaration that the fruits originate from a pest-free area that meets the requirements of paragraphs (a) and (b) of this section.

(3) *Safeguarding.* If fruits or vegetables are moved from a pest-free area into or through an area that is not free of that pest, the fruits or vegetables must be safeguarded during the time they are present in a non-pest-free area by being covered with insect-proof

mesh screens or plastic tarpaulins, including while in transit to the packinghouse and while awaiting packaging. If fruits or vegetables are moved through an area that is not free of that pest during transit to a port, they must be packed in insect-proof cartons or containers or be covered by insect-proof mesh or plastic tarpaulins during transit to the port and subsequent export to the United States. These safeguards described in this section must be intact upon arrival in the United States.

(Approved by the Office of Management and Budget under control numbers 0579–0049, 0579–0316 and 0579–0293)

§319.56–6 Trust fund agreements.

If APHIS personnel need to be physically present in an exporting country or region to facilitate the exportation of fruits or vegetables and APHIS services are to be funded by the national plant protection organization (NPPO) of the exporting country or a private export group, then the NPPO or the private export group must enter into a trust fund agreement with APHIS that is in effect at the time the fruits or vegetables are exported. Under the agreement, the NPPO of the exporting country or the private export group must pay in advance all estimated costs that APHIS expects to incur in providing inspection services in the exporting country. These costs will include administrative expenses incurred in conducting the services and all salaries (including overtime and the Federal share of employee benefits), travel expenses (including per diem expenses), and other incidental expenses incurred by the inspectors in performing services. The agreement must require the NPPO of the exporting country or region or a private export group to deposit a certified or cashier's check with APHIS for the amount of those costs, as estimated by APHIS. The agreement must further specify that, if the deposit is not sufficient to meet all costs incurred by APHIS, the NPPO of the exporting country or a private export group must deposit with APHIS, before the services will be completed, a certified or cashier's check for the amount of the remaining costs, as determined by APHIS. After a final audit at the conclusion of each shipping season, any

overpayment of funds would be returned to the NPPO of the exporting country or region or a private export group, or held on account.

§319.56–7 Territorial applicability and exceptions.

(a) The regulations in this subpart apply to importations of fruits and vegetables into any area of the United States, except as provided in this section.

(b) *Importations of fruits and vegetables into Guam.* (1) The following fruits and vegetables may be imported into Guam without treatment, except as may be required under §319.56–3(d), and in accordance with all the requirements of this subpart as modified by this section:

(i) All leafy vegetables and root crops from the Bonin Islands, Volcano Islands, and Ryukyu Islands.

(ii) All fruits and vegetables from Palau and the Federated States of Micronesia (FSM), except *Artocarpus* spp. (breadfruit, jackfruit, and chempedak), citrus, curacao apple, guava, Malay or mountain apple (*Syzygium* spp.), mango, and papaya, and except dasheen from the Yap district of FSM and from Palau, and bitter melon (*Momordica charantia*) from Palau. The excepted products are approved for entry into Guam after treatment in accordance with part 305 of this chapter.

(iii) *Allium* (without tops), artichokes, bananas, bell peppers, cabbage, carrots, celery, Chinese cabbage, citrus fruits, eggplant, grapes, lettuce, melons, okra, parsley, peas, persimmons, potatoes, rhubarb, squash (*Cucurbita maxima*), stone and pome fruits, string beans, sweetpotatoes, tomatoes, turnip greens, turnips, and watermelons from Japan and Korea.

(iv) Leafy vegetables, celery, and potatoes from the Philippine Islands.

(v) Carrots (without tops), celery, lettuce, peas, potatoes, and radishes (without tops) from Australia.

(vi) Arrowroot, asparagus, bean sprouts, broccoli, cabbage, carrots (without tops), cassava, cauliflower, celery, chives, cow-cabbage, dasheen, garlic, gingerroot, horseradish, kale, kudzu, leek, lettuce, onions, Portuguese cabbage, turnip, udo, water

259

chestnut, watercress, waterlily root, and yam bean root from Taiwan.

(vii) Lettuce from Papua New Guinea.

(viii) Carrots (without tops), celery, lettuce, loquats, onions, persimmons, potatoes, tomatoes, and stone fruits from New Zealand.

(ix) Asparagus, carrots (without tops), celery, lettuce, and radishes (without tops) from Thailand.

(x) Green corn on the cob.

(xi) All other fruits and vegetables approved for entry into any other part or port of the United States, and except any which are specifically designated in this subpart as not approved.

(2) An inspector in Guam may accept an oral application and issue an oral permit for products listed in paragraph (a) of this section, which is deemed to fulfill the requirements of § 319.56–3(b) of this subpart. The inspector may waive the documentation required in § 319.56–3 for such products whenever the inspector finds that information available from other sources meets the requirements under this subpart for the information normally supplied by such documentation.

(3) The provisions of § 319.56–11 do not apply to chestnuts and acorns imported into Guam, which are enterable into Guam without permit or other restriction under this subpart. If chestnuts or acorns imported under this paragraph are found infected, infested, or contaminated with any plant pest and are not subject to disposal under this subpart, disposition may be made in accordance with § 330.106 of this chapter.

(4) Baskets or other containers made of coconut fronds are not approved for use as containers for fruits and vegetables imported into Guam. Fruits and vegetables in such baskets or containers offered for importation into Guam will not be regarded as meeting § 319.56–3(a).

(c) *Importation of fruits and vegetables into the U.S. Virgin Islands.* (1) Fruits and vegetables grown in the British Virgin Islands may be imported into the U.S. Virgin Islands in accordance with § 319.56–3, except that:

(i) Such fruits and vegetables are exempt from the permit requirements of § 319.56–3(b); and

(ii) Mangoes grown in the British Virgin Islands are prohibited entry into the U.S. Virgin Islands.

(2) Okra produced in the West Indies may be imported into the U.S. Virgin Islands without treatment but are subject to inspection at the port of arrival.

[72 FR 39501, July 18, 2007, as amended at 75 FR 4252, Jan. 26, 2010]

§§ 319.56–8—319.56–9 [Reserved]

§ 319.56–10 Importation of fruits and vegetables from Canada.

(a) *General permit for fruits and vegetables grown in Canada.* Fruits and vegetables grown in Canada and offered for entry into the United States will be subject to the inspection, treatment, and other requirements of § 319.56–3(d), but may otherwise be imported into the United States without restriction under this subpart; provided, that:

(1) Consignments of *Allium* spp. consisting of the whole plant or above ground parts must be accompanied by a phytosanitary certificate issued by the national plant protection organization of Canada with an additional declaration stating that the articles are free from *Acrolepipsis assectella* (Zeller).

(2) Potatoes from Newfoundland and that portion of the Municipality of Central Saanich in the Province of British Columbia east of the West Saanich Road are prohibited importation into the United States in accordance with § 319.37–20.

(b) [Reserved]

(Approved by the Office of Management and Budget under control number 0579–0316)

[72 FR 39501, July 18, 2007, as amended at 83 FR 11865, Mar. 19, 2018]

§ 319.56–11 Importation of dried, cured, or processed fruits, vegetables, nuts, and legumes.

(a) Dried, cured, or processed fruits and vegetables (except frozen fruits and vegetables), including cured figs and dates, raisins, nuts, and dried beans and peas, may be imported without permit, phytosanitary certificate, or other compliance with this subpart, except as specifically provided otherwise in this section or elsewhere in this part.

(b) *Acorns and chestnuts*—(1) *From countries other than Canada and Mexico;*

treatment required. Acorns and chestnuts intended for purposes other than propagation, except those grown in and shipped from Canada and Mexico, must be imported into the United States under permit, and subject to all the requirements of §319.56–3, and must be treated in accordance with part 305 of this chapter.[2]

(2) *From Canada and Mexico.* Acorns and chestnuts grown in and shipped from Canada and Mexico for purposes other than propagation may be imported in accordance with paragraph (a) of this section.

(3) *For propagation.* Acorns and chestnuts from any country may be imported for propagation only in accordance with the applicable requirements in §§319.37–1 through 319.37–23.

(c) *Macadamia nuts.* Macadamia nuts in the husk or shell are prohibited importation into the United States unless the macadamia nuts were produced in, and imported from, St. Eustatius.

[72 FR 39501, July 18, 2007, as amended at 75 FR 4252, Jan. 26, 2010; 83 FR 11866, Mar. 19, 2018]

§319.56–12 Importation of frozen fruits and vegetables.

Frozen fruits and vegetables may be imported into the United States in accordance with §319.56–3. Such fruits and vegetables must be held in accordance with the requirements for importing frozen fruits and vegetables in part 305 of this chapter..

[72 FR 39501, July 18, 2007, as amended at 75 FR 4252, Jan. 26, 2010]

Subpart M—Wheat Diseases

SOURCE: 70 FR 8231, Feb. 18, 2005, unless otherwise noted. Redesignated at 84 FR 2429, Feb. 7, 2019

§319.59–1 Definitions.

Administrator. The Administrator of the Animal and Plant Health Inspection Service, United States Department of Agriculture, or any employee of the United States Department of Ag-

riculture delegated to act in his or her stead.

Animal and Plant Health Inspection Service (APHIS). The Animal and Plant Health Inspection Service of the U.S. Department of Agriculture.

Controlled import permit. A written or electronically transmitted authorization issued by APHIS for the importation into the United States of otherwise prohibited or restricted plant material for experimental, therapeutic, or developmental purposes, under controlled conditions as prescribed by the Administrator in accordance with §319.6.

From. An article is considered to be "from" any country or locality in which it was grown.

Grain. Wheat (*Triticum aestivum*), durum wheat (*Triticum durum*), and triticale (*Triticum aestivum* × *Secale cereale*) used for consumption or processing and not for planting.

Hay. Host crops cut and dried for feeding to livestock. Hay cut after reaching the dough stage may contain mature kernels of the host crop.

Host crops. Plants or plant parts, including grain, seed, or hay, of wheat (*Triticum aestivum*), durum wheat (*Triticum durum*), and triticale (*Triticum aestivum* × *Secale cereale*).

Inspector. Any individual authorized by the Administrator of APHIS or the Commissioner of the Bureau of Customs and Border Protection, Department of Homeland Security, to enforce the regulations in this subpart.

Karnal bunt. A plant disease caused by the fungus *Tilletia indica* (Mitra) Mundkur.

Plant. Any plant (including any plant part) for or capable of propagation, including a tree, a tissue culture, a plantlet culture, pollen, a shrub, a vine, a cutting, a graft, a scion, a bud, a bulb, a root, and a seed.

Seed. Wheat (*Triticum aestivum*), durum wheat (*Triticum durum*), and triticale (*Triticum aestivum* × *Secale cereale*) used for propagation.

Spp. (species). All species, clones, cultivars, strains, varieties, and hybrids, of a genus.

Straw. The vegetative material left after the harvest of host crops. Straw is generally used as animal feed or bedding, as mulch, or for erosion control.

---

[2] Acorns and chestnuts imported into Guam are subject to the requirements of §319.56–7(b).

United States. The States, the Commonwealth of the Northern Mariana Islands, the Commonwealth of Puerto Rico, the District of Columbia, Guam, the Virgin Islands of the United States, or any other territory or possession of the United States.

[70 FR 8231, Feb. 18, 2005, as amended at 70 FR 71212, Nov. 28, 2005; 78 FR 25571, May 2, 2013; 83 FR 11866, Mar. 19, 2018]

§ 319.59–2 General import prohibitions; exceptions.

(a) [Reserved]

(b) Articles listed in § 319.59–3 as prohibited importation pending risk evaluation, and articles regulated for Karnal bunt in § 319.59–4(a) may be imported for experimental, therapeutic, or developmental purposes under a controlled import permit issued in accordance with § 319.6 if:

(1) Imported at the National Plant Germplasm Inspection Station, Building 580, Beltsville Agricultural Center East, Beltsville, MD 20705, or through any USDA plant inspection station listed in accordance with § 319.37–8(a);

(2) Imported pursuant to a controlled import permit issued for such article and kept on file at the National Plant Germplasm Inspection Station;

(3) Imported under conditions of treatment, processing, growing, shipment, or disposal specified on the controlled import permit and found by the Administrator to be adequate to prevent the introduction into the United States of tree, plant, or fruit diseases, injurious insects, and other plant pests, and

(4) Imported with a controlled import tag or label securely attached to the outside of the container containing the article or securely attached to the article itself if not in a container, and with such tag or label bearing a controlled import permit number corresponding to the number of the controlled import permit issued for such article.

(c) The importation of any host crops (including seed and any other plant parts) that are for planting or capable of being planted is restricted under

Subpart H—Plants for Planting of this part.

[70 FR 8231, Feb. 18, 2005, as amended at 70 FR 71212, Nov. 28, 2005; 72 FR 43523, Aug. 6, 2007; 78 FR 25571, May 2, 2013; 83 FR 11866, Mar. 19, 2018; 84 FR 2429, Feb. 7, 2019]

§ 319.59–3 Articles prohibited importation pending risk evaluation.

The articles listed in paragraph (a) of this section from the countries and localities listed in paragraph (b) of this section are prohibited from being imported or offered for entry into the United States, except as provided in § 319.59–2(b), pending the completion of an evaluation by APHIS of the potential pest risks associated with the articles. The national plant protection organization of any listed country or locality may contact APHIS[1] to initiate the preparation of a risk evaluation. If supported by the results of the risk evaluation, APHIS will take action to remove that country or locality from the list in paragraph (b) of this section.

(a) The following articles of *Triticum* spp. (wheat) or of *Aegilops* spp. (barb goatgrass, goatgrass): Straw (other than straw, with or without heads, which has been processed or manufactured for use indoors, such as for decorative purposes or for use in toys); chaff; and products of the milling process (*i.e.*, bran, shorts, thistle sharps, and pollards) other than flour.

(b) Afghanistan, Algeria, Armenia, Australia, Azerbaijan, Bangladesh, Belarus, Bulgaria, Chile, China, Cyprus, Egypt, Estonia, Falkland Islands, Georgia, Greece, Guatemala, Hungary, India, Iran, Iraq, Israel, Italy, Japan, Kazakhstan, Kyrgyzstan, Latvia, Libya, Lithuania, Moldova, Morocco, Nepal, North Korea, Oman, Pakistan, Portugal, Romania, Russia, Spain, Tajikistan, Tanzania, Tunisia, Turkey, Turkmenistan, South Africa, South Korea, Ukraine, Uzbekistan, and Venezuela.

[70 FR 8231, Feb. 18, 2005, as amended at 70 FR 71212, Nov. 28, 2005; 83 FR 11866, Mar. 19, 2018]

---

[1] Requests should be submitted in writing to Phytosanitary Issues Management, PPQ, APHIS, 4700 River Road, Unit 140, Riverdale, MD 20737–1236.

§ 319.59–4 Karnal bunt.

(a) *Regulated articles.* The following are regulated articles for Karnal bunt:

(1) Conveyances, including trucks, railroad cars, and other containers used to move host crops from a region listed in paragraph (b)(1) of this section that test positive for Karnal bunt through the presence of bunted kernels;

(2) Plant parts, including grain, straw, or hay, of all varieties of wheat (*Triticum aestivum*), durum wheat (*Triticum durum*), and triticale (*Triticum aestivum* × *Secale cereale*) from a region listed in paragraph (b)(1) of this section, except for straw/stalks/seed heads for decorative purposes that have been processed or manufactured prior to movement and are intended for use indoors;

(3) *Tilletia indica* (Mitra) Mundkur;

(4) Mechanized harvesting equipment that has been used in the production of wheat, durum wheat, or triticale that has tested positive for Karnal bunt through the presence of bunted kernels; and

(5) Seed conditioning equipment and storage/handling equipment that has been used in the production of wheat, durum wheat, or triticale seed found to contain the spores of *Tilletia indica.*

(b)(1) Karnal bunt is known to occur in the following regions: Afghanistan, India, Iran, Iraq, Mexico, Nepal, Pakistan, and South Africa.

(2) The Administrator may recognize an area within a region listed in paragraph (b)(1) of this section as an area free of Karnal bunt whenever he or she determines that the area meets the requirements of the International Standard for Phytosanitary Measures (ISPM) No. 4, "Requirements for the establishment of pest free areas." The international standard was established by the International Plant Protection Convention of the United Nations' Food and Agriculture Organization and is incorporated by reference in § 300.5 of this chapter. APHIS will publish a notice in the FEDERAL REGISTER and maintain on an APHIS Web site a list of the specific areas that are approved as areas in which Karnal bunt is not known to occur in order to provide the public with current, valid information. Areas listed as being free from Karnal bunt are subject to audit by APHIS to verify that they continue to merit such listing.

(c) *Handling, inspection and phytosanitary certificates.* Unless otherwise prohibited under § 319.59–3 of this subpart, any articles described in paragraph (a)(2) of this section that are from a region listed in paragraph (b)(1) of this section may be imported into the United States subject to the following conditions:

(1) The articles must be from an area that has been recognized, in accordance with paragraph (b)(2) of this section, to be an area free of Karnal bunt, or the articles have been tested and found to be free of Karnal bunt;

(2) The articles have not been commingled prior to arrival at a U.S. port of entry with articles from areas where Karnal bunt is known to occur;

(3) The articles offered for entry must be made available to an inspector for examination and remain at the port until released, or authorized further movement pending release, by an inspector; and

(4) The articles must be accompanied by a phytosanitary certificate issued by the national plant protection organization of the region of origin that includes the following additional declaration: "These articles originated in an area where Karnal bunt is not known to occur, as attested to either by survey results or by testing for bunted kernels or spores."

(d) *Treatments.* (1) Prior to entry into the United States, the following articles must be cleaned by removing any soil and plant debris that may be present.

(i) All conveyances and mechanized harvesting equipment used for storing and handling wheat, durum wheat, or triticale that tested positive for Karnal bunt based on bunted kernels.

(ii) All grain storage and handling equipment used to store or handle seed that has tested spore positive or grain that has tested bunted-kernel positive.

(iii) All seed-conditioning equipment used to store or handle seed that has tested spore-positive.

(2) Articles listed in paragraphs (d)(1)(i) and (d)(1)(ii) of this section will require disinfection in addition to cleaning prior to entry into the United

States if an inspector or an official of the plant protection organization of the country of origin determines that disinfection is necessary to prevent the spread of Karnal bunt. Disinfection is required for all seed conditioning equipment covered under paragraph (d)(1)(iii) prior to entry into the United States.

(3) Items that require disinfection prior to entry into the United States must be disinfected in accordance with part 305 of this chapter.

(Approved by the Office of Management and Budget under control number 0579–0240)

[70 FR 8231, Feb. 18, 2005, as amended at 75 FR 4253, Jan. 26, 2010; 83 FR 11866, Mar. 19, 2018]

Subpart N—Packing Materials

SOURCE: Redesignated at 84 FR 2429, Feb. 7, 2019.

QUARANTINE

§ 319.69 Notice of quarantine.

(a) The following plants and plant products, when used as packing materials, are prohibited entry into the United States from the countries and localities named in this paragraph (a), exceptions to the prohibitions may be authorized in the case of specific materials which have been so prepared, manufactured, or processed that in the judgment of the inspector no pest risk is involved in their entry:

(1) Rice straw, hulls, and chaff; from all countries.

(2) Corn and allied plants (maize, sorghum, broomcorn, Sudan grass, napier grass, jobs-tears, teosinte, Polytoca, Sclerachne, Chionachne); all parts, from all countries except Mexico, and the countries of Central America, the West Indies, and South America.

(3) Cotton and cotton products (lint, waste, seed cotton, cottonseed, and cottonseed hulls); from all countries.

(4) Sugarcane; all parts of the plant including bagasse, from all countries.

(5) Bamboo; leaves and small shoots, from all countries.

(6) Leaves of plants; from all countries.

(7) Forest litter; from all countries.

(8) Organic decaying vegetative matter from all countries, unless the mat-

ter is expressly authorized to be used as a packing material in this part. Exceptions to the prohibitions in paragraphs (a)(1) through (7) of this section may be authorized in the case of specific materials which has been so prepared, manufactured, or processed that in the judgment of the inspector no pest risk is involved in their entry.

(b) The following plants and plant products when used as packing materials will be permitted entry into the United States from the countries and localities designated below only in accordance with the regulations in this subpart:

(1) Cereal straw, hulls, and chaff (such as oats, barley, and rye) from all countries, except rice straw, hulls, and chaff, which are prohibited importation from all countries by paragraph (a)(1) of this section, and except wheat straw, hulls, and chaff, which are restricted importation by § 319.59 of this part from any country or locality listed in § 319.59–2 of this part.

(2) Corn and allied plants (maize, sorghum, broomcorn, Sudan grass, napier grass, jobs-tears, teosinte, Polytoca, Sclerachne, Chionachne); all parts, from Mexico and the countries of Central America, the West Indies, and South America.

(3) Grasses and hay and similar indefinite dried or cured masses of grasses, weeds, and herbaceous plants; from all countries.

(c) The importation of plants and plant products that are prohibited or restricted under paragraphs (a) and (b) of this section may be authorized for experimental, therapeutic, or developmental purposes under conditions specified in a controlled import permit issued in accordance with § 319.6.

(d) This quarantine shall leave in full force and effect all other quarantines and orders.

(e) As used in this subpart, unless the context otherwise requires, the term *United States* means the States, the District of Columbia, Guam, Puerto Rico, and the Virgin Islands of the United States.

[24 FR 10788, Dec. 29, 1959, as amended at 26 FR 9333, Oct. 4, 1961; 36 FR 24917, Dec. 24, 1971; 60 FR 27682, May 25, 1995; 63 FR 31102, June 8, 1998; 78 FR 25571, May 2, 2013; 84 FR 29958, June 25, 2019]

§319.69a Administrative instructions and interpretation relating to the entry into Guam of plant materials specified in §319.69.

(a) Plants and products designated in §319.69(a)(1), (3), (4), and (5) and (b)(1) and (3) as prohibited or restricted entry into the United States from the countries and localities named may be imported into Guam as packing materials without prohibition or restriction under this subpart. Inspection of such importations may be made under the general authority of §330.105(a) of this chapter. If an importation is found infected, infested, or contaminated with any plant pest and is not subject to disposal under this part, disposition may be made in accordance with §330.106 of this chapter.

(b) Corn and allied plants listed in §319.69(a)(2) may be imported into Guam subject to the requirements of §§319.69-2, 319.69-3, and 319.69-4.

(c) Under §319.69(a) (6) and (7), coconut fronds and other parts of the coconut trees are prohibited entry into Guam as packing materials except as permitted in §319.37-11.

[24 FR 10788, Dec. 29, 1959, as amended at 60 FR 27682, May 25, 1995; 62 FR 65009, Dec. 10, 1997; 83 FR 11866, Mar. 19, 2018]

RULES AND REGULATIONS

§319.69-1 Definitions.

(a) *Packing materials.* The expression "packing material", as used in §319.69, includes any of the plants or plant products enumerated, when these are associated with or accompany any commodity or shipment to serve for filling, wrapping, ties, lining, mats, moisture retention, protection, or for any other purpose; and the word "packing", as used in the expression "packing materials", shall include the presence of such materials within, in contact with, or accompanying such commodity or shipment.[1]

---

[1] Since it is the packing materials themselves which constitute the danger and not the manner of use, it is intended that the definition shall include their presence within or accompanying a shipment regardless of their function or relation to a shipment or the character of the shipment.

(b) *Inspector.* An inspector of the U.S. Department of Agriculture.

[75 FR 17292, Apr. 6, 2010, as amended at 84 FR 29958, June 25, 2019]

§319.69-2 Freedom from pests.

All packing materials allowed entry under restriction shall be free from injurious insects and plant diseases.

§319.69-3 Entry inspection.

All packing materials shall be subject to inspection at time of entry.

§319.69-4 Disposition of materials found in violation.

If the inspector shall find packing materials associated with or accompanying any commodity or shipment being imported, or to have been imported, in violation of §319.69 or of the regulations in this subpart or shall find them infested or infected with injurious insects or plant diseases, the inspector may refuse entry to the shipment, or the inspector may seize and destroy or otherwise dispose of such packing material, or the inspector may require it to be replaced, or sterilized, or otherwise treated.

[24 FR 10788, Dec. 29, 1959, as amended at 70 FR 33326, June 7, 2005]

§319.69-5 Types of organic decaying vegetative matter authorized for packing.

The following types of organic decaying vegetative matter are authorized as safe for packing:

(a) Peat;

(b) Peat moss; and

(c) Osmunda fiber.

[84 FR 29958, June 25, 2019]

Subpart O—Coffee

SOURCE: 63 FR 65650, Nov. 30, 1998, unless otherwise noted. Redesignated at 84 FR 2429, Feb. 7, 2019.

§319.73-1 Definitions.

Administrator. The Administrator of the Animal and Plant Health Inspection Service, United States Department of Agriculture, or any employee of the United States Department of Agriculture delegated to act in his or her stead.

Inspector. Any individual authorized by the Administrator to enforce this subpart.

Sample. Unroasted coffee not for commercial resale. Intended use includes, but is not limited to, evaluation, testing, or market analysis.

United States. The States, District of Columbia, Guam, Northern Mariana Islands, Puerto Rico, and the Virgin Islands of the United States.

Unroasted coffee. The raw or unroasted seeds or beans of coffee intended for processing.

[63 FR 65650, Nov. 30, 1998, as amended at 83 FR 11866, Mar. 19, 2018]

§ 319.73-2 Products prohibited importation.

(a) To prevent the spread of the coffee berry borer *Hypothenemus hampei* (Ferrari) and the fungus *Hemileia vastatrix* (Berkely and Broome), which causes an injurious rust disease, the following articles are prohibited importation into Hawaii and Puerto Rico, except as provided in § 319.73-3 of this subpart:

(1) Unroasted coffee;

(2) Coffee leaves; and

(3) Empty sacks previously used for unroasted coffee.

(b) The importation of any coffee plants (including bare seeds, seeds in pulp, and any other plant parts) that are for planting or capable of being planted is restricted under Subpart H—Plants for Planting of this part.

[63 FR 65650, Nov. 30, 1998, as amended at 83 FR 11866, Mar. 19, 2018; 84 FR 2429, Feb. 7, 2019]

§ 319.73-3 Conditions for transit movement of certain products through Puerto Rico or Hawaii.

(a) *Mail.* Samples of unroasted coffee that are transiting Hawaii or Puerto Rico en route to other destinations and that are packaged to prevent the escape of any plant pests may proceed without action by an inspector. Packaging that would prevent the escape of plant pests includes, but is not limited to, sealed cartons, airtight containers, or vacuum packaging. Samples of unroasted coffee received by mail but not packaged in this manner are subject to inspection and safeguard by an inspector. These samples must be re-

turned to origin or forwarded to a destination outside Hawaii or Puerto Rico in a time specified by an inspector and in packaging that will prevent the escape of any plant pests. If this action is not possible, the samples must be destroyed.

(b) *Cargo.* Samples of unroasted coffee that are transiting Hawaii or Puerto Rico as cargo and that remain on the carrier may proceed to a destination outside Hawaii or Puerto Rico without action by an inspector. Samples may be transshipped in Puerto Rico or Hawaii only after an inspector determines that they are packaged to prevent the escape of any plant pests. Samples that are not packaged in this manner must be rewrapped or repackaged in a manner prescribed by an inspector to prevent the escape of plant pests before the transshipment will be allowed.

(c) Other mail, cargo, and baggage shipments of articles covered by § 319.73-2 arriving in Puerto Rico or Hawaii may not be unloaded or transshipped in Puerto Rico or Hawaii and are subject to inspection and other applicable requirements of the Plant Safeguard Regulations (part 352 of this chapter).

§ 319.73-4 Costs.

All costs of inspection, packing materials, handling, cleaning, safeguarding, treating, or other disposal of products or articles under this subpart will be borne by the owner, importer, or agent of the owner or importer, including a broker. The services of an inspector during regularly assigned hours of duty and at the usual places of duty will be furnished without cost to the importer.

Subpart P—Cut Flowers

SOURCE: 64 FR 38110, July 15, 1999, unless otherwise noted. Redesignated at 84 FR 2429, Feb. 7, 2019.

§ 319.74-1 Definitions.

Administrator. The Administrator of the Animal and Plant Health Inspection Service, United States Department of Agriculture, or any employee of the United States Department of Agriculture delegated to act in his or her stead.

Controlled import permit. A written or electronically transmitted authorization issued by APHIS for the importation into the United States of otherwise prohibited or restricted plant material for experimental, therapeutic, or developmental purposes, under controlled conditions as prescribed by the Administrator in accordance with §319.6.

Cut flower. The highly perishable commodity known in the commercial flower-producing industry as a cut flower, which is the severed portion of a plant, including the inflorescence and any parts of the plant attached to it, in a fresh state and not for planting. This definition does not include dried, bleached, dyed, or chemically treated decorative plant materials; filler or greenery, such as fern fronds and asparagus plumes, frequently packed with fresh cut flowers; or Christmas greenery, such as holly, mistletoe, and Christmas trees.

Inspector. Any individual authorized by the Administrator to enforce this subpart.

United States. All of the States, the District of Columbia, Guam, the Northern Mariana Islands, Puerto Rico, the Virgin Islands of the United States, and all other territories or possessions of the United States.

[64 FR 38110, July 15, 1999, as amended at 78 FR 25571, May 2, 2013; 83 FR 11866, Mar. 19, 2018]

§319.74–2 Conditions governing the entry of cut flowers.

(a) *Inspection.* All cut flowers imported into the United States must be made available to an inspector for examination at the port of first arrival and must remain at the port of first arrival until released, or authorized further movement, by an inspector.

(b) *Actions to prevent the introduction of plant pests; notice by an inspector.* If an inspector orders any disinfection, cleaning, treatment, reexportation, or other action with regard to imported cut flowers that are found to be infested with injurious plant pests or infected with diseases, the inspector will provide an emergency action notification (PPQ Form 523) to the importer, owner, or agent or representative of the importer or owner of the cut flow-

ers. The importer, owner, or agent or representative of the importer or owner must, within the time specified in the PPQ Form 523 and at his or her own expense, destroy the cut flowers, ship them to a point outside the United States, move them to an authorized site, and/or apply treatments, clean, or apply other safeguards to the cut flowers as prescribed by the inspector on the PPQ Form 523. Further, if the importer, owner, or agent or representative of the importer or owner fails to follow the conditions on PPQ Form 523 by the time specified on the form, APHIS will arrange for destruction of the cut flowers, and the importer, owner, or agent or representative of the importer or owner will be responsible for all costs incurred. Cut flowers that have been cleaned or treated must be made available for further inspection, cleaning, and treatment at the option of the inspector at any time and place indicated by the inspector before the requirements of this subpart will have been met. Neither the Department of Agriculture nor the inspector may be held responsible for any adverse effects of treatment on imported cut flowers.

(c) *Fumigation for agromyzids.* Cut flowers imported from any country or locality and found upon inspection to be infested with agromyzids (insects of the family Agromyzidae) must be fumigated at the time of importation with methyl bromide in accordance with part 305 of this chapter, with the following exceptions:

(1) Fumigation will not be required for cut flowers imported from Canada (including Labrador and Newfoundland) or Mexico because of the finding of agromyzids.

(2) Fumigation will not be required for cut flowers of *Chrysanthemum* spp. imported from Colombia or the Dominican Republic because of the finding of agromyzids, when such agromyzids are identified by an inspector to be only agromyzids of the species *Liriomyza trifolii* (Burgess).

(d) *Chrysanthemum white rust hosts.* (1) The following *Chrysanthemum*, *Leucanthemella*, and *Nipponanthemum* spp. are considered to be hosts of chrysanthemum white rust:

Accepted name of susceptible species	Synonyms	Common name
Chrysanthemum arcticum L.	Arctanthemum arcticum (L.) Tzvelev and Dendranthema arcticum (L.) Tzvelev.	Arctic chrysanthemum and arctic daisy.
Chrysanthemum boreale (Makino) Makino.	Chrysanthemum indicum L. var. boreale Makino and Dendranthema boreale (Makino) Ling ex Kitam.	
Chrysanthemum indicum L.	Dendranthema indicum (L.) Des Moul.	
Chrysanthemum japonense Nakai ..	Dendranthema japonense (Nakai) Kitam. and Dendranthema occidentali-japonense Kitam.	Nojigiku.
Chrysanthemum japonicum Makino	Chrysanthemum makinoi Matsum. & Nakai and Dendranthema japonicum (Makino) Kitam.	Ryuno-giku.
Chrysanthemum × morifolium Ramat.	Anthemis grandiflorum Ramat., Anthemis stipulacea Moench, Chrysanthemum sinense Sabine ex Sweet, Chrysanthemum stipulaceum (Moench) W. Wight, Dendranthema × grandiflorum (Ramat.) Kitam., Dendranthema × morifolium (Ramat.) Tzvelev, and Matricaria morifolia Ramat.	Florist's chrysanthemum, chrysanthemum, and mum.
Chrysanthemum pacificum Nakai ...	Ajania pacifica (Nakai) K. Bremer & Humphries and Dendranthema pacificum (Nakai) Kitam.	Iso-giku.
Chrysanthemum shiwogiku Kitam ..	Ajania shiwogiku (Kitam.) K. Bremer & Humphries and Dendranthema shiwogiku (Kitam.) Kitam.	Shio-giku.
Chrysanthemum yoshinaganthum Makino ex Kitam.	Dendranthema yoshinaganthum (Makino ex Kitam.) Kitam.	
Chrysanthemum zawadskii Herbich subsp. yezoense (Maek.) Y. N. Lee.	Chrysanthemum arcticum subsp. maekawanum Kitam, Chrysanthemum arcticum var. yezoense Maek. [basionym], Chrysanthemum yezoense Maek. [basionym], Dendranthema yezoense (F. Maek.) D. J. N. Hind, and Leucanthemum yezoense (Maek.) á. Löve & D. Löve.	
Chrysanthemum zawadskii Herbich subsp. zawadskii.	Chrysanthemum sibiricum Turcz. ex DC., nom. inval., Dendranthema zawadskii (Herbich) Tzvelev, and Dendranthema zawadskii var. zawadskii.	
Leucanthemella serotina (L.) Tzvelev.	Chrysanthemum serotinum L., Chrysanthemum uliginosum (Waldst. & Kit. ex Willd.) Pers., and Pyrethrum uliginosum (Waldst. & Kit. ex Willd.).	Giant daisy or high daisy.
Nipponanthemum nipponicum (Franch. ex Maxim.) Kitam.	Chrysanthemum nipponicum (Franch. ex Maxim.) Matsum. and Leucanthemum nipponicum Franch. ex Maxim.	Nippon daisy or Nippon-chrysanthemum.

(2) Chrysanthemum white rust is considered to exist in the following regions: Andorra, Argentina, Australia, Belarus, Bosnia and Herzegovina, Brazil, Brunei, Canary Islands, Chile, China, Colombia, Croatia, Ecuador, Iceland, Japan, Korea, Liechtenstein, Macedonia, Malaysia, Mexico, Moldova, Monaco, New Zealand, Norway, Peru, Republic of South Africa, Russia, San Marino, Switzerland, Taiwan, Thailand, Tunisia, Ukraine, Uruguay, Venezuela, Yugoslavia; the European Union (Austria, Belgium, Bulgaria, Cyprus, Czech Republic, Denmark, Estonia, Finland, France, Germany, Greece, Hungary, Ireland, Italy, Latvia, Lithuania, Luxembourg, Malta, Netherlands, Poland, Portugal, Romania, Slovakia, Slovenia, Spain, Sweden, and United Kingdom); and all countries, territories, and possessions of countries located in part or entirely between 90° and 180° East longitude.

(3) Cut flowers of any species listed in paragraph (d)(1) of this section may be imported into the United States from any region listed in paragraph (d)(2) of this section only under the following conditions:

(i) The flowers must be grown in a production site that is registered with the national plant protection organization (NPPO) of the country in which the production site is located or with the NPPO's designee, and the NPPO or its designee must provide a list of registered sites to APHIS.

(ii) Each shipment of cut flowers must be accompanied by a phytosanitary certificate or equivalent documentation, issued by the NPPO of the country of origin or its designee, that contains an additional declaration stating that the place of production as well as the consignment have been inspected and found free of *Puccinia horiana*.

(iii) Box labels and other documents accompanying shipments of cut flowers must be marked with the identity of the registered production site.

(iv) APHIS-authorized inspectors must also be allowed access to production sites and other areas necessary to monitor the chrysanthemum white rust-free status of the production sites.

(4) Cut flowers not meeting these conditions will be refused entry into the United States. The detection of chrysanthemum white rust in a shipment of cut flowers from a registered production site upon arrival in the United States will result in the prohibition of imports originating from the production site until such time when APHIS and the NPPO of the exporting country, can agree that the eradication measures taken have been effective and that the pest risk within the production site has been eliminated.

(e) *Irradiation.* Cut flowers and foliage that are required under this part to be treated or subjected to inspection to control one or more of the plant pests for which irradiation is an approved treatment under part 305 of this chapter may instead be treated with irradiation. Irradiation treatment must be conducted in accordance with the requirements of part 305 of this chapter. There is a possibility that some cut flowers could be damaged by such irradiation.

(f) *Refusal of entry.* If an inspector finds that imported cut flowers are so infested with a plant pest or infected with disease that, in the judgment of the inspector, they cannot be cleaned or treated, or if they contain soil or other prohibited contaminants, the entire lot may be refused entry into the United States.

(Approved by the Office of Management and Budget under control number 0579–0271)

[64 FR 38110, July 15, 1999, as amended at 71 FR 4464, Jan. 27, 2006; 72 FR 15811, Apr. 3, 2007; 75 FR 4253, Jan. 26, 2010]

§319.74–3 Importations for experimental or similar purposes.

Cut flowers may be imported for experimental, therapeutic, or developmental purposes under conditions specified in a controlled import permit issued in accordance with §319.6.

[78 FR 25571, May 2, 2013]

§319.74–4 Costs and charges.

The Animal and Plant Health Inspection Service, U.S. Department of Agriculture, will be responsible only for the costs of providing the services of an inspector during regularly assigned hours of duty and at the usual places of duty (provisions relating to costs for other services of an inspector are contained in 7 CFR part 354). The importer, owner, or agent or representative of the importer or owner of cut flowers is responsible for all additional costs of inspection, treatment, movement, storage, or destruction ordered by an inspector under this subpart, including the costs of any labor, chemicals, packing materials, or other supplies required.

Subpart Q—Khapra Beetle

SOURCE: Redesignated at 84 FR 2429, Feb. 7, 2019.

§319.75 Restrictions on importation of regulated articles; disposal of articles refused importation.

(a) The Secretary has determined that in order to prevent the entry into the United States of khapra beetle (*Trogoderma granarium* Everts) it is necessary to restrict the importation of certain articles from foreign countries and localities. Accordingly, no person shall import any regulated article unless in conformity with all of the applicable restrictions in this subpart.

(b) Any article refused importation for noncompliance with the requirements of this subpart shall be promptly removed from the United States or abandoned by the importer, and pending such action shall be subject to the immediate application of such safeguards against escape of plant pests as the inspector determines necessary to prevent the introduction into the United States of plant pests. If the article is not promptly safeguarded, removed from the United States, or abandoned by the importer for destruction, it may be seized, destroyed, or otherwise disposed of in accordance with section 414 of the Plant Protection Act (7 U.S.C. 7714).

(c) A regulated article may be imported without complying with other restrictions under this subpart if:

(1) Imported for experimental, therapeutic, or developmental purposes under the conditions specified in a controlled import permit issued in accordance with § 319.6;

(2) Imported at the National Plant Germplasm Inspection Station, Building 580, Beltsville Agricultural Research Center East, Beltsville, MD 20705, or through any USDA plant inspection station listed in § 319.37–8(a); and

(3) Imported with a controlled import tag or label securely attached to the outside of the container containing the article or securely attached to the article itself if not in a container, and with such tag or label bearing a controlled import permit number corresponding to the number of the controlled import permit issued for such article.

[46 FR 38334, July 27, 1981, as amended at 47 FR 3085, Jan. 22, 1982; 66 FR 21057, Apr. 27, 2001; 72 FR 43523, Aug. 6, 2007; 78 FR 25572, May 2, 2013; 79 FR 19811, Apr. 10, 2014; 83 FR 11866, Mar. 19, 2018]

§ 319.75–1 Definitions.

Terms used in the singular form in this subpart shall be construed as the plural, and vice-versa, as the case may demand. The following terms, when used in this subpart, shall be construed, respectively, to mean:

Administrator. The Administrator of the Animal and Plant Health Inspection Service, United States Department of Agriculture, or any employee of the United States Department of Agriculture delegated to act in his or her stead.

From. An article is considered to be "from" any country or locality in which it originated or any country(ies) or locality(ies) in which it was offloaded prior to arrival in the United States.

Import. (importation, imported). To import or move into the United States.

Inspector. Any individual authorized by the Administrator or the Commissioner of U.S. Customs and Border Protection, Department of Homeland Security, to enforce the regulations in this subpart.

Person. Any individual, corporation, company, society, association or other organized group.

Phytosanitary certificate of inspection. A document relating to a regulated article, which is issued by a plant protection official of the country in which the regulated article was grown, which is issued not more than 15 days prior to shipment of the regulated article from the country in which grown, which is addressed to the plant protection service of the United States (Plant Protection and Quarantine), which contains a description of the regulated article intended to be imported into the United States, which certifies that the article has been thoroughly inspected, is believed to be free from injurious plant diseases, injurious insect pests, and other plant pests, and is otherwise believed to be eligible for importation pursuant to the current phytosanitary laws and regulations of the United States.

Plant gum. Any of numerous colloidal polysaccharide substances of plant origin that are gelatinous when moist but harden on drying. Plant gums include but are not limited to acacia gum, guar gum, gum arabic, locust gum and tragacanth gum.

Plant pest. The egg, pupal, and larval stages as well as any other living stage of any insects, mites, nematodes, slugs, snails, protozoa, or other invertebrate animals, bacteria, fungi, other parasitic plants or reproductive parts thereof, viruses, or any organisms similar to or allied with any of the foregoing, or any infectious substances, which can directly or indirectly injure or cause disease or damage in any plants or parts thereof, or any processed, manufactured, or other products of plants.

Plant Protection and Quarantine. The organizational unit within the Animal and Plant Health Inspection Service, U.S. Department of Agriculture, delegated responsibility for enforcing provisions of the Plant Protection Act and related legislation, quarantines, and regulations.

Secretary. The Secretary of Agriculture, or any other officer or employee of the Department of Agriculture to whom authority to act in his/her stead has been or may hereafter be delegated.

United States. The States, District of Columbia, American Samoa, Guam,

Northern Mariana Islands, Puerto Rico, and the Virgin Islands of the United States.

[46 FR 38334, July 27, 1981, as amended at 47 FR 3085, Jan. 22, 1982; 49 FR 1876, Jan. 16, 1984; 50 FR 8704, 8706, Mar. 5, 1985; 66 FR 21057, Apr. 27, 2001; 78 FR 25572, May 2, 2013; 19811, Apr. 10, 2014; 79 FR 77841, Dec. 29, 2014; 83 FR 11866, Mar. 19, 2018]

§319.75–2 Regulated articles.[1]

(a) The following articles are regulated articles from all countries designated in accordance with paragraph (c) of this section as infested with khapra beetle and are subject to mandatory treatment in accordance with §319.75–4:

(1) Seeds of the plant family Cucurbitaceae[2] if in shipments greater than 2 ounces, if not for propagation;

(2) Goatskins, lambskins, and sheepskins (excluding goatskins, lambskins, and sheepskins which are fully tanned, blue-chromed, pickled in mineral acid, or salted and moist);

(3) Plant gums and plant gum seeds shipped as bulk cargo (in an unpackaged state);

(4) Used jute or burlap bagging not containing cargo;

(5) Used jute or burlap bagging that is used as a packing material (such as filler, wrapping, ties, lining, matting, moisture retention material, or protection material), and the cargo for which the used jute or burlap bagging is used as a packing material; and

---

[1] The importation of restricted articles may be subject to prohibitions or restrictions under other provisions of 7 CFR part 319. For example, fresh whole chilies (*Capsicum* spp.) and fresh whole red peppers (*Capsicum* spp.) from Pakistan are prohibited from being imported into the United States under the provisions of Subpart L—Fruits and Vegetables of this part, and the importation of any restricted articles that are for planting or capable of being planted is restricted under Subpart H—Plants for Planting of this part.

[2] Seeds of the plant family Cucurbitaceae include but are not limited to: *Benincasa hispida* (wax gourd), *Citrullus lanatus* (watermelon), *Cucumis melo* (muskmelon, cantaloupe, honeydew), *Cucumis sativus* (cucumber), *Cucurbita pepo* (pumpkin, squashes, vegetable marrow), *Lagenaria siceraria* (calabash, gourd), *Luffa cylindrica* (dishcloth gourd), *Mormordica charantia* (bitter melon), and *Sechium edule* (chayote).

(6) Whole chilies (*Capsicum* spp.), whole red peppers (*Capsicum* spp.), and cumin seeds (*Cuminum cyminum*) when packed in new jute or burlap bagging;

(b) The following articles are regulated articles from all countries designated in accordance with paragraph (c) of this section as infested with khapra beetle or that have the potential to be infested with khapra beetle and are prohibited entry into the United States in passenger baggage and personal effects. Commercial shipments must be accompanied by a phytosanitary certificate issued in accordance with §319.75–9 and containing an additional declaration stating: "The shipment was inspected and found free of khapra beetle (*Trogoderma granarium*)."

(1) Rice (*Oryza sativa*); and

(2) Chick peas (*Cicer* spp.), safflower seeds (*Carthamus tinctorius*), and soybeans (*Glycine max*).

(c) The Administrator will designate a country or an area under a specific jurisdictional authority as infested with khapra beetle when we receive official notification from the country or area that it is infested or when we intercept the pest in a commercial shipment from that country. The Administrator will publish the list of countries or areas under a specific jurisdictional authority found to be infested with khapra beetle on the Plant Protection and Quarantine Web site, *http://www.aphis.usda.gov/import_export/plants/manuals/ports/downloads/kb.pdf.* After a change is made to the list of infested countries or areas, we will publish a notice in the FEDERAL REGISTER informing the public that the change has occurred.

[79 FR 77841, Dec. 29, 2014, as amended at 80 FR 43010, July 21, 2015; 83 FR 11866, Mar. 19, 2018; 84 FR 2429, Feb. 7, 2019]

§319.75–3 Permits.

A regulated article may be imported only after issuance of a written permit or oral authorization by the Plant Protection and Quarantine Programs in accordance with §§319.7 through 319.7–5.

(Approved by the Office of Management and Budget under control number 0579–0049)

[79 FR 19811, Apr. 10, 2014]

§ 319.75–4   Treatments.

Prior to moving into the United States from the port of entry, a regulated article listed in § 319.75–2(a) shall be treated for possible infestation with khapra beetle in accordance with part 305 of this chapter.

[79 FR 77841, Dec. 29, 2014]

§ 319.75–5   Marking and identity.

(a) Any regulated article at the time of importation shall plainly and correctly bear on the outer container (if in a container) or on the regulated article (if not in a container) the following information:

(1) General nature and quantity of the contents,

(2) Country or locality of origin,

(3) Name and address of shipper, owner, or person shipping or forwarding the article,

(4) Name and address of consignee,

(5) Identifying shipper's mark and number, and

(b) Any regulated article shall be accompanied at the time of importation by an invoice or packing list indicating the contents of the shipment.

(Approved by the Office of Management and Budget under control number 0579–0049)

[46 FR 38334, July 27, 1981, as amended at 47 FR 3085, Jan. 22, 1982; 48 FR 57466, Dec. 30, 1983; 79 FR 19811, Apr. 10, 2014]

§ 319.75–6   Arrival notification.

Promptly upon arrival of any regulated article at a port of entry, the importer shall notify Plant Protection and Quarantine of the arrival by such means as a manifest, Customs entry document, commercial invoice, waybill, a broker's document, or a notice form provided for that purpose.

(Approved by the Office of Management and Budget under control number 0579–0049)

[46 FR 38334, July 27, 1981, as amended at 48 FR 57466, Dec. 30, 1983; 79 FR 19811, Apr. 10, 2014]

§ 319.75–7   Costs and charges.

The services of the inspector during regularly assigned hours of duty and at the usual places of duty shall be furnished without cost to the importer.[3] The importer shall be responsible for arrangements for treatments required under § 319.75–4. Any treatment required under § 319.75–4 for a regulated article shall be performed at the port of entry by a nongovernmental fumigator at the importer's expense, and shall be performed under the supervision of an inspector. Plant Protection and Quarantine will not be responsible for any costs or charges, other than those indicated in this section.

[46 FR 38334, July 27, 1981, as amended at 79 FR 19811, Apr. 10, 2014; 84 FR 2429, Feb. 7, 2019]

§ 319.75–8   Ports of entry.

Any regulated article shall be imported only at a Customs designated port of entry indicated in 19 CFR 101.3(b)(1) and found by the Administrator and specified on the permit issued pursuant to § 319.75–3 to have a nongovernmental fumigator available at the port to treat such regulated article pursuant to § 319.75–4. It is the responsibility of the importer to arrange with the nongovernmental fumigator for treatment of the article.

[46 FR 38334, July 27, 1981, as amended at 72 FR 43523, Aug. 6, 2007; 78 FR 25572, May 2, 2013; 79 FR 19811, Apr. 10, 2014; 83 FR 11866, Mar. 19, 2018]

§ 319.75–9   Inspection and phytosanitary certificate of inspection.

(a) Any fruit, vegetable, or other plant product designated as a regulated article and grown in a country maintaining an official system of inspection for the purpose of determining whether such article is free from injurious plant diseases, injurious insect pests, and other plant pests shall be accompanied by a phytosanitary certificate of inspection from the plant protection service of such country at the time of importation or offer for importation into the United States. Such certificate may cover more than one article and more than one container kept together during shipment and offer for importation.

---

[3] Provisions relating to costs for other services of an inspector are contained in 7 CFR part 354.

(b) Any fruit, vegetable, seed, or other plant product designated as a regulated article which is accompanied by a valid phytosanitary certificate of inspection is subject to inspection by an inspector at the time of importation into the United States for the purpose of determining whether such article is free of injurious plant diseases, injurious insect pests, and other plant pests, and whether such article is otherwise eligible to be imported into the United States.

(c) Any fruit, vegetable, seed, or other plant product designated as a regulated article and grown in a country not maintaining an official system of inspection for the purpose of determining whether such article is free from injurious plant diseases, or injurious insect pests, and other plant pests shall be inspected by an inspector at the time of importation into the United States for the purpose of determining whether such article is free of such diseases and pests and whether such article is otherwise eligible to be imported into the United States.

[50 FR 8707, Mar. 5, 1985, as amended at 79 FR 19811, Apr. 10, 2014; 83 FR 11866, Mar. 19, 2018]

Subpart R—Gypsy Moth Host Material from Canada

Source: 64 FR 45866, Aug. 23, 1999, unless otherwise noted. Redesignated at 84 FR 2429, Feb. 7, 2019.

§ 319.77-1 Definitions.

Animal and Plant Health Inspection Service (APHIS). The Animal and Plant Health Inspection Service of the U.S. Department of Agriculture.

Canadian infested area. Any area of Canada listed as a gypsy moth infested area in § 319.77-3 of this subpart.

Canadian noninfested area. Any area of Canada that is not listed as a gypsy moth infested area in § 319.77-3 of this subpart.

Certification of origin. A signed, accurate statement certifying the area in which a regulated article was produced or grown. The statement may be provided directly on the shipping documents accompanying shipments of commercial wood products from Can-

ada, or may be provided on a separate certificate.

Gypsy moth. The insect known as the gypsy moth, Lymantria dispar (Linnaeus), in any stage of development.

Import (imported, importation). To bring or move into the territorial limits of the United States.

Mobile home. Any vehicle, other than a recreational vehicle, designed to serve, when parked, as a dwelling or place of business.

Outdoor household articles. Articles associated with a household that are generally kept or used outside the home. Examples of outdoor household articles are awnings, barbeque grills, bicycles, boats, dog houses, firewood, garden tools, hauling trailers, outdoor furniture and toys, recreational vehicles and their associated equipment, and tents.

Phytosanitary certificate. A document issued by an official authorized by the national government of Canada that contains a description of the regulated article intended for importation into the United States and that certifies that the article has been thoroughly inspected or treated, is believed to be free from plant pests, and is otherwise believed to be eligible for importation pursuant to the current phytosanitary laws and regulations of the United States. A phytosanitary certificate must be addressed to the Animal and Plant Health Inspection Service and may be issued no more than 14 days prior to the shipment of the regulated article.

Recreational vehicles. Vehicles, including pickup truck campers, one-piece motor homes, and travel trailers, designed to serve as temporary places of dwelling.

United States. All of the States of the United States, the District of Columbia, Guam, the Northern Mariana Islands, Puerto Rico, the Virgin Islands of the United States, and all other territories and possessions of the United States.

U.S. infested area. Any area of the United States listed as a gypsy moth generally infested area in § 301.45-3 of this chapter.

U.S. noninfested area. Any area of the United States that is not listed as a gypsy moth generally infested area in § 301.45-3 of this chapter.

[64 FR 45866, Aug. 23, 1999, as amended at 65 FR 38175, June 20, 2000]

§ 319.77-2 Regulated articles.

In order to prevent the spread of gypsy moth from Canada into non-infested areas of the United States, the gypsy moth host materials listed in paragraphs (a) through (e) of this section are designated as regulated articles. Regulated articles may be imported into the United States from Canada only under the conditions described in § 319.77-4 of this subpart.

(a) Trees without roots (e.g., Christmas trees), unless they were greenhouse-grown throughout the year;

(b) Logs with bark attached;

(c) Pulpwood with bark attached;

(d) Bark and bark products;

(e) Outdoor household articles;

(f) Mobile homes and their associated equipment; and

(g) Stone and quarry products.

[64 FR 45866, Aug. 23, 1999, as amended at 71 FR 40878, July 19, 2006; 83 FR 11866, Mar. 19, 2018; 84 FR 29958, June 25, 2019]

§ 319.77-3 Gypsy moth infested areas in Canada.

The following areas in Canada are known to be infested with gypsy moth:

(a) *Province of New Brunswick*—(1) *Charlotte County.* That portion of Charlotte County that includes the following parishes: Campobello Island, Dumbarton, Dufferin, Grand Manan Island, St. Andrews, St. Croix, St. David, St. George, St. James, St. Patrick, and St. Stephen.

(2) *Kings County.* That portion of Kings County that includes the following parishes: Greenwich, Kars, and Springfield.

(3) *Queens County.* (i) That portion of Queens County that includes the following parishes: Canning, Cambridge, Gagetown, Johnston, and Wickham; and

(ii) That portion of Chipman Parish south or west of highway 10; and

(iii) That portion of Waterborough Parish west of highway 10 and south of highway 2.

(4) *Sunbury County.* That portion of Sunbury County that includes the following parishes: Blissville, Burton, Gladstone, Lincoln, and Sheffield.

(5) *York County.* (i) That portion of York County that includes the City of Fredericton and the following parishes: North Lake and McAdam; and

(ii) That portion of Queensbury parish south and east of the Scotch Lake Road beginning in the west at Bear Island on the St. John River and ending at the Parish border on the east.

(b) *Province of Nova Scotia*—(1) *Annapolis County.* The entire county.

(2) *Digby County.* The entire county.

(3) *Halifax County.* The area of the county bounded by a line beginning at the intersection of the Halifax/Lunenburg County border and the Atlantic Ocean; then north along the Halifax/Lunenburg County border to the Halifax/Hants County border; then east along the Halifax/Hants County border to route 354; then south along route 354 to route 568 (Beaverbank-Windsor Junction Road); then east along route 568 (Beaverbank-Windsor Junction Road) to route 416 (Fall River Road); then east and north along route 416 (Fall River Road) to route 2; then south along route 2 to route 102/118; then south along route 118 to route 107; then south along route 107 to route 7; then east along route 7 to route 328; then south along route 328 to the shoreline of Cole Harbour; then west along the seashore from Cole Harbour to the point of beginning.

(4) *Hants County.* The area of the county bounded by a line beginning at the intersection of the Hants/Kings County border and the shoreline of the Minas Basin; then southwest along the Hants/Kings County border to the Hants/Lunenburg County border; then southeast along the Hants/Lunenburg County border to the Hants/Halifax County border; then east along the Hants/Halifax County border to route 354; then north along route 354 to the Minas Basin; then west along the shoreline of the Minas Basin to the point of beginning.

(5) *Kings County.* The entire county.

(6) *Lunenberg County.* The entire county.

(7) *Queens County.* The entire county.

(8) *Shelburne County.* The entire county.

(9) *Yarmouth County.* The entire county.

(c) *Province of Ontario.* (1) That portion of the Province of Ontario that includes the following counties and regional municipalities: Brant, Bruce, Dufferin, Durham, Elgin, Essex, Frontenac, Grey, Haldimand-Norfolk, Haliburton, Halton, Hamilton-Wentworth, Hastings, Huron, Kent, Lambton, Lanark, Leeds-Granville, Lennox-Addington, Middlesex, Muskoka, Niagara, Northumberland, Ottawa-Carleton, Oxford, Parry Sound, Peel, Perth, Peterborough, Prescott-Russell, Prince Edward, Renfrew, Simcoe, Stormont-Dundas-Glengarry, Victoria, Waterloo, Wellington, and York; and

(2) That portion of Algoma District that includes the City of Sault Ste. Marie and the following townships: Bright, Bright Additional, Cobden, Denis, Garden River First Nation, Indian Reserve #7, Johnson, Korah, Laird, Lefroy, Lewis, Long, MacDonald, Parke, Plummer Additional, Prince, Tarbutt, Tarbutt Additional, Tarentorus, Thessalon, Thompson, Shedden, Spragge, and Striker; and

(3) That portion of Algoma District south of Highway 17 and east of the City of Sault Ste. Marie; and

(4) That portion of Manitoulin District that includes: Cockburn Island, Great Cloche Island, Manitoulin Island, St. Joseph Island, and all Indian Reserves; and

(5) That portion of Nipissing District that includes the City of North Bay; and

(6) That portion of Nipissing District south of the Ottawa and Mattawa rivers; and

(7) That portion of Nipissing District south of highway 17 and west of the City of North Bay; and

(8) That portion of Sudbury District that includes the City of Sudbury and the townships of Baldwin, Dryden, Dunlop, Graham, Hallam, Hymen, Indian Reserves #4, #5, and #6, Lorne, Louise; May, McKim, Nairn, Neelon, Porter, Salter, Shakespeare, Victoria, and Waters; and

(9) That portion of the Sudbury District south of Highway 17.

(d) *Province of Quebec.* (1) That portion of the Province of Quebec that includes the following regional county municipalities: Acton, Arthabaska, Asbestos, Beauce-Sartigan, Beauharnois-Salaberry, Bécancour, Bellechasse, Brome-Missisquoi, Champlain, Coaticook, Communauté Urbaine de Montréal, Communauté Urbaine de L'Outaouais, D'Autray, Desjardins, Deux-Montagnes, Drummond, Francheville, Joliette, L'Amiante, L'Assomption, L'Érable, L'île-d'Orléans, Lajemmerais, Laval, La Nouvelle-Beauce, La Rivière-du-Nord, La Vallée-du-Richelieu, Le Bas-Richelieu, Le Granit, Le Haut-Richelieu, Le Haut-Saint-Francois, Le Haut-Saint-Laurent, Le Haute-Yamaska, Le Val-Saint-Francois, Les Chutes-de-la-Chaudire, Les Collines-de-L'Outaouais, Les Etchemins, Les Jardins-de-Napierville, Les Maskoutains, Les Moulins, Lotbinière, Memphrémagog, Mirabel, Montcalm, Montmagny, Nicolet-Yamaska, Robert-Cliche, Roussillon, Rouville, Sherbrooke, Therese-de Blainville, and Vaudreuil-Soulanges; and

(2) That portion of the regional county municipality of Antoine-Llabelle that includes the following municipalities: Notre-Dame-du-Laus, Notre-Dame-de-Pontmain, and Saint-Aimé-du-Lac-des-Iles; and

(3) That portion of the regional county municipality of Argenteuil that includes the following municipalities: Brownsburg, Calumet, Carillon, Chatham, Grenville, Lachute, Saint-André-d'Argenteuil, and Saint-André-Est; and

(4) That portion of the regional county municipality of Communauté Urbaine De Québec that includes the following municipalities: Cap-Rouge, L'Ancienne-Lorette, Québec, Saint-Augustin-de-Desmaures, Sainte-Foy, Sillery, and Vanier; and

(5) That portion of the regional county municipality of La Vallée-de-la-Gatineau that includes the following municipalities: Denholm, Gracefield, Kazabazua, Lac-Sainte-Marie, Low, Northfield, and Wright; and

(6) That portion of the regional county municipality of Le Centre-de-la-Mauricie that includes the following municipalities: Charette, Notre-Dame-du-Mont-Carmel, Sainte-Elie, Shawinigan, and Shawinigan (Sud); and

(7) That portion of the regional county municipality of Les Laurentides that includes the following municipality: La Conception; and

(8) That portion of the regional county municipality of Les Pays-d'en-Haut that includes the following municipality: Mont-Rolland; and

(9) That portion of the regional county municipality of Maskinongé that includes the following municipalities: Louiseville, Maskinongé, Saint-Joseph-de-Maskinongé, Saint-Barnabé, Saint-Sévère, Saint-Léon-le-Grand, Saint-Paulin, Sainte-Ursule, Saint-Justin, Saint-édouard-de-Maskinongé, Sainte-Angèle-de-Prémont, and Yamachiche; and

(10) That portion of the regional county municipality of Matawinie that includes the following municipalities: Saint-Félix-de-Valois, Saint-Jean-de-Matha, Rawdon, and Chertsey; and

(11) That portion of the regional county municipality of Papineau that includes the following municipalities: Fassett, Lochaber, Lochaber-Partie-Ouest, Mayo, Montebello, Montpellier, Mulgrave-et-Derry, Notre-Dame-de-Bon-Secours-Partie-Nord, Papineauville, Plaisance, Ripon (Village et Canton), Saint-André-Avellin (Village et Paroise), Sainte-Angélique, Saint-Sixte, and Thurso; and

(12) That portion of the regional county municipality of Pontiac that includes the following municipalities: Bristol, Shawville, Clarendon, Portage-du-Fort, Bryson, Campbell's Bay, Grand-Calumet, Litchfield, Thorne, Alleyn-et-Cawood, Leslie-Clapham-et-Huddersfield, Fort-Coulonge, Mansfield-et-Pontefract, Waltham-et-Bryson, L'Isle-aux-Allumettes-Partie-Est, Chapeau, L'Isle-aux-Allumettes, Chichester, Sheen-Esher-Aberdeen-et-Malakoff, and Rapides-des-Joachims; and

(13) That portion of the regional county municipality of Portneuf that includes the following municipalities: Cap-Santé, Deschambault, Donnacona, Grondines, Neuville, and Pointe-aux-Trembles.

[65 FR 38175, June 20, 2000, as amended at 67 FR 59453, Sept. 23, 2002]

§ 319.77–4   Conditions for the importation of regulated articles.

(a) *Trees and shrubs.*[1] (1) Trees without roots (*e.g.*, Christmas trees) may be imported into the United States from any area of Canada without restriction under this subpart if they:

(i) Were greenhouse-grown throughout the year;

(ii) Are destined for a U.S. infested area and will not be moved through any U.S. noninfested areas; or

(iii) Are Christmas trees destined for a U.S. infested area and will not be moved through any U.S. noninfested areas other than noninfested areas in the counties of Aroostock, Franklin, Oxford, Piscataquis, Penobscot, and Somerset, ME (*i.e.*, areas in those counties that are not listed in 7 CFR 301.45–3).

(2) Trees without roots (e.g., Christmas trees) that are destined for a U.S. noninfested area or will be moved through a U.S. noninfested area may be imported into the United States from Canada only under the following conditions:

(i) If the trees originated in a Canadian infested area, they must be accompanied by an officially endorsed Canadian phytosanitary certificate that includes an additional declaration confirming that the trees have been inspected and found free of gypsy moth or that the trees have been treated for gypsy moth in accordance with part 305 of this chapter.

(ii) If the trees originated in a Canadian noninfested area, they must be accompanied by a certification of origin stating that they were produced in an area of Canada where gypsy moth is not known to occur.

(b) *Bark and bark products and logs and pulpwood with bark attached.*[2] (1) Bark and bark products or logs or pulpwood with bark attached that are destined for a U.S. infested area and that will not be moved through any U.S.

---

[1] Trees and Shrubs from Canada may be subject to additional restrictions under "Subpart I—Logs, Lumber, and Other Wood Articles" (§§ 319.40–1 through 319.40–11).

[2] Bark, bark products, and logs from Canada are also subject to restrictions under " Subpart I—Logs, Lumber, and Other Wood Articles" (§§ 319.40 through 319.40–11 of this part).

noninfested area other than noninfested areas in the counties of Aroostock, Franklin, Oxford, Piscataquis, Penobscot, and Somerset, ME (*i.e.*, areas in those counties that are not listed in § 301.45–3 of this chapter) may be imported from any area of Canada without restriction under this subpart.

(2) Bark and bark products or logs or pulpwood with bark attached that are destined for a U.S. noninfested area or will be moved through a U.S. noninfested area may be imported into the United States from Canada only under the following conditions:

(i) If the bark, bark products, logs, or pulpwood originated in a Canadian infested area, they must be either:

(A) Accompanied by an officially endorsed Canadian phytosanitary certificate that includes an additional declaration confirming that they have been inspected and found free of gypsy moth or that they have been treated for gypsy moth in accordance with part 305 of this chapter; or

(B) Destined for a specified U.S. processing plant or mill under compliance agreement with the Animal and Plant Health Inspection Service for specified handling or processing.

(ii) If the bark, bark products, logs, or pulpwood originated in a Canadian noninfested area, they must be accompanied by a certification of origin stating that they were produced in an area of Canada where gypsy moth is not known to occur.

(o) *Outdoor household articles and mobile homes and their associated equipment.* (1) Outdoor household articles and mobile homes and their associated equipment that are destined for a U.S. infested area and will not be moved through any U.S. noninfested areas may be imported from any area in Canada without restriction under this subpart.

(2) Outdoor household articles and mobile homes and their associated equipment that are being moved from a Canadian noninfested area may be imported into any area of the United States without restriction under this subpart.

(3) Outdoor household articles and mobile homes and their associated equipment that are being moved from a Canadian infested area into a U.S. noninfested area, or that will be moved through a U.S. noninfested area, may be imported into the United States only if they are accompanied by a statement, signed by their owner, stating that they have been inspected and found free of gypsy moth.

(d) *Stone and quarry products.* Stone and quarry products originating in a Canadian infested area may be imported into the United States only if they are destined for an infested area of the United States and will not be moved through any noninfested areas of the United States, and may be moved through the United States if they are moved only through infested areas.

(Approved by the Office of Management and Budget under control number 0579–0142)

[64 FR 45866, Aug. 23, 1999, as amended at 65 FR 38176, June 20, 2000; 69 FR 61589, Oct. 20, 2004; 70 FR 33326, June 7, 2005; 71 FR 40878, July 19, 2006; 83 FR 11866, Mar. 19, 2018; 84 FR 2429, Feb. 7, 2019; 84 FR 29958, June 25, 2019]

§ 319.77–5 **Disposition of regulated articles denied entry.**

Any regulated article that is denied entry into the United States because it does not meet the requirements of this subpart must be promptly safeguarded or removed from the United States. If the article is not promptly safeguarded or removed from the United States, it may be seized, destroyed, or otherwise disposed of in accordance with section 414 of the Plant Protection Act (7 U.S.C. 7714).

[64 FR 45866, Aug. 23, 1999, as amended at 66 FR 21057, Apr. 27, 2001]

PART 322—BEES, BEEKEEPING BY-PRODUCTS, AND BEEKEEPING EQUIPMENT

Subpart A—General Provisions

AUTHORITY: 7 U.S.C. 281; 7 U.S.C. 7701–7772 and 7781–7786; 7 CFR 2.22, 2.80, and 371.3.

SOURCE: 69 FR 61747, Oct. 21, 2004, unless otherwise noted.

Subpart A—General Provisions

§ 322.1   Definitions.

Administrator. The Administrator, Animal and Plant Health Inspection Service, or an individual authorized to act for the Administrator.

Animal and Plant Health Inspection Service (APHIS). The Animal and Plant Health Inspection Service of the United States Department of Agriculture.

Bee. Any member of the superfamily *Apoidea* in any life stage, including germ plasm.

Beekeeping byproduct. Material for use in hives, including, but not limited to, beeswax for beekeeping, pollen for bee feed, or honey for bee feed.

Beekeeping equipment. Equipment used to house and manage bees, including, but not limited to, bee boards, hive bodies, bee nests and nesting material, smokers, hive tools, gloves or other clothing, and shipping containers.

Beekeeping establishment. All of the facilities, including apiaries, honey houses, and other facilities, and land that comprise a proprietor's beekeeping business.

Brood. The larvae, pupae, or postovipositional ova (including embryos) of bees.

Destination State. The State, district, or territory of the United States that is the final destination of imported bees, beekeeping byproducts, or beekeeping equipment.

Germ plasm. The semen and preovipositional ova of bees.

Hive. A box or other shelter containing a colony of bees.

Honeybee. Any live bee of the genus Apis in any life stage except germ plasm.

Inspector. Any employee of the Animal and Plant Health Inspection Service, or other individual authorized by the Administrator to carry out the provisions of this part.

Office International des Epizooties (OIE). The organization in the Food and Agriculture Organization of the United Nations responsible for the International Animal Health Code, which includes a section regarding bee diseases in international trade.

Package bees. Queen honeybees with attendant adult honeybees placed in a shipping container, such as a tube or cage.

Queen. The actively reproducing adult female in a colony of bees.

Slumgum. Residue remaining after the beeswax rendering process. It is composed of beeswax mixed with debris or refuse that accumulates when wax cappings or comb are melted. The residue can include wax moth cocoons, dead bees, bee parts, and other detritus from the colony.

Undesirable species or subspecies of honeybees. Honeybee species or subspecies including, but not limited to, *Apis mellifera scutellata,* commonly known as the African honeybee, and its hybrids; *Apis mellifera capensis,* commonly known as the Cape honeybee; and *Apis cerana,* commonly known as the Oriental honeybee.

United States. The States, District of Columbia, American Samoa, Guam, Northern Mariana Islands, Puerto Rico, and the Virgin Islands of the United States.

§322.2 General requirements for interstate movement and importation.

(a) *Interstate movement.* (1) The following regions of the United States are considered pest-free areas for Varroa mite, tracheal mite, small hive beetle, and African honeybee: Hawaii.

(2) In order to prevent the introduction of Varroa mite, tracheal mite, small hive beetle, and African honeybee into the pest-free areas listed in paragraph (a)(1) of this section, interstate movement of honeybees into those areas is prohibited.

(b) *Importation.* In order to prevent the introduction into the United States of bee diseases and parasites, and undesirable species and subspecies of honeybees:

(1) You may import bees, honeybee germ plasm, and beekeeping byproducts into the United States only in accordance with this part.

(2) You may not import pollen derived from bee colonies and intended for use as bee feed into the United States.

(3)(i) You may not import used beekeeping equipment into the United States, unless that used beekeeping equipment either:

(A) Will be used solely for indoor display purposes and will not come into contact with indigenous bees; or

(B) Consists of bee boards that contain live brood of bees, other than honeybees, from a region listed in §322.4(c).

(ii) New, unused beekeeping equipment is eligible for importation into the United States if it complies with all applicable regulations in this chapter.

(c) *Movements not in compliance.* (1) Any honeybees, honeybee germ plasm, bees other than honeybees, beekeeping byproducts, or used beekeeping equipment not in compliance with this part that are imported into the United States will be either:

(i) Immediately exported from the United States by you at your expense; or

(ii) Destroyed by us at your expense.

(2) Pending exportation or destruction, we will immediately apply any necessary safeguards to the bees, beekeeping byproducts, or used beekeeping equipment to prevent the introduction of bee diseases and parasites, and undesirable species and subspecies of honeybees into the United States.

§322.3 Costs and charges.

We will furnish, without cost, the services of an inspector during normal business hours and at the inspector's places of duty. You will be responsible for all costs and charges arising from inspection outside of normal business hours or away from the inspector's places of duty.[1] You are also responsible for all costs and charges related to any exportation or destruction of shipments, in accordance with §322.2(c)(1). Further, if you import bees or germ plasm into a containment facility for research or processing, you will be responsible for all additional costs and charges associated with the importation.

_____

[1] Information on costs for services of an inspector are contained in part 354 of this chapter.

Subpart B—Importation of Adult Honeybees, Honeybee Germ Plasm, and Bees Other Than Honeybees From Approved Regions

§ 322.4　Approved regions.

(a) *Adult honeybees.* The following regions are approved for the importation of adult honeybees into the continental United States (not including Hawaii) under the conditions of this subpart: Australia, Canada, and New Zealand.

(b) *Honeybee germ plasm.* The following regions are approved for the importation of honeybee germ plasm into the United States under the conditions of this subpart: Australia, Bermuda, Canada, France, Great Britain, New Zealand, and Sweden.

(c) *Bees other than honeybees.* The following regions are approved for the importation of bees other than honeybees into the continental United States (not including Hawaii) under the conditions of this subpart: Canada.

(d) If the name of the region from which you want to import adult honeybees, honeybee germ plasm, or bees other than honeybees into the United States does not appear in paragraphs (a), (b), or (c), respectively, of this section, refer to subpart C of this part, "Importation of Restricted Organisms," for requirements.

(e) For information on approving other regions for the importation of adult honeybees, honeybee germ plasm, or bees other than honeybees into the United States, see § 322.12.

§ 322.5　General requirements.

(a) All shipments of bees and honeybee germ plasm imported into the United States under this subpart must be shipped directly to the United States from an approved region.

(b) *Adult honeybees.* (1) You may import adult honeybees under this subpart only from regions listed in § 322.4(a).

(2) The honeybees must be package bees or queens with attending adult bees.

(c) *Honeybee germ plasm.* You may import honeybee germ plasm under this subpart only from regions listed in § 322.4(b).

(d) *Bees other than honeybees.* (1) You may import live adult bees or live brood and essential nest substrate under this subpart only from regions listed in § 322.4(c).

(2) The live bees or brood must belong to one of the following species:

(i) Bumblebees of the species *Bombus impatiens*;

(ii) Bumblebees of the species *Bombus occidentalis*;

(iii) Alfalfa leafcutter bee (*Megachile rotundata*);

(iv) Blue orchard bee (*Osmia lignaria*); or

(v) Horn-faced bee (*Osmia cornifrons*).

(3) If you want to import species of bees other than those listed in paragraph (d)(2) of this section, refer to subpart C of this part, "Importation of Restricted Organisms," for requirements.

§ 322.6　Export certificate.

Each shipment of bees and honeybee germ plasm arriving in the United States from an approved region must be accompanied by an export certificate issued by the appropriate regulatory agency of the national government of the exporting region.

(a) *Adult honeybees.* (1) For adult honeybees, the export certificate must:

(i) Certify that the hives from which the honeybees in the shipment were derived were individually inspected by an official of the regulatory agency no more than 10 days prior to export;

(ii) Identify any diseases, parasites, or undesirable species or subspecies of honeybee found in the hive during that preexport inspection; and

(iii) Certify that the bees in the shipment were produced in the exporting region and are the offspring of bees or semen also produced in the exporting region.

(2) If the export certificate identifies a bee disease or parasite of concern to the United States, including, but not limited to, Thai sacbrood virus, *Tropilaelaps clareae,* and *Euvarroa sinhai,* or an undesirable species or subspecies of honeybee, including, but not limited to, the Cape honeybee (*Apis mellifera capensis*) and the Oriental honeybee (*Apis cerana*), as occurring in the

hive from which the shipment was derived, we will refuse the shipment's entry into the United States.

(b) *Honeybee germ plasm.* (1) For honeybee germ plasm, the export certificate must:

(i) Certify that the hives from which the germ plasm in each shipment was derived were individually inspected by an official of the regulatory agency no more than 10 days prior to export;

(ii) Identify any diseases, parasites, or undesirable species or subspecies of honeybee found in the hive during that preexport inspection; and

(iii) Certify that the bees in the hives from which the shipment was derived were produced in the exporting region and are the offspring of bees or semen also produced in the exporting region.

(2) If the export certificate identifies a bee disease or parasite of concern to the United States, including, but not limited to, Thai sacbrood virus, *Tropilaelaps clareae,* and *Euvarroa sinhai,* or an undesirable species or subspecies of honeybee, including, but not limited to, the Cape honeybee (*Apis mellifera capensis*) and the Oriental honeybee (*Apis cerana*), as occurring in the hive from which the shipment was derived, we will refuse the shipment's entry into the United States.

(c) *Bees other than honeybees.* For bees other than honeybees, the export certificate must certify that the bees in the shipment were produced in the exporting region and are the offspring of bees or semen also produced in the exporting region.

(Approved by the Office of Management and Budget under control number 0579-0207)

§ 322.7 Notice of arrival.

(a) At least 10 business days prior to the arrival in the United States of any shipment of bees or honeybee germ plasm imported into the United States under this subpart, you must notify APHIS of the impending arrival. Your notification must include the following information:

(1) Your name, address, and telephone number;

(2) The name and address of the receiving apiary;

(3) The name, address, and telephone number of the producer;

(4) The U.S. port where you expect the shipment to arrive. The port must be staffed by an APHIS inspector (see § 322.11);

(5) The date you expect the shipment to arrive at that U.S. port;

(6) The scientific name(s) of the organisms in the shipment;

(7) A description of the shipment (*i.e.,* package bees, queen bees, nest boxes, etc.); and

(8) The total number of organisms you expect to receive.

(b) You must provide the notification to APHIS through one of the following means:

(1) By mail to the Permit Unit, PPQ, APHIS, 4700 River Road Unit 133, Riverdale, MD 20737-1236; or

(2) By facsimile at (301) 734-8700; or

(3) By electronic mail to *Notification@usda.gov, or*

(4) Using a U.S. Government electronic information exchange system or other authorized method.

(Approved by the Office of Management and Budget under control number 0579-0207)

[69 FR 61747, Oct. 21, 2004, as amended at 81 FR 40150, June 21, 2016]

§ 322.8 Packaging of shipments.

(a) *Adult honeybees.* All shipments of adult honeybees imported into the United States under this subpart:

(1) Must be packaged to prevent the escape of any bees or bee pests;

(2) Must not include any brood, comb, pollen, or honey; and

(3) May include sugar water or crystallized sugar (*e.g.,* candy) for use as food during transit.

(b) *Bees other than honeybees*—(1) *Adult bees.* All adult bees other than honeybees imported into the United States must be packaged to prevent the escape of any bees or bee pests.

(2) *Live brood.* For live brood of bees other than honeybees, packages:

(i) Must be securely closed;

(ii) May not include any soil, except for that which is present in nest cells that include developing, immature bees;

(iii) May include only packing materials that were grown or produced in the exporting region and that meet all other applicable requirements of this chapter, such as the regulations pertaining to unmanufactured wood in part 319 of this chapter and the plant pest regulations in part 330 of this chapter; and

(iv) May consist of brood housed in new or used bee boards, provided the bee boards meet all applicable requirements of this part.

§ 322.9 Mailed packages.

(a) If you import a package of honeybees, honeybee germ plasm, or bees other than honeybees under this subpart through the mail or through commercial express delivery, you must mark all sides of the outside of that package with the contents of the shipment, i.e., "Live Bees," "Bee Germ Plasm," or "Live Bee Brood," and the name of the exporting region. The marking must be clearly visible using black letters at least 1 inch in height on a white background.

(b) If you import a package of honeybees, honeybee germ plasm, or bees other than honeybees under this subpart through commercial express delivery, you must provide an accurate description of the complete contents of the shipment, i.e., "Live Bees," "Bee Germ Plasm," or "Live Bee Brood," for the shipment's delivery manifest entry.

(c) In addition to the export certificate required in § 322.6, a package of honeybees, honeybee germ plasm, or bees other than honeybees imported under this subpart by commercial express delivery must be accompanied at the time of arrival in the United States by an invoice or packing list accurately indicating the complete contents of the shipment.

§ 322.10 Inspection; refusal of entry.

(a) Shipments of honeybees, honeybee germ plasm, and bees other than honeybees imported into the United States under this subpart will be inspected at the port of entry in the United States for:

(1) Proper documentation (see § 322.6);

(2) Timely notice of arrival (see § 322.7); and

(3) Adequate packaging (see § 322.8).

(b) If, upon inspection, any shipment fails to meet the requirements of this part, that shipment will be refused entry into the United States. In accordance with § 322.2(c), the inspector will offer you, or in your absence the

shipper, the opportunity to immediately export any refused shipments. If you, or in your absence the shipper, decline to immediately export the shipment, we will destroy the shipment at your expense.

§ 322.11 Ports of entry.

Shipments of honeybees, honeybee germ plasm, and bees other than honeybees imported under this subpart may enter the United States only at a port of entry staffed by an APHIS inspector.[2]

§ 322.12 Risk assessment procedures for approving countries.

(a) The national government of the region wishing to export must request that we perform a risk assessment for the importation into the United States of honeybees, honeybee germ plasm, or bees other than honeybees from that region.

(b) When we receive a request, we will evaluate the science-based risks associated with such importation. Our risk assessment will be based on information provided by the exporting region, information from topical scientific literature, and, if applicable, information we gain from a site visit to the exporting region. The risk assessment will include:

(1) Identification of all bee diseases, including fungi, bacteria, viruses, mycoplasmas, and protozoa, that occur in the exporting region but not in the United States or that are listed as significant for international trade by the Office International des Epizooties (OIE);

(2) Identification of all bee parasites, including mites, that occur in the exporting region but not in the United States or that are listed as significant for international trade by the OIE;

(3) Identification of all species and subspecies of honeybees that occur in the exporting region but not in the United States or that are listed as significant for international trade by the OIE, if applicable;

---

[2] To find out if a specific port is staffed by an APHIS inspector, or for a list of ports staffed by APHIS inspectors, contact Permit Unit, PPQ, APHIS, 4700 River Road Unit 133, Riverdale, MD 20737–1236; toll-free (877) 770–5990; fax (301) 734–8700.

(4) Identification of all pests of bee culture, such as the small hive beetle, that occur in the exporting region but not in the United States or that are listed as significant for international trade by the OIE;

(5) Evaluation of the probability of establishment, including pathway, entry, colonization, and spread potentials, of any diseases, parasites, undesirable species or subspecies of honeybees, or pests identified in accordance with paragraphs (b)(1), (2), (3), or (4) of this section;

(6) Evaluation of the potential consequences of establishment, including economic, environmental, and perceived social and political effects, of each disease, parasite, undesirable species or subspecies of honeybees, or pest identified in accordance with paragraphs (b)(1), (2), (3), or (4) of this section; and

(7) Consideration of the effectiveness of the regulatory system of the exporting region to control bee diseases, parasites, undesirable species and subspecies of honeybees, and pests that occur there and to prevent occurrences of new bee diseases, parasites, undesirable species and subspecies of honeybees, and pests.

(c) Based on the conclusions of the risk assessment, we will either:

(1) Publish in the FEDERAL REGISTER a notice of proposed rulemaking to allow honeybees, honeybee germ plasm, or bees other than honeybees to be imported into the United States from that region; or

(2) Deny the request in writing, stating the specific reasons for that action.

(d) We will publish a notice of availability of all completed risk assessments for public comment.

(Approved by the Office of Management and Budget under control number 0579–0207)

Subpart C—Importation of Restricted Organisms

§322.13 General requirements; restricted organisms.

(a) For the purposes of this part, the following are restricted organisms:

(1) Honeybee brood in the comb;

(2) Adult honeybees from any region other than those listed in §322.4(a);

(3) Honeybee germ plasm from any region other than those listed in §322.4(b); and

(4) Bees other than honeybees, in any life stage, from any region other than those listed in §322.4(c) or any species of bee other than those listed in §322.5(d)(2).

(b) Persons importing restricted organisms into the United States must be Federal, State, or university researchers; be at least 18 years of age; and be physically present during normal business hours at an address within the United States specified on the permit during any periods when articles are being imported or moved interstate under the permit. All such importations must be for research or experimental purposes and in accordance with this part.

[69 FR 61747, Oct. 21, 2004, as amended at 79 FR 19811, Apr. 10, 2014]

§322.14 Documentation; applying for a permit to import a restricted organism.

Any restricted organism imported into the United States must be accompanied by both a permit, in accordance with paragraph (a) of this section, and an invoice or packing list accurately indicating the complete contents of the shipment, in accordance with paragraph (b) of this section.

(a) *Permit.* You must submit a completed application for a permit to import restricted organisms at least 30 days prior to scheduling arrival of those organisms. You may import a restricted organism only if we approve your application and issue you a permit. Our procedures for reviewing permit applications are provided in §322.15. To apply for a permit, you must supply, either on a completed PPQ Form 526 or in some other written form, the following information: [3]

(1) *Applicant information.* Your name, title, organization, address, telephone

---

[3] Mail your completed application to Permit Unit, PPQ, APHIS, 4700 River Road Unit 133, Riverdale, MD 20737–1236. A PPQ Form 526 may be obtained by writing to the same address, calling toll-free (877) 770–5990, faxing your request to (301) 734–8700, or downloading the form from *http://www.aphis.usda.gov/ppq/ss/permits/pests/.*

number, facsimile number, and electronic mail address (provide all that are applicable).

(2) *Application type.* New permit, permit renewal, or amendment to existing permit (if a renewal or amendment, provide the current permit number).

(3) *Type of movement.* Select or write "Import into the United States."

(4) *Scientific name of organism.* Genus, species, subspecies or strain, and author (if known).

(5) *Type of organism.* Select or write "Bees and/or bee germ plasm."

(6) *Taxonomic classification.* Family of restricted organisms.

(7) *Life stage(s).* Semen, preovipositional eggs, embryos, postovipositional eggs, larvae, pupae, or adults. If adult queens, please specify.

(8) Number of shipments.

(9) Number of specimens per shipment.

(10) Is the organism established in the United States?

(11) Is the organism established in the destination State?

(12) Media or species of host material accompanying the organism (*e.g.*, pollen, honey, wax, nesting material).

(13) *Source of organism (include any that apply, and list region of origin).* Supplier (provide supplier's name and address), wild collected, or reared under controlled conditions.

(14) *Method of shipment.* Airmail, express delivery (list company name).

(15) Port(s) of entry.

(16) Approximate date(s) of arrival at the port of entry.

(17) *Destination.* Provide the address of the location where the organism will be received and maintained, including building and room numbers where applicable.

(18) *Intended use (include any that apply).* Select or write "Scientific Study."

(19) Has your facility been evaluated by APHIS? If yes, list date(s) of approval. Is your facility approved for the species of bees or bee germ plasm for which you are seeking a permit?

(20) Provide your signature and the date of your signature under the following certification: "I certify that all statements and entries I have made on this document are true and accurate to the best of my knowledge and belief. I understand that any intentional false statement or misrepresentation made on this document is a violation of law and punishable by a fine of not more than $10,000, or imprisonment, or not more than 5 years, or both. (18 U.S.C. 1001)." If you are required to have a sponsor for your permit application, your sponsor must also sign and date under the same certification.

(b) *Invoice.* Any restricted organism must be accompanied at the time of arrival in the United States by an invoice or packing list accurately indicating the complete contents of the shipment and the exporting region.

(Approved by the Office of Management and Budget under control number 0579–0207)

[69 FR 61747, Oct. 21, 2004, as amended at 79 FR 19811, Apr. 10, 2014]

§ 322.15 APHIS review of permit applications; denial or revocation of permits.

(a) *Review of permit applications to import restricted organisms*—(1) *Consultation.* During our review of your permit application, we may consult with any Federal officials; appropriate officials of any State, Territory, or other jurisdiction in the United States in charge of research or regulatory programs relative to bees; and any other qualified governmental or private research laboratory, institution, or individual. We will conduct these consultations to gain information on the risks associated with the importation of the restricted organisms.

(2) *Review by destination State.* We will transmit a copy of your permit application, along with our anticipated decision on the application, to the appropriate regulatory official in the destination State for review and recommendation. A State's response, which we will consider before taking final action on the permit application, may take one of the following forms:

(i) The State recommends that we issue the permit;

(ii) The State recommends that we issue the permit with specified additional conditions;

(iii) The State recommends that we deny the permit application and provides scientific, risk-based reasons supporting that recommendation; or

(iv) The State makes no recommendation, thereby concurring with our decision regarding the issuance of the permit. [4]

(b) *Results of review.* After a complete review of your application, we will either:

(1) Issue you a written permit with, if applicable, certain specific conditions listed for the importation of the restricted organisms you applied to import. You must initial each condition on the proposed permit and return the proposed permit conditions to the Permit Unit before we will issue you a signed valid permit; or

(2) Notify you that your application has been denied and provide reasons for the denial.

(c) *Denial of permit applications.* APHIS will deny an application for a permit to import a restricted organism regulated under this subpart when, in its opinion, such movement would involve a danger of dissemination of an exotic bee disease or parasite, or an undesirable species or subspecies of honeybee. Danger of such dissemination may be deemed to exist when:

(1) Existing safeguards against dissemination are inadequate and no adequate safeguards can be arranged; or

(2) The potential for disseminating an exotic bee disease or parasite, or an undesirable species or subspecies of honeybee, with the restricted organism outweighs the probable benefits that could be derived from the proposed movement and use of the restricted organism; or

(3) When you, as a previous permittee, failed to maintain the safeguards or otherwise observe the conditions prescribed in a previous permit and have failed to demonstrate your ability or intent to observe them in the future; or

(4) The proposed movement of the restricted organism is adverse to the conduct of an eradication, suppression, control, or regulatory program of APHIS.

(5) APHIS may also deny a permit to import restricted organisms:

[4] If a State regulatory official does not respond within 20 business days, we will conclude that the State has chosen to make no recommendation regarding the issuance of the permit.

(i) To a person who has previously failed to comply with any APHIS regulation, except:

(A) A permit revoked in an investigation concerning that failure has been reinstated on appeal, at the discretion of APHIS; or

(B) All measures ordered by APHIS to correct the failure, including but not limited to, payment of penalties or restitution, have been complied with to the satisfaction of APHIS.

(ii) To a person who has previously failed to comply with any international or Federal regulation or instruction concerning the importation of prohibited or restricted foreign agricultural products; or

(iii) If the application for a permit contains information that is found to be materially false, fraudulent, deceptive, or misrepresentative.

(d) *Cancellation of permits.* (1) APHIS may cancel any outstanding permit whenever:

(i) We receive information subsequent to the issuance of the permit of circumstances that would constitute cause for the denial of an application for permit under paragraph (c) of this section; or

(ii) You, as the permittee, fail to maintain the safeguards or otherwise observe the conditions specified in the permit or in any applicable regulations.

(2) Upon cancellation of a permit, you must either:

(i) Surrender all restricted organisms to an APHIS inspector; or

(ii) Destroy all restricted organisms under the supervision of an APHIS inspector.

(e) *Appealing the denial of permit applications or revocation of permits.* If your permit application has been denied or your permit has been revoked, APHIS will inform you in writing, including by electronic methods, as promptly as circumstances permit and will include the reasons for the denial or revocation. You may appeal the decision by writing to APHIS within 10 business days from the date you received the communication notifying you of the denial or revocation of the permit. Your appeal must state all facts and reasons upon which you are relying to show that your permit application was wrongfully denied or your permit was wrongfully revoked. APHIS will grant or deny the appeal in writing and will state in writing the reason for the decision. The denial or revocation will remain in effect during the resolution of the appeal.

(Approved by the Office of Management and Budget under control number 0579–0207)

[69 FR 61747, Oct. 21, 2004, as amended at 79 FR 19811, Apr. 10, 2014]

§ 322.16 Packaging of shipments.

(a) Restricted organisms must be packed in a container or combination of containers that will prevent the escape of the organisms and the leakage of any contained materials. The container must be sufficiently strong to prevent it from rupturing or breaking during shipment.

(b) The outer container must be clearly marked with the contents of the shipment, i.e., either "Live Bees," "Bee Germ Plasm," or "Live Bee Brood," and the name of the region of origin.

(c) Only approved packing materials may be used in a shipment of restricted organisms.

(1) The following materials are approved as packing materials: Absorbent cotton or processed cotton padding free of cottonseed; cages made of processed wood; cellulose materials; excelsior; felt; ground peat (peat moss); paper or paper products; phenolic resin foam; sawdust; sponge rubber; thread waste, twine, or cord; and vermiculite.

(2) Other materials, such as host material for the organism, soil, or other types of packing material, may be included in a container only if identified in the permit application and approved by APHIS on the permit.

§ 322.17 Mailed packages.

(a) If you import a restricted organism through the mail or through commercial express delivery, you must attach a special mailing label (APHIS Form 599), which APHIS will provide with your permit, to the package or container. The mailing label indicates that APHIS has authorized the shipment.

(b) You must address the package containing the restricted organism to the containment facility or apiary identified on the permit (post office boxes are not allowed).

(c) If the restricted organism arrives in the mail without the mailing label described in paragraph (a) of this section or addressed to a containment facility or apiary other than the one listed on the permit, an inspector will refuse to allow the organism to enter the United States.

§ 322.18 Restricted organisms in a commercial vehicle arriving at a land border port in the United States.

(a) If you import a restricted organism through a land border port in the United States by commercial vehicle (i.e., automobile or truck), then the person carrying the restricted organism must present the permit required by § 322.14 and an invoice or packing slip accurately indicating the complete contents of the shipment to the inspector at the land border port.

(b) The restricted organisms must be surrendered at the port of entry and can continue on to the destination identified on the permit only by a bonded carrier (commercial express delivery).

(c) If you fail to present a copy of the permit and an invoice or packing list accurately indicating the complete contents of the shipment at the port of entry, an inspector will refuse the organism's entry to the United States or confiscate and destroy the refused material.

(Approved by the Office of Management and Budget under control number 0579–0207)

§ 322.19 Inspection; refusal of entry.

(a) APHIS may inspect any restricted organism at the time of importation to determine if the organism meets all of the requirements of this part.

(b) If, upon inspection, any shipment fails to meet the requirements of the regulations, that shipment will be refused entry into the United States. In accordance with § 322.2(c), the inspector will offer the shipper the opportunity to immediately export any refused shipments. If the shipper declines to immediately export the shipment, we will destroy the shipment at his or her expense.

§322.20 Ports of entry.

A restricted organism may be imported only at a port of entry staffed by an APHIS inspector.[5] After a restricted organism has been cleared for importation at the port of entry, the organism can only be transported by a bonded commercial carrier immediately and directly from the port of entry to the containment facility or apiary identified on the permit. You may open the package containing the restricted organism only within the containment facility or apiary identified on the permit.

(Approved by the Office of Management and Budget under control number 0579–0207)

§322.21 Post-entry handling.

(a) Immediately following clearance at the port of entry, a restricted organism must move by a bonded commercial carrier directly to a containment facility or apiary that has been inspected and approved by APHIS.[6] We must inspect and approve the containment facility or apiary before we will issue a permit to import a restricted organism.

(b) *Inspection of premises.* Prior to issuing a permit to import restricted organisms, we will inspect the apiary or containment facility where you intend to contain the restricted organisms. In order to approve the apiary or containment facility, an inspector must determine that adequate safeguards are in place to prevent the release of diseases or parasites of bees, or of undesirable species or strains of honeybees. We will use the following criteria to determine whether adequate safeguards are in place:

(1) *Enclosed containment facilities.* (i) Will the facility's entryways, windows, and other structures, including water, air, and waste handling systems, con-

tain the restricted organisms, parasites and pathogens, and prevent the entry of other organisms and unauthorized visitors?

(ii) Does the facility have operational and procedural safeguards in place to prevent the escape of the restricted organisms, parasites, and pathogens, and to prevent the entry of other organisms and unauthorized visitors?

(iii) Does the facility have a means of inactivating or sterilizing restricted organisms and any breeding materials, pathogens, parasites, containers, or other material?

(2) *Containment apiaries.* (i) Is the apiary located in an area devoid of indigenous bees and sufficiently isolated to prevent contact between indigenous bees and imported restricted organisms? Is the area extending from the apiary to the nearest indigenous bees constantly unsuitable for foraging individuals of the imported restricted organisms?

(ii) Does the apiary have sufficient physical barriers to prevent the entry of unauthorized visitors?

(iii) Does the apiary have operational and procedural safeguards in place to prevent the escape of the restricted organisms, parasites, and pathogens, and to prevent the entry of other organisms and unauthorized visitors?

(iv) Does the apiary have a means of inactivating or sterilizing restricted organisms, and any hives, wax, pathogens, parasites, containers, or other materials?

(3) Containment apiaries for honeybees resulting from germ plasm imported from nonapproved regions.

(i) Does the apiary have sufficient physical barriers to prevent the entry of unauthorized visitors?

(ii) Are there sufficient physical barriers (*e.g.*, excluders) in hives in the apiary to prevent the escape of all adult queen and drone honeybees resulting from the germ plasm?

(iii) Does the apiary have operational and procedural safeguards in place to prevent the escape of all queen and drone honeybees resulting from the germ plasm?

(iv) Does the apiary have a means of destroying colonies of honeybees with undesirable characteristics that may result from imported germ plasm?

---

[5] To find out if a specific port is staffed by an APHIS inspector, or for a list of ports staffed by APHIS inspectors, contact Permit Unit, PPQ, APHIS, 4700 River Road Unit 133, Riverdale, MD 20737–1236; toll-free (877) 770–5990; fax (301) 734–8700.

[6] For a list of approved facilities, or to arrange to have a facility inspected by APHIS, contact Permit Unit, PPQ, APHIS, 4700 River Road Unit 133, Riverdale, MD 20737–1236; toll-free (877) 770–5990.

(c) *Holding in containment.* (1) If we issue a permit for importing restricted organisms into an approved containment facility or apiary, you may not remove or release the restricted organisms, or the progeny or germ plasm resulting from the restricted organisms, from the apiary or facility without our prior approval.

(2) You must allow us to inspect the apiary or facility and all documents associated with the importation or holding of restricted organisms at any time to determine whether safeguards are being maintained to prevent the release of the restricted organisms, their progeny and germ plasm, parasites, and pathogens.

(3) You must inform us immediately, but no later than 24 hours after detection, if restricted organisms escape from the facility

(d) *Release from containment apiary or facility.* (1) After rearing the restricted organisms in an approved containment facility or apiary through at least 4 months of active reproduction with no evidence of nonindigenous parasites or pathogens or of undesirable characteristics, you may submit a request to us for the release of the bees. The request must include:

(i) Inspection protocols;

(ii) Inspection frequencies;

(iii) Names and titles of inspectors;

(iv) Complete information, including laboratory reports, on detection of diseases and parasites in the population;

(v) Complete notes and observations on behavior, such as aggressiveness and swarming; and

(vi) Any other information or data relating to bee diseases, parasites, or adverse species or subspecies.

(2) Mail your request for release to the Permit Unit, PPQ, APHIS, 4700 River Road Unit 133, Riverdale, MD 20737–1236, or fax to (301) 734–8700.

(3) When we receive a complete request for release from containment, we will evaluate the request and determine whether the bees may be released. Our evaluation may include an environmental assessment or environmental impact statement prepared in accordance with the National Environmental Policy Act. We may conduct an additional inspection of the bees during our evaluation of the request. You will receive a written statement as soon as circumstances allow that approves or denies your request for release of the bees.

(Approved by the Office of Management and Budget under control number 0579–0207)

Subpart D—Transit of Restricted Organisms Through the United States

§ 322.22   General requirements.

(a) You may transit restricted organisms from any region through the United States to another region only in accordance with this part. For a list of restricted organisms, see § 322.13(a).

(b) You may ship restricted organisms only aboard aircraft to the United States for transit to another country.

(c) You may transload a shipment of restricted organisms only once during the shipment's entire transit through the United States and only at an airport in the continental United States. You may not transload restricted organisms in Hawaii. In Hawaii, the restricted organisms must remain on, and depart for another destination aboard, the same aircraft on which the shipment arrived at the Hawaiian airport.

§ 322.23   Documentation.

Each shipment of restricted organisms transiting the United States must be accompanied by a document issued by the appropriate regulatory authority of the national government of the region of origin stating that the shipment has been inspected and determined to meet the packaging requirements in § 322.24.

§ 322.24   Packaging of transit shipments.

(a) Restricted organisms transiting the United States must be packaged in securely closed and completely enclosed containers that prevent the escape of organisms and the leakage of any contained materials. The container must be sufficiently strong and durable to prevent it from rupturing or breaking during shipment.

(b) In addition to the requirements in paragraph (a) of this section, each pallet of cages containing honeybees transiting the United States must be covered by an escape-proof net that is secured to the pallet so that no honeybees can escape from underneath the net.

(c) The outside of the package must be clearly marked with the contents of the transit shipment, *i.e.*, either "Live Bees," "Bee Germ Plasm," or "Live Bee Brood," and the name of the exporting region.

§322.25 Notice of arrival.

At least 2 business days prior to the expected date of arrival of restricted organisms at a port in the continental United States for in-transit movement, you or your shipper must contact the port to give the following information:

(a) The name of each U.S. airport where the shipment will arrive;

(b) The name of the U.S. airport where the shipment will be transloaded (if applicable);

(c) The date of the shipment's arrival at each U.S. airport;

(d) The date of the shipment's departure from each U.S. airport;

(e) The names, phone numbers, and addresses of both the shipper and receiver;

(f) The number of units in the shipment (*i.e.*, number of queens or number of cages of package bees); and

(g) The name of the airline carrying the shipment.

(Approved by the Office of Management and Budget under control number 0579–0207)

§322.26 Inspection and handling.

(a) All shipments of restricted organisms transiting the United States are subject to inspection at the port in the United States for compliance with this part. If, upon inspection, a transit shipment of restricted articles is found not to meet the requirements of this part, we will destroy the shipment at your expense.

(b) *Transloading*—(1) *Adult bees.* You may transload adult bees from one aircraft to another aircraft at the port of arrival in the United States only under the supervision of an inspector. If the adult bees cannot be transloaded immediately to the subsequent flight, you

must store them within a completely enclosed building. Adult bees may not be transloaded from an aircraft to ground transportation for subsequent movement through the United States.

(2) *Bee germ plasm.* You may transload bee germ plasm from one aircraft to another at the port of arrival in the United States only under the supervision of an inspector.

§322.27 Eligible ports for transit shipments.

You may transit restricted organisms only through a port of entry staffed by an APHIS inspector.[7]

Subpart E—Importation and Transit of Restricted Articles

§322.28 General requirements; restricted articles.

(a) The following articles from any region are restricted articles:

(1) Dead bees of any genus;

(2) Beeswax for beekeeping; and

(3) Honey for bee feed.

(b) Restricted articles may only be imported into or transit the United States in accordance with this part.

§322.29 Dead bees.

(a) Dead bees imported into or transiting the United States must be either:

(1) Immersed in a solution containing at least 70 percent alcohol or a suitable fixative for genetic research;

(2) Immersed in liquid nitrogen; or

(3) Pinned and dried in the manner of scientific specimens.

(b) Dead bees are subject to inspection at the port of entry in the United States to confirm that the requirements of paragraph (a) of this section have been met.

§322.30 Export certificate.

Each shipment of restricted articles, except for dead bees, imported into or transiting the United States must be accompanied by an export certificate

---

[7] To find out if a specific port is staffed by an APHIS inspector, or for a list of ports staffed by APHIS inspectors, contact Permit Unit, PPQ, APHIS, 4700 River Road Unit 133, Riverdale, MD 20737–1236; toll-free (877) 770–5990; fax (301) 734–8700.

issued by the appropriate regulatory agency of the national government of the exporting region. The export certificate must state that the articles in the shipment have been treated as follows:

(a) *Beeswax.* Must have been liquefied, and slumgum and honey must be removed.

(b) *Honey for bee feed.* Heated to 212 °F (100 °C) for 30 minutes.

(Approved by the Office of Management and Budget under control number 0579–0207)

§ 322.31 Notice of arrival.

(a) At least 10 business days prior to the arrival in the United States of any shipment of restricted articles, you must notify APHIS of the impending arrival. Your notification must include the following information:

(1) Your name, address, and telephone number;

(2) The name and address of the recipient of the restricted articles;

(3) The name, address, and telephone number of the producer;

(4) The date you expect to receive the shipment;

(5) A description of the contents of the shipment (*i.e.*, dead bees, honey for bee feed, etc.); and

(6) The total number of restricted articles you expect to receive.

(b) You must provide the notification to APHIS through one of the following means:

(1) By mail to the Permit Unit, PPQ, APHIS, 4700 River Road Unit 133, Riverdale, MD 20737–1236; or

(2) By facsimile at (301) 734–8700; or

(3) By electronic mail to *Notification@usda.gov, or*

(4) Using a U.S. Government electronic information exchange system or other authorized method.

(Approved by the Office of Management and Budget under control number 0579–0207)

[69 FR 61747, Oct. 21, 2004, as amended at 81 FR 40150, June 21, 2016]

§ 322.32 Mailed packages.

(a) If you import a restricted article through the mail or through commercial express delivery, you must mark all sides of the outside of that package with the contents of the shipment and the name of the exporting region. The marking must be clearly visible using black letters at least 1 inch in height on a white background.

(b) If you import a restricted article through commercial express delivery, you must provide an accurate description of the complete contents of the shipment for the shipment's delivery manifest entry.

(c) In addition to the export certificate required in § 322.30 (if applicable), a restricted article that is imported by mail or commercial express delivery must be accompanied by an invoice or packing list accurately indicating the complete contents of the shipment.

(Approved by the Office of Management and Budget under control number 0579–0207)

§ 322.33 Restricted articles in a commercial bonded vehicle arriving at a land border port in the United States.

If you import a restricted article through a land border port in the United States by commercial vehicle (*i.e.*, automobile or truck), then the person carrying the package containing the restricted article or the driver of the vehicle must present the export certificate required by § 322.30 (if applicable) and an invoice or packing slip accurately indicating the complete contents of the shipment to the inspector at the land border port.

§ 322.34 Inspection; refusal of entry.

(a) You must present shipments of restricted articles to the inspector at the port of entry in the United States. Shipments of restricted articles must remain at the port of entry until released by the inspector.

(b) The inspector at the port will confirm that all shipments of restricted articles have proper documentation (see § 322.30) and that you provided notice of arrival for all shipments of restricted articles (see § 322.32).

(c) If, upon inspection, any shipment fails to meet the requirements of this part, that shipment will be refused entry into the United States. In accordance with § 322.2(c), the inspector will offer you, or in your absence the shipper, the opportunity to immediately export any refused shipments, or confiscate and destroy the refused shipments.

(Approved by the Office of Management and Budget under control number 0579–0207)

§322.35 Ports of entry.

A restricted article may be imported only at a port of entry staffed by an APHIS inspector. To find out if a specific port is staffed by an APHIS inspector, or for a list of ports staffed by APHIS inspectors, contact Permit Unit, PPQ, APHIS, 4700 River Road Unit 133, Riverdale, Maryland 20737–1236; toll-free (877) 770–5990; fax (301) 734–8700.

PART 330—FEDERAL PLANT PEST REGULATIONS; GENERAL; PLANT PESTS, BIOLOGICAL CONTROL ORGANISMS, AND ASSOCIATED ARTICLES; GARBAGE

Subpart A—General Provisions

AUTHORITY: 7 U.S.C. 1633, 7701–7772, 7781–7786, and 8301–8317; 21 U.S.C. 136 and 136a; 31 U.S.C. 9701; 7 CFR 2.22, 2.80, and 371.3.

SOURCE: 24 FR 10825, Dec. 29, 1959, unless otherwise noted.

Subpart A—General Provisions

SOURCE: Redesignated at 84 FR 2429, Feb. 7, 2019.

§330.100 Definitions.

The following terms, when used in this part, shall be construed, respectively, to mean:

Administrative instructions. Published documents relating to the enforcement of this part, and issued under authority thereof by the Administrator.

Administrator. The Administrator of the Animal and Plant Health Inspection Service (APHIS), United States Department of Agriculture, or any employee of APHIS to whom authority has been delegated to act in the Administrator's stead.

Animal and Plant Health Inspection Service (APHIS). The Animal and Plant Health Inspection Service of the United States Department of Agriculture.

Article. Any material or tangible object, including a living organism, that could harbor living plant pests or noxious weeds. The term includes associated articles such as soil and packaging.

Biocontainment facility. A physical structure or portion thereof, constructed and maintained in order to contain plant pests, biological control organisms, or associated articles.

Biological control organism. Any enemy, antagonist, or competitor used to control a plant pest or noxious weed.

Continental United States. The contiguous 48 States, Alaska, and the District of Columbia.

Continued curation permit. A permit issued prior to the expiration date for an import permit or interstate movement permit in order for a permittee to continue research or other actions listed on the import or interstate movement permit. Continued curation permits do not allow acquisition of additional organisms for research and other authorized activities and only address retention of existing organisms for authorized uses.

Department. The United States Department of Agriculture.

Deputy Administrator. The Deputy Administrator of the Plant Protection and Quarantine Programs or any employee of the Plant Protection and Quarantine Programs delegated to act in his or her stead.

Enter (entry). To move into, or the act of movement into, the commerce of the United States.

EPA. The Environmental Protection Agency of the United States.

Export (exportation). To move from, or the act of movement from, the United States to any place outside the United States.

Garbage. That material designated as "garbage" in § 330.400(b).

Hand-carry. Importation of an organism that remains in one's personal possession and in close proximity to one's person.

Import (importation). To move into, or the act of movement into, the territorial limits of the United States.

Inspector. Any individual authorized by the Administrator of APHIS or the Commissioner of U.S. Customs and Border Protection to enforce the regulations in this part.

Interstate movement. Movement from one State into or through any other State; or movement within the District of Columbia, Guam, the U.S. Virgin Islands, or any other territory or possession of the United States.

Living. Viable or potentially viable.

Means of conveyance. Any personal or public property used for or intended for use for the movement of any other property. This specifically includes, but is not limited to, automobiles, trucks, railway cars, aircraft, boats, freight containers, and other means of transportation.

Move (moved and movement). To carry, enter, import, mail, ship, or transport; to aid, abet, cause, or induce the carrying, entering, importing, mailing, shipping, or transporting; to offer to carry, enter, import, mail, ship, or transport; to receive to carry, enter, import, mail, ship, or transport; to release into the environment, or to allow any of those activities.

Noxious weed. Any plant or plant product that can directly or indirectly injure or cause damage to crops (including nursery stock or plant products), livestock, poultry, or other interests of agriculture, irrigation, navigation, the natural resources of the United States, the public health, or the environment.

Owner. The owner, or his or her agent, having possession of a plant pest, biological control organism, associated article, or any other means of conveyance, products, or article subject to the regulations in this part.

Permit. A written authorization, including by electronic methods, by the Administrator to move plant pests, biological control organisms, or associated articles under conditions prescribed by the Administrator.

Permittee. The person to whom APHIS has issued a permit in accordance with this part and who must comply with the provisions of the permit and the regulations in this part.

Person. Any individual, partnership, corporation, association, joint venture, or other legal entity.

Plant. Any plant (including any plant part) for or capable of propagation including trees, tissue cultures, plantlet cultures, pollen, shrubs, vines, cuttings, grafts, scions, buds, bulbs, roots, and seeds.

Plant pest. Any living stage of any of the following that can directly or indirectly injure, cause damage to, or cause disease in any plant or plant product: A protozoan, nonhuman animal, parasitic plant, bacterium, fungus, virus or viroid, infectious agent or other pathogen, or any article similar to or allied with any of the foregoing.

Plant product. Any flower, fruit, vegetable, root, bulb, seed, or other plant part that is not included in the definition of plant; or any manufactured or processed plant or plant part.

Plant Protection and Quarantine Programs. The Plant Protection and Quarantine Programs of the Animal and Plant Inspection Health Service.

Pure culture. A single species of invertebrate originating only from an identified/described population and free of disease and parasites, cryptic species, soil and other biological material except host material and substrate as APHIS deems appropriate. Examples of identified/described population are those originating from a specific laboratory colony or field collection from a specified geographic area, such as an entire country or States or provinces of a country.

Regulated garbage. That material designated as regulated garbage in §330.400(c) and (d).

Responsible individual. One or more individuals who a permittee designates to appropriately oversee and control the staff, facilities, and/or site(s) at the location(s) specified on the permit as the ultimate destination of the plant pest, biological control organism, or associated article, to ensure compliance with the permit conditions during all phases of the activities being performed with the regulated articles authorized under a permit issued in accordance with this part for the movement or curation of a plant pest, biological control organism, or associated article. For the duration of the permit, the individual(s) must serve as a primary contact for communication with APHIS. The permittee may designate him or herself as the responsible individual. The responsible individual(s) must be at least 18 years of age and to be able meet with and provide information to an APHIS representative within a reasonable time frame. In accordance with section 7734 of the Plant Protection Act (7 U.S.C. 7701 *et seq.*), the act, omission, or failure of any responsible individual will also be deemed the act, omission, or failure of a permittee.

Secure shipment. Shipment of a regulated plant pest, biological control organism, or associated article in a container or a means of conveyance of sufficient strength and integrity to prevent leakage of contents and to withstand shocks, pressure changes, and other conditions incident to ordinary handling in transportation.

Shelf-stable. The condition achieved in a product, by application of heat, alone or in combination with other ingredients and/or other treatments, of being rendered free of microorganisms capable of growing in the product at nonrefrigerated conditions (over 50 °F or 10 °C).

Soil. The unconsolidated material from the earth's surface that consists of rock and mineral particles and that supports or is capable of supporting biotic communities.

State. Any of the States of the United States, the Commonwealth of the Northern Mariana Islands, the Commonwealth of Puerto Rico, the District of Columbia, Guam, the U.S. Virgin Islands, and all other territories or possessions of the United States.

Sterilization (sterile, sterilized). A chemical or physical process that results in the death of all living organisms on or within the article subject to the process. Examples include, but are not limited to, autoclaving and incineration.

Taxon (taxa). Any recognized grouping or rank within the biological nomenclature of organisms, such as class, order, family, genus, species, subspecies, pathovar, biotype, race, forma specialis, or cultivar.

Transit. Movement from and to a foreign destination through the United States.

United States. All of the States and territories.

U.S. Customs and Border Protection (CBP). U.S. Customs and Border Protection within the Department of Homeland Security.

[84 FR 29958, June 25, 2019]

§330.101 Policy.

The purpose of the regulations in this part is to prevent the dissemination of plant pests into the United States, or interstate, by regulating the movement of plant pests into or through the United States, or interstate, and the movement of means of conveyance, earth, stone and quarry products, garbage, and certain other products and articles into or through the United States, or from any Territory or possession into or through any other Territory or possession or the continental United States. The Deputy Administrator shall employ procedures to carry out this purpose which will impose a minimum of impediment to foreign commerce and travel whenever practicable, consistent with proper precaution against plant pest dissemination. The same policy is to be applied in the case of interstate commerce and travel.

§ 330.102 Basis for certain regulations.

Under the authority of the Plant Protection Act, the Secretary may prohibit or restrict the importation, entry, exportation, or movement in interstate commerce of any plant, plant product, biological control organism, noxious weed, article (including baggage, mail, garbage, earth, stone, and quarry products) or means of conveyance if such actions are necessary to prevent the introduction into or the dissemination within the United States of a plant pest or noxious weed.

[66 FR 21058, Apr. 27, 2001]

§ 330.103 Documentation.

Any notifications, reports, and similar documentation not specified in the regulations in this part, but necessary to carry out the purpose of the regulations, will be prescribed in administrative instructions.

§ 330.104 Ports of entry.

Ports of entry for plant pests, means of conveyance, or other products or articles of any character whatsoever the entry or movement of which is regulated by the regulations in this part may be specified in administrative instructions or in the permits if permits are required by the regulations. The ports of entry shall be those named in 19 CFR 101.3(b)(1), except as otherwise provided by administrative instructions or by permits issued in accordance with this part, and except those ports of entry listed below.

LIST OF EXCEPTIONS TO CUSTOMS DESIGNATED PORTS OF ENTRY

State	Port of entry
[Reserved]	[Reserved]

[24 FR 10825, Dec. 29, 1959, as amended at 72 FR 43523, Aug. 6, 2007]

§ 330.105 Inspection.

(a) *Inspection of foreign arrivals.* In order to prevent the dissemination into the United States of plant pests and for the purpose of carrying out the regulations in this part, all plant pests; means of conveyance and their stores; baggage; mail; plants; plant products; soil; stone and quarry products under this part; garbage; and any other product or article of any character whatsoever which an inspector considers may be infested or infected by or contain a plant pest, arriving in the United States from any place outside thereof for entry into or movement through the United States shall be subject to inspection by an inspector at the port of first arrival, except that mail will be handled in accordance with the joint customs and postal regulations for inspecting and handling mail. No such plant pests; means of conveyance or their stores; baggage; mail; plants; plant products; soil; stone or quarry products under this part; garbage; or other products or articles which an inspector notifies the Customs authorities should be held for inspection shall be released by Customs officers for entry or onward movement until released by an inspector. The release of all means of conveyance, products and articles regulated under parts 319, 321, and 352 of this chapter shall be in accordance with the requirements of those parts and the applicable provisions in this part. Whenever it shall be deemed safe to modify the requirements of this section by exempting any class of means of conveyance, products or articles from the requirement that they be held for inspection and release of the inspector, the exemptions shall be specified in administrative instructions. Inspectors shall make local arrangements, in accordance with policies of the Plant Protection and Quarantine Programs, with the Collector of Customs for the release by Customs officers on behalf of the inspector of any class of means of conveyance, their stores, baggage, mail, or other products or articles when such arrangements do not increase unduly the danger of plant pest dissemination and will facilitate clearance of means of conveyance, baggage, mail, or other products or articles.

(b) *Inspection of domestic movements.* For the purpose of preventing the interstate movement of plant pests, provisions requiring inspection of means of conveyance and products or articles moving interstate may be issued as regulations in association with quarantines in part 301 or part 318 of this chapter or in this part.

NOTE: Notices appearing at 24 FR 4650, June 9, 1959, 24 FR 5363, July 2, 1959, 24 FR 6889, August 26, 1959, and 24 FR 7519, September 18, 1959, provide in part as follows: That means of conveyance subject to such inspection and release requirements and arriving at any port of entry outside the regularly assigned hours of duty of the Federal plant quarantine inspector, will be held for such inspection and release, until the regularly assigned hours of duty. However, notice is also hereby given that pursuant to the provisions of the Act of August 28, 1950 (7 U.S.C. 2260) such inspection service outside of the regularly assigned hours of duty may be made available to any interested person, upon a reimbursable basis and in accordance with applicable regulations, upon request to the Plant Quarantine Inspector in Charge at such port.

Information concerning regularly assigned hours of duty for Federal plant quarantine inspectors at each port where such inspection is available may be obtained locally by application to the Plant Quarantine Inspector in Charge at such port.

[24 FR 10825, Dec. 29, 1959, as amended at 62 FR 65009, Dec. 10, 1997; 84 FR 29960, June 25, 2019]

§330.106 Emergency measures.

(a) *Procedures to prevent pest dissemination.* Whenever inspection of any means of conveyance, stores, baggage, mail, plants, plant products, earth, stone and quarry products, garbage, or other products or articles of any character whatsoever, arriving in the United States from a place outside thereof, or moving interstate, discloses a plant pest, or provides a reason to believe such a pest is present (other than one moving under permit in accordance with any conditions in the permit and the provisions in this part) which is new to, or not theretofore known to be widely prevalent or distributed within and throughout the United States, the inspector shall employ procedures necessary to prevent the dissemination of the plant pest. Such procedures shall also be employed with respect to means of conveyance or products or articles of any character whatsoever which have moved into the United States or interstate and which the inspector has reason to believe were infested or infected by or contained any such plant pest at the time of such movement. The inspector may follow administrative instructions containing procedures prescribed for certain situations, or he may follow a procedure selected by him from administratively approved methods known to be effective. The procedure may involve seizure, quarantine, treatment in accordance with part 305 of this chapter, application of other remedial measures, exportation, return to shipping point of origin, destruction, or other disposal, but no means of conveyance, product, article, or plant pest owned by any person shall be destroyed, exported, or returned to shipping point of origin or ordered to be so handled, unless there is, in the opinion of the inspector, no less drastic action adequate to prevent the dissemination of the plant pest. In forming such an opinion that no less drastic action is adequate, the inspector shall be guided by applicable specific and general instructions received from officers of the Plant Protection and Quarantine Programs. In taking action with respect to any means of conveyance, product, article, plant pest, the inspector shall take cognizance of applicable requirements of the customs and postal laws and regulations.

(b) *Orders for remedial measures.* The inspector may order the owner of any means of conveyance, product, article, or plant pest, subject to disposal under paragraph (a) of this section, to treat, apply other remedial measures, destroy, or make other disposal thereof without cost to the Federal Government and in a manner specified in accordance with paragraph (a) of this section.

(c) *Failure to apply remedial measures.* If the measures required by the inspector are not applied promptly by the owner within the time limits specified by the inspector, the inspector shall apply measures necessary to prevent the dissemination of the plant pests.

(d) *Khapra beetle infestations of means of conveyance, or cargo or stores thereof;*

other infestations. As a means of preventing the dissemination into the United States, or interstate, of the khapra beetle (Trogoderma granarium Everts), the following procedures will be applicable when that insect is found, or there is reason to believe it is present, in a means of conveyance within paragraph (a) of this section, or in any cargo or stores in such a means of conveyance, or in any cargo or stores unloaded or landed, or being unloaded or landed, in the United States therefrom. These procedures will also apply with respect to other plant pests when the inspector finds they are necessary and sufficient to prevent the spread of such pests.

(1) *Infestation in storerooms and similar compartments of means of conveyance (except aircraft).* (i) When infestation is found only in stores or storerooms, galleys, pantries, or similar noncargo compartments of a means of conveyance, except aircraft, the inspector shall prescribe and supervise the application of such remedial measures as, in his opinion, will be effective under conditions that will not spread the infestation to other parts of the means of conveyance, or to adjacent piers or other installations. If, in the opinion of the inspector, fumigation is the only available safeguard to eliminate the infestation, he shall order the owner to arrange for immediate fumigation of the infested stores and portions of the means of conveyance.

(ii) If the means of conveyance is to leave the territorial limits of the United States directly for a port in another country within 24 hours of such order, the inspector may suspend compliance with the fumigation requirement pending departure from the United States. Pending fumigation or departure, the inspector may seal the openings of infested compartments, packages, or articles, if in his opinion the action is necessary to prevent plant pest dissemination while the means of conveyance remains in the teritorial limits of the United States, as authorized in § 330.110. The inspector may extend the 24-hour period to 48 hours, if, in his judgment, such extension is warranted by plans of the owner to remove the means of conveyance from the territorial limits of the

United States within the extended period, the inability of the contractor to begin fumigation within the 24-hour period, or other reason deemed valid by the inspector. Further extension shall be given only under authority of the Deputy Administrator. Pending compliance with the requirement of fumigation, or the departure from the territorial limits of the United States directly for a port in another country, no stores, laundry, furnishings or equipment, or other articles or products whether in cargo or stores, shall be unloaded from the means of conveyance except as authorized by the inspector and under conditions prescribed by him. The owner of an infested means of conveyance under notice for fumigation which leaves the territorial limits of the United States without fumigation should arrange for the eradication of the infestation before returning to the same or another port in the United States. Upon return to a port in the United States and unless the infestation has been eliminated to the satisfaction of the inspector, the means of conveyance shall be subject to fumigation immediately upon arrival in the United States. Unloading or landing of any product or article shall not be permitted pending compliance with the fumigation requirement, except as authorized by the inspector and under conditions prescribed by him.

(iii) If the means of conveyance is to remain at the port where the infestation was found or is to be moved to another port in the United States, the inspector shall prescribe and supervise the application of the remedial measures at the port where the infestation is found, as provided in this paragraph, or he may authorize the means of conveyance to be moved to another port for fumigation or the application of other remedial measures under safeguards prescribed by him.

(iv) In all instances where the inspector prescribed procedures concerned with the application of remedial measures which involve (*a*) withholding permission to discharge articles or products; (*b*) permission to discharge after such permission has been withheld; (*c*) discontinuance of discharging; or (*d*) resumption of discharging after it has been discontinued, the appropriate Customs officer shall be immediately notified in writing. The inspector shall also inform the Customs officers at the port where the infestation is found and at such other ports as may be necessary of the requirement for fumigation and/or permission to move coastwise to another U.S. port for fumigation or other remedial measures.

(2) *Infestation in cargo compartments of means of conveyance (except aircraft)*. When infestation is found in cargo compartments or in cargo of a means of conveyance, except aircraft, the inspector shall prescribe and supervise the application of such remedial measures as, in his opinion are necessary, with respect to the cargo and the portions of the means of conveyance which contain or contained or were contaminated by the infested cargo. If in the opinion of the inspector fumigation is the only available safeguard to eliminate the infestation, he shall order the owner to arrange for immediate fumigation of the infested portions of such means of conveyance and cargo. However, if such cargo compartments cannot be fumigated without fumigating the entire means of conveyance, the inspector may order the entire means of conveyance and cargo to be fumigated. The inspector shall notify the owner of the means of conveyance of such requirement and the owner shall arrange for immediate fumigation. Discharge of cargo shall be discontinued unless the inspector allows it to continue under safeguards to be prescribed by him. The provisions applicable to stores and storerooms in paragraph (d)(1) (ii) and (iii) of this section shall apply to cargo and cargo areas of such means of conveyance. Customs officers shall be informed as required in paragraph (d)(1)(iv) of this section.

(3) *Infestation in an aircraft*. If infestation is found in an aircraft, the inspector may apply seals as provided in §330.110, and he may require such temporary safeguards as he deems necessary, including the discontinuance of further unloading or landing of any products or articles except as authorized by him. Upon finding such infestation in an aircraft the inspector shall promptly notify the Plant Protection and Quarantine Programs of all circumstances and the temporary safeguards employed, and the Plant Protection and Quarantine Programs will specify the measures for eliminating the infestation which will not be deleterious to the aircraft or its operating components. Any insecticidal application required shall be approved by the Deputy Administrator for use in aircraft. If the aircraft is to depart from the territorial limits of the United States within 24 hours after the infestation is found, the inspector shall permit such departure in lieu of the application of other measures and shall prior to departure break any seals that would prevent access to the aircraft or safe operation thereof. Other seals shall remain intact at time of departure and shall be broken by the aircraft commander or a crew member upon his order only after the aircraft is beyond the territorial limits of the United States. Extension of the 24-hour period shall be given only under authority of the Deputy Administrator. The owner of the aircraft under notice of khapra beetle infestation which leaves the territorial limits of the United States before the infestation has been eradicated should arrange for eradication before returning the aircraft to the United States. Upon return to the United States, if the infestation is not eliminated to the satisfaction of the inspector, the aircraft shall be subject to the same disinfestation requirements and other safeguards immediately upon arrival in the United States. Customs officers shall be notified as required in paragraph (d)(1)(iv) of this section.

(4) *Precautions*. The owner of a means of conveyance required to be fumigated pursuant to this section shall arrange with a competent operator to apply the fumigant under the supervision of the inspector. The owner shall understand that if certain fumigants are used they may result in residues in or on foodstuffs which may render them unsafe for use as food items. He is hereby warned against such use unless as ascertains that the fumigated foodstuffs are fit for human consumption. It should also be understood by the owner that emergency measures prescribed by the inspector to safeguard against dissemination of infestation may have adverse effects on certain products and

articles, and that the acceptance of fumigation as a requirement is an alternative to the immediate removal of the infested means of conveyance and any products and articles thereon, from the territorial limits of the United States. Products or articles in a means of conveyance, or compartments thereof, which may be exposed to methyl bromide or other remedial measures and may be adversely affected thereby, may be removed from the means of conveyance or compartments thereof prior to the application of the remedial measures if in the opinion of the inspector this can be done without danger of plant pest dissemination and under conditions authorized by him, for additional inspection and/or application of effective remedial measures.

[24 FR 10825, Dec. 29, 1959, as amended at 25 FR 8989, Sept. 20, 1960; 32 FR 6339, Apr. 21, 1967; 36 FR 24917, Dec. 24, 1971; 66 FR 21058, Apr. 27, 2001; 69 FR 12265, Mar. 16, 2004; 75 FR 4253, Jan. 26, 2010]

§ 330.107 Costs.

All costs (including those incurred under § 330.106 of this part by the government or the owner) incident to the inspection, handling, cleaning, safeguarding, treating, or other disposal of means of conveyance or products, articles, or · plant pests under this part shall be borne by the owner. Services of the inspector during regularly assigned hours of duty at the usual places of duty shall be furnished without cost to the person requesting the services, unless a user fee is payable under § 354.3 of this chapter.

CROSS REFERENCE: See note following § 330.105.

[56 FR 14844, Apr. 12, 1991]

§ 330.108 Authority to issue administrative instructions.

The Deputy Administrator is authorized to issue the administrative instructions for which provision is made in the regulations in this part, for the purpose of preventing dissemination of plant pests into the United States or interstate. In addition, whenever the Deputy Administrator shall find that existing conditions as to pest risk involved in the movement of plant pests, means of conveyance, or other products

or articles to which the regulations in this part apply, make it safe to modify by making less stringent the restrictions contained in any of such regulations, he shall publish such findings in administrative instructions, specifying the manner in which the regulations shall be made less stringent whereupon such modification shall become effective.

§ 330.109 Caution.

In applying treatments or taking other measures prescribed in administrative instructions or by the inspector, it should be understood that inexactness or carelessness may result in injury or damage.

§ 330.110 Seals.

(a) *Use authorized; form.* Whenever, in the opinion of the inspector, it is necessary, as a safeguard in order to prevent the dissemination of plant pests into the United States, or interstate, seals may be applied to openings, packages, or articles requiring the security provided by such seals. The words "openings, packages, or articles" shall include any form of container, shelf, bin, compartment, or other opening, package, or article which the inspector may have occasion to seal in lieu of more drastic action or otherwise, as a safeguard against plant pest dissemination. The seals may be automatic metal seals or labels or tags and will be provided by the Plant Protection and Quarantine Programs. When they consist of a label or tag, they will be printed in black ink on yellow paper and read substantially as follows: "Warning! The opening, package, or article to which this seal is affixed is sealed under authority of law. This seal is not to be broken while within the territorial limits of the United States except by, or under instructions of, an inspector."

(b) *Breaking of seals.* Seals may be broken: (1) By an inspector; (2) by a Customs officer for Customs purposes, in which case the opening, package, or article will be resealed with Customs seals; (3) by the owner or his agent when the means of conveyance, product, or article has left the territorial limits of the United States; (4) by any person authorized by the inspector or the Deputy Administrator under conditions specified by the inspector or Deputy Administrator. No person shall break seals applied under authority of this section except as provided in this paragraph. The movement into or through the United States, or interstate, of any means of conveyance or product or article on which a seal, applied under this paragraph, has been broken in violation of this paragraph is hereby prohibited, except as authorized by an inspector.

(c) *Notice of sealing.* When an inspector seals any opening, product or article, he shall explain the purpose of such action to the owner or his representative and shall present him with a written notice of the conditions under which the seal may be broken, if requested to do so.

[25 FR 8990, Sept. 20, 1960, as amended at 36 FR 24917, Dec. 24, 1971]

§330.111 Advance notification of arrival of aircraft and watercraft.

The owner, operator, or other representative of any aircraft or watercraft entering the United States from a foreign country, or arriving in the continental United States from Hawaii or any territory or possession of the United States, shall provide every Plant Protection and Quarantine office (PPQ office) serving a port of arrival on the itinerary of the craft while in the United States with advance notification of intent to arrive at that port. This advance notification of arrival shall:

(a) Reach the appropriate PPQ office not less than 12 hours before the craft's estimated time of arrival at the port;

(b) Be communicated by radio, wire, telephone, or any other means; and

(c) Include the following information:

(1) The name or other identifying feature of the individual craft;

(2) The date and estimated time of arrival at the port;

(3) The location of arrival, providing the most site-specific data available, such as the dock, pier, wharf, berth, mole, anchorage, gate, or facility, and;

(4) The names of all foreign and non-Continental U.S. ports where any cargo, crew, or passenger destined for the continental United States has boarded the craft since its most recent arrival at a port in the United States.

(d) If the craft's estimated time of arrival changes by more than one hour, the PPQ office that serves the port of arrival must be notified and provided with updated information immediately.

(e) If the craft's site of arrival changes after a PPQ office has received advance notification of arrival, both that PPQ office and the newly affected PPQ office shall be notified of this change immediately. This applies, too, to site-specific changes involving watercraft.

(f) If the craft's point of arrival is an anchorage, the PPQ office shall be notified, as soon as possible after the craft's arrival at the anchorage, of the specific site, such as berth, mole, pier, to which the craft will be moving, as well as of its estimated time of arrival at that site.

(g) Aircraft and watercraft meeting any of the following conditions are exempt from the provisions in this section, and need not provide advance notification of arrival:

(1) The craft is not regularly used to carry passengers or cargo for a fee;

(2) The aircraft is making a flight scheduled in the Official Airline Guide, North American Edition, or the Official Airline Guide, Worldwide Edition, unless the scheduled time of arrival changes by more than one hour or the plane is diverted to another landing port;

(3) An inspector has precleared the aircraft in Hawaii, a territory or possession of the United States, or a foreign port, having determined that the aircraft contained only articles that are not prohibited or restricted importation into the United States under the provisions of 7 CFR chapter III and 9 CFR chapter I; or

(4) Personnel of the United States armed forces, including the U.S. Coast Guard, in Hawaii, a territory or possession of the United States, or a foreign port, have precleared an aircraft, having determined that the aircraft contained only articles that are not prohibited or restricted importation into the United States under the provisions of 7 CFR chapter III and 9 CFR chapter I.

(5) The owner, operator, or other representative of the aircraft or watercraft not leaving the United States has been informed in writing by a PPQ inspector that notification of intended arrival is not required at subsequent ports in the United States.

(Approved by the Office of Management and Budget under control number 0579–0054)

[52 FR 49344, Dec. 31, 1987]

Subpart B—Movement of Plant Pests, Biological Control Organisms, and Associated Articles

SOURCE: 84 FR 29960, June 25, 2019, unless otherwise noted.

§ 330.200 Scope and general restrictions.

(a) *Restrictions.* No person shall import, move interstate, transit, or release into the environment plant pests, biological control organisms, or associated articles, unless the importation, interstate movement, transit, or release into the environment of the plant pests, biological control organisms, or associated articles is:

(1) Authorized under an import, interstate movement, or continued curation permit issued in accordance with § 330.201; or

(2) Authorized in accordance with other APHIS regulations in this chapter; or

(3) Explicitly granted an exception from permitting requirements in this subpart; or

(4) Authorized under a general permit issued by the Administrator.

(b) *Plant pests regulated by this subpart.* For the purposes of this subpart, and except for an organism that has undergone genetic engineering as defined in § 340.3 of this chapter, APHIS will consider an organism to be a plant pest if the organism directly or indirectly injures, causes damage to, or causes disease in a plant or plant product, or if the organism is an unknown risk to plants or plant products, but is similar to an organism known to directly or indirectly injure, cause damage to, or cause disease in a plant or plant product. Plant pests that have undergone genetic engineering, as defined in § 340.3 of this chapter, are subject to the regulations of part 340 of this chapter.

(c) *Biological control organisms regulated by this subpart.* For the purposes of this subpart, biological control organisms include:

(1) Invertebrate predators and parasites (parasitoids) used to control invertebrate plant pests;

(2) Invertebrate competitors used to control invertebrate plant pests;

(3) Invertebrate herbivores used to control noxious weeds;

(4) Microbial pathogens used to control invertebrate plant pests;

(5) Microbial pathogens used to control noxious weeds;

(6) Microbial parasites used to control plant pathogens; and

(7) Any other types of biological control organisms, as determined by APHIS.

(d) *Biological control organisms not regulated by this subpart.* Paragraph (c) of this section notwithstanding, biological control organisms that have undergone genetic engineering, as defined in § 340.3 of this chapter, as well as products that are currently under an EPA experimental use permit, a Federal Insecticide Fungicide and Rodenticide Act (FIFRA) section 18 emergency exemption, or products that are currently registered with EPA as a microbial pesticide product, are not regulated under this subpart. Additionally, biological control organisms that are pesticides that are not registered with EPA, but are being transferred, sold, or distributed in accordance with EPA's regulations in 40 CFR 152.30, are not regulated under this subpart for their interstate movement or importation. However, an importer desiring to import a shipment of biological control organisms subject to FIFRA must submit to the EPA Administrator a Notice of Arrival of Pesticides and Devices as required by CBP regulations at 19 CFR 12.112. The Administrator will provide notification to the importer indicating the disposition to be made of shipment upon its entry into the customs territory of the United States.

[84 FR 29960, June 25, 2019, as amended at 85 FR 29832, May 18, 2020]

§330.201 Permit requirements.

(a) *Types of permits.* APHIS issues import permits, interstate movement permits, continued curation permits, and transit permits for plant pests, biological control organisms, and associated articles.[1]

(1) *Import permit.* Import permits are issued to persons for secure shipment from outside the United States into the territorial limits of the United States. When import permits are issued to individuals, these individuals must be 18 years of age or older and have a physical address within the United States. When import permits are issued to corporate persons, these persons must maintain an address or business office in the United States with one or more designated individuals for service of process.

(2) *Interstate movement permit.* Interstate movement permits are issued to persons for secure shipment from any State into or through any other State. When interstate movement permits are issued to individuals, these individuals must be 18 years of age or older and have a physical address within the United States. When interstate movement permits are issued to corporate persons, these persons must maintain an address or business office in the United States with a designated individual for service of process.

(3) *Continued curation permits.* Continued curation permits are issued in conjunction with and prior to the expiration date for an import permit or interstate movement permit, in order for the permittee to continue the actions listed on the import permit or interstate movement permit. When continued curation permits are issued to individuals, these individuals must be 18 years of age or older and have a physical address within the United States. When continued curation permits are issued to corporate persons, these persons must maintain an address or business office in the United States with one or more designated individuals for service of process.

(4) *Transit permits.* Transit permits are issued for secure shipments through the United States. Transit permits are issued in accordance with part 352 of this chapter.

(b) *Applying for a permit.* Permit applications must be submitted by the applicant in writing or electronically through one of the means listed at *http://www.aphis.usda.gov/plant_health/permits/index.shtml* in advance of the action(s) proposed on the permit application.

(c) *Completing a permit application.* A permit application must be complete before APHIS will evaluate it in order to determine whether to issue the permit requested. To facilitate timely processing, applications should be submitted as far in advance as possible of the date of the proposed permit activity. Guidance regarding how to complete a permit application, including guidance specific to the various information blocks on the application, is available at *http://www.aphis.usda.gov/plant_health/permits/index.shtml.*

(d) *APHIS action on permit applications.* APHIS will review the information on the application to determine whether it is complete. In order to consider an application complete, APHIS may request additional information that it determines to be necessary in order to assess the risk to plants and plant products that may be posed by the actions proposed on the application. When it is determined that an application is complete, APHIS will commence review of the information provided.

(1) *State or Tribal consultation and comment; consultation with other individuals.* APHIS will share a copy of the permit application, and the proposed permit conditions, with the appropriate State or Tribal regulatory officials, and may share the application and the proposed conditions with other persons or groups to provide comment.

(2) *Initial assessment of sites and facilities.* Prior to issuance of a permit, APHIS will assess all sites and facilities that are listed on the permit application, including private residences,

---

[1] Persons contemplating the shipment of plant pests, biological control organisms, or associated articles to places outside the United States should make arrangements directly, or through the recipient, with the country of destination for the export of the plant pests, biological control organisms, or associated articles into that country.

biocontainment facilities, and field locations where the organism[2] or associated article will be held or released. As part of this assessment, all sites and facilities are subject to inspection. All facilities must be determined by APHIS to be constructed and maintained in a manner that prevents the dissemination or dispersal of plant pests, biological control organisms, or associated articles from the facility. The applicant must provide all information requested by APHIS regarding this assessment, and must allow all inspections requested by APHIS during normal business hours (8 a.m. to 4:30 p.m., Monday through Friday, excluding holidays). Failure to do so constitutes grounds for denial of the permit application.

(3) *Issuance of a permit.* APHIS may issue a permit to an applicant if APHIS concludes that the actions indicated in the permit application are not likely to introduce or disseminate a plant pest, biological control organism, or noxious weed within the United States in a manner that exposes plants and plant products to unacceptable risk. Issuance will occur as follows:

(i) Prior to issuing the permit, APHIS will notify the applicant in writing or electronically of all proposed permit conditions. The applicant must agree in writing or electronically that he or she, and all his or her employees, agents, and/or officers, will comply with all permit conditions and all provisions of this subpart. If the organism or associated article will be contained in a private residence, the applicant must state in this agreement that he or she authorizes APHIS to conduct unscheduled assessments of the residence during normal business hours if a permit is issued.

(ii) APHIS will issue the permit after it receives and reviews the applicant's agreement. The permit will be valid for no more than 3 years. During that period, the permittee must abide by all permitting conditions, and the use of the organism or associated article must conform to the intended use on the permit. Moreover, the use of organisms derived from a regulated parent

organism during that period must conform to the intended use specified on the permit for the parent organism.

(iii) All activities carried out under the permit must cease on or before the expiration date for the permit, unless, prior to that expiration date, the permittee has submitted a new permit application and a new permit has been issued to authorize continuation of those actions.

(iv) At any point following issuance of a permit but prior to its expiration date, an inspector may conduct unscheduled assessments of the site or facility in which the organisms or associated articles are held, to determine whether they are constructed and are being maintained in a manner that prevents the dissemination of organisms or associated articles from the site or facility. The permittee must allow all such assessments requested by APHIS during normal business hours. Failure to allow such assessments constitutes grounds for revocation of the permit.

(4) *Denial of a permit application.* APHIS may deny an application for a permit if:

(i) APHIS concludes that the actions proposed in the permit application would present an unacceptable risk to plants and plant products because of the introduction or dissemination of a plant pest, biological control organism, or noxious weed within the United States; or

(ii) The actions proposed in the permit application would be adverse to the conduct of an APHIS eradication, suppression, control, or regulatory program; or

(iii) A State or Tribal executive official, or a State or Tribal plant protection official authorized to do so, objects to the movement in writing and provides specific, detailed information that there is a risk the movement will result in the dissemination of a plant pest or noxious weed into the State, APHIS evaluates the information and agrees, and APHIS determines that such plant pest or noxious weed risk cannot be adequately addressed or mitigated; or

(iv) The applicant does not agree to observe all of the proposed permit conditions that APHIS has determined are

---

[2] Includes biological control organisms and plant pests.

necessary to mitigate identified risks; or

(v) The applicant does not provide information requested by APHIS as part of an assessment of sites or facilities, or does not allow APHIS to inspect sites or facilities associated with the actions listed on the permit application; or

(vi) APHIS determines that the applicant has not followed prior permit conditions, or has not adequately demonstrated that they can meet the requirements for the current application. Factors that may contribute to such a determination include, but are not limited to:

(A) The applicant, or a partnership, firm, corporation, or other legal entity in which the applicant has a substantial interest, financial or otherwise, has not complied with any permit that was previously issued by APHIS.

(B) Issuing the permit would circumvent any order denying or revoking a previous permit issued by APHIS.

(C) The applicant has previously failed to comply with any APHIS regulation.

(D) The applicant has previously failed to comply with any other Federal, State, or local laws, regulations, or instructions pertaining to plant health.

(E) The applicant has previously failed to comply with the laws or regulations of a national plant protection organization or equivalent body, as these pertain to plant health.

(F) APHIS has determined that the applicant has made false or fraudulent statements or provided false or fraudulent records to APHIS.

(G) The applicant has been convicted or has pled *nolo contendere* to any crime involving fraud, bribery, extortion, or any other crime involving a lack of integrity.

(5) *Withdrawal of a permit application.* Any permit application may be withdrawn at the request of the applicant. If the applicant wishes to withdraw a permit application, he or she must provide the request in writing to APHIS. APHIS will provide written notification to the applicant as promptly as circumstances allow regarding reception of the request and withdrawal of the application.

(6) *Cancellation of a permit.* Any permit that has been issued may be canceled at the request of the permittee. If a permittee wishes a permit to be canceled, he or she must provide the request in writing to APHIS–PPQ. Whenever a permit is canceled, APHIS will notify the permittee in writing regarding such cancellation.

(7) *Revocation of a permit.* APHIS may revoke a permit for any of the following reasons:

(i) After issuing the permit, APHIS obtains information that would have otherwise provided grounds for it to deny the permit application; or

(ii) APHIS determines that the actions undertaken under the permit have resulted in or are likely to result in the introduction into or dissemination within the United States of a plant pest or noxious weed in a manner that presents an unacceptable risk to plants or plant products; or

(iii) APHIS determines that the permittee, or any employee, agent, or officer of the permittee, has failed to comply with a provision of the permit or the regulations under which the permit was issued.

(8) *Amendment of permits*—(i) *Amendment at permittee's request.* If a permittee determines that circumstances have changed since the permit was initially issued and wishes the permit to be amended accordingly, he or she must request the amendment, either through APHIS' online portal for permit applications, or by contacting APHIS directly via phone or email. The permittee may have to provide supporting information justifying the amendment. APHIS will review the amendment request, and may amend the permit if only minor changes are necessary. Requests for more substantive changes may require a new permit application. Prior to issuance of an amended permit, the permittee may be required to agree in writing that he or she, and his or her employees, agents, and/or officers will comply with the amended permit and conditions.

(ii) *Amendment initiated by APHIS.* APHIS may amend any permit and its conditions at any time, upon determining that the amendment is needed to address newly identified considerations concerning the risks presented by the organism or the activities being conducted under the permit. APHIS may also amend a permit at any time to ensure that the permit conditions are consistent with all of the requirements of this part. As soon as circumstances allow, APHIS will notify the permittee of the amendment to the permit and the reason(s) for it. Depending on the nature of the amendment, the permittee may have to agree in writing or electronically that he or she, and his or her employees, agents, and/or officers, will comply with the permit and conditions as amended before APHIS will issue the amended permit. If APHIS requests such an agreement, and the permittee does not agree in writing that he or she, and his or her employees, agents, and/or officers, will comply with the amended permit and conditions, the existing permit will be revoked.

(9) *Suspension of permitted actions.* APHIS may suspend authorization of actions authorized under a permit if it identifies new factors that cause it to reevaluate the risk associated with those actions. APHIS will notify the permittee in writing of this suspension explaining the reasons for it and stating the actions for which APHIS is suspending authorization. Depending on the results of APHIS' evaluation, APHIS will subsequently contact the permittee to remove the suspension, amend the permit, or revoke the permit.

(10) *Appeals.* Any person whose application has been denied, whose permit has been revoked or amended, or whose authorization for actions authorized under a permit has been suspended, may appeal the decision in writing to the Administrator within 10 business days after receiving the written notification of the denial, revocation, amendment, or suspension. The appeal shall state all of the facts and reasons upon which the person relies to show that the application was wrongfully denied, permit revoked or amended, or authorization for actions under a permit suspended. The Administrator shall grant or deny the appeal, stating the reasons for the decision as promptly as circumstances allow.

(Approved by the Office of Management and Budget Under Control Number 0579–0054)

§ 330.202 Biological control organisms.

(a) *General conditions for importation, interstate movement, and release of biological control organisms.* Except as provided in paragraph (b) of this section, no biological control organism regulated under this subpart may be imported, moved in interstate commerce, or released into the environment unless a permit has been issued in accordance with § 330.201 authorizing such importation, interstate movement, or release, and the organism is moved or released in accordance with this permit and the regulations in this subpart. The regulations in 40 CFR parts 1500 through 1508, part 1b of this title, and part 372 of this chapter may require APHIS to request additional information from an applicant regarding the proposed release of a biological control organism as part of its evaluation of a permit application. Further information regarding the types of information that may be requested, and the manner in which this information will be evaluated, is found at *http://www.aphis.usda.gov/plant_health/permits/index.shtml.*

(b) *Exceptions from permitting requirements for certain biological control organisms.* APHIS has determined that certain biological control organisms have become established throughout their geographical or ecological range in the continental United States, such that the additional release of pure cultures derived from field populations of taxa of such organisms into the environment of the continental United States will present no additional plant pest risk (direct or indirect) to plants or plant products. Lists of biological control organisms for invertebrate plant pests and for weeds are maintained on the PPQ Permits and Certifications website at *https://www.aphis.usda.gov/aphis/resources/permits.*

(1) *Importation and interstate movement of listed organisms.* Pure cultures of organisms excepted from permit requirements, unless otherwise indicated, may be imported or moved interstate within the continental United States without further restriction under this subpart.

(2) *Release of listed organisms.* Pure cultures of organisms on the list may be released into the environment of the

continental United States without further restriction under this subpart.

(c) *Additions to the list of organisms granted exceptions from permitting requirements for their importation, interstate movement, or release.* Any person may request that APHIS add a biological control organism to the list referred to in paragraph (b) of this section by submitting a petition to APHIS via email to *pest.permits@usda.gov* or through any means listed at *http://www.aphis.usda.gov/plant_health/permits/index.shtml.* The petition must include the following information:

(1) Evidence indicating that the organism is indigenous to the continental United States throughout its geographical or ecological range, or evidence indicating that the organism has produced self-replicating populations within the continental United States for an amount of time sufficient, based on the organism's taxon, to consider that taxon established throughout its geographical or ecological range in the continental United States; or

(2) Evidence that the organism's geographical or ecological range includes an extremely limited area of or none of the continental United States based on its inability to maintain year to year self-replicating populations despite repeated introductions over a sufficient range of time; or

(3) The petition would include evidence that the organism cannot establish anywhere in the continental United States; or

(4) Results from a field study where data were collected from representative habitats occupied by the biological control organism. Studies must include sampling for any direct or indirect impacts on target and non-target hosts of the biological control organism in these habitats. Supporting scientific literature must be cited; or

(5) Any other data, including published scientific reports, that suggest that subsequent releases of the organism into the environment of the continental United States will present no additional plant pest risk (direct or indirect) to plants or plant products.

(d) *APHIS review of petitions*—(1) *Evaluation.* APHIS will review the petition to determine whether it is complete. If APHIS determines that the petition is complete, it will conduct an evaluation of the petition to determine whether there is sufficient evidence that the organism exists throughout its geographical or ecological range in the continental United States and that subsequent releases of pure cultures of field populations of the organism into the environment of the continental United States will present no additional plant pest risk (direct or indirect) to plants or plant products.

(2) *Notice of availability of the petition.* If APHIS determines that there is sufficient evidence that the organism exists throughout its geographical or ecological range in the continental United States and that subsequent releases of pure cultures of the organism into the environment of the continental United States will present no additional plant pest risk to plants or plant products, APHIS will publish a notice in the FEDERAL REGISTER announcing the availability of the petition and requesting public comment on that document.

(3) *Notice of determination.* (i) If no comments are received, or if the comments received do not lead APHIS to reconsider its determination, APHIS will publish in the FEDERAL REGISTER a subsequent notice describing the comments received and stating that the organism has been added to the list referred to in paragraph (b) of this section.

(ii) If the comments received lead APHIS to reconsider its determination, APHIS will publish in the FEDERAL REGISTER a subsequent notice describing the comments received and stating its reasons for determining not to add the organism to the list referred to in paragraph (b) of this section.

(e) *Removal of organisms from the list of exempt organisms.* Any biological control organism may be removed from the list referred to in paragraph (b) of this section if information emerges that would have otherwise led APHIS to deny the petition to add the organism to the list. Whenever an organism is removed from the list, APHIS will publish a notice in the FEDERAL REGISTER announcing that action and the basis for it.

(Approved by the Office of Management and Budget under control number 0579–0187)

§ 330.203 Soil.

(a) *Requirements.* The Administrator has determined that, unless it has been sterilized, soil is an associated article, and is thus subject to the permitting requirements of § 330.201, unless its movement:

(1) Is regulated pursuant to other APHIS regulations in this chapter; or

(2) Does not require such a permit under the provisions of paragraph (b)(1) or (c)(1) of this section.

(b) *Conditions governing the importation of soil*—(1) *Permit.* Except as provided in § 319.37-10 of this chapter and except for soil imported from areas of Canada not regulated by the national plant protection organization of Canada for a soil-borne plant pest, soil may be imported into the United States if an import permit has been issued in accordance with § 330.201 and if the soil is imported under the conditions specified on the permit.

(2) *Additional conditions for the importation of soil via hand-carry.* In addition to the condition of paragraph (b)(1) of this section, soil may be hand-carried into the United States only if the importation meets the conditions of § 330.205.

(3) *Additional conditions for the importation of soil intended for the extraction of plant pests.* In addition to the condition of paragraph (b)(1) of this section, soil may be imported into the United States for the extraction of plant pests if the soil will be imported directly to an APHIS-approved biocontainment facility.

(4) *Additional conditions for the importation of soil contaminated with plant pests and intended for disposal.* In addition to the condition of paragraph (b)(1) of this section, soil may be imported into the United States for the disposal of plant pests if the soil will be imported directly to an APHIS-approved disposal facility.

(5) *Exemptions.* The articles listed in this paragraph (b) are not soil, provided that they are free of organic material. Therefore, they may be imported into the United States without an import permit issued in accordance with § 330.201, unless the Administrator has issued an order stating otherwise. All such articles are, however, subject to inspection at the port of first arrival, subsequent reinspection at other locations, other remedial measures deemed necessary by an inspector to remove any risk the items pose of disseminating plant pests or noxious weeds, and any other restrictions of this chapter:

(i) Consolidated material derived from any strata or substrata of the earth. Examples include clay (laterites, bentonite, china clay, attapulgite, tierrafino), talc, chalk, slate, iron ore, and gravel.

(ii) Sediment, mud, or rock from saltwater bodies of water.

(iii) Cosmetic mud and other commercial mud products.

(iv) Stones, rocks, and quarry products.

(c) *Conditions governing the interstate movement of soil*—(1) *General conditions.* Except for soil moved in accordance with paragraphs (c)(2) through (5) of this section, soil may be moved interstate within the United States without prior issuance of an interstate movement permit in accordance with § 330.201 or further restriction under this subpart. However, all soil moved interstate is subject to any movement restrictions and remedial measures specified for such movement referenced in part 301 of this chapter.

(2) *Conditions for the interstate movement within the continental United States of soil intended for the extraction of plant pests.* Soil may be moved in interstate commerce within the continental United States with the intent of extracting plant pests, only if an interstate movement permit has been issued for its movement in accordance with § 330.201, and if the soil will be moved directly to an APHIS-approved biocontainment facility in a secure manner that prevents its dissemination into the outside environment.

(3) *Conditions for the interstate movement within the continental United States of soil infested with plant pests and intended for disposal.* Soil may be moved in interstate commerce within the continental United States with the intent of disposing of plant pests, only if an interstate movement permit has been issued for its movement in accordance with § 330.201, and the soil will be moved directly to an APHIS-approved disposal facility in a secure manner

that prevents its dissemination into the outside environment.

(4) *Conditions for the interstate movement of soil samples from an area quarantined in accordance with part 301 of this chapter for chemical or compositional testing or analysis.* Soil samples may be moved for chemical or compositional testing or analysis from an area that is quarantined in accordance with part 301 of this chapter without prior issuance of an interstate movement permit in accordance with § 330.201 or further restriction under this chapter, provided that the soil is moved to a laboratory that has entered into and is operating under a compliance agreement with APHIS, is abiding by all terms and conditions of the compliance agreement, and is approved by APHIS to test and/or analyze such samples.

(5) *Additional conditions for interstate movement of soil to, from, or between Hawaii, the territories, and the continental United States.* In addition to all general conditions for interstate movement of soil, soil may be moved in interstate commerce to, from, or between Hawaii, the territories, and the continental United States only if an interstate movement permit has been issued for its movement in accordance with § 330.201. In addition, soil moved to, from, or between Hawaii, the territories, and the continental United States with the intent of extracting plant pests is subject to the conditions of paragraph (c)(2) of this section, while soil infested with plant pests and intended for disposal is subject to the conditions of paragraph (c)(3) of this section.

(d) *Conditions governing the transit of soil through the United States.* Soil may transit through the United States only if a transit permit has been issued for its movement in accordance with part 352 of this chapter.

(Approved by the Office of Management and Budget Under Control Number 0579–0054)

§ **330.204 Exceptions to permitting requirements for the importation or interstate movement of certain plant pests.**

Pursuant to section 7711 of the Plant Protection Act (7 U.S.C. 7701 *et seq.*), the Administrator has determined that certain plant pests may be moved interstate within the continental United States without restriction. The list of all such plant pests is on the PPQ Permits and Certifications website at *https://www.aphis.usda.gov/aphis/resources/permits.* Plant pests listed as being excepted from permitting requirements, unless otherwise indicated, may be moved interstate within the continental United States without further restriction under this subpart.

(a) *Categories.* In order to be included on the list, a plant pest must:

(1) Be from field populations or lab cultures derived from field populations of a taxon that is established throughout its entire geographical or ecological range within the continental United States; or

(2) Be commercially available and raised under the regulatory purview of other Federal agencies.

(b) *Petition process to add plant pests to the list*—(1) *Petition.* Any person may petition APHIS to have an additional plant pest added to the list of plant pests that may be imported into or moved in interstate commerce within the continental United States without restriction. To submit a petition, the person must provide, in writing, information supporting the placement of a particular pest in one of the categories listed in paragraph (a) of this section.

(i) Information that the plant pest belongs to a taxon that is established throughout its entire geographical or ecological range within the United States must include scientific literature, unpublished studies, or data regarding:

(A) The biology of the plant pest, including characteristics that allow it to be identified, known hosts, and virulence;

(B) The geographical or ecological range of the plant pest within the continental United States; and

(C) The areas of the continental United States within which the plant pest is established.

(ii) Information that the plant pest is commercially available and raised under the regulatory purview of another Federal agency must include a citation to the relevant law, regulation, or order under which the agency exercises such oversight.

307

(2) *APHIS review.* APHIS will review the information contained in the petition to determine whether it is complete. In order to consider the petition complete, APHIS may require additional information to determine whether the plant pest belongs to one of the categories listed in paragraph (a) of this section. When it is determined that the information is complete, APHIS will commence review of the petition.

(3) *Action on petitions to add pests.* (i) If, after review of the petition, APHIS determines there is insufficient evidence that the plant pest belongs to one of the categories listed in paragraph (a) of this section, APHIS will deny the petition, and notify the petitioner in writing regarding this denial.

(ii) If, after review of the petition, APHIS determines that the plant pest belongs to one of the categories in paragraph (a) of this section, APHIS will publish a notice in the FEDERAL REGISTER that announces the availability of the petition and any supporting documentation to the public, that states that APHIS intends to add the plant pest to the list of plant pests that may be imported into or moved in interstate commerce within the continental United States without restriction, and that requests public comment. If no comments are received on the notice, or if, based on the comments received, APHIS determines that its conclusions regarding the petition have not been affected, APHIS will publish in the FEDERAL REGISTER a subsequent notice stating that the plant pest has been added to the list.

(c) *Petition process to have plant pests removed from the list*—(1) *Petition.* Any person may petition to have a plant pest removed from the list of plant pests that may be imported into or moved interstate within the continental United States without restriction by writing to APHIS. The petition must contain independently verifiable information demonstrating that APHIS' initial determination that the plant pest belongs to one of the categories in paragraph (a) of the section should be changed, or that additional information is now available that would have caused us to change the initial decision.

(2) *APHIS review.* APHIS will review the information contained in the petition to determine whether it is complete. In order to consider the petition complete, APHIS may require additional information supporting the petitioner's claim. When it is determined that the information is complete, APHIS will commence review of the petition.

(3) *APHIS action on petitions to remove pests.* (i) If, after review of the petition, APHIS determines that there is insufficient evidence to suggest that its initial determination should be changed, APHIS will deny the petition, and notify the petitioner in writing regarding this denial.

(ii) If, after review of the petition, APHIS determines that there is a sufficient basis to suggest that its initial determination should be changed, APHIS will publish a notice in the FEDERAL REGISTER that announces the availability of the petition, and that requests public comment regarding removing the plant pest from the list of plant pests that may be imported into or move in interstate commerce within the continental United States without restriction. If no comments are received on the notice, or if the comments received do not affect APHIS' conclusions regarding the petition, APHIS will publish a subsequent notice in the FEDERAL REGISTER stating that the plant pest has been removed from the list.

(d) *APHIS-initiated changes to the list.* (1) APHIS may propose to add a plant pest to or remove a pest from the list of plant pests that may be imported into or move in interstate commerce within the continental United States without restriction, if it determines that there is sufficient evidence that the plant pest belongs to one of the categories listed in paragraph (a) of the section, or if evidence emerges that leads APHIS to reconsider its initial determination that the plant pest was or was not in one of the categories listed in paragraph (a) of this section. APHIS will publish a notice in the FEDERAL REGISTER announcing this proposed addition or removal, making available any supporting documentation that it prepares, and requesting public comment.

(2) If no comments are received on the notice or if the comments received do not affect the conclusions of the notice, APHIS will publish a subsequent notice in the FEDERAL REGISTER stating that the plant pest has been added to or removed from the list.

(Approved by the Office of Management and Budget Under Control Number 0579–0187)

§330.205 Hand-carry of plant pests, biological control organisms, and soil.

Plant pests, biological control organisms, and soil may be hand-carried into the United States only in accordance with the provisions of this section.

(a) *Authorization to hand-carry*—(1) *Application for a permit; specification of "hand-carry" as proposed method of movement.* A person must apply for an import permit for the plant pest, biological control organism, or soil, in accordance with §330.201, and specify hand-carry of the organism or article as the method of proposed movement.

(2) *Specification of individual who will hand-carry.* The application must also specify the individual or individuals who will hand-carry the plant pest, biological control organism, or soil into the United States. If APHIS authorizes this individual or these individuals to hand-carry, the authorization may not be transferred to nor actions under it performed by individuals other than those identified on the permit application.

(b) *Notification of intent to hand-carry.* After the permittee has obtained an import permit but no less than 20 days prior to movement, the permittee must provide APHIS through APHIS' online portal for permit applications or by fax with the names of the designated hand carrier, or carriers, assigned to that movement. Additional conditions for hand-carry are available on the APHIS website.[3]

(c) *Denial, amendment, or cancellation of authorization to hand-carry.* APHIS may deny a request to hand-carry, or amend or cancel any hand-carry authorization at any time, if it deems such action necessary to prevent the

introduction or dissemination of plant pests or noxious weeds within the United States.

(d) *Appeal of denial, amendment, or cancellation.* Any person whose request to hand-carry has been denied, or whose authorization to hand-carry has been amended or canceled, may appeal the decision in writing to APHIS.

§330.206 Packaging requirements.

Shipments in which plant pests, biological control organisms, and associated articles are imported into, moved in interstate commerce, or transited through the United States must meet the general packaging requirements of this section, as well as all specific packaging requirements on the permit itself.

(a) *Packaging requiremspents.* All shipments must consist of an outer shipping container and at least two packages within the container. Both the container and inner packages must be securely sealed to prevent the dissemination of the enclosed plant pests, biological control organisms, or associated articles.

(1) *Outer shipping container.* The outer shipping container must be rigid, impenetrable and durable enough to remain closed and structurally intact in the event of dropping, lateral impact with other objects, and other shocks incidental to handling.

(2) *Inner packages.* The innermost package or packages within the shipping container must contain all of the organisms or articles that will be moved. As a safeguard, the innermost package must be placed within another, larger package. All packages within the shipping container must be constructed or safeguarded so that they will remain sealed and structurally intact throughout transit. The packages must be able to withstand changes in pressure, temperature, and other climatic conditions incidental to shipment.

(b) *Packing material.* Packing materials may be placed in the inner packages or shipping container for such purposes as cushioning, stabilizing,

---

[3] *https://www.aphis.usda.gov/plant_health/permits/organism/downloads/HandCarryPolicy.pdf.*

water absorption or retention, nourishment or substrate for regulated articles, etc. Packing material for importation must be free of plant pests, noxious weeds, biological control organisms not listed on the permit or associated articles, and, as such, must be new, or must have been sterilized or disinfected prior to reuse. Packing material must be suited for the enclosed organism or article, as well as any medium in which the organism or article will be maintained.

(c) *Requirements following receipt of the shipment at the point of destination.* (1) Packing material, including media and substrates, must be destroyed by incineration, be decontaminated using autoclaving or another approved method, or otherwise be disposed of in a manner specified in the permit itself.

(2) Shipping containers may be reused, provided that the container has not been contaminated with plant pests, noxious weeds, biological control organisms, or associated articles. Shipping containers that have been in contact with or otherwise contaminated with any of these items must be sufficiently sterilized or disinfected prior to reuse, or otherwise disposed of.

(d) *Costs.* Permittees who fail to meet the requirements of this section may be held responsible for all costs incident to inspection, rerouting, repackaging, subsequent movement, and any treatments.

§ 330.207 Cost and charges.

The inspection services of APHIS inspectors during regularly assigned hours of duty and at the usual places of duty will be furnished without cost. APHIS will not be responsible for any costs or charges incidental to inspections or compliance with the provisions of this subpart, other than for the inspection services of the inspector.

Subpart C—Movement of Soil, Stone, and Quarry Products [Reserved]

Subpart D—Garbage

SOURCE: 71 FR 49314, Aug. 23, 2006, unless otherwise noted. Redesignated at 84 FR 2429, Feb. 7, 2019.

§ 330.400 Regulation of certain garbage.

(a) *Certain interstate movements and imports*—(1) *Interstate movements of garbage from Hawaii and U.S. territories and possessions to other States.* Hawaii, Puerto Rico, American Samoa, the Commonwealth of the Northern Mariana Islands, the Federated States of Micronesia, Guam, the U.S. Virgin Islands, Republic of the Marshall Islands, and the Republic of Palau are hereby quarantined, and the movement of garbage therefrom to any other State is hereby prohibited except as provided in this subpart in order to prevent the introduction and spread of exotic plant pests and diseases.

(2) *Imports of garbage.* In order to protect against the introduction of exotic animal and plant pests and diseases, the importation of garbage from all foreign countries except Canada is prohibited except as provided in § 330.401(b).

(b) *Definitions—Agricultural waste.* By-products generated by the rearing of animals and the production and harvest of crops or trees. Animal waste, a large component of agricultural waste, includes waste (*e.g.,* feed waste, bedding and litter, and feedlot and paddock runoff) from livestock, dairy, and other animal-related agricultural and farming practices.

Approved facility. A facility approved by the Administrator, Animal and Plant Health Inspection Service, upon his determination that it has equipment and uses procedures that are adequate to prevent the dissemination of plant pests and livestock or poultry diseases, and that it is certified by an appropriate Government official as currently complying with the applicable laws for environmental protection.

Approved sewage system. A sewage system approved by the Administrator, Animal and Plant Health Inspection Service, upon his determination that the system is designed and operated in such a way as to preclude the discharge of sewage effluents onto land surfaces or into lagoons or other stationary waters, and otherwise is adequate to prevent the dissemination of plant pests and livestock or poultry diseases, and that is certified by an appropriate

Government official as currently complying with the applicable laws for environmental protection.

Carrier. The principal operator of a means of conveyance.

Garbage. All waste material that is derived in whole or in part from fruits, vegetables, meats, or other plant or animal (including poultry) material, and other refuse of any character whatsoever that has been associated with any such material.

Incineration. To reduce garbage to ash by burning.

Interstate. From one State into or through any other State.

Sterilization. Cooking garbage at an internal temperature of 212 °F for 30 minutes.

Stores. The food, supplies, and other provisions carried for the day-to-day operation of a conveyance and the care and feeding of its operators.

Yard waste. Solid waste composed predominantly of grass clippings, leaves, twigs, branches, and other garden refuse.

§330.401 Garbage generated onboard a conveyance.

(a) *Applicability.* This section applies to garbage generated onboard any means of conveyance during international or interstate movements as provided in this section and includes food scraps, table refuse, galley refuse, food wrappers or packaging materials, and other waste material from stores, food preparation areas, passengers' or crews' quarters, dining rooms, or any other areas on the means of conveyance. This section also applies to meals and other food that were available for consumption by passengers and crew on an aircraft but were not consumed.

(1) Not all garbage generated onboard a means of conveyance is regulated for the purposes of this section. Garbage regulated for the purposes of this section is defined as "regulated garbage" in paragraphs (b) and (c) of this section.

(2) Garbage that is commingled with regulated garbage is also regulated garbage.

(b) *Garbage regulated because of movements outside the United States or Canada.* For purposes of this section, garbage on or removed from a means of conveyance is regulated garbage, if, when the garbage is on or removed from the means of conveyance, the means of conveyance has been in any port outside the United States and Canada within the previous 2-year period. There are, however, two exceptions to this provision. These exceptions are as follows:

(1) *Exception 1: Aircraft.* Garbage on or removed from an aircraft is exempt from requirements under paragraph (d) of this section if the following conditions are met when the garbage is on or removed from the aircraft:

(i) The aircraft had previously been cleared of all garbage and of all meats and meat products, whatever the country of origin, except meats that are shelf-stable; all fresh and condensed milk and cream from countries designated in 9 CFR 94.1 as those in which foot-and-mouth disease exists; all fresh fruits and vegetables; and all eggs; and the items previously cleared from the aircraft as prescribed by this paragraph have been disposed of according to the procedures for disposing of regulated garbage, as specified in paragraphs (d)(2) and (d)(3) of this section.

(ii) After the garbage and stores referred to in paragraph (b)(1)(i) of this section were removed, the aircraft has not been in a non-Canadian foreign port.

(2) *Exception 2: Other conveyances.* Garbage on or removed in the United States from a means of conveyance other than an aircraft is exempt from requirements under paragraph (d) of this section if the following conditions are met when the garbage is on or removed from the means of conveyance:

(i) The means of conveyance is accompanied by a certificate from an inspector stating the following:

(A) That the means of conveyance had previously been cleared of all garbage and of all meats and meat products, whatever the country of origin, except meats that are shelf-stable; all fresh and condensed milk and cream from countries designated in 9 CFR 94.1 as those in which foot-and-mouth disease exists; all fresh fruits and vegetables; and all eggs; and the items previously cleared from the means of conveyance as prescribed by this paragraph have been disposed of according

311

to the procedures for disposing of regulated garbage, as specified in paragraphs (d)(2) and (d)(3) of this section.

(B) That the means of conveyance had then been cleaned and disinfected in the presence of the inspector; and

(ii) Since being cleaned and disinfected, the means of conveyance has not been in a non-Canadian foreign port.

(c) *Garbage regulated because of certain movements to or from Hawaii, territories, or possessions.* For purposes of this section, garbage on or removed from a means of conveyance is regulated garbage, if at the time the garbage is on or removed from the means of conveyance, the means of conveyance has moved during the previous 1-year period, either directly or indirectly, to the continental United States from any territory or possession or from Hawaii, to any territory or possession from any other territory or possession or from Hawaii, or to Hawaii from any territory or possession. There are, however, two exceptions to this provision. These exceptions are as follows:

(1) *Exception 1: Aircraft.* Garbage on or removed from an aircraft is exempt from requirements under paragraph (d) of this section if the following two conditions are met when the garbage is on or removed from the aircraft:

(i) The aircraft had been previously cleared of all garbage and all fresh fruits and vegetables, and the items previously cleared from the aircraft as prescribed by this paragraph have been disposed of according to the procedures for disposing of regulated garbage, as specified in paragraphs (d)(2) and (d)(3) of this section.

(ii) After the garbage and stores referred to in paragraph (c)(1)(i) of this section were removed, the aircraft has not moved to the continental United States from any territory or possession or from Hawaii; to any territory or possession from any other territory or possession or from Hawaii; or to Hawaii from any territory or possession.

(2) *Exception 2: Other conveyances.* Garbage on or removed from a means of conveyance other than an aircraft is exempt from requirements under paragraph (d) of this section if the following two conditions are met when

the garbage is on or removed from the means of conveyance:

(i) The means of conveyance is accompanied by a certificate from an inspector stating that the means of conveyance had been cleared of all garbage and all fresh fruits and vegetables; and the items previously cleared from the means of conveyance as prescribed by this paragraph have been disposed of according to the procedures for disposing of regulated garbage, as specified in paragraphs (d)(2) and (d)(3) of this section.

(ii) After being cleared of the garbage and stores referred to in paragraph (c)(2)(i) of this section, the means of conveyance has not moved to the continental United States from any territory or possession or from Hawaii; to any territory or possession from any other territory or possession or from Hawaii; or to Hawaii from any territory or possession.

(d) *Restrictions on regulated garbage.* (1) Regulated garbage may not be disposed of, placed on, or removed from a means of conveyance except in accordance with this section.

(2) Regulated garbage is subject to general surveillance for compliance with this section by inspectors and to disposal measures authorized by the Plant Protection Act and the Animal Health Protection Act to prevent the introduction and dissemination of pests and diseases of plants and livestock.

(3) All regulated garbage must be contained in tight, covered, leak-proof receptacles during storage on board a means of conveyance while in the territorial waters, or while otherwise within the territory of the United States. All such receptacles shall be contained inside the guard rail if on a watercraft. Such regulated garbage shall not be unloaded from such means of conveyance in the United States unless such regulated garbage is removed in tight, covered, leak-proof receptacles under the direction of an inspector to an approved facility for incineration, sterilization, or grinding into an approved sewage system, under direct supervision by such an inspector, or such regulated garbage is removed for other handling in such manner and under such supervision as may, upon request

in specific cases, be approved by the Administrator as adequate to prevent the introduction and dissemination of plant pests and animal diseases and sufficient to ensure compliance with applicable laws for environmental protection. *Provided that,* a cruise ship may dispose of regulated garbage in landfills at Alaskan ports only, if and only if the cruise ship does not have prohibited or restricted meat or animal products on board at the time it enters Alaskan waters for the cruise season, and only if the cruise ship, except for incidental travel through international waters necessary to navigate safely between ports, remains in Canadian and U.S. waters off the west coast of North America, and calls only at continental U.S. and Canadian ports during the entire cruise season.

(i) Application for approval of a facility or sewage system may be made in writing by the authorized representative of any carrier or by the official having jurisdiction over the port or place of arrival of the means of conveyance to the Administrator, Animal and Plant Health Inspection Service, U.S. Department of Agriculture, Washington, DC 20250. The application must be endorsed by the operator of the facility or sewage system.

(ii) Approval will be granted if the Administrator determines that the requirements set forth in this section are met. Approval may be denied or withdrawn at any time, if the Administrator determines that such requirements are not met, after notice of the proposed denial or withdrawal of the approval and the reasons therefor, and an opportunity to demonstrate or achieve compliance with such requirements, has been afforded to the operator of the facility or sewage system and to the applicant for approval. However, approval may also be withdrawn without such prior procedure in any case in which the public health, interest, or safety requires immediate action, and in such case, the operator of the facility or sewage system and the applicant for approval shall promptly thereafter be given notice of the withdrawal and the reasons therefor and an opportunity to show cause why the approval should be reinstated.

(e) The Plant Protection and Quarantine Programs and Veterinary Services, Animal, and Plant Health Inspection Service, will cooperate with other Federal, State, and local agencies responsible for enforcing other statutes and regulations governing disposal of the regulated garbage to the end that such disposal shall be adequate to prevent the dissemination of plant pests and livestock or poultry diseases and comply with applicable laws for environmental protection. The inspectors, in maintaining surveillance over regulated garbage movements and disposal, shall coordinate their activities with the activities of representatives of the Environmental Protection Agency and other Federal, State, and local agencies also having jurisdiction over such regulated garbage

§ 330.402 **Garbage generated in Hawaii.**

(a) *Applicability.* This section applies to garbage generated in households, commercial establishments, institutions, and businesses prior to interstate movement from Hawaii, and includes used paper, discarded cans and bottles, and food scraps. Such garbage includes, and is commonly known as, municipal solid waste.

(1) Industrial process wastes, mining wastes, sewage sludge, incinerator ash, or other wastes from Hawaii that the Administrator determines do not pose risks of introducing animal or plant pests or diseases into the continental United States are not regulated under this section.

(2) The interstate movement from Hawaii to the continental United States of agricultural wastes and yard waste (other than incidental amounts (less than 3 percent) that may be present in municipal solid waste despite reasonable efforts to maintain source separation) is prohibited.

(3) Garbage generated onboard any means of conveyance during interstate movement from Hawaii is regulated under § 330.401.

(b) *Restrictions on interstate movement of garbage.* The interstate movement of garbage generated in Hawaii to the continental United States is regulated as provided in this section.

(1) The garbage must be processed, packaged, safeguarded, and disposed of using a methodology that the Administrator has determined is adequate to prevent the introduction or dissemination of plant pests into noninfested areas of the United States.

(2) The garbage must be moved under a compliance agreement in accordance with § 330.403. APHIS will only enter into a compliance agreement when the Administrator is satisfied that the Agency has first satisfied all its obligations under the National Environmental Policy Act and all applicable Federal and State statutes to fully assess the impacts associated with the movement of garbage under the compliance agreement.

(3) All such garbage moved interstate from Hawaii to any of the continental United States must be moved in compliance with all applicable laws for environmental protection.

§ 330.403 Compliance agreement and cancellation.

(a) Any person engaged in the business of handling or disposing of garbage in accordance with this subpart must first enter into a compliance agreement with the Animal and Plant Health Inspection Service (APHIS). Compliance agreement forms (PPQ Form 519) are available without charge from local USDA/APHIS/Plant Protection and Quarantine offices, which are listed in telephone directories.

(b) A person who enters into a compliance agreement, and employees or agents of that person, must comply with the following conditions and any supplemental conditions which are listed in the compliance agreement, as deemed by the Administrator to be necessary to prevent the dissemination into or within the United States of plant pests and livestock or poultry diseases:

(1) Comply with all applicable provisions of this subpart;

(2) Allow inspectors access to all records maintained by the person regarding handling or disposal of garbage, and to all areas where handling or disposal of garbage occurs;

(3)(i) If the garbage is regulated under § 330.401, remove garbage from a means of conveyance only in tight, covered, leak-proof receptacles;

(ii) If the garbage is regulated under § 330.402, transport garbage interstate in packaging approved by the Administrator;

(4) Move the garbage only to a facility approved by the Administrator; and

(5) At the approved facility, dispose of the garbage in a manner approved by the Administrator and described in the compliance agreement.

(c) Approval for a compliance agreement may be denied at any time if the Administrator determines that the applicant has not met or is unable to meet the requirements set forth in this subpart. Prior to denying any application for a compliance agreement, APHIS will provide notice to the applicant thereof, and will provide the applicant with an opportunity to demonstrate or achieve compliance with requirements.

(d) Any compliance agreement may be canceled, either orally or in writing, by an inspector whenever the inspector finds that the person who has entered into the compliance agreement has failed to comply with this subpart. If the cancellation is oral, the cancellation and the reasons for the cancellation will be confirmed in writing as promptly as circumstances allow. Any person whose compliance agreement has been canceled may appeal the decision, in writing, within 10 days after receiving written notification of the cancellation. The appeal must state all of the facts and reasons upon which the person relies to show that the compliance agreement was wrongfully canceled. As promptly as circumstances allow, the Administrator will grant or deny the appeal, in writing, stating the reasons for the decision. A hearing will be held to resolve any conflict as to any material fact. Rules of practice concerning a hearing will be adopted by the Administrator. This administrative remedy must be exhausted before a person can file suit in court challenging the cancellation of a compliance agreement.

(e) Where a compliance agreement is denied or canceled, the person who entered into or applied for the compliance agreement may be prohibited, at the discretion of the Administrator, from handling or disposing of regulated garbage.

.(Approved by the Office of Management and Budget under control numbers 0579–0015, 0579–0054, and 0579–0292)

PART 331—POSSESSION, USE, AND TRANSFER OF SELECT AGENTS AND TOXINS

Sec.

AUTHORITY: 7 U.S.C. 8401; 7 CFR 2.22, 2.80, and 371.3.

SOURCE: 70 FR 13278, Mar. 18, 2005, unless otherwise noted.

§ 331.1 Definitions.

Administrator. The Administrator, Animal and Plant Health Inspection Service, or any person authorized to act for the Administrator.

Animal and Plant Health Inspection Service (APHIS). The Animal and Plant Health Inspection Service of the U.S. Department of Agriculture.

Attorney General. The Attorney General of the United States or any person authorized to act for the Attorney General.

Biological agent. Any microorganism (including, but not limited to, bacteria, viruses, fungi, or protozoa), or infectious substance, or any naturally occurring, bioengineered, or synthesized component of any such microorganism or infectious substance, capable of causing:

(1) Death, disease, or other biological malfunction in a human, an animal, a plant, or another living organism;

(2) Deterioration of food, water, equipment, supplies, or material of any kind; or

(3) Deleterious alteration of the environment.

Centers for Disease Control and Prevention (CDC). The Centers for Disease Control and Prevention of the U.S. Department of Health and Human Services.

Diagnosis. The analysis of specimens for the purpose of identifying or confirming the presence or characteristics of a select agent or toxin, provided that such analysis is directly related to protecting the public health or safety, animal health or animal products, or plant health or plant products.

Entity. Any government agency (Federal, State, or local), academic institution, corporation, company, partnership, society, association, firm, sole proprietorship, or other legal entity.

HHS Secretary. The Secretary of the Department of Health and Human Services or his or her designee, unless otherwise specified.

HHS select agent and/or toxin. A biological agent or toxin listed in 42 CFR 73.3.

Import. To move into, or the act of movement into, the territorial limits of the United States.

Information security. Protecting information and information systems from unauthorized access, use, disclosure, disruption, modification, or destruction in order to provide:

(1) *Integrity,* which means guarding against improper information modification or destruction, and includes ensuring information authenticity;

(2) *Confidentiality,* which means preserving authorized restrictions on access and disclosure, including means for protecting personal privacy and proprietary information; and

(3) *Availability,* which means ensuring timely and reliable access to and use of information.

Interstate. From one State into or through any other State, or within the District of Columbia, Guam, the Virgin Islands of the United States, or any other territory or possession of the United States.

Permit. A written authorization by the Administrator to import or move interstate select agents or toxins, under conditions prescribed by the Administrator.

PPQ. The Plant Protection and Quarantine Programs of the Animal and Plant Health Inspection Service.

Principal investigator. The one individual who is designated by the entity to direct a project or program and who is responsible to the entity for the scientific and technical direction of that project or program.

Recombinant nucleic acids. (1) Molecules that are constructed by joining nucleic acid molecules and that can replicate in a living cell; or

(2) Molecules that result from the replication of those described in paragraph (1) of this definition.

Responsible official. The individual designated by an entity with the authority and control to ensure compliance with the regulations in this part.

Security barrier. A physical structure that is designed to prevent entry by unauthorized persons.

Select agent and/or toxin. A biological agent or toxin listed in § 331.3.

Specimen. Samples of material from humans, animals, plants, or the environment, or isolates or cultures from such samples, for diagnosis, verification, or proficiency testing.

State. Any of the several States of the United States, the Commonwealth of the Northern Mariana Islands, the Commonwealth of Puerto Rico, the District of Columbia, Guam, the Virgin Islands of the United States, or any other territory or possession of the United States.

Synthetic nucleic acids. (1) Molecules that are chemically or by other means synthesized or amplified, including those that are chemically or otherwise modified but can base pair with naturally occurring nucleic acid molecules (i.e., synthetic nucleic acids); or

(2) Molecules that result from the replication of those described in paragraph (1) of this definition.

Toxin. The toxic material or product of plants, animals, microorganisms (including, but not limited to, bacteria, viruses, fungi, or protozoa), or infectious substances, or a recombinant or synthesized molecule, whatever their origin and method of production, and includes:

(1) Any poisonous substance or biological product that may be engineered as a result of biotechnology produced by a living organism; or

(2) Any poisonous isomer or biological product, homolog, or derivative of such a substance.

United States. All of the States.

USDA. The U.S. Department of Agriculture.

Validated inactivation procedure. A procedure, whose efficacy is confirmed by data generated from a viability testing protocol, to render a select agent non-viable but allows the select agent to retain characteristics of interest for future use; or to render any nucleic acids that can produce infectious forms of any select agent virus non-infectious for future use.

Verification. The demonstration of obtaining established performance (*e.g.,* accuracy, precision, and the analytical sensitivity and specificity) specifications for any procedure used for diagnosis.

Viability testing protocol. A protocol to confirm the validated inactivation procedure by demonstrating the material is free of all viable select agent.

[70 FR 13278, Mar. 18, 2005, as amended at 77 FR 61074, Oct. 5, 2012; 79 FR 26830, May 12, 2014; 82 FR 6204, Jan. 19, 2017]

§ 331.2 Purpose and scope.

This part implements the provisions of the Agricultural Bioterrorism Protection Act of 2002 setting forth the requirements for possession, use, and transfer of select agents and toxins. The biological agents and toxins listed in this part have the potential to pose a severe threat to plant health or plant products.

§ 331.3 PPQ select agents and toxins.

(a) Except as provided in paragraphs (d) and (e) of this section, the Administrator has determined that the biological agents and toxins listed in this section have been determined to have the potential to pose a severe threat to plant health or to plant products.

(b) PPQ select agents and toxins:

Coniothyrium glycines, (formerly *Phoma glycinicola, Pyrenochaeta glycines*);

Peronosclerospora   *philippinensis*
  (*Peronosclerospora sacchari*);
Ralstonia solanacearum;
Rathayibacter toxicus;j
Sclerophthora rayssiae;
Synchytrium endobioticum;
Xanthomonas oryzae.

(c) Genetic elements, recombinant and/or synthetic nucleic acids, and recombinant and/or synthetic organisms:

(1) Nucleic acids that can produce infectious forms of any of the select agent viruses listed in paragraph (b) of this section.

(2) Recombinant and/or synthetic nucleic acids that encode for the functional forms of any toxin listed in paragraph (b) of this section if the nucleic acids:

(i) Can be expressed *in vivo* or *in vitro*; or

(ii) Are in a vector or recombinant host genome and can be expressed *in vivo* or *in vitro*.

(3) Select agents and toxins listed in paragraph (b) of this section that have been genetically modified.

(d) Select agents or toxins that meet any of the following criteria are excluded from the requirements of this part:

(1) Any select agent or toxin that is in its naturally occurring environment, provided that the agent or toxin has not been intentionally introduced, cultivated, collected, or otherwise extracted from its natural source.

(2) Nonviable select agents or nontoxic toxins.

(3) A select agent or toxin that has been subjected to decontamination or a destruction procedure when intended for waste disposal.

(4) A select agent or regulated nucleic acid that can produce infectious forms of any select agent virus that has been subjected to a validated inactivation procedure that is confirmed through a viability testing protocol. Surrogate strains that are known to possess equivalent properties with respect to inactivation can be used to validate an inactivation procedure; however, if there are known strain-to-strain variations in the resistance of a select agent to an inactivation procedure, then an inactivation procedure validated on a lesser resistant strain must also be validated on the more resistant strains.

(5) Material containing a select agent that is subjected to a procedure that removes all viable select agent cells, spores, or virus particles if the material is subjected to a viability testing protocol to ensure that the removal method has rendered the material free of all viable select agent.

(6) A select agent or regulated nucleic acids that can produce infectious forms of any select agent virus not subjected to a validated inactivation procedure or material containing a select agent not subjected to a procedure that removes all viable select agent cells, spores, or virus particles if the material is determined by the Administrator to be effectively inactivated or effectively removed. To apply for a determination an individual or entity must submit a written request and supporting scientific information to APHIS. A written decision granting or denying the request will be issued.

(7) A PPQ select toxin identified in an original food sample or clinical sample.

(8) Waste generated during the delivery of patient care by health care professionals from a patient diagnosed with an illness or condition associated with a select agent, where that waste is decontaminated or transferred for destruction by complying with State and Federal regulations within 7 calendar days of the conclusion of patient care.

(9) Any subspecies of *Ralstonia solanacearum* except race 3, biovar 2 and all subspecies of *Sclerophthora rayssiae* except var. *zeae*, provided that the individual or entity can identify that the agent is within the exclusion category.

(e) An attenuated strain of a select agent or a select toxin modified to be less potent or toxic may be excluded from the requirements of this part based upon a determination by the Administrator that the attenuated strain or modified toxin does not pose a severe threat to plant health or plant products.

(1) To apply for exclusion, an individual or entity must submit a written request and supporting scientific information. A written decision granting or denying the request will be issued. An

exclusion will be effective upon notification to the applicant. Exclusions will be listed on the National Select Agent Registry Web site at *http:// www.selectagents.gov/.*

(2) If an excluded attenuated strain or modified toxin is subjected to any manipulation that restores or enhances its virulence or toxic activity, the resulting select agent or toxin will be subject to the requirements of this part.

(3) An individual or entity may make a written request to the Administrator for reconsideration of a decision denying an application for the exclusion of an attenuated strain of a select agent or a select toxin modified to be less potent or toxic. The written request for reconsideration must state the facts and reasoning upon which the individual or entity relies to show the decision was incorrect. The Administrator will grant or deny the request for reconsideration as promptly as circumstances allow and will state, in writing, the reasons for the decision.

(f) Any select agent or toxin seized by a Federal law enforcement agency will be excluded from the requirements of this part during the period between seizure of the agent or toxin and the transfer or destruction of such agent or toxin provided that:

(1) As soon as practicable, the Federal law enforcement agency transfers the seized agent or toxin to an entity eligible to receive such agent or toxin or destroys the agent or toxin by a recognized sterilization or inactivation process.

(2) The Federal law enforcement agency safeguards and secures the seized agent or toxin against theft, loss, or release, and reports any theft, loss, or release of such agent or toxin.

(3) The Federal law enforcement agency reports the seizure of the select agent or toxin to APHIS or CDC. The seizure must be reported within 24 hours by telephone, facsimile, or e-mail. This report must be followed by submission of APHIS/CDC Form 4 within 7 calendar days after seizure of the select agent or toxin. A copy of the completed form must be maintained for 3 years.

(4) The Federal law enforcement agency reports the final disposition of the select agent or toxin to APHIS or CDC by submission of APHIS/CDC Form 4. A copy of the completed form must be maintained for 3 years.

[70 FR 13278, Mar. 18, 2005, as amended at 73 FR 61331, Oct. 16, 2008; 77 FR 61075, Oct. 5, 2012; 79 FR 26830, May 12, 2014; 82 FR 6204, Jan. 19. 2017; 83 FR 48201, Sept. 24, 2018]

§ 331.4   [Reserved]

§ 331.5   Exemptions.

(a) Diagnostic laboratories and other entities that possess, use, or transfer a select agent or toxin that is contained in a specimen presented for diagnosis or verification will be exempt from the requirements of this part for such agent or toxin contained in the specimen, provided that:

(1) Unless directed otherwise by the Administrator, within 7 calendar days after identification of the select agent or toxin, the select agent or toxin is transferred in accordance with § 331.16 or destroyed on-site by a recognized sterilization or inactivation process.

(2) The agent or toxin is secured against theft, loss, or release during the period between identification of the agent or toxin and transfer or destruction of such agent or toxin, and any theft, loss, or release of such agent or toxin is reported.

(3) The identification of the agent or toxin is reported to APHIS or CDC, the specimen provider, and to other appropriate authorities when required by Federal, State, or local law by telephone, facsimile, or email. This report must be followed by submission of APHIS/CDC Form 4 to APHIS or CDC within 7 calendar days after identification.

(b) In addition to the exemption provided in paragraph (a) of this section, the Administrator may grant a specific exemption upon a showing of good cause and upon his or her determination that such exemption is consistent with protecting plant health or plant products. An individual or entity may request in writing an exemption from the requirements of this part. If granted, such exemptions are valid for a maximum of 3 years; thereafter, an individual or entity must request a new exemption. If a request for exemption is denied, an individual or entity may request reconsideration in writing to the Administrator. The request for reconsideration must state all of the facts and reasons upon which the individual or entity relies to show that the exemption was wrongfully denied. The Administrator will grant or deny the request for reconsideration as promptly as circumstances allow and will state, in writing, the reasons for the decision.

[70 FR 13278, Mar. 18, 2005, as amended at 82 FR 6204, Jan. 19, 2017]

§ 331.6 [Reserved]

§ 331.7 Registration and related security risk assessments.

(a) Unless exempted under § 331.5, an individual or entity shall not possess, use, or transfer any select agent or toxin without a certificate of registration issued by the Administrator.

(b) As a condition of registration, each entity is required to be in compliance with the requirements of this part for select agents and toxins listed on the registration regardless of whether the entity is in actual possession of the select agent or toxin. With regard to toxins, the entity registered for possession, use, or transfer of a toxin must be in compliance with the requirements of this part regardless of the amount of toxins currently in its possession.

(c) As a condition of registration, each entity must designate an individual to be its responsible official. While most registrants are likely to be entities, in the event that an individual applies for and is granted a certificate of registration, the individual will be considered the responsible official.

(d)(1) As a condition of registration, the following must be approved by the Administrator or the HHS Secretary based on a security risk assessment by the Attorney General:

(i) The individual or entity;

(ii) The responsible official; and

(iii) Unless otherwise exempted under this section, any individual who owns or controls the entity.

(2) Federal, State, or local governmental agencies, including public accredited academic institutions, are exempt from the security risk assess-

ments for the entity and the individual who owns or controls such entity.

(3) An individual will be deemed to own or control an entity under the following conditions:[1]

(i) For a private institution of higher education, an individual will be deemed to own or control the entity if the individual is in a managerial or executive capacity with regard to the entity's select agents or toxins or with regard to the individuals with access to the select agents or toxins possessed, used, or transferred by the entity.

(ii) For entities other than institutions of higher education, an individual will be deemed to own or control the entity if the individual:

(A) Owns 50 percent or more of the entity, or is a holder or owner of 50 percent or more of its voting stock; or

(B) Is in a managerial or executive capacity with regard to the entity's select agents or toxins or with regard to the individuals with access to the select agents or toxins possessed, used, or transferred by the entity.

(4) An entity will be considered to be an institution of higher education if it is an institution of higher education as defined in section 101(a) of the Higher Education Act of 1965 (20 U.S.C. 1001(a)), or is an organization described in 501(c)(3) of the Internal Revenue Code of 1986, as amended (26 U.S.C. 501(c)(3)).

(5) To obtain a security risk assessment, an individual or entity must submit the information necessary to conduct a security risk assessment to the Attorney General.

(e) To apply for a certificate of registration for only PPQ select agents or toxins, or for PPQ and VS select agents or toxins, an individual or entity must submit the information requested in the registration application package (APHIS/CDC Form 1) to APHIS. To apply for a certificate of registration for overlap select agents or toxins, overlap select agents or toxins and any combination of PPQ or VS select agents or toxins, or HHS select agents or toxins and any combination of PPQ

---

[1] These conditions may apply to more than one individual.

or VS select agents or toxins, an individual or entity must submit the information requested in the registration application package (APHIS/CDC Form 1) to APHIS or CDC, but not both.

(f) Prior to the issuance of a certificate of registration, the responsible official must promptly provide notification of any changes to the application for registration by submitting the relevant page(s) of the registration application.

(g) The issuance of a certificate of registration may be contingent upon inspection or submission of additional information, such as the security plan, biosafety plan, incident response plan, or any other documents required to be prepared under this part.

(h) A certificate of registration will be valid for one physical location (a room, a building, or a group of buildings) where the responsible official will be able to perform the responsibilities required in this part, for specific select agents or toxins, and for specific activities.

(i) A certificate of registration may be amended to reflect changes in circumstances (e.g., replacement of the responsible official or other personnel changes, changes in ownership or control of the entity, changes in the activities involving any select agents or toxins, or the addition or removal of select agents or toxins).

(1) Prior to any change, the responsible official must apply for an amendment to a certificate of registration by submitting the relevant page(s) of the registration application. [2]

(2) The responsible official will be notified in writing if an application to amend a certificate of registration has been approved. Approval of an amendment may be contingent upon an inspection or submission of additional information, such as the security plan, biosafety plan, incident response plan, or any other documents required to be prepared under this part.

(3) No change may be made without such approval.

(j) An entity must immediately notify APHIS or CDC if it loses the services of its responsible official. In the event that an entity loses the services of its responsible official, an entity may continue to possess or use select agents or toxins only if it appoints as the responsible official another individual who has been approved by the Administrator or the HHS Secretary following a security risk assessment by the Attorney General and who meets the requirements of this part.

(k) A certificate of registration will be terminated upon the written request of the entity if the entity no longer possesses or uses any select agents or toxins and no longer wishes to be registered.

(l) A certificate of registration will be valid for a maximum of 3 years.

[70 FR 13278, Mar. 18, 2005, as amended at 82 FR 6205, Jan. 19, 2017]

§ 331.8 Denial, revocation, or suspension of registration.

(a) An application may be denied or a certificate of registration revoked or suspended if:

(1) The individual or entity, the responsible official, or an individual who owns or controls the entity is within any of the categories described in 18 U.S.C. 175b;

(2) The individual or entity, the responsible official, or an individual who owns or controls the entity is reasonably suspected by any Federal law enforcement or intelligence agency of:

(i) Committing a crime set forth in 18 U.S.C. 2332b(g)(5); or

(ii) Knowing involvement with an organization that engages in domestic or international terrorism (as defined in 18 U.S.C. 2331) or with any other organization that engages in intentional crimes of violence; or

(iii) Being an agent of a foreign power as defined in 50 U.S.C. 1801;

(3) The individual or entity does not meet the requirements of this part; [3] or

(4) It is determined that such action is necessary to protect plant health or plant products.

_____

[2] Depending on the change, a security risk assessment by the Attorney General may also be required (e.g., replacement of the responsible official, changes in ownership or control of the entity, new researchers or graduate students, etc.).

[3] If registration is denied for this reason, we may provide technical assistance and guidance.

(b) Upon revocation or suspension of a certificate of registration, the individual or entity must:

(1) Immediately stop all use of each select agent or toxin covered by the revocation or suspension order;

(2) Immediately safeguard and secure each select agent or toxin covered by the revocation or suspension order from theft, loss, or release; and

(3) Comply with all disposition instructions issued by the Administrator for each select agent or toxin covered by the revocation or suspension.

(c) Denial of an application for registration and revocation or suspension of registration may be appealed under §331.20. However, any denial of an application for registration or revocation or suspension of a certificate of registration will remain in effect until a final agency decision has been rendered.

§331.9 Responsible official.

(a) An individual or entity required to register under this part must designate an individual to be the responsible official. The responsible official must:

(1) Be approved by the Administrator or the HHS Secretary following a security risk assessment by the Attorney General.

(2) Be familiar with the requirements of this part.

(3) Have authority and responsibility to act on behalf of the entity.

(4) Ensure compliance with the requirements of this part.

(5) Have a physical (and not merely a telephonic or audio/visual) presence at the registered entity to ensure that the entity is in compliance with the select agent regulations and be able to respond in a timely manner to onsite incidents involving select agents and toxins in accordance with the entity's incident response plan.

(6) Ensure that annual inspections are conducted of each registered space where select agents or toxins are stored or used in order to ensure compliance with the requirements of this part. The results of each inspection must be documented, and any deficiencies identified during an inspection must be corrected and the corrections documented.

(7) Ensure that individuals are provided the contact information for the USDA Office of Inspector General Hotline and the HHS Office of Inspector General Hotline so that they may anonymously report any biosafety/biocontainment or security concerns related to select agents and toxins.

(8) Investigate to determine the reason for any failure of a validated inactivation procedure or any failure to remove viable select agent from material. If the responsible official is unable to determine the cause of a deviation from a validated inactivation procedure or a viable select agent removal method; or receives any report of any inactivation failure after the movement of material to another location, the responsible official must report immediately by telephone or email the inactivation or viable agent removal method failure to APHIS or CDC.

(9) Review, and revise as necessary, each of the entity's validated inactivation procedures or viable select agent removal methods. The review must be conducted annually or after any change in principal investigator, change in the validated inactivation procedure or viable select agent removal method, or failure of the validated inactivation procedure or viable select agent removal method. The review must be documented and training must be conducted if there are any changes to the validated inactivation procedure, viable select agent removal method, or viability testing protocol.

(b) An entity may designate one or more individuals to serve as an alternate responsible official who acts for the responsible official in his/her absence. These individuals must have the authority and control to ensure compliance with the regulations when acting as the responsible official.

(c) The responsible official must report the identification and final disposition of any select agent or toxin contained in a specimen for diagnosis or verification.

(1) The identification of the select agent or toxin must be immediately reported by telephone, facsimile, or e-mail. The final disposition of the agent or toxin must be reported by submission of APHIS/CDC Form 4 within 7 calendar days after identification. A

copy of the completed form must be maintained for 3 years.

(2) Less stringent reporting may be required during agricultural emergencies or outbreaks, or in endemic areas.

[70 FR 13278, Mar. 18, 2005, as amended at 77 FR 61075, Oct. 5, 2012; 82 FR 6205, Jan. 19, 2017]

§ 331.10 Restricting access to select agents and toxins; security risk assessments.

(a) An individual or entity required to register under this part may not provide an individual access to a select agent or toxin, and an individual may not access a select agent or toxin, unless the individual is approved by the Administrator or the HHS Secretary following a security risk assessment by the Attorney General.

(b) An individual will be deemed to have access at any point in time if the individual has possession of a select agent or toxin (e.g., carries, uses, or manipulates) or the ability to gain possession of a select agent or toxin.

(c) Each individual with access to select agents or toxins must have the appropriate education, training, and/or experience to handle or use such agents or toxins.

(d) To apply for access approval, each individual must submit the information necessary to conduct a security risk assessment to the Attorney General.

(e) A person with valid approval from the HHS Secretary or Administrator to have access to select agents or toxins may request, through his or her Responsible Official, that the HHS Secretary or Administrator provide their approved access status to another registered individual or entity for a specified period of time. A responsible official must immediately notify the responsible official of the visiting entity if the person's access to select agents or toxins has been terminated.

(f) An individual's security risk assessment may be expedited upon written request by the responsible official and a showing of good cause (e.g., agricultural emergencies, national security, or a short-term visit by a prominent researcher). A written decision granting or denying the request will be issued.

(g) An individual's access approval may be denied, limited, or revoked if:

(1) The individual is within any of the categories described in 18 U.S.C. 175b;

(2) The individual is reasonably suspected by any Federal law enforcement or intelligence agency of committing a crime set forth in 18 U.S.C. 2332b(g)(5); knowing involvement with an organization that engages in domestic or international terrorism (as defined in 18 U.S.C. 2331) or with any other organization that engages in intentional crimes of violence; or being an agent of a foreign power as defined in 50 U.S.C. 1801; or

(3) It is determined that such action is necessary to protect plant health or plant products.

(h) An individual may appeal the Administrator's decision to deny, limit, or revoke access approval under § 331.20.

(i) Access approval is valid for a maximum of 3 years.

(j) The responsible official must immediately notify APHIS or CDC when an individual's access to select agents or toxins is terminated by the entity and the reasons therefore.

[70 FR 13278, Mar. 18, 2005, as amended at 77 FR 61075, Oct. 5, 2012; 82 FR 6205, Jan. 19, 2017]

§ 331.11 Security.

(a) An individual or entity required to register under this part must develop and implement a written security plan. The security plan must be sufficient to safeguard the select agent or toxin against unauthorized access, theft, loss, or release.

(b) The security plan must be designed according to a site-specific risk assessment and must provide graded protection in accordance with the risk of the select agent or toxin, given its intended use. A current security plan must be submitted for initial registration, renewal of registration, or when requested.

(c) The security plan must:

(1) Describe procedures for physical security, inventory control, and information systems control;

(2) Contain provisions for the control of access to select agents and toxins, including the safeguarding of animals (including arthropods) or plants intentionally or accidentally exposed to or infected with a select agent, against unauthorized access, theft, loss or release.

(3) Contain provisions for routine cleaning, maintenance, and repairs;

(4) Establish procedures for removing unauthorized or suspicious persons;

(5) Describe procedures for addressing loss or compromise of keys, keycards, passwords, combinations, etc. and protocols for changing access permissions or locks following staff changes;

(6) Contain procedures for reporting unauthorized or suspicious persons or activities, loss or theft of select agents or toxins, release of select agents or toxins, or alteration of inventory records;

(7) Contain provisions for ensuring that all individuals with access approval from the Administrator or the HHS Secretary understand and comply with the security procedures;

(8) Describe procedures for how the Responsible Official will be informed of suspicious activity that may be criminal in nature and related to the entity, its personnel, or its select agents or toxins; and describe procedures for how the entity will notify the appropriate Federal, State, or local law enforcement agencies of such activity.

(9) Contain provisions for information security that:

(i) Ensure that all external connections to systems which manage security for the registered space are isolated or have controls that permit only authorized and authenticated users;

(ii) Ensure that authorized and authenticated users are only granted access to select agent and toxin related information, files, equipment (e.g., servers or mass storage devices), and applications as necessary to fulfill their roles and responsibilities, and that access is modified when the user's roles and responsibilities change or when their access to select agents and toxins is suspended or revoked;

(iii) Ensure that controls are in place that are designed to prevent malicious code (such as, but not limited to, computer viruses, worms, spyware) from compromising the confidentiality, integrity, or availability of information systems which manage access to spaces registered under this part or records as specified in §331.17;

(iv) Establish a robust configuration management practice for information systems to include regular patching and updates made to operating systems and individual applications; and

(v) Establish procedures that provide backup security measures in the event that access control systems, surveillance devices, and/or systems that manage the requirements of §331.17 are rendered inoperable.

(10) Contain provisions and policies for shipping, receiving, and storage of select agents and toxins, including documented procedures for receiving, monitoring, and shipping of all select agents and toxins. These provisions must provide that an entity will properly secure containers on site and have a written contingency plan for unexpected shipments.

(d) An individual or entity must adhere to the following security requirements or implement measures to achieve an equivalent or greater level of security:

(1) Allow access only to individuals with access approval from the Administrator or the HHS Secretary;

(2) Allow individuals not approved for access by the Administrator or the HHS Secretary to conduct routine cleaning, maintenance, repairs, and other activities not related to select agents or toxins only when continuously escorted by an approved individual if the potential to access to select agents or toxins exists;

(3) Provide for the control of select agents and toxins by requiring freezers, refrigerators, cabinets, and other containers where select agents or toxins are stored to be secured against unauthorized access (e.g., card access system, lock boxes);

(4) Inspect all suspicious packages before they are brought into or removed from an area where select agents or toxins are used or stored;

(5) Establish a protocol for intra-entity transfers under the supervision of an individual with access approval from the Administrator or the HHS Secretary, including chain-of-custody

documents and provisions for safeguarding against theft, loss, or release; and

(6) Require that individuals with access approval from the Administrator or the HHS Secretary refrain from sharing with any other person their unique means of accessing a select agent or toxin (*e.g.*, keycards or passwords);

(7) Require that individuals with access approval from the Administrator or the HHS Secretary immediately report any of the following to the responsible official:

(i) Any loss or compromise of keys, passwords, combinations, etc.;

(ii) Any suspicious persons or activities;

(iii) Any loss or theft of select agents or toxins;

(iv) Any release of a select agent or toxin;

(v) Any sign that inventory or use records for select agents or toxins have been altered or otherwise compromised; and

(vi) Any loss of computer, hard drive or other data storage device containing information that can be used to gain access to select agents or toxins; and

(8) Separate areas where select agents and toxins are stored or used from the public areas of the building.

(e) Entities must conduct complete inventory audits of all affected select agents and toxins in long-term storage when any of the following occur:

(1) Upon the physical relocation of a collection or inventory of select agents or toxins for those select agents or toxins in the collection or inventory;

(2) Upon the departure or arrival of a principal investigator for those select agents and toxins under the control of that principal investigator; or

(3) In the event of a theft or loss of a select agent or toxin, all select agents and toxins under the control of that principal investigator.

(f) [Reserved]

(g) In developing a security plan, an individual or entity should consider the document entitled, " Security Plan Guidance." This document is available on the National Select Agent Registry at *http://www.selectagents.gov/*.

(h) The plan must be reviewed annually and revised as necessary. Drills or exercises must be conducted at least annually to test and evaluate the effectiveness of the plan. The plan must be reviewed and revised, as necessary, after any drill or exercise and after any incident. Drills or exercises must be documented to include how the drill or exercise tested and evaluated the plan, any problems that were identified and corrective action(s) taken, and the names of registered entity personnel participants.

[70 FR 13278, Mar. 18, 2005, as amended at 77 FR 61075, Oct. 5, 2012; 79 FR 26830, May 12, 2014; 82 FR 6205, Jan. 19, 2017; 83 FR 48202, Sept. 24, 2018]

§ 331.12　Biocontainment.

(a) An individual or entity required to register under this part must develop and implement a written biocontainment plan that is commensurate with the risk of the select agent or toxin, given its intended use.[4] The biocontainment plan must contain sufficient information and documentation to describe the biocontainment procedures for the select agent or toxin, including any animals (including arthropods) or plants intentionally or accidentally exposed to or infected with a select agent. The current biocontainment plan must be submitted for initial registration, renewal of registration, or when requested. The biocontainment plan must include the following provisions:

(1) The hazardous characteristics of each agent or toxin listed on the entity's registration and the biocontainment risk associated with laboratory procedures related to the select agent or toxin;

(2) Safeguards in place with associated work practices to protect entity personnel, the public, and the environment from exposure to the select agent or toxin including, but not limited to: Personal protective equipment and other safety equipment; containment equipment including, but not limited to, biological safety cabinets, animal caging systems, and centrifuge safety containers; and engineering controls and other facility safeguards;

---

[4] Technical assistance and guidance may be obtained by contacting APHIS.

(3) Written procedures for each validated method used for disinfection, decontamination, or destruction, as appropriate, of all contaminated or presumptively contaminated materials including, but not limited to: Cultures and other materials related to the propagation of select agents or toxins, items related to the analysis of select agents and toxins, personal protective equipment, arthropod containment systems, extracted plant and/or arthropod tissues, laboratory surfaces and equipment, and effluent material; and

(4) Procedures for the handling of select agents and toxins in the same spaces with non-select agents and toxins to prevent unintentional contamination.

(b) The biocontainment procedures must be sufficient to contain the select agent or toxin (*e.g.*, physical structure and features of the entity, and operational and procedural safeguards).

(c) In developing a biocontainment plan, an individual or entity should consider the following:

(1) "Containment Facilities and Safeguards for Exotic Plant Pathogens and Pests" (Robert P. Kahn and S.B. Mathur eds., 1999); and

(2) "A Practical Guide to Containment: Greenhouse Research with Transgenic Plants and Microbes" (Patricia L. Traynor ed., 2001).

(d) [Reserved]

(e) The plan must be reviewed annually and revised as necessary. Drills or exercises must be conducted at least annually to test and evaluate the effectiveness of the plan. The plan must be reviewed and revised, as necessary, after any drill or exercise and after any incident. Drills or exercises must be documented to include how the drill or exercise tested and evaluated the plan, any problems that were identified and corrective action(s) taken, and the names of registered entity personnel participants.

[70 FR 13278, Mar. 18, 2005, as amended at 77 FR 61076, Oct. 5, 2012; 79 FR 26830, May 12, 2014; 82 FR 6205, Jan. 19, 2017]

§ 331.13 Restricted experiments.

(a) An individual or entity may not conduct or possess products resulting from the following experiments unless approved by and conducted in accordance with the conditions prescribed by the Administrator:

(1) Experiments that involve the deliberate transfer of, or selection for, a drug or chemical resistance trait to select agents that are not known to acquire the trait naturally, if such acquisition could compromise the control of disease agents in humans, veterinary medicine, or agriculture.

(2) Experiments involving the deliberate formation of synthetic or recombinant nucleic acids containing genes for the biosynthesis of select toxins lethal for vertebrates at an $LD[50]<100$ ng/kg body weight.

(b) The Administrator may revoke approval to conduct any of the experiments in paragraph (a) of this section, or revoke or suspend a certificate of registration, if the individual or entity fails to comply with the requirements of this part.

(c) To apply for approval to conduct any of the experiments in paragraph (a) of this section, an individual or entity must submit a written request and supporting scientific information to the Administrator. A written decision granting or denying the request will be issued.

[70 FR 13278, Mar. 18, 2005, as amended at 77 FR 61076, Oct. 5, 2012; 79 FR 26830, May 12, 2014]

§ 331.14 Incident response. [5]

(a) An individual or entity required to register under this part must develop and implement a written incident response plan [6] based upon a site specific risk assessment. The incident response plan must be coordinated with any entity-wide plans, kept in the workplace, and available to employees for review. The current incident response plan must be submitted for initial registration, renewal of registration, or when requested.

(b) The incident response plan must fully describe the entity's response procedures for the theft, loss, or release of

---

[5] Nothing in this section is meant to supersede or preempt incident response requirements imposed by other statutes or regulations.

[6] Technical assistance and guidance may be obtained by contacting APHIS.

a select agent or toxin; inventory discrepancies; security breaches (including information systems); severe weather and other natural disasters; workplace violence; bomb threats and suspicious packages; and emergencies such as fire, gas leak, explosion, power outage, and other natural and manmade events.

(c) The response procedures must account for hazards associated with the select agent or toxin and appropriate actions to contain such select agent or toxin, including any animals (including arthropods) or plants intentionally or accidentally exposed to or infected with a select agent.

(d) The incident response plan must also contain the following information:

(1) The name and contact information (*e.g.*, home and work) for the individual or entity (*e.g.*, responsible official, alternate responsible official(s), biosafety officer, etc.);

(2) The name and contact information for the building owner and/or manager, where applicable;

(3) The name and contact information for tenant offices, where applicable;

(4) The name and contact information for the physical security official for the building, where applicable;

(5) Personnel roles and lines of authority and communication;

(6) Planning and coordination with local emergency responders;

(7) Procedures to be followed by employees performing rescue or medical duties;

(8) Emergency medical treatment and first aid;

(9) A list of personal protective and emergency equipment, and their locations;

(10) Site security and control;

(11) Procedures for emergency evacuation, including type of evacuation, exit route assignments, safe distances, and places of refuge; and

(12) Decontamination procedures.

(e) [Reserved]

(f) The plan must be reviewed annually and revised as necessary. Drills or exercises must be conducted at least annually to test and evaluate the effectiveness of the plan. The plan must be reviewed and revised, as necessary, after any drill or exercise and after any incident. Drills or exercises must be documented to include how the drill or exercise tested and evaluated the plan, any problems that were identified and corrective action(s) taken, and the names of registered entity personnel participants.

[70 FR 13278, Mar. 18, 2005, as amended at 77 FR 61076, Oct. 5, 2012; 82 FR 6206, Jan. 19, 2017]

§ 331.15 Training.

(a) An individual or entity required to register under this part must provide information and training on biocontainment, biosafety, security (including security awareness), and incident response to:

(1) Each individual with access approval from the Administrator or HHS Secretary. The training must address the particular needs of the individual, the work they will do, and the risks posed by the select agents or toxins. The training must be accomplished prior to the individual's entry into an area where a select agent is handled or stored, or within 12 months of the date the individual was approved by the Administrator or the HHS Secretary for access, whichever is earlier.

(2) Each individual not approved for access to select agents and toxins by the Administrator or HHS Secretary before that individual enters areas under escort where select agents or toxins are handled or stored (*e.g.*, laboratories, growth chambers, animal rooms, greenhouses, storage areas, shipping/receiving areas, production facilities, etc.). Training for escorted personnel must be based on the risk associated with accessing areas where select agents and toxins are used and/or stored. The training must be accomplished prior to the individual's entry into where select agents or toxins are handled or stored (*e.g.*, laboratories, growth chambers, animal rooms, greenhouses, storage areas, shipping/receiving areas, production facilities, etc.).

(b) [Reserved]

(c) Refresher training must be provided annually for individuals with access approval from the HHS Secretary or Administrator or at such time as the registered individual or entity significantly amends its security, incident response, or biocontainment plans.

(d) The responsible official must ensure a record of the training provided to each individual with access to select agents and toxins and each escorted individual (e.g., laboratory workers, visitors, etc.) is maintained. The record must include the name of the individual, the date of the training, a description of the training provided, and the means used to verify that the employee understood the training.

(e) The responsible official must ensure and document that individuals are provided the contact information of the USDA Office of Inspector General Hotline and the HHS Office of Inspector General Hotline so that they may anonymously report any safety or security concerns related to select agents and toxins.

[77 FR 61076, Oct. 5, 2012, as amended at 82 FR 6206, Jan. 19, 2017]

§ 331.16 Transfers.

(a) Except as provided in paragraph (c) of this section, a select agent or toxin may only be transferred to an individual or entity registered to possess, use, or transfer that agent or toxin. A select agent or toxin may only be transferred under the conditions of this section and must be authorized by APHIS or CDC prior to the transfer.[7]

(b) A transfer may be authorized if:

(1) The sender:

(i) Has at the time of transfer a certificate of registration that covers the particular select agent or toxin to be transferred and meets all the requirements of this part;

(ii) Meets the exemption requirements for the particular select agent or toxin to be transferred; or

(iii) Is transferring the select agent or toxin from outside of the United States and meets all import requirements.

(2) At the time of transfer, the recipient has a certificate of registration that includes the particular select agent or toxin to be transferred and meets all of the requirements of this part.

---

[7]The requirements of this section do not apply to transfers within a registered entity (*i.e.*, the sender and the recipient are covered by the same certificate of registration).

(c) On a case-by-case basis, the Administrator may authorize a transfer of a select agent or toxin not otherwise eligible for transfer under this part under conditions prescribed by the Administrator.

(d) To obtain authorization for a transfer, APHIS/CDC Form 2 must be submitted.

(e) After authorization is provided by APHIS or CDC, the packaging of the select agent(s) and toxin(s) is performed by an individual approved by the HHS Secretary or Administrator to have access to select agents and toxins and is in compliance with all applicable laws concerning packaging.

(f) The sender must comply with all applicable laws governing shipping.

(g) Transportation in commerce starts when the select agent(s) or toxin(s) are packaged for shipment and ready for receipt by a courier transporting select agent(s) or toxin(s) and ends when the package is received by the intended recipient who is an individual approved by the HHS Secretary or Administrator to have access to select agents and toxins, following a security risk assessment by the Attorney General.

(h) The recipient must submit a completed APHIS/CDC Form 2 within 2 business days of receipt of a select agent or toxin.

(i) The recipient must immediately notify APHIS or CDC if the select agent or toxin has not been received within 48 hours after the expected delivery time or if the package containing the select agent or toxin has been damaged to the extent that a release of the select agent or toxin may have occurred.

(j) An authorization for a transfer shall be valid only for 30 calendar days after issuance, except that such an authorization becomes immediately null and void if any facts supporting the authorization change (*e.g.*, change in the certificate of registration for the sender or recipient, change in the application for transfer).

[70 FR 13278, Mar. 18, 2005, as amended at 77 FR 61077, Oct. 5, 2012; 82 FR 6206, Jan. 19, 2017]

§ 331.17 Records.

(a) An individual or entity required to register under this part must maintain complete records relating to the activities covered by this part. Such records must include:

(1) An accurate, current inventory for each select agent (including viral genetic elements, recombinant and/or synthetic nucleic acids, and organisms containing recombinant and/or synthetic nucleic acids) held in long-term storage (placement in a system designed to ensure viability for future use, such as in a freezer or lyophilized materials), including:

(i) The name and characteristics (*e.g.*, strain designation, GenBank Accession number, etc.);

(ii) The quantity acquired from another individual or entity (*e.g.*, containers, vials, tubes, etc.), date of acquisition, and the source;

(iii) Where stored (*e.g.*, building, room, and freezer or other storage container);

(iv) When moved from storage and by whom and when returned to storage and by whom;

(v) The select agent used, purpose of use, and, when applicable, final disposition;

(vi) Records created under § 331.16 (Transfers);

(vii) For intra-entity transfers (sender and the recipient are covered by the same certificate of registration), the select agent, the quantity transferred, the date of transfer, the sender, and the recipient; and

(viii) Records created under § 331.19 (Notification of theft, loss, or release);

(2) An accurate, current accounting of any animals or plants intentionally or accidentally exposed to or infected with a select agent (including number and species, location, and appropriate disposition);

(3) An accurate, current inventory for each toxin held, including:

(i) The name and characteristics;

(ii) The quantity acquired from another individual or entity (*e.g.*, containers, vials, tubes, etc.), date of acquisition, and the source;

(iii) The initial and current quantity amount (*e.g.*, milligrams, milliliters, grams, etc.);

(iv) The toxin used and purpose of use, quantity, date(s) of the use and by whom;

(v) Where stored (*e.g.*, building, room, and freezer or other storage container);

(vi) When moved from storage and by whom and when returned to storage and by whom, including quantity amount;

(vii) Records created under § 331.16 (Transfers);

(viii) For intra-entity transfers (sender and the recipient are covered by the same certificate of registration), the toxin, the quantity transferred, the date of transfer, the sender, and the recipient;

(ix) Records created under § 331.19 (Notification of theft, loss, or release);

(x) If destroyed, the quantity of toxin destroyed, the date of such action, and by whom.

(4) A current list of all individuals that have been granted access approval by the Administrator or the HHS Secretary;

(5) Information about all entries into areas containing select agents or toxins, including the name of the individual, name of the escort (if applicable), and the date and time of entry;

(6) Accurate, current records created under § 331.9(c) (Responsible official), § 331.11 (Security), § 331.12 (Biocontainment), § 331.14 (Incident response), and § 331.15 (Training);

(7) A written explanation of any discrepancies; and

(8) For select agents or material containing select agents or regulated nucleic acids that can produce infectious forms of any select agent virus that have been subjected to a validated inactivation procedure or a procedure for removal of viable select agent:

(i) A written description of the validated inactivation procedure or viable select agent removal method used, including validation data;

(ii) A written description of the viability testing protocol used;

(iii) A written description of the investigation conducted by the entity responsible official involving an inactivation or viable select agent removal failure and the corrective actions taken;

(iv) The name of each individual performing the validated inactivation or viable select agent removal method;

(v) The date(s) the validated inactivation or viable select agent removal method was completed;

(vi) The location where the validated inactivation or viable select agent removal method was performed; and

(vii) A certificate, signed by the principal investigator, that includes the date of inactivation or viable select agent removal, the validated inactivation or viable select agent removal method used, and the name of the principal investigator. A copy of the certificate must accompany any transfer of inactivated or select agent removed material.

(b) The individual or entity must implement a system to ensure that all records and databases created under this part are accurate and legible, have controlled access, and that their authenticity may be verified.

(c) The individual or entity must promptly produce upon request any information that is related to the requirements of this part but is not otherwise contained in a record required to be kept by this section. The location of such information may include, but is not limited to, biocontainment certifications, laboratory notebooks, institutional biosafety and/or animal use committee minutes and approved protocols, and records associated with occupational health and suitability programs. All records created under this part must be maintained for 3 years.

[70 FR 13278, Mar. 18, 2005, as amended at 77 FR 61077, Oct. 5, 2012; 82 FR 6206, Jan. 19, 2017]

§331.18 Inspections.

(a) Without prior notification, APHIS must be allowed to inspect any site at which activities regulated under this part are conducted and must be allowed to inspect and copy any records relating to the activities covered by this part.

(b) Prior to issuing a certificate of registration to an individual or entity, APHIS may inspect and evaluate their premises and records to ensure compliance with this part.

§331.19 Notification of theft, loss, or release.

(a) An individual or entity must immediately notify APHIS or CDC upon discovery of the theft or loss of a select agent or toxin. Thefts or losses must be reported even if the select agent or toxin is subsequently recovered or the responsible parties are identified.

(1) The theft or loss of a select agent or toxin must be reported by telephone, facsimile, or e-mail. The following information must be provided:

(i) The name of the select agent or toxin and any identifying information (*e.g.*, strain or other characterization information);

(ii) An estimate of the quantity stolen or lost;

(iii) An estimate of the time during which the theft or loss occurred;

(iv) The location (building, room) from which the theft or loss occurred; and

(v) The list of Federal, State, or local law enforcement agencies to which the individual or entity reported, or intends to report, the theft or loss.

(2) A completed APHIS/CDC Form 3 must be submitted within 7 calendar days.

(b) An individual or entity must notify APHIS or CDC immediately upon discovery of a release of a select agent or toxin outside of the primary barriers of the biocontainment area.

(1) The release of a select agent or toxin must be reported by telephone, facsimile, or e-mail. The following information must be provided:

(i) The name of the select agent or toxin and any identifying information (*e.g.*, strain or other characterization information);

(ii) An estimate of the quantity released;

(iii) The time and duration of the release;

(iv) The location (building, room) from which the release occurred; and

(v) The number of individuals potentially exposed at the entity;

(vi) Actions taken to respond to the release; and

(vii) Hazards posed by the release.

(2) A completed APHIS/CDC Form 3 must be submitted within 7 calendar days.

[70 FR 13278, Mar. 18, 2005, as amended at 77 FR 61077, Oct. 5, 2012]

§ 331.20 Administrative review.

(a) An individual or entity may appeal a denial, revocation, or suspension of registration under this part. The appeal must be in writing, state the factual basis for the appeal, and be submitted to the Administrator within 30 calendar days of the decision.

(b) An individual may appeal a denial, limitation, or revocation of access approval under this part. The appeal must be in writing, state the factual basis for the appeal, and be submitted to the Administrator within 180 calendar days of the decision.

(c) The Administrator's decision constitutes final agency action.

[77 FR 61077, Oct. 5, 2012]

PART 340—MOVEMENT OF ORGANISMS MODIFIED OR PRODUCED THROUGH GENETIC ENGINEERING

Sec.
340.1 Applicability of this part.
340.2 Scope of this part.
340.3 Definitions.
340.4 Regulatory status review.
340.5 Permits.
340.6 Record retention, compliance, and enforcement.
340.7 Confidential business information.
340.8 Costs and charges.

AUTHORITY: 7 U.S.C. 7701–7772 and 7781–7786; 31 U.S.C. 9701; 7 CFR 2.22, 2.80, and 371.3.

SOURCE: 85 FR 29832, May 18, 2020, unless otherwise noted.

§ 340.1 Applicability of this part.

(a) The regulations in this part apply to those organisms described in § 340.2, but not to any organism that is exempt from this part under paragraph (b), (c), or (d) of this section.

(b) The regulations in this part do not apply to plants that have been modified such that they contain either a single modification of a type listed in paragraphs (b)(1) through (3) of this section, or additional modifications as determined by the Administrator, and

described in paragraph (b)(4) of this section.

(1) The genetic modification is a change resulting from cellular repair of a targeted DNA break in the absence of an externally provided repair template; or

(2) The genetic modification is a targeted single base pair substitution; or

(3) The genetic modification introduces a gene known to occur in the plant's gene pool, or makes changes in a targeted sequence to correspond to a known allele of such a gene or to a known structural variation present in the gene pool.

(4) The Administrator may propose to exempt plants with additional modifications, based on what could be achieved through conventional breeding. Such proposals may be Agency-initiated, and follow the process in paragraph (b)(4)(i) of this section, or in response to a request made in accordance with paragraph (b)(4)(ii) of this section.

(i) *APHIS-initiated proposals for exemptions.* APHIS will publish a notice in the FEDERAL REGISTER of the proposal by the Administrator to exempt plants with additional modifications. The notice will make available any supporting documentation, and will request public comment. After reviewing the comments, APHIS will publish a subsequent notice in the FEDERAL REGISTER announcing its final determination.

(ii) *Other parties' requests for exemptions.* Any person may request that the Administrator exempt plants developed with additional modifications that could be achieved through conventional breeding. To submit a request, the person must provide, in writing, information supporting the modification(s). Supporting information must include the following:

(A) A description of the modification(s);

(B) The factual grounds demonstrating that the proposed modification(s) could be achieved through conventional plant breeding;

(C) Copies of scientific literature, unpublished studies, or other data that support the request; and

(D) Any information known to the requestor that would be unfavorable to the request.

(iii) *Timeframe for Agency review of requests for additional exemptions.* After APHIS receives all information required under paragraph (b)(4)(ii) of this section, APHIS will complete its review of the request and render a determination within 12 months, except in circumstances that could not reasonably have been anticipated.

(iv) *Denial of requests.* If APHIS disagrees with the conclusions of the request or determines that there is insufficient evidence that the modification could be achieved through conventional breeding methods, APHIS will deny the request and notify the requestor in writing regarding this denial.

(v) *Agreement with requests.* If APHIS initially determines that the modification could be achieved through conventional breeding methods, APHIS will publish a notice in the FEDERAL REGISTER and request public comments in accordance with the process set forth in paragraph (b)(4)(i) of this section. After reviewing the comments, APHIS will publish a subsequent notice in the FEDERAL REGISTER announcing its final determination.

(vi) *website posting.* A list specifying the additional modifications will be posted on the APHIS website at *https://www.aphis.usda.gov/aphis/ourfocus/biotechnology.*

(c) The regulations in this part do not apply to a plant with:

(1) A plant-trait-mechanism of action combination that has previously undergone an analysis by APHIS in accordance with §340.4 and has been determined by APHIS not to be regulated under this part, or

(2) A plant-trait-mechanism of action combination found in a plant that APHIS determined to be deregulated in response to a petition submitted prior to October 1, 2021, pursuant to §340.6 as that section was set forth prior to August 17, 2020. All plants determined by APHIS to be deregulated pursuant to §340.6 as that section was set forth prior to August 17, 2020 will retain their nonregulated status under these regulations.

(d) The regulations in this part do not apply to plants determined by APHIS not to require regulation under this part pursuant to the "Am I Regulated" process. All plants determined by APHIS not to require regulation under this part pursuant to the "Am I Regulated" process will retain their nonregulated status under these regulations.

(e) Developers may request confirmation from APHIS that a plant is not within the scope of this part. APHIS will provide a written response (confirmation letter) within 120 days of receiving a sufficiently detailed confirmation request, except in circumstances that could not reasonably have been anticipated.

(Approved by the Office of Management and Budget under control number 0579–0471)

§340.2 Scope of this part.

Except under a permit issued by the Administrator in accordance with §340.5, no person shall move any GE organism that:

(a) Is a plant that has a plant-trait-mechanism of action combination that has not been evaluated by APHIS in accordance with §340.4 or that, as a result of such evaluation, is subject to the regulations; or

(b) Meets the definition of a *plant pest* in §340.3; or

(c) Is not a plant but has received deoxyribonucleic acid (DNA) from a plant pest, as defined in §340.3, and the DNA from the donor organism either is capable of producing an infectious agent that causes plant disease or encodes a compound that is capable of causing plant disease; or

(d) Is a microorganism used to control plant pests, or an invertebrate predator or parasite (parasitoid) used to control invertebrate plant pests, and could pose a plant pest risk; or

(e) Is a plant that encodes a product intended for pharmaceutical or industrial use.

§340.3 Definitions.

Terms used in the singular form in this part shall be construed as the plural, and vice versa, as the case may demand. The following terms, when used in this part, shall be construed, respectively, to mean:

Access. The ability during regular business hours to enter, or pass to and from, a location, inspect, and/or obtain or make use or copies of any records, data, or samples necessary to evaluate compliance with this part and all conditions of a permit issued in accordance with § 340.5.

Administrator. The Administrator of the Animal and Plant Health Inspection Service (APHIS) or any other employee of APHIS to whom authority has been or may be delegated to act in the Administrator's stead.

Agent. A person who is designated by the responsible person to act in whole or in part on behalf of the permittee to maintain control over an organism under permit during its movement and to ensure compliance with all applicable permit conditions and the requirements in this part. Multiple agents may be associated with a single responsible person or permit. Agents may be, but are not limited to, brokers, farmers, researchers, or site cooperators. An agent must be at least 18 years of age and be a legal resident of the United States.

Animal and Plant Health Inspection Service (APHIS). An agency of the United States Department of Agriculture (USDA).

Article. Any material or tangible object that could harbor plant pests.

Contained facility. A structure for the storage and/or propagation of living organisms designed with physical barriers capable of preventing the escape of the organisms. Examples include but are not limited to laboratories, growth chambers, fermenters, and containment greenhouses.

Donor organism. The organism from which genetic material is obtained for transfer to the recipient organism.

Environment. All the land, air, and water; and all living organisms in association with land, air, and water.

Gene pool. Germplasm within which sexual recombination is possible as a result of hybridization, including via methods such as embryo culture or bridging crosses.

Genetic engineering. Techniques that use recombinant, synthesized, or amplified nucleic acids to modify or create a genome.

Import (importation). To move into, or the act of movement into, the territorial limits of the United States.

Inspector. Any individual authorized by the Administrator or by the Commissioner of Customs and Border Protection, Department of Homeland Security, to enforce the regulations in this part.

Interstate. From one State into or through any other State or within the District of Columbia, the Commonwealth of the Northern Mariana Islands, the Commonwealth of Puerto Rico, Guam, the Virgin Islands of the United States, or any other territory or possession of the United States.

Mechanism of action (MOA). The biochemical process(es) through which genetic material determines a trait.

Move (moving, movement). To carry, enter, import, mail, ship, or transport; aid, abet, cause, or induce the carrying, entering, importing, mailing, shipping, or transporting; to offer to carry, enter, import, mail, ship, or transport; to receive to carry, enter, import, mail, ship, or transport; to release into the environment; or to allow any of the above activities to occur.

Organism. Any active, infective, or dormant stage of life form of an entity characterized as living, including vertebrate and invertebrate animals, plants, bacteria, fungi, mycoplasmas, mycoplasma-like organisms, as well as entities such as viroids, viruses, or any entity characterized as living, related to the foregoing.

Permit. A written authorization, including by electronic methods, by the Administrator to move organisms regulated under this part and associated articles under conditions prescribed by the Administrator.

Person. Any individual, partnership, corporation, company, society, association, or other organized group.

Plant. Any plant (including any plant part) for or capable of propagation, including a tree, a tissue culture, a plantlet culture, pollen, a shrub, a vine, a cutting, a graft, a scion, a bud, a bulb, a root, or a seed.

Plant pest. Any living stage of a protozoan, nonhuman animal, parasitic plant, bacterium, fungus, virus or viroid, infectious agent or other pathogen, or any article similar to or allied with any of the foregoing, that can directly or indirectly injure, cause damage to, or cause disease in any plant or plant product.

Plant pest risk. The potential for direct or indirect injury to, damage to, or disease in any plant or plant product resulting from introducing or disseminating a plant pest, or the potential for exacerbating the impact of a plant pest.

Plant product. (1) Any flower, fruit, vegetable, root, bulb, seed, or other plant part that is not included in the definition of plant; or

(2) Any manufactured or processed plant or plant part.

Recipient organism. The organism whose nucleic acid sequence will be modified through the use of genetic engineering.

Release into the environment (environmental release). The use of an organism outside the physical constraints of a contained facility.

Responsible person. The individual responsible for maintaining control over a GE organism under permit during its movement and for ensuring compliance with all conditions contained in any applicable permit as well as with other requirements in this part and in the Plant Protection Act (7 U.S.C. 7701 *et seq.*). This individual must sign the permit application, and must be at least 18 years of age, and must be a legal resident of the United States.

Secure shipment. Shipment in a container or a means of conveyance of sufficient strength and integrity to withstand leakage of contents, shocks, pressure changes, and other conditions incident to ordinary handling in transportation.

State. Any of the several States of the United States, the Commonwealth of the Northern Mariana Islands, the Commonwealth of Puerto Rico, the District of Columbia, Guam, the Virgin Islands of the United States, or any other territories or possessions of the United States.

State or Tribal regulatory official. State or Tribal official with responsibilities for plant health, or any other duly designated State or Tribal official, in the State or on the Tribal lands where the movement is to take place.

Trait. An observable (able to be seen or otherwise identified) characteristic of an organism.

Unauthorized release. The intentional or accidental movement of an organism under a permit issued pursuant to this part in a manner not authorized by the permit; or the intentional or accidental movement without a permit of an organism that is subject to the regulations in this part.

§340.4 Regulatory status review.

(a)(1) Any person may submit a request to APHIS for a regulatory status review, pursuant to paragraph (b)(3) of this section.

(2) Any person may request re-review of a GE plant previously found to be subject to this part after an initial review was conducted, provided that the request is supported by new, scientifically valid evidence bearing on the plant pest risk associated with movement of the plant.

(3) APHIS may also initiate a regulatory status review or re-review of a GE plant to identify whether it is subject to regulation under this part.

(4) Information submitted in support of a request for a regulatory status review or re-review must meet the requirements listed in paragraphs (a)(4)(i) through (iii) of this section.

(i) A description of the comparator plant(s), to include genus, species, and any relevant subspecies information;

(ii) The genotype of the modified plant, including a detailed description of the differences in genotype between the modified and unmodified plant; and

(iii) A detailed description of the new trait(s) of the modified plant.

(iv) Detailed information on how to meet the above-listed requirements can be found on the APHIS website at *https://www.aphis.usda.gov/aphis/ ourfocus/biotechnology.* If APHIS proposes revisions to the detailed information on the APHIS website, APHIS will make the proposed revisions available for notice and public comment prior to implementation.

(b)(1) When APHIS receives a request for a regulatory status review of a GE plant, APHIS will conduct an initial review to determine whether there is a plausible pathway by which the GE plant, or any sexually compatible relatives that can acquire the engineered trait from the GE plant, would pose an increased plant pest risk relative to the plant pest risk posed by the respective non-GE or other appropriate comparator(s), based on the following factors:

(i) The biology of the comparator plant(s) and its sexually compatible relatives;

(ii) The trait and mechanism-of-action of the modification(s); and

(iii) The effect of the trait and mechanism-of-action on:

(A) The distribution, density, or development of the plant and its sexually compatible relatives;

(B) The production, creation, or enhancement of a plant pest or a reservoir for a plant pest;

(C) Harm to non-target organisms beneficial to agriculture; and

(D) The weedy impacts of the plant and its sexually compatible relatives.

(2) APHIS will complete the initial review within 180 days of receiving a request for a regulatory status review that meets the requirements specified in paragraph (a)(4) of this section, except in circumstances that could not reasonably have been anticipated. If APHIS does not identify a plausible pathway by which the GE plant or its sexually compatible relatives would pose an increased plant pest risk relative to the comparator(s) in the initial review, the GE plant is not subject to the regulations in this part. APHIS will post the plant, trait, and general description of the MOA on its website.

(b)(3)(i) If APHIS does identify a plausible pathway by which the GE plant or its sexually compatible relatives would pose an increased plant pest risk relative to the comparator(s) in the initial review, the requestor may apply for a permit and/or request that APHIS conduct an evaluation of the factor(s) of concern identified in the initial review to determine the likelihood and consequence of the plausible increased plant pest risk. APHIS may request additional information as needed to evaluate the factor(s) of concern.

(ii) For those GE plants for which such an evaluation is conducted, APHIS will publish the results of the evaluation in the FEDERAL REGISTER

and will solicit and review comments from the public. Except in circumstances that could not reasonably have been anticipated, APHIS will complete these steps within 15 months of receiving a request for a regulatory status review that meets the requirements specified in paragraph (a)(4) of this section.

(iii) If APHIS finds that the GE plant and its sexually compatible relatives are unlikely to pose an increased plant pest risk relative to their comparator(s), the GE plant is not subject to this part. APHIS will publish its evaluation of the plant-trait-MOA combination in a subsequent FEDERAL REGISTER document and will also post it on the APHIS website. If APHIS does not make such a finding, the GE plant will remain regulated under this part, and its movement will be allowed only under permit in accordance with § 340.5.

(c) This section is applicable beginning April 5, 2021 for GE corn, soybean, cotton, potato, tomato, and alfalfa, and on October 1, 2021 for all GE plants.

(Approved by the Office of Management and Budget under control number 0579–0471)

§ 340.5 Permits.

(a) *Permit requirement.* A permit from APHIS is required for the movement of all GE organisms subject to the regulations under this part.

(b) *Permit application requirements.* All applications for permits must be submitted in accordance with the requirements of this section. The responsible person must apply for and obtain a permit through APHIS' website. The application must also include the following information:

(1) *General information requirements for all permit applications.* All permit applications must include the name, title, and contact information of the responsible person and agent (if any); the country (or countries) and locality (or localities) where the organism was collected, developed, manufactured, reared, cultivated, and cultured (as applicable); the organism's genus, species and any relevant subspecies and common name information; the intended activity (*i.e.,* importation, interstate

movement, or release into the environment of the GE organism); and information on the intended trait and the genotype of the intended trait. All permit applications must be signed by the responsible person.

(2) *Information requirements for permit applications for interstate movement or importation.* Applications for permits for interstate movement or importation of GE organisms must include the following additional information:

(i) The origin and destination of the GE organism, including information on the addresses and contact details of the sender and recipient, if different from the responsible person;

(ii) The quantity of the GE organism, the method of shipment, and means of ensuring the security of the shipment against unauthorized release of the organism; and

(iii) The manner in which packaging material, shipping containers, and any other material accompanying the organism will be disposed of to prevent unauthorized release.

(3) *Information requirements for permit applications for release into the environment.* Applications for permits for release of GE organisms into the environment must include information on all proposed environmental release sites, including land area (size), Global Positioning System coordinates, addresses, and land use history of the site and adjacent areas; and the name and contact information of a person at each environmental release site, if different from the responsible person. In the event that additional release sites are requested after the issuance of a permit, APHIS will evaluate and amend permits as appropriate, in accordance with paragraph (l) of this section.

(c) *Exemption for GE Arabidopsis thaliana.* A permit for interstate movement is not required for GE *Arabidopsis thaliana*, provided that it is moved as a secure shipment, the modified genetic material is stably integrated into the plant genome, and the modified material does not include the complete infectious genome of a plant pest.

(d) *Exemption for GE disarmed Agrobacterium species.* A permit for importation or interstate movement is not required for any GE disarmed *Agrobacterium* species, provided that it

is moved as a secure shipment, the modified genetic material is stably integrated into the genome, and the modified material does not include the complete infectious genome of a plant pest.

(e) *Exemption for Drosophila melanogaster.* A permit for importation or interstate movement is not required for GE *Drosophila melanogaster*, provided that it is moved as a secure shipment and that any introduced genetic material is not designed to propagate through a population by biasing the inheritance rate.

(f) *Exemption for certain microbial pesticides.* A permit is not required for the movement of any GE microorganism product that is currently registered with the Environmental Protection Agency (EPA) as a microbial pesticide, so long as the microorganism is not a plant pest as defined in §340.3.

(g) *Exemption of certain plant-incorporated protectants.* A permit is not required for the movement of any GE plant modified solely to contain a plant-incorporated protectant that is currently registered with EPA as a pesticide product pursuant to the Federal Insecticide, Fungicide, and Rodenticide Act (7 U.S.C. 136 *et seq., FIFRA*) or that is currently exempted from FIFRA pursuant to 40 CFR 174.21.

(h) *Administrative actions—(1) Review of permit applications.* APHIS will review the permit application to determine whether it is complete. APHIS will notify the applicant orally or in writing if the application is incomplete, and the applicant will be provided the opportunity to revise the application. Once an application is complete, APHIS will review it to determine whether to approve or deny the application.

(2) *APHIS assignment of permit conditions.* If a permit application is approved, the Administrator will issue a permit with conditions as described in paragraph (i) of this section. Prior to issuance of a permit, the responsible person must agree in writing, in a manner prescribed by the Administrator, that the responsible person and all agents of the responsible person are aware of, understand, and will comply with the permit conditions. Failure to

335

comply with this provision will be grounds for the denial of a permit.

(3) *Inspections.* All premises associated with the permit are subject to inspection before and after permit issuance, and all materials associated with the movement are subject to sampling after permit issuance. The responsible person and agents must provide inspectors access to premises, facilities, release locations, storage areas, waypoints, materials, equipment, means of conveyance, documents, and records related to the movement of organisms permitted under this part. Failure to provide access for inspection prior to the issuance of a permit will be grounds for the denial of a permit. Failure to provide access for inspection following permit issuance will be grounds for withdrawal of the permit.

(4) *State or Tribal review and comment.* The Administrator will submit for notification and review a copy of the permit application, without confidential business information (CBI), and any permit conditions to the appropriate State or Tribal regulatory official. Timely comments received from the State or Tribal regulatory official will be considered by the Administrator prior to permit issuance.

(5) *Approval or denial of a permit.* Except in circumstances that could not reasonably have been anticipated, APHIS will approve or deny the permit within:

(i) 45 days of receipt of a complete application for a permit for interstate movement or for importation; or

(ii) 120 days of receipt of a complete application for a permit for release into the environment.

(iii) The 120-day period may be extended if preparation of an environmental assessment or environmental impact statement is necessary.

(i) *Permit conditions.* The standard conditions listed in this paragraph (i) will be assigned to all permits issued under this section. The Administrator may assign supplemental permit conditions as deemed necessary to ensure confinement of the GE organism. Prior to issuance of a permit or an amended permit, the responsible person will be required to agree in writing or electronically that he or she and his or her agents will comply with the conditions of the permit, as described in this paragraph (i). If the responsible person does not agree to the conditions, the amendment will be denied.

(1) The organism under permit must be maintained and disposed of in a manner so as to prevent its unauthorized release, spread, dispersal, and/or persistence in the environment.

(2) The organism under permit must be kept separate from other organisms, except as specifically allowed in the permit.

(3) The organism under permit must be maintained only in areas and premises specified in the permit.

(4) The identity of the organism under permit must be maintained and verifiable at all times.

(5) Authorized activities may be engaged in only while the permit is valid; the duration for which the permit is valid will be listed on the permit itself.

(6) Records related to activities carried out under the permit must be maintained by the responsible person and must be of sufficient accuracy, quality, and completeness to demonstrate compliance with all permit conditions and requirements under this part. APHIS must be allowed access to all records, to include visual inspection and reproduction (*e.g.,* photocopying, digital reproduction). The responsible person must submit reports and notices to APHIS, containing the information specified within the permit, at the times specified in the permit. At a minimum:

(i) Following an environmental release, environmental release reports must be submitted for all authorized release locations where the release occurred. Environmental release reports must contain details of sufficient accuracy, quality, and completeness to identify the location, shape, and size of the release and the organism(s) released into the environment. In the event no release occurs at an authorized location, an environmental release report of no environmental release must be submitted for all authorized locations where an environmental release did not occur. Unauthorized releases must be reported in accordance with paragraph (i)(9) of this section.

(ii) When the environmental release is of a plant, reports of volunteer monitoring activities and findings must be submitted for all authorized release locations where an environmental release occurred. If no monitoring activities are conducted, a volunteer monitoring report of no monitoring must be submitted indicating why no volunteer monitoring was done.

(7) Inspectors must be allowed access, during regular business hours, to all locations related to the permitted activities.

(8) The organism under permit must undergo the application of measures determined by the Administrator to be necessary to prevent its unauthorized release, spread, dispersal, and/or persistence in the environment.

(9) In the event of a possible or actual unauthorized release, the responsible person must contact APHIS as described in the permit within 24 hours of discovery and must subsequently supply a statement of facts in writing no later than 5 business days after discovery.

(10) The responsible person for a permit remains the responsible person for the permit unless a transfer of responsibility is approved by APHIS. The responsible person must contact APHIS to initiate any transfer. The new responsible person assumes all responsibilities for ensuring compliance with the existing permit and permit conditions and for meeting the requirements of this part.

(j) *Denial or withdrawal of a permit.* Permit applications may be denied, or permits withdrawn, in accordance with this paragraph.

(1) *Denial of permits.* The Administrator may deny, either orally or in writing, any application for a permit. If the denial is oral, the Administrator will then communicate, as promptly as circumstances allow, the denial, and the reasons for it, in writing. The Administrator may deny a permit application if:

(i) The Administrator concludes that the proposed actions, *e.g.*, movements under permit, may not prevent the unauthorized release, spread, dispersal, and/or persistence in the environment of the organism; or

(ii) The Administrator determines that the responsible person or any agent of the responsible person has failed to comply with any material provision of this part, any other regulations issued pursuant to the Plant Protection Act (7 U.S.C. 7701 *et seq.*) or the Plant Protection Act itself;

(iii) In addition, no permit will be issued if the responsible person and his or her agents do not agree in writing, in accordance with paragraph (h)(2) of this section, to comply with the permit conditions or, in accordance with paragraph (h)(3) of this section, to allow inspection by APHIS.

(2) *Withdrawal of permits.* The Administrator may withdraw, either orally or in writing, any permit that has been issued. If the withdrawal is oral, the Administrator will communicate, as promptly as circumstances allow, the withdrawal, and the reasons for it, in writing. The Administrator may withdraw a permit if:

(i) Following issuance of the permit, the Administrator receives information that would have provided grounds for APHIS to deny the original permit application;

(ii) The Administrator determines that actions taken under the permit have resulted in the unauthorized release, spread, dispersal, and/or persistence in the environment of the organism under permit; or

(iii) The Administrator determines that the responsible person or any agent of the responsible person has failed to comply at any time with any material provision of this part or with any other regulations issued pursuant to the Plant Protection Act (7 U.S.C. 7701 *et seq.*). This includes failure to comply with the conditions of any permit issued.

(k) *Appeal of denial or withdrawal of permit.* Any person whose permit application has been denied or whose permit has been withdrawn may appeal the decision in writing to the Administrator.[1]

---

[1] The Office of the Administrator, as established in §371.2 of this chapter, will review appeals involving the denial or withdrawal of a permit. Appeals may be sent to Office of the Administrator, United States Department of Agriculture, Jamie L. Whitten

Continued

The applicant must submit in writing an acknowledgment of the denial or withdrawal, and a statement of intent to appeal, within 10 days after receiving written notification of the denial or withdrawal. The applicant may request additional time to prepare the appeal. The appeal must state all of the facts and reasons upon which the person relies to assert that the permit was wrongfully denied or withdrawn. The Administrator will grant or deny the appeal in writing, stating the reasons for the decision as promptly as circumstances allow. If there is a conflict as to any material fact, a hearing shall be held to resolve such conflict.

(1) *Amendment of permits*—(1) *Amendment at responsible person's request.* If the responsible person determines that circumstances have changed since the permit was initially issued and wishes the permit to be amended accordingly, the responsible person must request the amendment by contacting APHIS directly. The responsible person will have to provide supporting information justifying the amendment. APHIS will review the amendment request, and will amend the permit if APHIS determines that relatively minor changes are necessary. Requests for more substantive changes will require a new permit application. Prior to issuance of an amended permit, the responsible person will be required to agree in writing or electronically that he or she and his or her agents will comply with the conditions of the amended permit. If the responsible person does not agree to the conditions, the amendment will be denied.

(2) *Amendment initiated by APHIS.* APHIS may amend any permit and its conditions at any time, upon determining that the amendment is needed to address plant pest risks presented by the organism or the activities allowed under the permit. APHIS will notify the responsible person of the amendment to the permit and, as soon as circumstances allow, the reason(s) for it. The responsible person may have to agree in writing or electronically that he or she and his or her agents will comply with the conditions of the

Building, Room 312-E, 1400 Independence Ave. SW, Washington, DC 20250.

amended permit before APHIS will issue it. If APHIS requests such an agreement, and the responsible person does not accept it, the existing permit will be withdrawn.

(m) *Shipping under a permit.* (1) All shipments of organisms under permit must be secure shipments. Organisms under permit must be shipped in accordance with the regulations in 49 CFR part 178.

(2) The container must be accompanied by a document that includes the names and contact details for the sender and recipient.

(3) For any organism to be imported into the United States, the outmost container must bear information regarding the nature and quantity of the contents; the country (or countries) and locality (localities) where collected, developed, manufactured, reared, cultivated, and cultured (as applicable); the name and address of the shipper, owner, or person shipping or forwarding the organism; the name, address, and telephone number of the consignee; the identifying shipper's mark and number; and the permit number authorizing the importation. For organisms imported under permits by mail, the container must also be addressed to a plant inspection station listed in the USDA Plants for Planting Manual, which can be accessed at: *https://www.aphis.usda.gov/import_export/plants/manuals/ports/downloads/plants_for_planting.pdf.* All imported containers of organisms under permits must be accompanied by an invoice or packing list indicating the contents of the shipment.

(4) Following the completion of the shipment, all packaging material, shipping containers, and any other material accompanying the organism will be devitalized consistent with supplemental permit conditions, or disposed of to prevent unauthorized release.

(n) *Applicability date:* This section is applicable beginning April 5, 2021.

(Approved by the Office of Management and Budget under control number 0579-0471)

§ 340.6 Record retention, compliance, and enforcement.

(a) *Recordkeeping.* Responsible persons and their agents are required to establish, keep, and make available to APHIS the following records:

(1) Records and reports required under § 340.5(i);

(2) Addresses and any other information (e.g., GPS coordinates, maps) needed to identify all locations where the organism under permit was stored or used, including all contained facilities and environmental release locations;

(3) A copy of the APHIS permit authorizing the permitted activity; and

(4) Legible copies of contracts (including amendments to contracts) between the responsible person and agents that conduct activities subject to this part for the responsible person, and copies of documents relating to agreements made without a written contract.

(b) *Record retention.* Records indicating that an organism under permit that was imported or moved interstate reached its intended destination must be retained for at least 2 years. All other records related to a permit must be retained for 5 years following the expiration of the permit, unless a longer retention period is determined to be needed by the Administrator and is documented in the supplemental permit conditions.

(c) *Compliance and enforcement.* (1) Responsible persons and their agents must comply with all of the requirements of this part. Failure to comply with any of the requirements of this part may result in any or all of the following:

(i) Denial of a permit application or withdrawal of a permit in accordance with § 340.5(j);

(ii) Application of remedial measures in accordance with the Plant Protection Act (7 U.S.C. 7701 *et seq.*); and

(iii) Criminal and/or civil penalties in accordance with the Plant Protection Act (7 U.S.C. 7701 *et seq.*).

(2) Prior to the issuance of a complaint seeking a civil penalty, the Administrator may enter into a stipulation, in accordance with § 380.10 of this chapter.

(d) *Liability for acts of an agent.* For purposes of enforcing this part, the act, omission, or failure of any agent for a responsible person may be deemed also to be the act, omission, or failure of the responsible person.

(Approved by the Office of Management and Budget under control number 0579–0471)

§ 340.7 Confidential business information.

Persons including confidential business information (CBI) in any document submitted to APHIS under this part should do so in the following manner. If there are portions of a document deemed to contain confidential business information, those portions must be identified, and each page containing such information must be marked "CBI Copy." A second copy of the document must be submitted with all such CBI deleted, and each page where the CBI was deleted must be marked "CBI Deleted." In addition, any person submitting CBI must justify how each piece of information requested to be treated as CBI is a trade secret or, if not a trade secret, is either commercial or financial information that is privileged or confidential.

§ 340.8 Costs and charges.

The services of the inspector related to carrying out this part and provided during regularly assigned hours of duty and at the usual places of duty will be furnished by APHIS without cost to the responsible person.[1] The U.S. Department of Agriculture will not be responsible for any costs or charges incidental to inspections or compliance with the provisions of this part, other than for the services of the inspector.

PART 351—IMPORTATION OF PLANTS OR PLANT PRODUCTS BY MAIL

AUTHORITY: 7 U.S.C. 7711–7714, 7721, 7754, and 7755; 7 CFR 2.22, 2.80, and 371.3.

[1] The Department's provisions relating to overtime charges for an inspector's services are set forth in part 354 of this chapter.

351.7 Regulations governing importation by mail of plant material for immediate export.

CROSS REFERENCE: For customs regulations governing importation of plants and plant products, see 19 CFR part 12.

AUTHORITY: 7 U.S.C. 7701–7772 and 7781–7786; 7 CFR 2.22, 2.80, and 371.3.

§ 351.1 Joint treatment generally.

The entry into the United States of certain plants, plant products, and soil is prohibited or restricted through various orders, quarantines, and regulations promulgated by the Administrator of the Animal and Plant Health Inspection Service (APHIS) under the authority of the Plant Protection Act (7 U.S.C. 7701–7772). To assist in enforcing the aforementioned orders, quarantines, and regulations, the Plant Protection and Quarantine Programs of APHIS have made provisions with the U.S. Postal and Customs Services to ensure closer inspection of prohibited or restricted imported articles.

[66 FR 21059, Apr. 27, 2001]

§ 351.2 Location of inspectors.

Inspectors of the Plant Protection and Quarantine Programs and customs officers are stationed at the following locations:

Anchorage, Alaska, Arlington, Va., Atlanta, Ga., Baltimore, Md., Baton Rouge, La., Blaine, Wash., Boston, Mass., Brownsville, Tex., Buffalo, N.Y., Calexico, Calif., Chantilly, Va., Charleston, S.C., Charlotte Amalie, St. Thomas, V.I., Chicago, Ill., Christiansted, St. Croix, V.I., Cleveland, Ohio., Corpus Christi, Tex., Dallas, Tex., Del Rio, Tex., Detroit, Mich., Douglas, Ariz., Dover, Del., Duluth, Minn., Eagle Pass, Tex., El Paso, Tex., Galveston, Tex., Hidalgo, Tex., Hilo, Hawaii, Hoboken, N.J., Honolulu, Hawaii, Houston, Tex., Jacksonville, Fla., Jamaica, L.I., N.Y., Key West, Fla., Laredo, Tex., McGuire AFB, N.J., Memphis, Tenn., Miami, Fla., Milwaukee, Wis., Mobile, Ala., New Orleans, La., New York, N.Y., Newport News, Va., Nogales, Ariz., Norfolk, Va., Pensacola, Fla., Philadelphia, Pa., Port Arthur, Tex., Port Canaveral, Fla., Port Everglades, Fla., Portland, Oreg., Presidio, Tex., Progreso, Tex., Ramey AFB, P.R., Roma, Tex., Rouses Point, N.Y., St. Paul, Minn., San Antonio, Tex., San Diego, Calif., San Francisco, Calif., San Juan, P.R., San Luis, Ariz., San Pedro, Calif., San Ysidro, Calif., Savannah, Ga., Seattle, Wash., Tampa, Fla.,

Toledo, Ohio, Washington, DC, West Palm Beach, Fla., Wilmington, N.C.

[28 FR 5203, May 24, 1963, as amended at 36 FR 24917, Dec. 24, 1971]

§ 351.3 Procedure on arrival.

All parcel post or other mail packages from foreign countries which, either from examination or external evidence, are found or are believed to contain plants or plant products, shall be dispatched for submission, or actually submitted, to the plant quarantine inspector at the most accessible location listed in § 351.2. The inspector shall pass upon the contents under the Plant Quarantine Act and Federal Plant Pest Act and with the cooperation of the customs and postal officers either

(a) Release the package from further plant quarantine examination and endorse his decision thereon; or

(b) Divert it to the Plant Quarantine Station at Washington, DC, Brownsville, Tex., Hoboken, N.J., Honolulu, Hawaii, Jamaica, L.I., N.Y., Laredo, Tex., Miami, Fla., New Orleans, La., San Francisco, Calif., San Juan, P.R., San Pedro, Calif., or Seattle, Wash., for whatever disposition is deemed warranted. If so diverted, the plant quarantine inspector shall attach to the package the yellow and green special mailing tag addressed to the proper quarantine station. A package so diverted shall be accompanied by customs card Form 3511 and transmitted to the appropriate Customs office for referral to the Plant Quarantine Station. Envelopes containing customs card Form 3511 addressed to the collector of customs, New York, N.Y., shall contain a notation that the material is to be referred to the Plant Protection and Quarantine Programs, Hoboken, N.J.

[24 FR 9923, Dec. 9, 1959, as amended at 28 FR 5204, May 24, 1963, as amended at 36 FR 24917, Dec. 24, 1971]

§ 351.4 Records.

The customs officers at Washington, DC, Brownsville, Tex., Hoboken, N.J., Honolulu, Hawaii, Jamaica, L.I., N.Y., Laredo, Tex., Miami, Fla., New Orleans, La., San Francisco, Calif., San Juan, P.R., San Pedro, Calif., or Seattle, Wash., shall keep a record of such packages as may be delivered to representatives of the Department of Agriculture, and upon the return thereof shall prepare a mail entry to accompany the dutiable package and deliver it to the postmaster for delivery or onward dispatch or in appropriate cases subject the shipment to formal customs entry procedure.

[28 FR 5204, May 24, 1963]

§351.5 Return or destruction.

Where the plant quarantine inspector requires the entire shipment to be returned to the country of origin as a prohibited importation (in which event he shall endorse his action thereon) and delivers the shipment to the collector of customs, the collector shall in turn deliver it to the postmaster for dispatch to the country of origin. If, upon examination, the plant material is deemed dangerous to plant life, the collector of customs shall permit the plant quarantine inspector to destroy immediately both the container and its contents. In either case the plant quarantine inspector shall notify the addressee of the action taken and the reason therefor. If the objectionable plant material forms only a portion of the contents of the mail package and in the judgment of the inspector the package can safely be delivered to the addressee, after removing and destroying the objectionable material, such procedure is authorized. In the latter case the inspector shall place in the package a memorandum (Form AQI-387) informing the addressee of the action taken by the inspector and describing the matter which has been seized and destroyed and the reasons therefor.

[24 FR 9923, Dec. 9, 1959, as amended at 36 FR 24917, Dec. 24, 1971]

§351.6 Packages in closed mail dispatches.

The foregoing instructions shall be followed in the treatment of packages containing plants or plant products received in closed mail dispatches made up for transmission directly to a post office located at a customs port at which no plant quarantine inspector is stationed. Such packages (accompanied by customs card Form 3511) shall be forwarded by the collector of customs through the postmaster to the most accessible location listed in §351.2 for appropriate treatment in the manner hereinbefore provided. This procedure shall also be followed in respect to such packages which are forwarded to unlisted post offices from the post office of original receipt, without having received plant quarantine examination. Packages discovered at post offices where no customs officer is located shall be forwarded by the postmaster under his official penalty envelope addressed to the collector of customs at the most accessible location listed for appropriate treatment as prescribed herein.

[24 FR 9923, Dec. 9, 1959, as amended at 36 FR 24917, Dec. 24, 1971]

§351.7 Regulations governing importation by mail of plant material for immediate export.

To collectors of customs and others concerned:

(a) Shipments of plant material may be imported by mail free of duty for immediate exportation by mail subject to the following regulations, which have been approved by the Department of Agriculture and the Post Office Department:

(1) Each shipment shall be dispatched in the mails from abroad, accompanied by a yellow and green special mail tag bearing the serial number of the permit for entry for immediate exportation or immediate transportation and exportation, issued by the U.S. Department of Agriculture, and also the postal form of customs declaration.

(2) Upon arrival, the shipment shall be detained by, or redispatched to, the postmaster at Washington, DC, Brownsville, Tex., Hoboken, N.J., Honolulu, Hawaii, Jamaica, L.I., N.Y., Laredo, Tex., Miami, Fla., New Orleans, La., San Francisco, Calif., San Juan, P.R., San Pedro, Calif., or Seattle, Wash., as may be appropriate, according to the address on the yellow and green tag, and there submitted to the customs officer and the Federal quarantine inspector. The merchandise shall under no circumstances be permitted to enter the commerce of the United States.

(3) After inspection by the customs and quarantine officers, and with their

approval, the addressee, or his authorized agent, shall repack and readdress the mail parcel under customs supervision; affix to the parcel the necessary postage, and comply with other mailing requirements, after which the parcel shall be delivered to the postmaster for exportation by mail pursuant to 19 CFR 9.11(a). The contents of the original parcel may be subdivided and exported in separate parcels in like manner.

(4) It will not be necessary to issue a customs mail entry nor to require formal entry of the shipments.

(5) The mail shipments referred to shall be accorded special handling only at the points specified in paragraph (a)(2) of this section.

(6) The foregoing procedure shall not affect the movement of plant material in the international mails in transit through the United States.

[24 FR 9923, Dec. 9, 1959, as amended at 28 FR 5204, May 24, 1963]

PART 352—PLANT QUARANTINE SAFEGUARD REGULATIONS

Authority: 7 U.S.C. 7701–7772 and 7781–7786; 21 U.S.C. 136 and 136a; 31 U.S.C. 9701; 7 CFR 2.22, 2.80, and 371.3.

Source: 25 FR 1929, Mar. 5, 1960, unless otherwise noted.

§ 352.1 Definitions.

(a) This part may be cited by the short title: "Safeguard Regulations." This title shall be understood to include both the regulations and administrative instructions in this part.

(b) Words used in the singular form in this part shall be deemed to import the plural and vice versa as the case may demand. For purposes of this part, unless the context otherwise requires, the following terms shall be construed, respectively, to mean:

Administrative instructions. Published documents set forth in this part relating to the enforcement of this part, and issued under authority thereof by the Deputy Administrator.

Biological control organism. Any enemy, antagonist, or competitor used to control a plant pest or noxious weed.

Brought in for temporary stay where unloading or landing is not intended. Brought in by carrier but not intended to be unloaded or landed from such carrier. This phrase includes movement (i) departing from the United States on the same carrier directly from the point of arrival therein; and (ii) transiting a part of the United States before departure therefrom, and applies whether movement under Customs procedure is as residue cargo or follows some form of Customs entry.

Carrier; means of conveyance. Automobile, truck, animal-drawn vehicle, railway car, aircraft, ship, or other means of transportation.

Customs. The U.S. Customs Service, Department of the Treasury, or, with reference to Guam, the Customs Office of the Government of Guam.

Deputy Administrator. The Deputy Administrator of the Plant Protection and Quarantine Programs or any employee of the Plant Protection and Quarantine Programs delegated to act in his or her stead.

Foreign trade zone. A formally prescribed area containing various physical facilities located in or adjacent to ports of entry under the jurisdiction of the United States and established, operated, and maintained as a foreign trade zone pursuant to the Foreign-Trade Zones Act of June 18, 1934 (48 Stat. 998–1003; 19 U.S.C. 81a–81u), as amended, wherein foreign merchandise, as well as domestic merchandise, may be deposited for approved purposes. Movement into and from such area is subject to applicable customs, plant quarantine, and other Federal requirements.

Immediate (export, trans-shipment, or transportation and exportation). The period which, in the opinion of the inspector, is the shortest practicable interval of time between the arrival of an incoming carrier and the departure of the outgoing carrier transporting a consignment of prohibited or restricted products or articles.

Inspector. Any individual authorized by the Administrator of APHIS or the Commissioner of Customs and Border Protection, Department of Homeland Security, to enforce the regulations in this part.

Intended for unloading and entry at a port other than the port of first arrival. Brought in by carrier at a port for movement to the port of entry under residue cargo procedure of Customs.

Noxious weed. Any plant or plant product that can directly or indirectly injure or cause damage to crops (including nursery stock or plant products), livestock, poultry, or other interests of agriculture, irrigation, navigation, the natural resources of the United States, the public health, or the environment.

Other product or article. Any product or article of any character whatsoever (other than plants, plant products, soil, plant pests, and means of conveyance), which an inspector considers may be infested or infected by or contain a plant pest.

Owner. The owner, or his agent (including the operator of a carrier), having responsible custody of a plant, plant product, plant pest, soil, or other product or article subject to this part.

Person. Any individual, partnership, corporation, association, joint venture, society, or other legal entity.

Plant pest. Any living stage of any of the following that can directly or indirectly injure, cause damage to, or cause disease in any plant or plant product: A protozoan, nonhuman animal, parasitic plant, bacterium, fungus, virus or viroid, infectious agent or other pathogen, or any article similar to or allied with any of the plant pests listed in this definition.

Plant Protection Act. Title IV of Public Law 106–224, 114 Stat. 438, 7 U.S.C. 7701–7772, which was enacted June 20, 2000.

Plant Protection and Quarantine Programs. The Plant Protection and Quarantine Programs, Animal and Plant Health Inspection Service, of the U.S. Department of Agriculture.

Plants and plant products. Nursery stock, other plants, plant parts, roots, bulbs, seeds, fruits, nuts, vegetables, and other plant products, and any product constituted, in whole or in part, of plant material which has not been so manufactured or processed as to eliminate pest risk.

Port. Any place designated by the President, Secretary of the Treasury, or Congress at which a Customs officer is assigned with authority to accept entries of merchandise, to collect duties, and to enforce the various provisions of the Customs and Navigation laws in force at that place.

Port of arrival. Any port in the United States at which a prohibited or restricted product or article arrives.

Port of entry. A port at which a specified shipment or means of conveyance is accepted for entry or admitted without entry into the United States.

Residue cargo. Shipments authorized by Customs to be transported under the Customs bond of the carrier on which the shipments arrive, without entry being filed, for direct export from the first port of arrival, or to another port for entry or for direct export at that port without entry being required.

Safeguard. A procedure for handling, maintaining, or disposing of prohibited or restricted products and articles subject to this part so as to eliminate the risk of plant pest dissemination which the prohibited or restricted products and articles may present.

Ship. Any means of transportation by water.

Soil. The unconsolidated material from the earth's surface that consists of rock and mineral particles and that supports or is capable of supporting biotic communities.

Stores and furnishings. Plants and plant products for use on board a carrier; e.g. as food or decorative material.

United States. The States, the District of Columbia, Guam, Puerto Rico, and the Virgin Islands of the United States, and the territorial waters of the United States adjacent to those land areas.

Unloaded or landed for transportation and exportation. Brought in by carrier and transferred to another carrier for transportation to another port for exportation, whether or not some form of Customs entry is made.

Unloaded or landed for transshipment and exportation. Brought in by carrier and transferred to another carrier for exportation from the same port, whether or not some form of Customs entry is made.

[25 FR 1929, Mar. 5, 1960, as amended at 36 FR 24917, Dec. 24, 1971; 37 FR 10554, May 25, 1972; 62 FR 65009, Dec. 10, 1997; 66 FR 21059, Apr. 27, 2001; 71 FR 49325, Aug. 23, 2006; 84 FR 29966, June 25, 2019]

§ 352.2 Purpose; relation to other regulations; applicability; preemption of State and local laws.

(a) Importations of plants, plant products, plant pests, biological control organisms, noxious weeds, soil, and other products and articles that may be infested or infected by or contain plant pests, biological control organisms, or noxious weeds are exempt from the prohibitions or restrictions contained in parts 319, 330, and 360 of this chapter if they meet one of the conditions in paragraphs (a)(1) through (a)(4) of this section and are moved into the United States and handled in compliance with this part. *Provided:* That these exemptions do not apply to cotton and covers imported into the United States from any country for exportation or transshipment and exportation or transportation and exportation as provided in §§ 319.8 through 319.8–26 of this chapter. Moreover, the applicable provisions of §§ 330.100 through 330.109 and 330.400 of this chapter also apply to products and articles subject to this part.

(1) They are brought in temporarily where loading and landing is not intended;

(2) They are unloaded or landed for transshipment and exportation;

(3) They are unloaded or landed for transportation and exportation; or

(4) They are intended for unloading and entry at a port other than the port of arrival.

(b) Prohibited or restricted products and articles offered for and refused entry into the United States under parts 319, 330, or 360 of this chapter are subject to the applicable provisions in this part regarding their subsequent handling in this country.

(c)(1) The provisions in this part apply whether the controls over arrival, temporary stay, unloading, landing, transshipment and exportation, or transportation and exportation, or other movement or possession in the United States and Guam are maintained by entry or other procedures of the U.S. Customs Service, Department of the Treasury, or the Customs Office of the Government of Guam, respectively. Such provisions will apply to arrivals in the United States, including arrivals in a foreign trade zone in the United States to which admission is sought in accordance with the Customs regulations in 19 CFR chapter I. Prohibited or restricted products and articles that have entered the United States, been exported pursuant to this part, and returned to the United States are subject to the applicable requirements of this part upon reentry.

(2) Any restrictions and requirements under this part with respect to the arrival, temporary stay, unloading, landing, transshipment, exportation, transportation and exportation, or other movement or possession in the United States of any product or article shall apply to any person who, respectively, brings into, maintains, unloads, lands, transships, exports, transports and exports, or otherwise moves or possesses in the United States such product or article, whether he is the person who was required to have a permit for the product or article or a subsequent custodian of such product or article, and failure to comply with all applicable restrictions and requirements under this part by any such person shall be deemed to be a violation of this part.

(d) Under section 436 of the Plant Protection Act (7 U.S.C. 7756), a State or political subdivision of a State may not regulate in foreign commerce any article, means of conveyance, plant, biological control organism, plant pest, noxious weed, or plant product in order to control a plant pest or noxious weed, to eradicate a plant pest or noxious weed, or to prevent the introduction or dissemination of a biological control organism, plant pest, or noxious weed.

[25 FR 1929, Mar. 5, 1960, as amended at 62 FR 65009, Dec. 10, 1997; 66 FR 21059, Apr. 27, 2001; 75 FR 68952, Nov. 10, 2010; 84 FR 29966, June 25, 2019]

§352.3 Enforcement and administration.

(a) Plants, plant products, plant pests, biological control organisms, noxious weeds, soil, and other products and articles subject to the regulations in this part that are unloaded, landed, or otherwise brought or moved into or through the United States in violation of this part may be seized, destroyed, or otherwise disposed of in accordance with section 414 of the Plant Protection Act (7 U.S.C. 7714). Any person who unloads, lands, or otherwise brings or moves into or through the United States any regulated plants, plant products, plant pests, biological control organisms, noxious weeds, soil, or other products or articles in violation of this part will be subject to prosecution under the applicable provisions of law.

(b) Whenever the Deputy Administrator of the Plant Protection and Quarantine Programs shall find that existing conditions of danger of plant pest escape or dissemination involved in the arrival, unloading, landing, or other movement, or possession in the United States of plants, plant products, plant pests, biological control organisms, noxious weeds, soil, or other products or articles subject to the regulations in this part, make it safe to modify by making less stringent the restrictions contained in any such regulation, he shall publish such findings in administrative instructions, specifying the manner in which the regulations shall be made less stringent with respect thereto, whereupon such modification shall become effective; or he may, upon request in specific cases, when the public interests will permit, authorize arrival, unloading, landing, or other movement, or possession in the United States under conditions that are less stringent than those contained in the regulations in this part.

(c) The Deputy Administrator also may set forth and publish, in administrative instructions, requirements and conditions for any class of products or articles supplemental to the regulations in this part, and may promulgate interpretations of this part.

(d) The Deputy Administrator shall employ procedures to carry out the purposes of this part which will impose a minimum of impediment to foreign commerce, consistent with proper precaution against plant pest, noxious weed, or biological control organism dissemination.

[25 FR 1929, Mar. 5, 1960, as amended at 66 FR 21059, Apr. 27, 2001; 75 FR 68952, Nov. 10, 2010; 84 FR 29966, June 25, 2019]

§352.4 Documentation.

(a) *Manifest.* Immediately upon the arrival of a carrier in the United States the owner shall make available to the inspector for examination a complete manifest or other documentation from which the inspector may determine whether there are on board any prohibited or restricted products or articles subject to this part, other than accompanied baggage and mail.

(b) *Other documentation.* Any notifications, reports, and similar documentation not specified in the regulations in this part, but necessary to carry out the purpose of the regulations, will be prescribed in administrative instructions.

(c) *Procedure after examination of documents.* After examination of the carrier cargo manifest or other documentation the inspector may notify the owner and the Customs officer that certain products or articles on board the carrier are subject to this part and may not be unloaded or landed for any purpose pending plant quarantine inspection. In such case the owner shall not unload or land such products or articles without authorization by an inspector.

§352.5 Permit; requirement, form and conditions.

(a) *General.* (1) Permits are required for the arrival, unloading or landing, or other movement into or through the United States of plants, plant products, plant pests, biological control organisms, noxious weeds, and soil subject to this part. The permit may consist of a general authorization as set out in paragraph (b), (c), or (d) of this section or § 352.11, or it may be a specific permit. A specific permit may be formal or oral except as a formal permit is required by paragraph (c) or (e) of this section. The Deputy Administrator may in administrative instructions require specific or formal permits for any class of products or articles subject to this part.

(2) A formal permit may be issued in prescribed form, in letter form, or a combination thereof. A rubber stamp impression or other endorsement made by the inspector on pertinent Customs documents covering the products or articles involved may constitute the formal permit in appropriate cases.

(b) *Permit for prohibited or restricted products or articles brought in for temporary stay where unloading or landing in the United States is not intended.* No permit other than the authorization contained in this paragraph shall be required for bringing into the United States any plants, plant products, plant pests, biological control organisms, noxious weeds, or soil subject to this part for temporary stay where unloading or landing in the United States is not intended, e.g., in connection with residue cargo movement under Customs procedure, or in connection with Customs entry for exportation or for transportation and exportation. This authorization also includes transshipment of products and articles under this paragraph from a carrier directly to another carrier of the same company when necessitated by an emergency or operating requirement and effected in accordance with safeguards prescribed in writing or orally by the inspector under § 352.10.

(c) *Permit for prohibited or restricted products or articles unloaded or landed for immediate transshipment and exportation, or immediate transportation and exportation.* When in the opinion of the inspector it is unnecessary to specify in a formal permit the safeguards required to prevent plant pest dissemination, plants, plant products, plant pests, biological control organisms, noxious weeds, or soil subject to this

part may be unloaded or landed for immediate transshipment and exportation or for immediate transportation and exportation, as provided in § 352.10, with the approval of the inspector and no further permit than the authorization contained in this paragraph; otherwise a formal permit shall be required for such unloading or landing.

(d) *Permit for restricted products or articles moving as residue cargo from port of first arrival to port of entry.* Restricted plants, plant products, plant pests, biological control organisms, noxious weeds, or soil subject to this part arriving in the United States for movement under residue cargo procedures of Customs from a port of first arrival to another port for Customs entry into the United States may be allowed to so move without permit other than the authorization contained in this paragraph, if the inspector finds that apparently they can meet the applicable requirements of parts 319, 330, and 360 of this chapter at the port where entry is to be made; otherwise a formal permit shall be required for such movement. Such restricted products and articles shall become subject to the applicable permit and other requirements of parts 319, 330, and 360 of this chapter upon arrival at the port where Customs entry is to be made and shall not be unloaded or landed unless they comply with the applicable requirements.

(e) *Formal permits required for certain prohibited or restricted products or articles brought into a foreign trade zone.* A formal permit must be obtained to bring any prohibited or restricted plants, plant products, plant pests, biological control organisms, noxious weeds, or soil subject to the provisions in this part, into a foreign trade zone for storage, manipulation, or other handling, except for immediate transshipment and exportation or for immediate transportation and exportation. Special conditions to safeguard such storage, manipulation, or other possession or handling may be specified in the permit, and when so specified shall be in addition to any other applicable requirements of this part or the safeguards prescribed by the inspector or otherwise under this part.

[25 FR 1929, Mar. 5, 1960, as amended at 62 FR 65009, Dec. 10, 1997; 66 FR 21059, Apr. 27, 2001; 75 FR 68952, Nov. 10, 2010; 84 FR 29966, June 25, 2019]

§ 352.6 Application for permit and approval or denial thereof.

(a) *Plants and plant products (including noxious weeds).* Except as otherwise provided in this paragraph, any person desiring to unload or land, or otherwise move into or through the United States, any plants or plant products for which a specific permit is required by § 352.5, shall in the case of prohibited plants or plant products, and should in the case of restricted plants or plant products, in advance of arrival in the United States of the plants or plant products, submit an application for a permit to the Plant Protection and Quarantine Programs,[1] stating such of the following information as is relevant: The name and address of the importer, the approximate quantity and kind of plants and plant products it is desired to import under this part, the country where grown, the United States port of arrival, the United States port of export, the proposed routing from the port of arrival to the port of exportation, means of transportation to be employed (*i.e.*, mail, air mail, express, air express, freight, air freight, baggage), and the name and address of the agent representing the importer. Applications may be made on forms provided for the purpose by the Plant Protection and Quarantine Programs, or orally, or by letter, telegram, or other means of communication furnishing all the information required by this paragraph. Applications need not be made for shipments handled under general authorizations set forth in § 352.5 (b), (c), or (d), or in § 352.11.

(b) *Plant pests.* Any person desiring to unload or land, or otherwise move into or through the United States, any plant pest for which a specific permit is required by § 352.5 shall, in advance of the arrival of the plant pests in the United States, submit an application to the Plant Protection and Quarantine Programs for a permit in accordance with § 330.201 of this chapter.

(c) *Soil.* Any person desiring to bring into or unload or land, or otherwise move into or through the United States, any soil for which a specific permit is required by § 352.5 shall, in advance of the arrival of the soil in the United States, submit an application for permit to the Plant Protection and Quarantine Programs as specified by § 330.203 of this chapter.

(d) *Constructive oral application.* If a permit has not been issued in advance of arrival, application for any required permit (other than a formal permit) shall be considered to have been made orally to the inspector at the port of arrival by presentation of the shipment for entry or its listing on the manifest or other documentation, but this shall not excuse failure to make timely application as required by this section. Express application is required for a formal permit.

(e) *Approval or denial of permits.* Upon approval of the application, the permit will be issued. Any conditions necessary to eliminate danger of plant pest, noxious weed, or biological control organism dissemination may be specified in the permit, or otherwise as provided in § 352.10. Permits will be denied if, in the opinion of the Deputy Administrator, it is not possible to prescribe conditions adequate to prevent danger of plant pest, noxious weed, or biological control organism dissemination by the plants, plant products, plant pests, or soil involved.

(Approved by the Office of Management and Budget under control number 0579–0049)

[25 FR 1929, Mar. 5, 1960, as amended at 36 FR 24917, Dec. 24, 1971; 37 FR 10554, May 25, 1972; 48 FR 57466, Dec. 30, 1983; 59 FR 67611, Dec. 30, 1994; 75 FR 68952, Nov. 10, 2010; 84 FR 29966, June 25, 2019]

§ 352.7 Notice of arrival.

Immediately upon arrival of any shipment of plants or plant products (including noxious weeds) subject to this part and covered by a specific permit, the importer shall submit to an

---

[1] Application for such permits should be addressed to the Animal and Plant Health Inspection Service, Plant Protection and Quarantine, Port Operations, Permit Unit, 4700 River Road, Unit 136, Riverdale, Maryland 20737–1236.

inspector notice of such arrival using a form provided for that purpose (Form PPQ–368) and, where relevant, the proposed routing to the proposed U.S. port of exit. Forms will be submitted using a U.S. Government electronic information exchange system or other authorized method. Notice of arrival shall not be required for other products or articles subject to this part since other available documentation meets the requirement for this notice.

(Approved by the Office of Management and Budget under control number 0579–0049)

[81 FR 40150, June 21, 2016]

§ 352.8 Marking requirements.

Prohibited and restricted products and articles subject to this part shall be adequately marked or otherwise identified by documentation to indicate their nature.

§ 352.9 Ports. =

The arrival, unloading, landing, or possession of plants, plant products, plant pests, biological control organisms, noxious weeds, soil, or other products or articles subject to this part shall not be allowed at points within the United States other than at the ports specified in the Customs Regulations in 19 CFR 1.1 and 19 CFR 6.13, and Agana, Guam, or such other ports as may be named in permits or administrative instructions. Restrictions on the ports which may be used for particular types of handling of any products or articles subject to this part may be specified generally in administrative instructions or in permits in specific cases. When ports are specified in permits or otherwise, the arrival, unloading, landing, or possession of the products or articles involved at other ports will not be allowed except as the inspector may authorize changes in the ports specified.

[25 FR 1929, Mar. 5, 1960, as amended at 75 FR 68952, Nov. 10, 2010; 84 FR 29966, June 25, 2019]

§ 352.10 Inspection; safeguards; disposal.

(a) *Inspection and release.* Prohibited and restricted products and articles subject to this part shall be subject to inspection at the port of first arrival in accordance with § 330.105(a) of this chapter and shall not be released by Customs officers for unloading, landing, or other onward movement or entry until released by an inspector or a Customs officer on behalf of an inspector in accordance with the procedure prescribed in § 330.105(a) of this chapter. If diversion or change of Customs entry is not permitted for any movements authorized under this part, the inspector at the original port of Customs entry shall appropriately endorse Customs documents to show that fact. However, the inspector at the U.S. port of export may approve diversion or change of Customs entry to permit movement to a different foreign country, or entry into the United States, subject to all other applicable requirements under this part or parts 319, 330, or 360 of this chapter. If diversion or change of Customs entry is desired at a Customs port in the United States where there is no inspector, the owner may apply to the Plant Protection and Quarantine Programs[2] for information as to applicable conditions. If diversion or change of Customs entry is desired at port, confirmation will be given by the Plant Protection and Quarantine Programs to the appropriate Customs officers and Plant Protection and Quarantine Programs inspectors.

(b) *Safeguards.* (1) The unloading, landing, retention on board as stores and furnishings or cargo, transshipment and exportation, transportation and exportation, onward movment to the port of entry as residue cargo or under a Customs entry for immediate transportation, and other movement or possession within the United States of prohibited or restricted products and articles under this part shall be subject to such safeguards as may be prescribed in the permits and this part and any others which, in the opinion of the inspector, are necessary and are specified by him to prevent plant pest, noxious weed, or biological control organism dissemination . In the case of prohibited or restricted products or articles subject to this part which are unloaded or landed

_____

[2] The Deputy Administrator, Plant Protection and Quarantine Programs, Animal and Plant Health Inspection Service, U.S. Department of Agriculture, Washington, DC 20250.

for transshipment and exportation or transportation and exportation, or for onward movement to the port of entry as residue cargo or under a Customs entry for immediate transportation, this shall include necessary safeguards with respect to any movement within the port area between the point of arrival and the point of temporary storage, other handling, or point of departure, including a foreign trade zone. Prohibited and restricted products and articles subject to this part which are unloaded or landed for transshipment and exportation or transportation and exportation, or for onward movement as residue cargo or under a Customs entry for immediate transportation, shall be transshipped, or transported and exported from the United States, or moved onward immediately. This shall mean the shortest practicable interval of time commensurate with the risk of plant pest, noxious weed, or biological control organism disseminatio required to transfer the products or articles from one carrier to another and to move them onward or from the United States. If, in the opinion of the inspector, considerations of risk of plant pest or noxious weed dissemination require, such movement shall be made without regard to the noncompetitive or competitive relations of the carriers concerned, and the inspector shall promptly report to the Plant Protection and Quarantine Programs the circumstances when the emergency is so acute that subsequent movement is required on a carrier of a company other than the one bringing the products or articles to the United States or on which onward movement was contemplated by the shipper or forwarding carrier. Prohibited or restricted plants, plant products, plant pests, biological control organisms, and soil which were intended for entry into the United States under parts 319, 330, or 360 of this chapter, or for movement into or through the United States under this part, and which were refused such entry or movement before unloading or landing, or which were refused such entry or movement after unloading or landing and are immediately reloaded on the same carrier, may be retained on board pending removal from the United States or other disposal, but shall be subject to the safeguards specified under this section. Prohibited or restricted products and articles which were refused entry or movement under said parts after unloading or landing and which are not immediately reloaded in accordance with this section shall be subject to such safeguard action as the inspector deems necessary to carry out the purposes of this part.

(2) Safeguards prescribed by an inspector under this section shall be prescribed to the owner by the inspector in writing except that the inspector may prescribe the safeguards orally when, in his opinion, the circumstances and related Customs procedures do not require written notice to the owner of the safeguards to be followed by the owner. In prescribing safeguards, the relevant requirements of parts 319, 330, or 360 of this chapter and this part shall be considered. The safeguards prescribed shall be the minimum required to preventplant pest, noxious weed, or biological control organism dissemination. Destruction or exportation shall be required only when no less drastic measures are deemed by the inspector to be adequate to prevent plant pest, noxious weed, or biological control organism dissemination . The inspector may follow administrative instructions promulgated for certain situations, or he may follow a procedure selected by him from administratively approved methods known to be effective in similar situations. In the case of aircraft that are contaminated with insect pests, only an insecticidal formulation, approved for use in aircraft, may be so applied as an emergency measure. If the application is not effective against the insect pests or if other pests must be safeguarded against, the inspector shall report the circumstances promptly to the Plant Protection and Quarantine Programs and receive instructions as to safeguards that will not have a deleterious effect on the structure of the aircraft or its operating equipment. In prescribing safeguards consideration will be given to such factors as:

(i) The nature and habits of the plant pests or biological control organisms known to be, or likely to be, present with the plants, plant products, soil, or other products or articles.

(ii) Nature of the plants, plant products, plant pests, or biological control organisms, soil, or other products or articles.

(iii) Nature of containers or other packaging and adequacy thereof to preventplant pest, noxious weed, or biological control organism dissemination.

(iv) Climatic conditions as they may have a bearing on plant pest or biological control organism dispersal, and refrigeration if provided.

(v) Routing pending exportation.

(vi) Presence of soil.

(vii) Construction or physical condition and type of carrier.

(viii) Facilities for treatment in accordance with part 305 of this chapter, or for incineration or other destruction.

(ix) Availability of transportation facilities for immediate exportation.

(x) Any other related factor which should be considered, such as intent to export to an adjacent or nearby country.

(c) *Disposal.* (1) If prohibited or restricted products or articles subject to this part are not safeguarded in accordance with measures prescribed under this part, or cannot be adequately safeguarded to prevent plant pest, noxious weed, or biological control organism dissemination, they shall be seized, destroyed, or otherwise disposed of according to law. Whenever disposal action is to be taken by the inspector he shall notify the local Customs officer in advance.

(2) When a shipment of any products or articles subject to this part has been handled in accordance with all conditions and safeguards prescribed in this part and in the permit and by the inspector, the inspector shall inform the local Customs officer concerned of the release of such products or articles, in appropriate manner.

[25 FR 1929, Mar. 5, 1960, as amended at 36 FR 24917, Dec. 24, 1971; 37 FR 10554, May 25, 1972; 62 FR 65009, Dec. 10, 1997; 66 FR 21059, Apr. 27, 2001; 75 FR 4253, Jan. 26, 2010; 75 FR 68952, Nov. 10, 2010; 84 FR 29966, June 25, 2019]

§ 352.11 **Mail.**

(a) *Transit mail.* (1) Plants, plant products, plant pests, biological control organisms, noxious weeds, soil, or other products or articles which arrive in the United States in closed dispatches by international mail or international parcel post and which are in transit through the United States to another country shall be allowed to move through the United States without further permit than the authorization contained in this section. Notice of arrival shall not be required as other documentation meets the requirement for this notice.

(2) Inspectors ordinarily will not inspect transit mail or parcel post, whether transmitted in open mail or in closed dispatches. They may do so if it comes to their attention that any such mail or parcel post contains prohibited or restricted products or articles which require safeguard action. Inspection and disposal in such cases will be made in accordance with this part and part 330 of this chapter, and in conformity with regulations and procedures of the Post Office Department for handling transit mail and parcel post.

(b) *Importation for exportation.* Plants and plant products to be imported for exportation, by mail, will be handled under permit in accordance with Part 351 of this chapter.

[25 FR 1929, Mar. 5, 1960, as amended at 75 FR 68952, Nov. 10, 2010; 84 FR 29966, June 25, 2019]

§ 352.12 **Baggage.**

Products or articles subject to this part which are contained in baggage shall be subject to the requirements of this part in the same manner as cargo.

§ 352.13 **Certain   conditions   under which change of Customs entry or diversion is permitted.**

When plants, plant products, plant pests, biological control organisms, noxious weeds, soil, or other products or articles released for exportation, transshipment and exportation, or transportation and exportation, under this part, have met all applicable permit and other requirements for importation, including inspection and treatment, as provided in part 319, 330, or 360 of this chapter, the form of Customs entry may be changed and the shipment may be diverted at any time to permit delivery of the products and articles to a destination in the United States, so far as the requirements in this part are involved. The Customs officer concerned at the original port of Customs entry shall be informed by the inspector that such release has been made and that such change of entry or diversion is approved under this part by appropriate endorsement of Customs documents.

[25 FR 1929, Mar. 5, 1960, as amended at 62 FR 65009, Dec. 10, 1997; 66 FR 21059, Apr. 27, 2001; 75 FR 68952, Nov. 10, 2010; 84 FR 29967, June 25, 2019]

§352.14 Costs.

All costs incident to the inspection, handling, safeguarding, or other disposal of prohibited or restricted products or articles under the provisons in this part shall be borne by the owner. Services of the inspector during regularly assigned hours of duty at the usual places of duty shall be furnished without cost to the person requesting the services, unless a user fee is payable under §354.3 of this chapter.

[56 FR 14844, Apr. 12, 1991]

§352.15 Caution.

In applying safeguards or taking other measures prescribed under the provisions in this part, it should be understood that inexactness or carelessness may result in injury or damage. It should also be understood by the owners that emergency measures prescribed by the inspector to safeguard against plant pest, noxious weed, or biological control organism dissemination may have adverse effects on certain products and articles and that they will take the calculated risk of such adverse effects of authorized measures.

[25 FR 1929, Mar. 5, 1960, as amended at 75 FR 68952, Nov. 10, 2010; 84 FR 29967, June 25, 2019]

§§352.16–352.28 [Reserved]

§352.29 Administrative instructions: Avocados from Mexico.

Avocados from Mexico may be moved through the United States to destinations outside the United States only in accordance with this section.

(a) *Permits.* Before moving the avocados through the United States, the owner must obtain a formal permit in accordance with §352.6 of this part.

(b) *Ports.* The avocados may enter the United States only at the following ports: Galveston or Houston, Texas; the border ports of Nogales, Arizona, or Brownsville, Eagle Pass, El Paso, Hidalgo, or Laredo, Texas; or at other ports within that area of the United States specified in paragraph (f) of this section.

(c) *Notice of arrival.* At the port of arrival, the owner must provide notification of the arrival of the avocados in accordance with §352.7 of this part.

(d) *Inspection.* The owner must make the avocados available for examination by an inspector. The avocados may not be moved from the port of arrival until released by an inspector.

(e) *Shipping requirements.* The avocados must be moved through the United States either by air or in a refrigerated truck or refrigerated rail car or in refrigerated containers on a truck or rail car. If the avocados are moved in refrigerated containers on a truck or rail car, an inspector must seal the containers with a serially numbered seal at the port of arrival. If the avocados are removed in a refrigerated truck or refrigerated rail car, an inspector must seal the truck or rail car with a serially numbered seal at the port of arrival. If the avocados are transferred to another vehicle or container in the United States, an inspector must be present to supervise the transfer and must apply a new serially numbered seal. The avocados must be moved through the United States under Customs bond.

(f) *Shipping areas.* Avocados moved by truck or rail car may transit only that area of the United States bounded on the west and south by a line extending from El Paso, Texas, to Salt Lake City, Utah, to Portland, Oregon, and due west from Portland; and on the east and south by a line extending from Brownsville, Texas, to Galveston, Texas, to Kinder, Louisiana, to Memphis, Tennessee, to Louisville, Kentucky, and due east from Louisville. All cities on these boundary lines are included in this area. If the avocados are moved by air, the aircraft may not land outside this area. Avocados that enter the United States at Nogales, Arizona, must be moved to El Paso,

Texas, by the route specified on the formal permit.

[52 FR 27671, July 23, 1987, as amended at 54 FR 43167, Oct. 23, 1989]

§ 352.30 Untreated oranges, tangerines, and grapefruit from Mexico.

The following provisions shall apply to the movement into or through the United States under this part of untreated oranges, tangerines, and grapefruit from Mexico in transit to foreign countries via United States ports on the Mexican border.

(a) *Untreated fruit; general*—(1) *Permit and notice of arrival required.* The owner shall, in advance of shipment of untreated oranges, tangerines, or grapefruit from Mexico via United States ports to any foreign country, procure a formal permit as provided in § 352.6, or application for permit may be submitted to the inspector at the port in the United States through which the shipment will move. Notice of arrival of such fruit shall be submitted as required by § 352.7.

(2) *Origin: period of entry.* Such fruit may enter from any State in Mexico throughout the year, in accordance with requirements of this section and other applicable provisions in this part.

(3) *Cleaning refrigerated containers prior to return to the United States from Canada.* Refrigerated containers that have been used to transport untreated oranges, tangerines, or grapefruit from Mexico through the United States to Canada shall be carefully swept and freed from all fruit, as well as boxes and rubbish, by the carrier involved prior to reentry into the United States.

(4) *Inspection; safeguards.* (i) Each shipment under paragraph (a) of this section shall be subject to such inspections and safeguards as are required by this section and such others as may be prescribed by the inspector pursuant to § 352.10.

(ii) Untreated oranges, tangerines, and grapefruit arriving from Mexico at authorized ports in the United States for movement to a foreign country shall be loaded into refrigerated containers and preinspected by an inspector for freedom of citrus leaves before entry into the United States or be accompanied by an acceptable certificate

from an inspector as to such freedom. Refrigerated containers loaded with untreated oranges, tangerines, and grapefruit that are not free of such leaves will be denied entry into the United States.

(iii) All refrigerated containers used to transport untreated fruit from Mexico through the United States to a foreign country under this paragraph (a) shall be subject to any treatment in accordance with part 305 of this chapter at the port of first arrival and elsewhere as may be required by the inspector, pursuant to this part, in order to prevent plant pest dissemination.

(b) *Additional conditions for overland movement of certain untreated fruit.* Untreated oranges, tangerines, and grapefruit from Mexico may move overland through the United States to a foreign country only in accordance with the following additional conditions:

(1) *Ports of entry.* Such fruit may enter only at Nogales, AZ, or Eagle Pass, El Paso, or Laredo, TX.

(2) *General transit conditions.* The following conditions apply to all shipments of untreated oranges, tangerines, and grapefruit from Mexico transiting the United States for movement to a foreign country:

(i) The fruit must be packed in insect-proof boxes or crates that prevent the escape or entry of adult, larval, or pupal fruit flies.[3]

(ii) Boxes or cartons of fruit must be enclosed in sealed, refrigerated containers of the type commonly used by the maritime or commercial trucking industry. An official seal must be applied to the container at the port of entry. The seal must not be removed except by an inspector, or after the shipment has left the United States.

(iii) The temperature in the refrigerated containers in which the fruit is transported must be maintained at 60 °F or lower.

(iv) If the seal on the containers in which such fruit is shipped is found to

---

[3] If there is a question as to whether packaging is adequate, send a request for approval of the packaging, together with a sample of the packaging, to the Animal and Plant Health Inspection Service, Plant Protection and Quarantine, Center for Plant Health Science and Technology, 1730 Varsity Drive, Suite 400, Raleigh, NC 27606.

have been broken, for any reason, before the container leaves the United States, or if the cooling system in the containers fails at any point during transit, an inspector at the port of entry must be contacted immediately.

(v) A transportation and exportation permit must be issued by an inspector for each shipment. This permit can be obtained from APHIS headquarters.[4]

(vi) If untreated fruit is transloaded to another container while in the United States, the transloading must be supervised by an inspector and a replacement official seal must be applied to the container to which the fruit is moved.

(vii) Shipments of such fruit must move by direct route, in Customs bond and under official seal, without diversion or change of entry en route, from the port of entry to the port of exit or to an approved port in the United States for export to another foreign country.

(viii) Shipments of such fruit may not traverse the counties of Cameron, Hidalgo, Starr, or Willacy, TX. Shipments of such fruit may only traverse areas listed under each type of carrier listed below.

(3) *Truck movement.* Trucks may haul refrigerated containers of such fruit from Mexico to shipside, or to approved refrigerated storage pending lading aboard ship, in Corpus Christi, Galveston, or Houston, TX, or alongside railway carriers or aircraft at the ports named in paragraph (b)(2) of this section for movement to a foreign country. Shipments of such fruit via truck may traverse only the territory within the United States bounded on the west by a line starting at Laredo, TX, on to El Paso, TX, to Salt Lake City, UT, and then to Portland, OR, and on the east by a line drawn from Laredo, TX to Hebbronville, TX, to Corpus Christi, TX, to Galveston, TX, to Kinder, LA, to Memphis, TN, and then to Louisville, KY, and routes directly northward.

(4) *Rail movement.* Shipments must move by direct route from the port of entry to the port of exit or to an approved North Atlantic port in the United States for export to another foreign country, as follows: The fruit may be entered at Nogales, AZ, only for direct rail routing to El Paso, TX, after which it shall traverse only the territory bounded on the west by a line drawn from Laredo, TX, to El Paso, TX, to Salt Lake City, UT, and then to Portland, OR, and on the east by a line drawn from Laredo, TX, to Hebbronville, TX, to Corpus Christi, TX, to Galveston, TX, to Kinder, LA, to Memphis, TN, and then to Louisville, KY, and routes directly northward. Such fruit may also enter the United States from Mexico at any port listed in paragraph (b)(1) of this section, for direct eastward rail movement, without diversion en route, for reentry into Mexico.

(5) *Air cargo movement.* Shipments of such fruit may move by direct route as air cargo, without change of entry while in the United States en route from the port of entry, to Canada. If an emergency occurs en route to the port of export that will require transshipment to another carrier, an inspector at the port of entry must be contacted immediately.

(c) *Additional conditions for movement of certain untreated fruit by water route.* Untreated oranges, tangerines, and grapefruit from Mexico may move from Mexico to a foreign country by water route through the United States under this section only in accordance with the following additional conditions:

(1) *Ports of entry.* Such oranges, tangerines, and grapefruit may enter only at New York, Boston, or such other North Atlantic ports in the United States as may be named in permits, for exportation, or at Galveston, Texas, for exportation by water route.

(2) *Routing through North Atlantic ports.* Such fruit entering via North Atlantic ports in the United States shall move by direct water route to New York or Boston, or to such other North Atlantic ports as may be named in the permit only for immediate direct export by water route to any foreign country, or for immediate transportation and exportation in Customs bond by direct rail route to Canada.

---

[4] To obtain this permit, contact the Animal and Plant Health Inspection Service, Plant Protection and Quarantine, Permit Unit, 4700 River Road Unit 133, Riverdale, MD 20737.

(d) *Restriction on diversion or change of Customs entry.* Diversion or change of Customs entry shall not be permitted with movements authorized under paragraph (b) (4) or (5) or paragraph (c) of this section and the inspector at the original port of Customs entry shall appropriately endorse the Customs documents to show that fact: *Provided,* That the inspector at such port of entry may, when consistent with the purposes of this part, approve diversion or change of Customs entry to permit movement to a different foreign country or entry into the United States subject to all other applicable requirements under this part or part 319 of this chapter. If diversion or change of Customs entry is desired at a Customs port in the United States where there is no inspector, the owner may apply to the Plant Protection and Quarantine Programs for information as to applicable conditions. If diversion or change of entry is approved at such a port, confirmation will be given by the Plant Protection and Quarantine Programs to appropriate Customs officers and Plant Protection and Quarantine Programs inspectors.

(e) *Untreated fruit from certain municipalities in Mexico.* Oranges, tangerines, and grapefruit in transit to foreign countries may be imported from certain municipalities in Mexico that meet the criteria of § 319.56–5 for freedom from fruit flies in accordance with the applicable conditions in part 319 of this chapter.

(f) *Treated fruit.* Oranges, tangerines, and grapefruit from Mexico that have been treated in Mexico in accordance with part 305 of this chapter may be moved through the United States ports for exportation in accordance with the regulations in part 319 of this chapter.

(g) *Costs.* Costs shall be borne by the owner of the fruit as provided in § 352.14. This includes all costs for preinspection and convoying of loaded trucks and supervision of transloading from trucks to approved carriers or storage in United States ports when augmented inspection service has to be provided for such preinspection, convoying, and supervision.

(Approved by the Office of Management and Budget under control number 0579–0303)

[25 FR 1929, Mar. 5, 1960, as amended at 36 FR 24917, Dec. 24, 1971; 37 FR 10554, May 25, 1972; 55 FR 23066, June 6, 1990; 56 FR 13066, Mar. 29, 1991; 67 FR 46578, July 16, 2002; 71 FR 49325, Aug. 23, 2006; 72 FR 39528, July 18, 2007; 75 FR 4253, Jan. 26, 2010; 84 FR 29967, June 25, 2019]

PART 353—EXPORT CERTIFICATION

AUTHORITY: 7 U.S.C. 7701–7772 and 7781–7786; 21 U.S.C. 136 and 136a; 7 CFR 2.22, 2.80, and 371.3.

SOURCE: 61 FR 15368, Apr. 8, 1996, unless otherwise noted.

§ 353.1　Definitions.

Administrator. The Administrator, Animal and Plant Health Inspection Service, or any person authorized to act for the Administrator.

Agent. An individual who meets the eligibility requirements set forth in § 353.6, and who is designated by the Animal and Plant Health Inspection Service to conduct phytosanitary field inspections of seed crops to serve as a basis for the issuance of phytosanitary certificates.

Animal and Plant Health Inspection Service (APHIS). The Animal and Plant Health Inspection Service of the U.S. Department of Agriculture.

Consignment. One shipment of plants or plant products, from one exporter, to one consignee, in one country, on one means of conveyance; or any mail shipment to one consignee.

Export certificate for processed plant products. A certificate (PPQ Form 578 or an approved electronic equivalent) issued by an inspector, describing the plant health condition of processed or manufactured plant products based on inspection of submitted samples and/or by virtue of the processing received.

Family. An inspector or agent and his or her spouse, their parents, children, and first cousins.

Industry-issued certificate. A certificate issued by a representative of the concerned agricultural or forestry industry under the terms of a written agreement with the Animal and Plant Health Inspection Service, giving assurance that a plant product has been handled, processed, or inspected in a manner required by a foreign government. An industry-issued certificate includes an ISPM 15 quality/treatment mark.

Inspector. An employee of the Animal and Plant Health Inspection Service, or a State or county plant regulatory official designated by the Secretary of Agriculture to inspect and certify to shippers and other interested parties, as to the phytosanitary condition of plant products inspected under the Act.

Non-government facility. A laboratory, research facility, inspection service, or other entity that is maintained, at least in part, for the purpose of providing laboratory testing or phytosanitary inspection services and that is not operated by the Federal Government or by the government of a State or a subdivision of a State.

Office of inspection. The office of an inspector of plants and plant products covered by this part.

Phytosanitary certificate. A certificate (PPQ Form 577 or an approved electronic equivalent) issued by an inspector, giving the phytosanitary condition of domestic plants or unprocessed or unmanufactured plant products based on inspection of the entire lot or representative samples drawn by a Federal or State employee authorized to conduct such sampling.

Phytosanitary certificate for reexport. A certificate (PPQ Form 579 or an approved electronic equivalent) issued by an inspector, giving the phytosanitary condition of foreign plants and plant products legally imported into the United States and subsequently offered for reexport. The certificate certifies that, based on the original foreign phytosanitary certificate and/or additional inspection or treatment in the United States, the plants and plant products are considered to conform to the current phytosanitary regulations of the receiving country and have not been subjected to the risk of infestation or infection during storage in the United States. Plants and plant products which transit the United States under Customs bond are not eligible to receive the phytosanitary certificate for reexport.

Plant pests. Any living stage of any insects, mites, nematodes, slugs, snails, protozoa, or other invertebrate animals, bacteria, fungi, other parasitic plants or reproductive parts thereof, viruses, or any organisms similar to or allied with any of the foregoing, or any infectious substances, which can directly or indirectly injure or cause disease or damage in any plants or parts thereof, or other products of plants.

Plant products. Products derived from nursery stock, other plants, plant parts, roots, bulbs, seeds, fruits, nuts, and vegetables, including manufactured or processed products.

Plants and plant products. Nursery stock, other plants, plant parts, roots, bulbs, seeds, fruits, nuts, vegetables and other plant products, including manufactured or processed products.

Reference Manual A. The *Reference Manual for Administration, Procedures, and Policies of the National Seed Health System,* published by the National Seed Health System (NSHS). Reference Manual A describes the structure, administration, procedures, policies, and working practices of the NSHS and also contains relevant documentation, forms, and references for the NSHS. Reference Manual A is incorporated by reference at §300.3 of this chapter, and is available by writing to Phytosanitary Issues Management, Operational Support, PPQ, APHIS, 4700 River Road Unit 140, Riverdale, MD 20737–1236, and on the APHIS Web site at *http://www.aphis.usda.gov/ppq/pim/accreditation.*

Reference Manual B. The *Reference Manual for Seed Health Testing and Phytosanitary Field Inspection Methods,* published by the National Seed Health System (NSHS). Reference Manual B contains the detailed seed health testing, seed sampling, and inspection procedures for the NSHS. Reference Manual B is incorporated by reference at § 300.4 of this chapter, and is available by writing to Phytosanitary Issues Management, Operational Support, PPQ, APHIS, 4700 River Road Unit 140, Riverdale, MD 20737–1236, and on the APHIS Web site at *http:// www.aphis.usda.gov/ppq/pim/accreditation*.

State. Any of the States of the United States, the District of Columbia, American Samoa, Guam, the Northern Mariana Islands, Puerto Rico, or the Virgin Islands of the United States.

The Act. Title IV of Public Law 106–224, 114 Stat. 438, 7 U.S.C. 7701–7772, which was enacted June 20, 2000.

[61 FR 15368, Apr. 8, 1996, as amended at 64 FR 1105, Jan. 8, 1999; 64 FR 72264, Dec. 27, 1999; 65 FR 50131, Aug. 17, 2000; 66 FR 21059, Apr. 27, 2001; 66 FR 37116, July 17, 2001; 66 FR 37400, July 18, 2001; 67 FR 8466, Feb. 25, 2002; 72 FR 35917, July 2, 2007; 81 FR 40151, June 21, 2016]

§ 353.2 Purpose and administration.

The export certification program does not require certification of any exports, but does provide certification of plants and plant products as a service to exporters. After assessing the phytosanitary condition of the plants or plant products intended for export, relative to the receiving country's regulations, an inspector issues an internationally recognized phytosanitary certificate, a phytosanitary certificate for reexport, or an export certificate for processed plant products if warranted. APHIS also enters into written agreements with industry to allow the issuance of industry-issued certificates giving assurance that a plant product has been handled, processed, or inspected in a manner required by a foreign government. An industry-issued certificate includes an ISPM 15 quality/ treatment mark.

[61 FR 15368, Apr. 8, 1996, as amended at 64 FR 72264, Dec. 27, 1999; 72 FR 35917, July 2, 2007; 81 FR 40151, June 21, 2016]

§ 353.3 Where service is offered.

(a) Information concerning the location of inspectors who may issue certificates for plants and plant products may be obtained by contacting one of the following regional offices:

Region	States
Northeastern, Blason II, 1st Floor, 505 South Lenola Road, Moorestown, NJ 08057.	CT, ME, MA, NH, RI, VT, NY, NJ, PA, MD, DE, VA, WI, MN, IL, IN, OH, MI, WV.
Southeastern, 3505 25th Avenue, Building 1, North, Gulfport, MS 39501.	FL, AL, GA, KY, MS, TN, NC, SC, PR, US VI.
Central, 3505 Boca Chica Blvd., Suite 360, Brownsville, TX 78521–4065.	TX, OK, NE, AR, KS, LA, IA, MO, ND, SD.
Western, 9580 Micron Avenue, Suite I, Sacramento, CA 95827.	HI, CA, CO, ID, MT, UT, WY, WA, OR, NV, NM, AZ, AK.

(b) Inspectors who may issue phytosanitary certificates for terrestrial plants listed in 50 CFR part 17 or 23 are available only at a port designated for export in 50 CFR part 24, or at a nondesignated port if allowed by the U.S. Department of the Interior pursuant to section 9 of the Endangered Species Act of 1973, as amended (16 U.S.C. 1538). The following locations are designated in 50 CFR part 24 as ports for export of terrestrial plants listed in 50 CFR part 17 or 23:

(1) Any terrestrial plant listed in 50 CFR part 17 or 23:

Nogales, AZ
Los Angeles, CA
San Diego, CA
San Francisco, CA
Miami, FL
Orlando, FL
Honolulu, HI
New Orleans, LA
Hoboken, NJ (Port of New York)
Jamaica, NY
San Juan, PR
Brownsville, TX
El Paso, TX
Houston, TX
Laredo, TX
Seattle, WA

(2) Any plant of the family Orchidaceae (orchids) listed in 50 CFR part 17 or 23:

Hilo, HI
Chicago, IL

(3) Roots of American ginseng (*Panax quinquefolius*) listed in 50 CFR 23.23:

Atlanta, GA
Chicago, IL
Baltimore, MD
St. Louis, MO
Milwaukee, WI

(4) Any plant listed in 50 CFR 17.12 or 23.23 and offered for exportation to Canada:

Detroit, MI
Buffalo, NY
Rouses Point, NY
Blaine, WA

(5) Any logs and lumber from trees listed in 50 CFR 17.12 or 23.23:

Mobile, AL
Savannah, GA
Baltimore, MD
Gulfport, MS
Wilmington and Morehead City, NC
Portland, OR
Philadelphia, PA
Charleston, SC
Norfolk, VA
Vancouver, WA

(6) Plants of the species *Dionaea muscipula* (Venus flytrap):

Wilmington, NC

§ 353.4 Products covered.

Plants and plant products when offered for export or re-export.

§ 353.5 Application for certification.

(a) To request the services of an inspector, a written application (PPQ Form 572) shall be made as far in advance as possible, and shall be filed in the office of inspection at the port of certification. Forms will be submitted using a U.S. Government electronic information exchange system or other authorized method.

(b) Each application shall be deemed filed when delivered to the proper office of inspection. When an application is filed, a record showing the date and time of filing shall be made in such office.

(c) Only one application for any consignment shall be accepted, and only one certificate for any consignment shall be issued.

(Approved by the Office of Management and Budget under control number 0579–0052)

[61 FR 15368, Apr. 8, 1996, as amended at 64 FR 72264, Dec. 27, 1999; 72 FR 35917, July 2, 2007; 81 FR 40151, June 21, 2016]

§ 353.6 Inspection.

Inspections shall be performed by agents, by inspectors, or by employees of a State plant protection agency who are authorized by the agency to perform field inspections in accordance with this part and who have successfully completed training in accordance with paragraph (a)(2)(iii) of this section. Employees of a State plant protection agency who are not agents may perform field inspections only under the supervision of an inspector.

(a) *Agent.* (1) Agents may conduct phytosanitary field inspections of seed crops in cooperation with and on behalf of those State plant regulatory agencies electing to use agents and maintaining a Memorandum of Understanding with the Animal and Plant Health Inspection Service in accordance with the regulations. The Memorandum of Understanding must state that agents shall be used in accordance with the regulations in this part. Agents are not authorized to issue Federal phytosanitary certificates, but are only authorized to conduct the field inspections of seed crops required as a basis for determining phytosanitary condition prior to the issuance of a phytosanitary certificate for the crops.

(2) To be eligible for designation as an agent, an individual must:

(i) Have the ability to recognize, in the crops he or she is responsible for inspecting, plant pests, including symptoms and/or signs of disease-causing organisms, of concern to importing countries.

(ii) Have a bachelor's degree in the biological sciences, and a minimum of 1 year's experience in identifying plant pests endemic to crops of commercial importance within the cooperating State, or a combination of higher education in the biological sciences and experience in identifying such plant pests, as follows:

0 years education and 5 years experience;
1 year education and 4 years experience;
2 years education and 3 years experience;
3 years education and 2 years experience; or
4 years education and 1 year experience.

The years of education and experience do not have to be acquired consecutively.

(iii) Successfully complete annual training provided by the State plant regulatory agency. The required training must include instruction in inspection procedures, identification of plant pests of quarantine importance to importing countries, methods of collection and submission of specimens (organisms and/or plants or plant parts) for identification, and preparation and submission of inspection report forms approved by the State plant regulatory agency.

(iv) Have access to Federal or State laboratories for the positive identification of plants pests detected.

(3) No agents shall inspect any plants or plant products in which they or a member of their family are directly or indirectly financially interested.

(b) *Inspector.* (1) An employee of the Animal and Plant Health Inspection Service, or a State or county regulatory official designated by the Secretary of Agriculture to inspect and certify to shippers and other interested parties, as to the phytosanitary condition of plants and plant products inspected under the Act.

(2) To be eligible for designation as an inspector, a State or county plant regulatory official must:

(i) Have a bachelor's degree in the biological sciences, and a minimum of 1 year's experience in Federal, State or county plant regulatory activities, or a combination of higher education in the biological sciences and experience in State plant regulatory activities, as follows:

0 years education and 5 years experience;
1 year education and 4 years experience;
2 years education and 3 years experience;
3 years education and 2 years experience; or
4 years education and 1 year experience.

The years of education and experience do not have to be acquired consecutively.

(ii) Successfully complete, as indicated by receipt of a passing grade, the Animal and Plant Health Inspection Service training course on phytosanitary certification.

(3) No inspectors shall inspect any plants or plant products in which they or a member of their family are directly or indirectly financially interested.

(c) *Applicant responsibility.* (1) When the services of an agent or an inspector are requested, the applicant shall make the plant or plant product accessible for inspection and identification and so place the plant or plant product to per-

mit physical inspection of the lot for plant pests.

(2) The applicant must furnish all labor involved in the inspection, including the moving, opening, and closing of containers.

(3) Certificates may be refused for failure to comply with any of the foregoing provisions.

§ 353.7 Certificates.

(a) *Phytosanitary certificate (PPQ Form 577).* (1) For each consignment of domestic plants or unprocessed plant products for which certification is requested, the inspector shall sign and issue a separate certificate based on the findings of the inspection.

(2) The original certificate shall immediately upon its issuance be delivered or mailed to the applicant or a person designated by the applicant.

(3) One copy of each certificate shall be filed in the office of inspection at the port of certification, and one forwarded to the Administrator.

(4) The Administrator may authorize inspectors to issue certificates on the basis of inspections made by cooperating Federal, State, and county agencies. The Administrator may also authorize inspectors to issue a certificate on the basis of a laboratory test or an inspection performed by a non-government facility accredited in accordance with § 353.8.

(5) Inspectors may issue new certificates on the basis of inspections for previous certifications when the previously issued certificates can be canceled before they have been accepted by the phytopathological authorities of the country of destination involved.

(b) *Export certificate for processed plant products (PPQ Form 578).* (1) For each consignment of processed plant products for which certification is requested, the inspector shall sign and issue a certificate based on the inspector's findings after inspecting submitted samples and/or by virtue of processing received.

(2) The original certificate shall immediately upon its issuance be delivered or mailed to the applicant or a person designated by the applicant.

(3) One copy of each certificate shall be filed in the office of inspection at the port of certification.

(4) The Administrator may authorize inspectors to issue certificates on the basis of inspections made by cooperating Federal, State, and county agencies. The Administrator may also authorize inspectors to issue a certificate on the basis of a laboratory test or an inspection performed by a non-government facility accredited in accordance with §353.8.

(5) Inspectors may issue new certificates on the basis of inspections/processing used for previous certifications.

(c) *Phytosanitary certificate for reexport (PPQ Form 579).* (1) For each consignment of foreign origin plants or unprocessed plant products for which certification is requested, the inspector shall sign and issue a certificate based on the original foreign phytosanitary certificate and/or additional inspection or treatment in the United States after determining that the consignment conforms to the current phytosanitary regulations of the receiving country and has not been subjected to the risk of infestation or infection during storage in the United States.

(2) The original certificate shall immediately upon its issuance be delivered or mailed to the applicant or a person designated by the applicant.

(3) One copy of each certificate shall be filed in the office of inspection at the port of certification, and one forwarded to the Administrator.

(4) The Administrator may authorize inspectors to issue certificates on the basis of inspections made by cooperating Federal, State, and county agencies. The Administrator may also authorize inspectors to issue a certificate on the basis of a laboratory test or an inspection performed by a non-government facility accredited in accordance with §353.8.

(5) Inspectors may issue new certificates on the basis of inspections for previous certifications when the previously issued certificates can be canceled before they have been accepted by the phytopathological authorities of the country of destination involved.

(d) *Industry-issued certificate.* A certificate issued under the terms of a written agreement between the Animal and Plant Health Inspection Service and an agricultural or forestry company or association giving assurance that a plant product has been handled, processed, or inspected in a manner required by a foreign government. An industry-issued certificate includes an ISPM 15 quality/treatment mark. The certificate may be issued by the individual who signs the agreement or his/her delegate.

(1) *Contents of written agreement.* In each written agreement, APHIS shall agree to cooperate and coordinate with the signatory agricultural or forestry company or association to facilitate the issuance of industry-issued certificates and to monitor activities under the agreement, and the concerned agricultural or forestry company or association agrees to comply with the requirements of the agreement. Each agreement shall specify the articles subject to the agreement and any measures necessary to prevent the introduction and dissemination into specified foreign countries of specified injurious plant pests. These measures could include such treatments as refrigeration, heat treatment, kiln drying, etc., and must include all necessary preshipment inspections and subsequent sign-offs and product labeling as identified by Plant Protection and Quarantine (PPQ), APHIS, based on the import requirements of the foreign country.

(2) *Termination of agreement.* An agreement may be terminated by any signatory to the agreement by giving written notice of termination to the other party. The effective date of the termination will be 15 days after the date of actual receipt of the written notice. Any agreement may be immediately withdrawn by the Administrator if he or she determines that articles covered by the agreement were moved in violation of any requirement of this chapter or any provision of the agreement. If the withdrawal is oral, the decision to withdraw the agreement and the reasons for the withdrawal of the agreement shall be confirmed in writing as promptly as circumstances permit. Withdrawal of an agreement may be appealed in writing to the Administrator within 10 days after receipt of the written notification of the withdrawal. The appeal shall state all of the facts and reasons upon which the appellant relies to show that

the agreement was wrongfully withdrawn. The Administrator shall grant or deny the appeal, in writing, stating the reasons for granting or denying the appeal as promptly as circumstances permit. If there is a conflict as to any material fact and the person from whom the agreement is withdrawn requests a hearing, a hearing shall be held to resolve the conflict. Rules of practice concerning the hearing shall be adopted by the Administrator. No written agreement will be signed with an individual or a company representative of the concerned agricultural or forestry company or association who has had a written agreement withdrawn during the 12 months following such withdrawal, unless the withdrawn agreement was reinstated upon appeal.

(Approved by the Office of Management and Budget under control number 0579–0052 and 0579–0147)

[61 FR 15368, Apr. 8, 1996, as amended at 64 FR 1105, Jan. 8, 1999; 64 FR 72265, Dec. 27, 1999; 66 FR 37116, July 17, 2001; 72 FR 35917, July 2, 2007]

§ 353.8 Accreditation of non-government facilities.

(a) The Administrator may accredit a non-government facility to perform specific laboratory testing or phytosanitary inspection services if the Administrator determines that the non-government facility meets the criteria of paragraph (b) of this section.[1]

(1) A non-government facility's compliance with the criteria of paragraph (b) of this section shall be determined through an assessment of the facility and its fitness to conduct the laboratory testing or phytosanitary inspection services for which it seeks to be accredited. If, after evaluating the results of the assessment, the Administrator determines that the facility meets the accreditation criteria, the facility's application for accreditation will be approved.

(2) The Administrator may deny accreditation to, or withdraw the accreditation of, any non-government facility

to conduct laboratory testing or phytosanitary inspection services upon a determination that the facility does not meet the criteria for accreditation or maintenance of accreditation under paragraph (b) of this section and has failed to take the remedial action recommended to correct identified deficiencies.

(i) In the case of a denial, the operator of the facility will be informed of the reasons for the denial and may appeal the decision in writing to the Administrator within 10 days after receiving notification of the denial. The appeal must include all of the facts and reasons upon which the person relies to show that the facility was wrongfully denied accreditation. The Administrator will grant or deny the appeal in writing as promptly as circumstances permit, stating the reason for his or her decision. If there is a conflict as to any material fact, a hearing will be held to resolve the conflict. Rules of practice concerning the hearing will be adopted by the Administrator.

(ii) In the case of withdrawal, before such action is taken, the operator of the facility will be informed of the reasons for the proposed withdrawal. The operator of the facility may appeal the proposed withdrawal in writing to the Administrator within 10 days after being informed of the reasons for the proposed withdrawal. The appeal must include all of the facts and reasons upon which the person relies to show that the reasons for the proposed withdrawal are incorrect or do not support the withdrawal of the accreditation of the facility. The Administrator will grant or deny the appeal in writing as promptly as circumstances permit, stating the reason for his or her decision. If there is a conflict as to any material fact, a hearing will be held to resolve the conflict. Rules of practice concerning the hearing will be adopted by the Administrator. However, withdrawal shall become effective pending final determination in the proceeding when the Administrator determines that such action is necessary to protect the public health, interest, or safety. Such withdrawal will be effective upon oral or written notification, whichever is earlier, to the operator of

---

[1] A list of accredited non-government facilities may be obtained by writing to Phytosanitary Issues Management, PPQ, APHIS, 4700 River Road, Unit 140, Riverdale, MD 20737–1236.

the facility. In the event of oral notification, written confirmation will be given as promptly as circumstances allow. This withdrawal will continue in effect pending the completion of the proceeding, and any judicial review thereof, unless otherwise ordered by the Administrator.

(3) The Administrator will withdraw the accreditation of a non-government facility if the operator of the facility informs APHIS in writing that the facility wishes to terminate its accredited status.

(4) A non-government facility whose accreditation has been denied or withdrawn may reapply for accreditation using the application procedures in paragraph (b) of this section. If the facility's accreditation was denied or withdrawn under the provisions of paragraph (a)(2) of this section, the facility operator must include with the application written documentation specifying what actions have been taken to correct the conditions that led to the denial or withdrawal of accreditation.

(5) All information gathered during the course of a non-government facility's assessment and during the term of its accreditation will be treated by APHIS with the appropriate level of confidentiality, as set forth in the U.S. Department of Agriculture's administrative regulations in §1.11 of this title.

(b) *Criteria for accreditation of nongovernment facilities.* (1) Specific standards for accreditation in a particular area of laboratory testing or phytosanitary inspection are set forth in this part and may be obtained by writing to APHIS. If specific standards for accreditation in a particular area of laboratory testing or phytosanitary inspection have not been promulgated by APHIS, and the Administrator determines that accreditation in that area is practical, APHIS will develop appropriate standards applicable to accreditation in the area for which the nongovernment facility is seeking accreditation and publish a notice of proposed rulemaking in the FEDERAL REGISTER to inform the public and other interested persons of the opportunity to comment on and participate in the development of those standards.

(2) The operator of a non-government facility seeking accreditation to conduct laboratory testing or phytosanitary inspection shall submit an application to the Administrator. The application must be completed and signed by the operator of the facility or his or her authorized representative and must contain the following:

(i) Legal name and full address of the facility;

(ii) Name, address, and telephone and fax number of the operator of the facility or his or her authorized representative;

(iii) A description of the facility, including its physical plant, primary function, scope of operation, and, if applicable, its relationship to a larger corporate entity; and

(iv) A description of the specific laboratory testing or phytosanitary inspection services for which the facility is seeking accreditation.

(3) Upon receipt of the application, APHIS will review the application to identify the scope of the assessment that will be required to adequately review the facility's fitness to conduct the laboratory testing or phytosanitary inspection services for which it is seeking accreditation. Before the assessment of the facility begins, the applicant's representative must agree, in writing, to fulfill the accreditation procedure, especially to receive the assessment team, to supply any information needed for the evaluation of the facility, and to enter into a trust fund agreement as provided by paragraph (c) of this section to pay the fees charged to the applicant facility regardless of the result of the assessment and to pay the charges of subsequent maintenance of the accreditation of the facility. Once the agreement has been signed, APHIS will assemble an assessment team and commence the assessment as soon as circumstances permit. The assessment team will measure the facility's fitness to conduct the laboratory testing or phytosanitary inspection services for which it is seeking accreditation against the specific standards identified by the Administrator for those services by reviewing the facility in the following areas:

361

(i) *Physical plant.* The facility's physical plant (e.g., laboratory space, office space, greenhouses, vehicles, etc.) must meet the criteria identified in the accreditation standards as necessary to properly conduct the laboratory testing or phytosanitary inspection services for which it seeks accreditation.

(ii) *Equipment.* The facility's personnel must possess or have unrestricted access to the equipment (e.g., microscopes, computers, scales, triers, etc.) identified in the accreditation standards as necessary to properly conduct the laboratory testing or phytosanitary inspection services for which it seeks accreditation. The calibration and monitoring of that equipment must be documented and conform to prescribed standards.

(iii) *Methods of testing or inspection.* The facility must have a quality manual or equivalent documentation that describes the system in place at the facility for the conduct of the laboratory testing or phytosanitary inspection services for which the facility seeks accreditation. The manual must be available to, and in use by, the facility personnel who perform the services. The methods and procedures followed by the facility to conduct the laboratory testing or phytosanitary inspection services for which it seeks accreditation must be commensurate with those identified in the accreditation standards and must be consistent with or equivalent to recognized international standards for such testing or inspection.

(iv) *Personnel.* The management and facility personnel accountable for the laboratory testing or phytosanitary inspection services for which the facility is seeking accreditation must be identified and must possess the training, education, or experience identified in the accreditation standards as necessary to properly conduct the testing or inspection services for which the facility seeks accreditation, and that training, education, or experience must be documented.

(4) To retain accreditation, the facility must agree to:

(i) Observe the specific standards applicable to its area of accreditation;

(ii) Be assessed and evaluated on a periodic basis by means of proficiency testing or check samples;

(iii) Demonstrate on request that it is able to perform the tests or inspection services representative of those for which it is accredited;

(iv) Resolve all identified deficiencies;

(v) Notify APHIS as soon as possible, but no more than 10 days following its occurrence, of any change in key management personnel or facility staff accountable for the laboratory testing or phytosanitary inspection services for which the facility is accredited; and

(vi) Report to APHIS as soon as possible, but no more than 10 days following its occurrence, any change involving the location, ownership, physical plant, equipment, or other conditions that existed at the facility at the time accreditation was granted.

(c) *Fees and trust fund agreement.* The fees charged by APHIS in connection with the initial accreditation of a nongovernment facility and the maintenance of that accreditation shall be adequate to recover the costs incurred by the government in the course of APHIS' accreditation activities. To cover those costs, the operator of the facility seeking accreditation must enter into a trust fund agreement with APHIS under which the operator of the facility will pay in advance all estimated costs that APHIS expects to incur through its involvement in the pre-accreditation assessment process and the maintenance of the facility's accreditation. Those costs shall include administrative expenses incurred in those activities, such as laboratory fees for evaluating check test results, and all salaries (including overtime and the Federal share of employee benefits), travel expenses (including per diem expenses), and other incidental expenses incurred by the APHIS in performing those activities. The operator of the facility must deposit a certified or cashier's check with APHIS for the amount of the costs, as estimated by APHIS. If the deposit is not sufficient to meet all costs incurred by APHIS, the operator of the facility must deposit another certified or cashier's check with APHIS for the amount of the remaining costs, as determined by

APHIS, before APHIS' services will be completed. After a final audit at the conclusion of the pre-accreditation assessment, any overpayment of funds will be returned to the operator of the facility or held on account until needed for future activities related to the maintenance of the facility's accreditation.

(Approved by the Office of Management and Budget under control number 0579–0130)

[64 FR 1105, Jan. 8, 1999, as amended at 66 FR 37400, July 18, 2001]

§ 353.9 Standards for accreditation of non-government facilities to perform laboratory seed health testing and seed crop phytosanitary inspection.

(a) *Application for accreditation, certification of accreditation, and monitoring of accredited facilities.* A facility may apply to be accredited to perform laboratory seed health testing or seed crop phytosanitary inspection, or to renew such accreditation, by submitting an application in accordance with § 353.8(b)(2) of this part. If there are portions of the application deemed to contain trade secret or confidential business information (CBI), each page of the application containing such information should be marked "CBI Copy." The application must be accompanied by a copy of the facility's quality manual and a nonrefundable application fee of $1,000. The applicant must make additional deposits to cover the costs of gaining and maintaining accreditation into a trust fund established in accordance with § 353.8(c) of this part upon request by the Administrator.

(1) Upon determining that a facility is eligible for accreditation, the Administrator will issue the facility a certificate of accreditation. Accreditation will be for a period of 3 years from the date of issuance of the certificate of accreditation and may be renewed by submitting a new application and application fee in accordance with this paragraph.

(2) The Administrator may deny or withdraw accreditation in accordance with § 353.8(a)(2) of this part. A facility may appeal denial of accreditation in accordance with § 353.8(a)(2)(i) of this part, and may appeal withdrawal of ac-

creditation in accordance with § 353.8(a)(2)(ii) of this part.

(3) A facility that has been denied accreditation or had its accreditation withdrawn may not reapply within 60 days of the date the facility was notified in writing that accreditation was denied or withdrawn.

(4) After a facility is accredited, the facility must allow APHIS access to the facility and all of its equipment and records for the purpose of conducting unannounced audits to determine the facility's continuing eligibility for accreditation. Such audits will occur at least once a year and may be performed more frequently at the discretion of the Administrator.

(b) *Standards for accreditation.* A facility that, in accordance with § 353.8(b)(2) of this part, applies to be accredited to perform laboratory seed health testing or seed crop phytosanitary inspection will be evaluated for accreditation against these standards:

(1) *Physical plant.* The facility's physical plant (e.g., laboratory space, office space, greenhouses, vehicles, etc.) must:

(i) Have laboratory and office spaces enclosed by walls and locking doors to prevent unauthorized access;

(ii) Conform to all State and local zoning and other ordinances; and

(iii) Provide a work area that is dedicated to laboratory functions and has sufficient space to conduct the required tests and store the materials and samples required for the tests in a manner that prevents contamination by other samples in the laboratory and from other sources.

(2) The facility must have access to all equipment required to conduct the laboratory testing or seed crop phytosanitary inspections for which it is accredited. Specific test methodologies, materials, and the calibration and monitoring of the equipment must conform to Reference Manual B, which is incorporated by reference at § 300.4 of this chapter. The general requirements for each test category are as follows:

(i) *Seed crop phytosanitary inspections.* Seed crop phytosanitary inspection may also include related activities such as collection of seed samples for later laboratory testing, visual inspection of seed just prior to export, and inspection of greenhouses or growth chambers where plants are grown for seed production, as well as visual inspection of seed crops. In the field, inspectors must use accurate field maps, hand lenses, and secure containers for the collection, storage, and transportation of samples. Inspectors must have direct access to a laboratory that is fully equipped to carry out any necessary diagnostic tests needed for field samples.

(ii) *Direct visual examination.* Visual examination of seed requires a stereo microscope. Visual examination of tissue requires a compound light microscope. Visual examination of loosely attached or accompanying material requires a centrifuge and shaker.

(iii) *Incubation.* Required equipment includes incubation chambers, laminar flow hoods, media preparation equipment, scales, pH meters, distilled and sterile water, gas burners, an autoclave, and the appropriate media for the specified tests.

(iv) *Grow-out tests.* Grow-out tests require a greenhouse, growth chamber, or an outdoor quarantine location, and access to a laboratory that is fully equipped to carry out any required diagnostic tests.

(v) *Serological tests.* These tests require grinding, extraction, and sample purification equipment; fluorescent microscopes; plate readers; spectrophotometers; and the appropriate assay materials; or the appropriate equipment to use field ready test kits.

(vi) *DNA probes.* To conduct these tests, a laboratory must be equipped with polymerase chain reaction (PCR) equipment, including thermal cyclers, electrophoresis and gel blotting equipment, and the reagents and DNA polymerases necessary to conduct the PCR.

(3) *Methods of testing and inspection.* The facility must conduct its laboratory seed health testing and seed crop phytosanitary inspection procedures in accordance with Reference Manual B. The facility must have a quality manual documenting its quality system for laboratory seed health testing and seed crop phytosanitary inspection proce-

dures. The quality system must follow the general guidelines described in ANSI/ASQC Q9001–1994, *American National Standard: Quality Systems-Model for Quality Assurance in Design, Development, Production, Installation and Servicing.* Acceptable models for quality systems for accredited facilities are also described in detail in Reference Manual A, which is incorporated by reference at § 300.3 of this chapter. The personnel who perform the testing and inspection services must comply with the quality manual, and management must enforce this compliance. The facility must maintain documented procedures for identification, collection, indexing, access, filing, storage, maintenance, and disposition of quality system records. The facility must maintain quality system records to demonstrate conformance to the quality manual and the effective operation of the quality system.

(4) *Personnel.* There must be a selection procedure and a training system to ensure technical competence of all staff members. The education, technical knowledge, and experience required to perform assigned test and inspection functions must be documented and clearly defined. In addition:

(i) Evaluation of plant or tissue samples must be undertaken by a plant pathologist or by laboratory technicians under the supervision of a plant pathologist, who may provide such supervision either on-site, or from a remote location. Where personnel are required to be trained at a facility to evaluate the particular types of plants or tissue samples handled by the facility, the training program must be evaluated by APHIS and determined to be effective.

(ii) All staff must have access to and be familiar with the reference materials, guides, and manuals required for the routine performance of the tests and inspections they conduct.

(Approved by the Office of Management and Budget under control number 0579–0130)

[66 FR 37400, July 18, 2001, as amended at 67 FR 8466, Feb. 25, 2002]

PART 354—OVERTIME SERVICES RELATING TO IMPORTS AND EXPORTS; AND USER FEES

Sec.
354.1 Overtime work at border ports, sea ports, and airports.
354.2 Administrative instructions prescribing commuted traveltime.
354.3 User fees for certain international services.
354.4 User fees for certain domestic services.
354.5 Penalties for nonpayment or late payment of user fees.

AUTHORITY: 7 U.S.C. 7701–7772, 7781–7786, and 8301–8317; 21 U.S.C. 136 and 136a; 49 U.S.C. 80503; 7 CFR 2.22, 2.80, and 371.3.

§ 354.1 Overtime work at border ports, sea ports, and airports.

(a)(1) Any person, firm, or corporation having ownership, custody, or control of plants, plant products, animals, animal byproducts, or other commodities or articles subject to inspection, laboratory testing, certification, or quarantine under this chapter and subchapter D of chapter I, title 9 CFR, who requires the services of an employee of the Animal and Plant Health Inspection Service or U.S. Customs and Border Protection on a Sunday or holiday, or at any other time outside the regular tour of duty of that employee, shall sufficiently in advance of the period of Sunday, holiday, or overtime service request the Animal and Plant Health Inspection Service or U.S. Customs and Border Protection inspector in charge to furnish the service during the overtime or Sunday or holiday period, and shall pay the Government at the rate listed in the following table, except as provided in paragraphs (a)(1)(i), (ii), and (iii), and (a)(3) of this section:

OVERTIME FOR INSPECTION, LABORATORY TESTING, CERTIFICATION, OR QUARANTINE OF PLANT, PLANT PRODUCTS, ANIMALS, ANIMAL PRODUCTS OR OTHER REGULATED COMMODITIES

Outside the employee's normal tour of duty	Overtime rates (per hour)		
	Nov. 2, 2015– Sept. 30, 2016	Oct. 1, 2016– Sept. 30, 2017	Beginning Oct. 1, 2017
Monday through Saturday and holidays	$75	$75	$75
Sundays	99	99	100

(i) For any services performed on a Sunday or holiday, or at any time after 5 p.m. or before 8 a.m. on a weekday, in connection with the arrivals in or departure from the United States of a private aircraft or vessel, the total amount payable shall not exceed $25 for all inspection services performed by the U.S. Customs and Border Protection, Public Health Service, and the Department of Agriculture;

(ii) Owners and operators of aircraft will be provided service without reimbursement during regularly established hours of service on a Sunday or holiday; and

(iii) The overtime rate to be charged owners or operators of aircraft at airports of entry or other places of inspection as a consequence of the operation of the aircraft, for work performed outside of the regularly established hours of service is listed in the following table:

OVERTIME FOR COMMERCIAL AIRLINE INSPECTION SERVICES [1]

Outside the employee's normal tour of duty	Overtime rates (per hour)		
	Nov. 2, 2015– Sept. 30, 2016	Oct. 1, 2016– Sept. 30, 2017	Beginning Oct. 1, 2017
Monday through Saturday and holidays	$64	$65	$65
Sundays	85	86	86

[1] These charges exclude administrative overhead costs.

(2) Except as provided in paragraph (a)(3) of this section, a minimum charge of 2 hours shall be made for any Sunday or holiday or unscheduled overtime duty performed by an employee on a day when no work was scheduled for him or her, or which is performed by an employee on his or her regular workday beginning either at least 1 hour before his or her scheduled tour of duty or which is not in direct continuation of the employee's regular tour of duty. In addition, each such period of Sunday or holiday or unscheduled overtime work to which the 2-hour minimum charge provision applies may include a commuted traveltime period (CTT) the amount of which shall be prescribed in administrative instructions to be issued by the Administrator, Animal and Plant Health Inspection Service or U.S. Customs and Border Protection for the areas in which the Sunday or holiday or overtime work is performed and such period shall be established as nearly as may be practicable to cover the time necessarily spent in reporting to and returning from the place at which the employee performs such Sunday or holiday or overtime duty. With respect to places of duty within the metropolitan area of the employee's headquarters, such CTT period shall not exceed 3 hours. It shall be administratively determined from time to time which days constitute holidays. The circumstances under which such CTT periods shall be charged and the percentage applicable in each circumstance are as reflected in the following table:

	Actual time [1] charge—no minimum	2-hour guarantee charge	Commuted [2] traveltime (CTT) charge
CHARGES FOR INSPECTION WITHIN METROPOLITAN AREA OF EMPLOYEE'S HEADQUARTERS			
Work beginning before daily tour begins:			
8 to 59 minutes	Yes	No	None.
60 to 119 minutes.	....	Yes	½ CTT.
120 minutes or more.	....	Yes	Full CTT.
Work beginning after daily tour ends:			
Direct continuation.	Yes	No	None.

	Actual time [1] charge—no minimum	2-hour guarantee charge	Commuted [2] traveltime (CTT) charge
Break-in-service of:			
2–29 minutes	....	Yes	None.
30–60 minutes.	....	Yes	½ CTT.
61 minutes or more.	....	Yes	Full CTT.
CHARGES FOR INSPECTION SERVICES PERFORMED OUTSIDE METROPOLITAN AREA OF EMPLOYEE'S HEADQUARTERS			
Work beginning before daily tour begins:			
8 to 59 minutes	Yes	No	½ CTT.
60 minutes or more.	....	Yes	Full CTT.
Work beginning after daily tour ends:			
Direct continuations.	Yes	No	½ CTT.
2–59 minutes	No	Yes	½ CTT.
60 minutes or more.	No	Yes	Full CTT.
CHARGES FOR CALL OUT INSPECTION SERVICE ON HOLIDAY OR NONWORKDAY			
Work beginning at any time.	No	Yes	Full CTT.

[1] Actual time charged when work is contiguous with the daily tour will be in quarter hour multiples, with service time of 8 minutes or more rounded up to the next quarter hour and any time of less than 8 minutes will be disregarded.

[2] The full CTT allowance will be the amount of commuted traveltime prescribed for the place at which the inspections are performed. See § 354.2. One-half CTT is ½ of the full CTT period.

(3) The overtime rate and all other charges, including minimum and commute compensation charges, to be billed for services provided by an employee of U.S. Customs and Border Protection shall be charged according to the provisions of this section, 5 CFR part 551, or 19 CFR 24.16.

(b) The Animal and Plant Health Inspection Service or U.S. Customs and Border Protection inspector in charge of honoring a request to furnish inspection, laboratory testing, quarantine or certification service, shall assign employees to such Sunday or holiday or overtime duty with due regard to the work program and availability of employees for duty.

(c) As used in this section—

(1) The term *private aircraft* means any civilian aircraft not being used to transport persons or property for compensation or hire, and

(2) The term *private vessel* means any civilian vessel not being used (i) to transport persons or property for compensation or hire, or (ii) in fishing operations or in processing of fish or fish products.

(d)(1) Any principal, or any person, firm, partnership, corporation, or other legal entity acting as an agent or broker by requesting Sunday, holiday, or overtime services of an Animal and Plant Health Inspection Service or U.S. Customs and Border Protection inspector on behalf of any other person, firm, partnership, corporation, or other legal entity (principal), and who has not previously requested such service from an Animal and Plant Health Inspection Service or U.S. Customs and Border Protection inspector, must pay the inspector before service is provided.

(2) Since the payment must be collected before service can be provided, the Animal and Plant Health Inspection Service or U.S. Customs and Border Protection inspector will estimate the amount to be paid. Any difference between the inspector's estimate and the actual amount owed to the Animal and Plant Health Inspection Service or U.S. Customs and Border Protection will be resolved as soon as reasonably possible following the delivery of service, with the Animal and Plant Health Inspection Service or U.S. Customs and Border Protection either returning the difference to the agent, broker, or principal, or billing the agent, broker, or principal for the difference.

(3) The prepayment must be in some guaranteed form, such as money order, certified check, or cash. Prepayment in guaranteed form will continue until the Animal and Plant Health Inspection Service or U.S. Customs and Border Protection determines that the agent, broker, or principal has established an acceptable credit history.

(4) For security reasons, cash payments will be accepted only from 7 a.m. to 5 p.m., and only at a location designated by the Animal and Plant Health Inspection Service or U.S. Customs and Border Protection inspector.

(e)(1) Any principal, or any person, firm, partnership, corporation, or other legal entity requesting Sunday, holiday, or overtime services of an Animal and Plant Health Inspection Service or U.S. Customs and Border Protection inspector, and who has a debt to the Animal and Plant Health Inspection Service or U.S. Customs and Border Protection more than 60 days delinquent, must pay the inspector before service is provided.

(2) Since the payment must be collected before service can be provided, the Animal and Plant Health Inspection Service or U.S. Customs and Border Protection inspector will estimate the amount to be paid. Any difference between the inspector's estimate and the actual amount owed to the Animal and Plant Health Inspection Service or U.S. Customs and Border Protection will be resolved as soon as reasonably possible following the delivery of service, with the Animal and Plant Health Inspection Service or U.S. Customs and Border Protection either returning the difference to the agent, broker, or principal, or billing the agent, broker, or principal for the difference.

(3) The prepayment must be in some guaranteed form, such as money order, certified check, or cash. Prepayment in guaranteed form will continue until the debtor pays the delinquent debt.

(4) For security reasons, cash payments will be accepted only from 7 a.m. to 5 p.m., and only at a location designated by the Animal and Plant Health Inspection Service or U.S. Customs and Border Protection inspector.

(f) Reimbursable Sunday, holiday, or overtime services will be denied to any principal, or any person, firm, partnership, corporation, or other legal entity who has a debt to the Animal and Plant Health Inspection Service or U.S. Customs and Border Protection more than 90 days delinquent. Services will be denied until the delinquent debt is paid.

[49 FR 1173, Jan. 10, 1984, as amended at 49 FR 12186, Mar. 29, 1984; 49 FR 19441, May 8, 1984; 52 FR 16822, May 6, 1987; 53 FR 52975, Dec. 30, 1988; 54 FR 13506, Apr. 4, 1989; 55 FR 3198, Jan. 31, 1990; 55 FR 41059, Oct. 9, 1990; 56 FR 1082, Jan. 11, 1991; 58 FR 32434, June 10, 1993; 67 FR 48523, July 25, 2002; 68 FR 51882, Aug. 29, 2003; 80 FR 59566, Oct. 2, 2015]

§ 354.2 Administrative instructions prescribing commuted traveltime.

Each period of overtime and holiday duty, as defined in § 354.1 shall, in addition, include a commuted traveltime period for the respective ports, stations, and areas in which employees are located. The prescribed commuted traveltime periods are set forth below:

COMMUTED TRAVELTIME ALLOWANCES

[In hours]

Location covered	Served from—	Metropolitan area	
		Within	Outside
Alabama:			
Birmingham (including Birmingham Municipal Airport)	Alabaster	....	2
Birmingham (including Birmingham Municipal Airport)	Pelham	....	2
Chickasaw	Mobile	....	2
Huntsville		1	
Mobile		2	
Montgomery		1	
Undesignated ports		....	3
Alaska:			
Anchorage		1	
Seward	Anchorage	....	6
Undesignated ports		....	3
Arizona:			
Davis-Monthan AFB, Tucson	Nogales	....	4
Douglas		1	
Douglas	Nogales	....	6
Fort Huachuca Army Base, Sierra Vista	Douglas or Nogales	....	3
Fort Huachuca Army Base, Sierra Vista	Tucson	....	4
Nogales		2	
Phoenix		2	
Phoenix	Nogales	....	6
Phoenix	Tucson	....	5
San Luis		2	
Sasabe	Nogales	....	4
Tucson		1	
Tucson	Nogales	....	3
Yuma International airport		1	
Undesignated ports		....	3
Arkansas:			
Dardanelle	Conway	....	3
Dardanelle	Little Rock	....	3
Eaker AFB	Blytheville	....	1
Fort Smith	Conway	....	5
Fort Smith	Little Rock	....	6
Helena	Blytheville	....	5
Helena	Little Rock	....	5
Little Rock		1	
Little Rock	Conway	....	2
Little Rock AFB		2	
Little Rock AFB	Conway	....	2
Osceola	Blytheville	....	2
Pine Bluff	Conway	....	4
Pine Bluff	Little Rock	2	
Undesignated ports		....	3
West Memphis	Blytheville	....	3
Bahamas:			
Nassau		1	
Freeport		1	
Bermuda:			
Ferry Reach		1	
California:			
Andrade	Calexico	....	2
Antioch	San Francisco	....	4
Bakersfield	Shafter	1	
Beale AFB	Sacramento	....	4
Burbank	Los Angeles	....	3
Calexico		1	
Camp Pendleton, USMC, Oceanside	San Diego	....	3
Castle AFB	Merced	....	1
Edwards AFB	Ontario	....	4
El Segundo	Los Angeles	....	2
El Toro MCAS	Los Angeles	....	3
Fairfield	Stockton	....	4
Fresno	Shafter	....	5
Fresno	Stockton	....	5
Hanford	Shafter	....	5

COMMUTED TRAVELTIME ALLOWANCES—Continued

[In hours]

Location covered	Served from—	Metropolitan area	
		Within	Outside
Lemoore	Shafter	....	5
George AFB	Los Angeles	....	4
Hamilton AFB, Novato	Travis AFB	....	3
Los Angeles (including San Pedro, Los Angeles Harbor, Los Angeles International Airport, Long Beach Harbor, and Long Beach Municipal Airport).		2	
March AFB	Los Angeles	....	4
March AFB	Ontario	....	3
Martinez	San Francisco	....	3
Mather Field AFB	Travis AFB	....	3
Mather Airfield	Sacramento	3	
Mather AFB	Stockton	....	3
McClellan AFB	Sacramento	3	
McClellan AFB	Stockton		4
McClellan AFB	Travis AFB	....	3
Merced/Atwater (Old Castle AFB)	Stockton		3
Moffett Field NAS, Sunnyside	San Francisco		3
Moffett Field NAS, Sunnyside	San Jose	2	
Monterey	San Jose	....	5
Monterey	San Francisco	....	6
Moss Beach Landing	San Jose		4
Norton AFB	Los Angeles		4
Ontario	Los Angeles	....	3
Palm Springs International Airport	Ontario		4
Pittsburg	San Francisco	....	4
Port Chicago	San Francisco	....	3
Port Hueneme	Port Hueneme	1	
Port Hueneme	San Pedro	....	4
Redwood City	San Francisco	....	2
Richmond	San Francisco	....	3
Rodeo	San Francisco	....	3
Sacramento		1	
Sacramento	San Francisco	....	5
Sacramento	San Jose	....	6
Sacramento	Stockton	....	3
Sacramento	Travis AFB	....	2
Sacramento International Airport	Sacramento	3	
Sacramento Metropolitan Airport	Stockton		4
Sacramento Metropolitan Airport	Travis AFB	....	3
Sacramento Seaport	Sacramento	2	
San Bernardino International Airport (Old Norton AFB)	Ontario		2
San Diego (including Mexican border at San Ysidro; Brown, Gillespie, and Lindbergh Fields, Imperial Beach; North Island, Miramar and Naval and Civilian Maritime within the San Diego Unified Port District).		2	
San Diego	Los Angeles	....	6
San Francisco (including Alameda, Oakland, San Francisco International Airport, and Oakland International Airport).		2	
San Francisco	San Jose	....	4
San Jose		2	
San Jose	Sacramento	....	5
San Jose	San Francisco	....	4
San Jose	Stockton	....	5
San Luis Obispo Seaport	Port Hueneme	....	5
Santa Barbara Airport	Port Hueneme	....	2
Seal Beach	Los Angeles	....	2
Southern California International Airport (Old George AFB).	Ontario	....	3
Stockton		1	
Stockton	Sacramento	....	3
Stockton	San Jose	....	4½
Tecate	San Ysidro	....	3
Travis AFB		1	
Travis AFB	San Francisco		4
Vallejo	San Francisco	....	2
Undesignated ports		....	3
Canada:			
Vancouver, BC (including Richmond)	Blaine	....	4

COMMUTED TRAVELTIME ALLOWANCES—Continued

[In hours]

Location covered	Served from—	Metropolitan area	
		Within	Outside
Colorado:			
Denver (including Stapleton International Airport)		2	
Ent AFB (Peterson Field)	Denver		5
Stapleton International Airport	Ft. Collins		4
Connecticut:			
Bridgeport	Groton		4
Bridgeport	Wallingford		2
Bridgeport	Warwick, RI		6
Bridgeport	Windsor Locks		4
Groton (including New London)	Wallingford		4
Groton (including New London)	Warwick, RI		4
Groton (including New London)	Windsor Locks		4
Groton (including New London)		2	
New Haven	Groton		3
New Haven	Wallingford, CT		1
New Haven	Warwick, RI		6
New Haven	Windsor Locks		3
Windsor Locks (including Bradley Field)		1	
Windsor Locks (including Bradley Field)	Boston, MA		6
Windsor Locks (including Bradley Field)	Groton		4
Windsor Locks (including Bradley Field)	Hadley, MA		2
Windsor Locks (including Bradley Field)	Wallingford		3
Windsor Locks (including Bradley Field)	Warwick, RI		6
Undesignated ports			3
Delaware:			
Dover		1	
Dover	Wilmington		3½
Wilmington (including NCCA, Delaware City, and Claymont).		2	
Wilmington (including NCCA, Delaware City, and Claymont).	Dover		3
Wilmington	Philadelphia, PA		3
Wilmington	Baltimore, MD		5
Wilmington	Bridgeton		3
Wilmington	Trenton		3
District of Columbia:			
Washington, DC Metropolitan area (including Arlington and Alexandria, VA; Andrews AFB, MD; and Washington Navy Yard).	Andrews AFB, MD	2	
Washington, DC Metropolitan area (including Arlington and Alexandria, VA; Andrews AFB, MD; and Washington Navy Yard).	Baltimore, MD	2	
Washington, DC Metropolitan area (including Arlington and Alexandria, VA; Andrews AFB, MD; and Washington Navy Yard).	Beltsville, MD	2	
Washington, DC Metropolitan area (including Arlington and Alexandria, VA; Andrews AFB, MD; and Washington Navy Yard).	Dulles International Airport, VA	2½	
Florida:			
Apalachicola	Panama City		3
Apalachicola	Pensacola		6
Boca Grande	Tampa		5
Eglin AFB	Panama City		3
Eglin AFB	Pensacola		3
Fort Lauderdale		2	
Fort Meyers	Tampa		5
Fort Myers	Fort Myers	2	
Fort Myers	Palmetto		5
Fort Pierce		1	
Fort Pierce	West Palm Beach		3
Jacksonville		2	
Key West		1	
Marathon	Key West		2
Miami		2	
Orlando		2	
Orlando	Port Canaveral		3
Orlando	Ocoee		2
Panama City		1	
Panama City	Mobile, AL	7	

COMMUTED TRAVELTIME ALLOWANCES—Continued

[In hours]

Location covered	Served from—	Metropolitan area	
		Within	Outside
Panama City	Pensacola	....	5
Patrick AFB		1	
Pensacola		1	
Pensacola	Mobile, AL	....	3
Pensacola	Panama City	....	5
Pompano Beach	Fort Lauderdale	....	3
Port Canaveral		1	
Port Everglades		2	
Port St. Joe	Panama City	....	2
Port St. Joe	Pensacola	....	6
St. Petersburg/Clearwater	Tampa	2	
Tampa		2	
West Palm Beach		1	
Undesignated ports		....	3
Georgia:			
Atlanta		2	
Brunswick		2	
Brunswick	Savannah	....	4
Columbus	Atlanta	....	4
Marietta	Atlanta	....	2
St. Mary's	Brunswick	....	3
St. Mary's	Jacksonville, FL	....	3
Savannah		2	
Undesignated ports		....	3
Hawaii:			
Barbers Point NAS	Honolulu	....	2
Barking Sands NAS	Lihue	....	3
Hakalau	Hilo	....	2
Hilo		1	
Honolulu		2	
Kaanapali, Lahaina, Maui	Kahului, Maui	....	2
Kaanapali, Lahaina (Maui)	Honolulu	....	6
Kahului, Maui		1	
Kahului, Maui	Honolulu	....	4
Kailua, Kona	Hilo	....	5
Kailua, Kona	Keahole	1	
Kalapana	Hilo	....	2
Kaneohi MCAS	Honolulu	....	2
Kapaa	Lihue	....	2
Kapahi	Lihue	....	2
Keaau	Hilo	....	1
Keahole		1	
Keauhou	Honolulu	....	5
Keauhou	Keahole	....	2
Kurtistown	Hilo	....	2
Lihue Airport	Lihue	1	
Lihue, Kauai		1	
Lihue, Kauai	Honolulu	....	5
Mahaiula	Keahole	....	2
Mt. View	Hilo	....	2
Napili-Kapalua	Maui	....	3
Nawiliwili	Lihue	1	
Pepeekeo	Hilo	....	2
Poipu	Lihue	....	3
Port Allen	Lihue	....	3
Princeville	Lihue	....	3
South Kohala	Hilo	....	4
Umauma	Hilo	....	2
Wahiawa, Oahu	Honolulu	....	2
Wailea-Makena	Maui	....	2
West Loch, Pearl Harbor	Honolulu	....	2
Undesignated ports		....	3
Idaho:			
Mountain Home AFB	Caldwell	....	3
Mountain Home AFB	Twin Falls	....	4
Illinois:			
Chicago		3	
Undesignated ports		3	
Indiana:			
Burns Harbor (including Gary)	Frankfort	....	5

COMMUTED TRAVELTIME ALLOWANCES—Continued

[In hours]

Location covered	Served from—	Metropolitan area	
		Within	Outside
Evansville	Franklin	....	6
Gary	Hanna	....	2
Indianapolis	Frankfort	....	3
Indianapolis	Franklin	....	2
Indianapolis		1	
Mount Vernon	Franklin	....	6
Undesignated ports		....	3
Iowa:			
Davenport	Des Moines	....	6
Des Moines		1	
Soioux City	Des Moines	....	6
Undesignated ports	Des Moines	....	6
Kansas:			
Johnson County Industrial	Kansas City, MO	....	2
Topeka	Kansas City, MO	....	3
Topeka	Wichita	....	6
Wichita		1½	
Kentucky:			
Covington	Lexington	....	4
Fort Campbell	Brentwood, TN	....	4
Fort Campbell	Jackson, TN	....	5
Greater Cincinnati Airport	Erlanger, KY	2	
Greater Cincinnati Airport	Louisville, KY	1	
Greater Cincinnati Airport	Louisville, KY	....	4
Louisville	Erlanger	....	4
Louisville	Lexington	....	4
Undesignated ports		....	3
Louisiana:			
Barksdale AFB, Shreveport	Baton Rouge	....	6
Barksdale AFB, Shreveport	Monroe	....	4
Barksdale AFB	Shreveport	....	1½
Baton Rouge (including Port Allen)		2	
Buras	New Orleans	....	4
Burnside	Baton Rouge	....	3
Cameron	Lake Charles	....	3
Carlyss	Lake Charles	....	2
Clifton Ridge	Lake Charles	....	2
Convent	Baton Rouge	....	3
Donaldsonville	Baton Rouge	....	3
England Air Park	Baton Rouge	....	5
England Air Park	Shreveport	....	5
England Air Park	Monroe	....	4
Fouchon	New Orleans	....	5
Geismar	Baton Rouge	....	2
Hackberry	Lake Charles	....	2
Lake Charles	Baton Rouge	....	5
Lake Charles	Port Arthur, TX	....	3
Lake Charles		1	
Morgan City	New Orleans	....	4
New Orleans		2	
Ostrica	New Orleans	....	4
Plaquemine	Baton Rouge	....	2
Port of Tallulah	Baton Rouge	....	6
Port of Tallulah	West Monroe	....	3
St. Gabriel	Baton Rouge	....	2
St. James	Baton Rouge	....	3
Uncle Sam	Baton Rouge	....	3
Venice	New Orleans	....	4
Points on the Mississippi River above the St. Charles-Jefferson Parish boundary to and including Gramercy, LA: any point below Chalmette, LA, on the east bank; and Belle Chasse, LA, and points to and including Port Sulphur on the west bank.	New Orleans	....	3
Undesignated ports		....	3
Maine:			
Bangor		1	
Bangor	Augusta	....	3
Bath	Portland	....	2
Brunswick NAS	Portland	....	2

COMMUTED TRAVELTIME ALLOWANCES—Continued

[In hours]

Location covered	Served from—	Metropolitan area	
		Within	Outside
Brunswick NAS	Augusta	....	2
Bucksport	Bangor	....	2
Cousins Island	Portland	....	1
Eastport	Bangor	....	6
Harpswell	Portland	....	2
Kittery	Portland	....	3
Portland		1	
Portland	Augusta	....	3
Portland	Manchester, NH	....	6
Searsport	Augusta	....	4
Searsport	Bangor	....	3
Wiscasset	Portland	....	3
Winterport	Bangor	....	2
Undesignated ports		....	3
Maryland:			
Aberdeen Proving Ground	Baltimore	....	3
Andrews AFB	Baltimore	2	
Andrews AFB	Baltimore	....	3½
Andrews AFB	Beltsville	2	
Andrews AFB	Dulles International Airport, VA	2½	
Annapolis	Baltimore	....	3
Baltimore		3	
Baltimore	Andrews AFB	....	3½
Cambridge	Baltimore	....	4
Cambridge	Dover, DE	....	4
Piney Point	Baltimore	....	5
Piney Point	Beltsville	....	4
Salisbury	Dover, DE	....	3
Salisbury	Baltimore	....	4
Undesignated ports	Dover, DE, or Dulles International Airport, VA.	....	3
(For other points in Maryland, see DC listing)			
Massachusetts:			
Boston		3	
Boston	New Bedford	....	5
Fall River	Boston	....	5
Fall River	New Bedford	....	2
Fall River	Warwick, RI	....	3
Gloucester	Boston	....	4
New Bedford		1	
New Bedford	Boston	....	5
New Bedford	Warwick, RI	....	3
Otis ANG/CGNS	Boston	....	6
Otis ANG/CGNS	New Bedford	....	3
Otis ANG/CGNS	Warwick, RI	....	5
Plymouth	Warwick, RI	....	3
Provincetown	Warwick, RI	....	6
Sandwich	Boston	....	6
Sandwich	New Bedford	....	3
Sandwich	Warwick, RI	....	4
Somerset	Boston	....	4
Somerset	New Bedford	....	2
Westover AFB	Boston	....	6
Westover AFB	Hadley	....	1½
Westover AFB	Windsor Locks, CT	....	2
Woods Hole	Boston	....	6
Woods Hole	New Bedford	....	3
Woods Hole	Warwick, RI	....	5
Undesignated ports	Boston, New Bedford, MA and Warwick, RI.	....	3
Undesignated ports	Windsor Locks, CT	....	3
Mexico:			
Camargo	Roma, TX	1	
Ciudad Acuna	Del Rio, TX	1	
Ciudad Acuna	Eagle Pass, TX	....	3
Ciudad Acuna	Laredo, TX	....	6
Ciudad Acuna	Pleasanton, TX	....	6
Ciudad Juarez	El Paso, TX	1	
Matamoros	Brownsville, TX	1	

COMMUTED TRAVELTIME ALLOWANCES—Continued

[In hours]

Location covered	Served from—	Metropolitan area	
		Within	Outside
Mier	Roma, TX	1	
Nuevo Cd. Guerrero	Roma, TX	1	
Nuevo Laredo	Del Rio, TX	....	4
Nuevo Laredo	Eagle Pass, TX	....	5
Nuevo Laredo	Laredo, TX	1	
Nuevo Laredo	Pharr, TX	....	6
Nuevo Laredo	Pleasanton, TX	....	5
Nuevo Progreso	Progreso, TX	1	
Ojinaga	El Paso, TX	....	6
Ojinago	Presidio	1	
Piedras Negras	Eagle Pass, TX	1	
Piedras Negras	Laredo, TX	....	5
Piedras Negras	Pharr, TX	....	10
Piedras Niegros	Pleasanton, TX	....	5
Reynosa Eagle	Pass, TX	....	12
Reynosa	Hidalgo, TX	1	
Reynosa	Laredo, TX	....	5
Reynosa	Mission, TX	....	1
Reynosa	Pharr, TX	....	1
San Jeronimo	Presidio, TX	....	6
San Jeronimo	Santa Theresa, NM	....	1
Michigan:			
Battle Creek	Grand Rapids	....	3
Bay City	Mt. Pleasant	....	3
Detroit (including Detroit Metropolitan Airport, and Willow Run Airport).	Romulus/Detroit	3	
Kent County Airport	Grand Rapids	1	
Muskegon	Grand Rapids	....	2
Pontiac	Romulus	....	4
Saginaw	Mt. Pleasant	....	3
Selfridge AFB	Port Huron	....	3
Minnesota:			
Duluth		2	
Duluth	Minneapolis	....	6
Minneapolis-St. Paul		2	
Mississippi:			
Greenville	Jackson	....	5
Gulfport		1	
Jackson		1	
Keesler AFB	Gulfport	....	1
Natchez	Brookhaven	....	4
Pascagoula	Gulfport	....	2
Pascagoula	Mobile, AL	....	3
Port Bienville	Gulfport	....	2
Vicksburg	Jackson	....	3
Undesignated ports		....	3
Missouri:			
Kansas City		1	
Kansas City	St. Charles	....	6
Kansas City International Airport		1	
Kansas City International Airport	St. Louis	....	6
Richards-Gebaur AFB	Kansas City	2	
Rosecrans AFB	Kansas City	....	3
St. Louis and St. Louis International Airport	St. Louis	2	
Whiteman, AFB	Kansas City	....	4
Montana:			
Butte International Airport	Billings	....	6
Great Falls International Airport	Great Falls	1	
Nebraska:			
Omaha (including Offutt AFB)	Lincoln	....	3
Nevada:			
Las Vegas		1	
Reno		1	
New Hampshire:			
Keene Airport, Keene	Groton	....	6
Lebanon	Manchester	....	5
Manchester		2	
Newington	Manchester	....	4
Newington	Portland, ME	....	3

COMMUTED TRAVELTIME ALLOWANCES—Continued

[In hours]

Location covered	Served from—	Metropolitan area	
		Within	Outside
Pease AFB	Manchester	....	4
Pease AFB	Portland, ME	....	3
Portsmouth	Manchester	....	4
Portsmouth	Portland, ME	....	3
Undesignated ports	Manchester	....	3
New Jersey:			
Atlantic City	McGuire AFB	....	3
Atlantic City	Mullica Hill	....	2½
Burlington	Trenton	....	1
Coast Guard Station, Cape May	Mullica Hill	....	4
Deepwater	Mullica Hill	....	2
Hammonton	Mullica Hill	....	2
Hammonton	Trenton	....	3
Lakehurst NAS	McGuire AFB	....	2
Leonardo	McGuire AFB	....	4
Leonardo	Trenton	....	4
McGuire AFB	Mullica Hill	....	3
McGuire AFB	Trenton	....	2
McGuire AFB, Wrightstown		2	
Paulsboro	Mullica Hill	....	1½
Paulsboro	Philadelphia, PA	....	3
Salem	McGuire AFB	....	3
Salem	Trenton	....	3
Trenton		2	
New Mexico:			
Albuquerque		1	
Columbus	Deming	....	1½
Columbus	El Paso, TX	....	6
Columbus	Las Cruces	....	4
Holloman AFB, Alamogordo	El Paso, TX	....	4
Santa Teresa	El Paso, TX	....	1½
Santa Teresa	Las Cruces	....	1½
Undesignated ports		....	3
New York:			
Alexandria Bay	Oneida	....	5
Buffalo		2	
Chateaugay (including Churubusco and Cannon Corners)	Rouses Point	....	3
Corning	Avoca	2	
Corning	Big Flats	1	
Farmingdale	Westhampton Beach	3	
Islip	Westhampton Beach	2	
Jamaica, Long Island		2	
Lewiston	Buffalo	....	2
Massena	Rouses Point	....	5
New York	Buffalo	3	
Niagara Falls	Buffalo	....	2
Ogdensburg	Rouses Point	....	6
Oswego	Buffalo	....	6
Oswego	Canandaigua	4	
Oswego	Oneida	3	
Plattsburgh	Rouses Point	....	3
Rochester	Avoca	3	
Rochester	Buffalo	....	4
Rochester	Canandaigua	2	
Rooseveltown	Rouses Point	....	5
Rouses Point (including Champlain)	Rouses Point	2	
Syracuse		1	
Syracuse	Buffalo	....	6
Syracuse	Canandaigua	3	
Syracuse	Oneida	2	
Watertown	Oneida	....	4
Westhampton ANG	Westhampton	1	
Undesignated ports	Buffalo or Rouses Point	....	3
North Carolina:			
Camp Lejeune	Morehead City	....	2
Camp Lejeune	Wilmington	....	3
Charlotte		2	
Charlotte	Burlington	....	5
Charlotte	Greensboro	....	4

COMMUTED TRAVELTIME ALLOWANCES—Continued

[In hours]

Location covered	Served from—	Metropolitan area	
		Within	Outside
Charlotte International Airport	Laurinburg	....	5
Cherry Point	Goldsboro	....	4
Cherry Point	Morehead City	....	2
Elizabeth City		1	
Fort Bragg	Fayetteville	2	
Greensboro		1½	
Greensboro	Charlotte	....	4
Greensboro	Fayetteville	....	4½
Greensboro	Laurinburg	....	6
Morehead City		1	
Morehead City	Clinton	....	4
Morehead City	Goldsboro	....	4
New River MCAS	Morehead City	....	3
Pope AFB	Clinton	....	2½
Pope AFB	Fayetteville	1½	
Pope AFB	Goldsboro	....	4
Raleigh	Clinton	....	4
Raleigh	Fayetteville	....	4
Raleigh	Goldsboro	....	4
Seymour-Johnson AFB	Raleigh	2	
Sunny Point Army Terminal, Southport	Goldsboro	1	
Wilmington	Wilmington	....	2
Undesignated ports		1	
		....	3
Ohio:			
Akron	Cleveland	....	2
Ashtabula	Cleveland	....	3
Cincinnati	Columbus	....	6
Cincinnati	Dayton	....	3
Cincinnati	Toledo	....	6
Cincinnati	Washington Court House	....	3
Cleveland		2	
Cleveland	Toledo	....	5
Columbus		2	
Columbus	Cleveland	....	6
Columbus	Dayton	....	4
Columbus	Toledo	....	6
Columbus	Washington Court House	....	3
Dayton		2	
Dayton	Columbus	....	4
Dayton	Toledo	....	6
Dayton	Washington Court House	....	3
Fairport Harbor	Cleveland	....	2
Greater Cincinnati Airport (Boone County, KY)	Erlanger, KY	2	
Lorraine	Cleveland	....	2
Lunken Airport	Erlanger, KY	....	1
Rickenbacker AFB	Cleveland	....	6
Rickenbacker AFB	Washington Court House	....	3
Toledo		2	
Toledo	Cleveland	....	5
Toledo	Romulus, MI	....	4
Undesignated ports		....	3
Oklahoma:			
Altus AFB	Oklahoma City	....	6
Altus AFB	Elk City	....	3
Oklahoma City	Tulsa	....	6
Port of Muskogee	Tulsa	....	2
Port Arrow	Tulsa	....	1
Port of Catoosa	Tulsa	....	1
Port of Rogers Terminal	Tulsa	....	1
Port of Verdigris	Tulsa	....	1
Tulsa		1	
Tulsa International Airport	Oklahoma City	....	6
Tulsa International Airport	Tulsa	1	
Tinker AFB, Oklahoma City		1	
Will Rogers World Airport, Oklahoma City		1	
Undesignated ports	Oklahoma City	....	3
Undesignated ports		....	3
Oregon:			
Astoria	Portland	....	5

COMMUTED TRAVELTIME ALLOWANCES—Continued

[In hours]

Location covered	Served from—	Metropolitan area	
		Within	Outside
Portland		2	
The Dalles	Ellensburg, WA	....	6
Undesignated ports		....	3
Pennsylvania:			
Allentown-Bethlehem Easton Airport	Carlisle	....	5
Allentown-Bethlehem	Dallas	....	5
Chester	Philadelphia	....	3
Chester	Wilmington, DE	....	1
Easton Airport	Gap	....	5
Erie	Buffalo, NY, or Cleveland, OH	....	4
Erie	Meadville	....	3
Erie	Mercer	....	4
Erie	Pittsburgh	....	6
Greater Pittsburgh International Airport	Cleveland, OH	....	6
Greater Pittsburgh International Airport	Meadville	....	5
Greater Pittsburgh International Airport	Mercer	....	4
Greater Pittsburgh International Airport	Pittsburgh	3	
Harrisburg International Airport	Carlisle	....	2
Harrisburg International Airport	Dallas	....	6
Harrisburg International Airport	Gap	....	4
Harrisburg International Airport	Philadelphia	....	5
Harrisburg International Airport	Williamsport	....	6
Lehigh Valley International Airport, Allentown	Gap	....	4
Lehigh Valley International Airport, Allentown	Sweet Valley	....	4
Marcus Hook	Philadelphia	....	3
Marcus Hook	Wilmington, DE	....	1
Philadelphia		3	
Philadelphia	Bridgeton	....	3
Philadelphia	McGuire Air Force Base	....	3
Philadelphia	Trenton	....	3
Philadelphia	Wilmington, DE	....	3
Tullytown	Philadelphia	....	3
Wilkes-Barre/Scranton International Airport	Dallas	....	2
Wilkes-Barre/Scranton Airport	Williamsport	....	6
Willow Grove NAS	Philadelphia	....	3
Undesignated ports	Dallas or Gap	....	3
Puerto Rico:			
Aguadilla	Mayaguez	....	2
Aguirre	Ponce	....	3
Arecibo	San Juan	....	3
Borinquen Field	Mayaguez	....	2
Fajardo	Roosevelt Road	....	1
Fajardo	San Juan	....	3
Guanica	Mayaguez	....	2
Guanica	Ponce	....	2
Guayama	Ponce	....	3
Guayanilla	Ponce	....	2
Humacao and Yabucoa	San Juan	....	4
Mayaguez (including points from Ramey to Cabo Rojo)	San Juan	....	5
Mayaguez and El Mani Airport		1	
Ponce (including all subports in the Ponce customs district).	San Juan	....	4
Ponce and Mercedita Airport		1	
Roosevelt Roads		1	
San Juan		2	
Tallaboa (Penuelas)	Ponce	....	2
Yabucoa	Roosevelt Roads	....	3
Undesignated ports		....	3
Rhode Island:			
Davisville NSD	Boston, MA	....	4
Davisville NSD	Warwick	....	2
Melville	Warwick	....	3
Newport	Boston, MA	....	5
Newport	Warwick	....	4
Portsmouth	Warwick	....	3
Providence	Boston, MA	....	4
Providence	Warwick	....	2
Quonset Point	Boston, MA	....	4
Quonset Point	Warwick	....	2

COMMUTED TRAVELTIME ALLOWANCES—Continued

[In hours]

Location covered	Served from—	Metropolitan area	
		Within	Outside
Saunderstown	Warwick	....	3
Tiverton	Warwick	....	3
Warwick		2	
Warwick	Groton, CT	....	4
Undesignated ports		....	3
South Carolina:			
Beaufort-Port Royal	Charleston	....	4
Beaufort-Port Royal	Savannah	....	3
Beaufort-Port Royal	Yamassee	....	3
Charleston		2	
Columbia	Charleston	....	4
Columbia		1	
Georgetown	Charleston	....	3
Greenville-Spartanburg		1	
McEntire NG Air Base	Columbia	....	2
McEntire NG Air Base, Eastover	Charleston	....	4
Myrtle Beach AFB	Charleston	....	4
Myrtle Beach AFB	Conway	....	1
Shaw AFB	Columbia and Florence	....	2
Shaw AFB, Sumter	Charleston	....	4
Undesignated ports		....	3
South Dakota:			
Ellsworth AFB	Pierre	....	6
Tennessee:			
Knoxville		1	
Memphis		2	
Memphis	Jackson	....	4
Memphis International Airport	Batesville, MS	....	2½
Millington	Jackson	....	4
Millington	Memphis	....	2
Nashville		2	
Nashville	Jackson	....	6
Undesignated ports		....	3
Texas:			
Aransas Pass	Corpus Christi	....	2½
Barbour's Cut	Houston	2	
Bayport	Houston	2	
Baytown	Houston	2	
Beaumont	Port Arthur	....	2
Beaumont	Lake Charles	....	3
Brownsville		1	
Brownsville	Pharr	....	3
Carswell Field, Fort Worth	Dallas-Fort Worth International Airport	....	3
Columbia	Laredo	....	2
Corpus Christi	Corpus Christi	2	
Corpus Christi NAS	Corpus Christi	....	2
Dallas		1	
Dallas (including Love Field)	Dallas-Fort Worth International Airport	....	1
Dallas-Fort Worth International Airport		1	
Dallas-Fort Worth International Airport	Decatur	....	2
Del Rio		1	
Del Rio	Eagle Pass	....	3
Donna	Hidalgo	....	2
Dyess AFB	Abilene	1	
Eagle Pass		1	
El Paso		1	
Fabens	El Paso	....	2
Falcon Heights	Roma	....	1½
Freeport	Galveston or Houston	....	3
Fort Hood	College Station	....	5
Fort Hood	Waco	....	3
Galveston		1	
Galveston	Houston	....	4
Gregory	Corpus Christi	....	2
Houston (including Houston Intercontinental Airport)	Bellville	....	4
Houston (including Houston Intercontinental Airport)	Bryan	....	4
Houston (including Houston Intercontinental Airport)	Georgetown	....	8
Houston (including Houston Intercontinental Airport)	Pleasanton	....	8
Ingleside and Harbor Island (Port Aransas)	Corpus Christi	....	3

COMMUTED TRAVELTIME ALLOWANCES—Continued

[In hours]

Location covered	Served from—	Metropolitan area	
		Within	Outside
Harlingen		1	
Harlingen	McAllen		2
Hidalgo	Brownsville		3
Houston (Including Houston Intercontinental Airport)		2	
Kelly AFB	San Antonio		2
La Feria	Hidalgo		2
Laredo		1	
Laughlin AFB	Del Rio		1
Meacham Field	Dallas-Fort Worth International Airport		2
Mercedes	Hidalgo		2
Orange	Port Arthur		2
Pharr (Includes Hidalgo and McAllen International Airport		1½	
Point Comfort		1	
Point Comfort	Victoria		2
Port Arthur		1	
Port Arthur	Lake Charles, LA		3
Port Isabel	Brownsville		2
Presidio		1	
Progreso		1	
Progreso	Brownsville or Hidalgo		2
Progreso	Harlingen		1½
Randolph AFB	San Antonio		2
Robert Grey Army Airfield	College Station		5
Robert Grey Army Airfield	San Antonio		6
Robert Grey Army Airfield	Waco		3
Rockport	Corpus Christi		3
Roma	Brownsville		5
Roma	Laredo		4
Roma (Includes Rio Grande City)		1	
Roma	Pharr		3
Sabine Pass	Port Arthur		1
San Antonio		1	
Texas City	Galveston		1
Weslaco	Hidalgo		2
Undesignated ports			3
Utah:			
Salt Lake City International Airport		2	
Hill Air Force Base, Ogden	Salt Lake City		4
Vermont:			
Alburg	Berlin		5
Alburg	Rouses Point, NY	1	
Alburg	St. Albans		2
Battleboro	Berlin		5
Battleboro	St. Albans		6
Burlington	Montpelier		3
Burlington	St. Albans	1	
Derby Line	Berlin		6
Derby Line	St. Albans		5
Highgate Springs	Berlin		4
Highgate Springs	St. Albans		2
Montpelier		1	
Richford	Berlin		5
Richford	St. Albans		3
St. Albans (including Highgate Springs and Morses Line)	Rouses Point, NY		3
Undesignated ports	Montpelier		3
Undesignated ports	Rouses Point, NY		3
Undesignated ports	St. Albans		3
Virgin Islands:			
Alexander Hamilton Airport, St. Croix		1	
Charlotte Amalie, St. Thomas		1	
Christiansted, St. Croix		1	
Cruz Bay, St. John, USVI	St. Thomas, USVI		3
Frederiksted, St. Croix		1	
Undesignated ports			3
Virginia:			
Alexandria or Arlington	Andrews AFB, MD	2	
Alexandria or Arlington	Dulles International Airport	2½	
Alexandria or Arlington	Beltsville, MD	2	
Dulles International Airport		2	

COMMUTED TRAVELTIME ALLOWANCES—Continued

[In hours]

Location covered	Served from—	Metropolitan area	
		Within	Outside
Dulles International Airport	Baltimore, MD		3
Dulles International Airport	Beltsville, MD	2½	
Dulles International Airport	Fredericksburg		3
Hopewell	Norfolk		5
Norfolk Metropolitan Area (including Chesapeake, Hampton, Newport News, Portsmouth and Virginia Beach).		2	
Quantico MCAS	Dulles International Airport		3
Quantico MCAS	Fredericksburg	1½	
Richmond	Norfolk		5
Undesignated ports			3
(For other points in Virginia, see District of Columbia listing)			
Washington:			
Aberdeen	Seattle, maritime port		4
Anacortes	Blaine		3
Ault Field	Blaine		4
Bangor NSO	Seattle, maritime port		4
Bellingham	Blaine		2
Blaine		1	
Brewster	Ellensberg		6
Brewster	Spokane		6
Brewster	Wenatchee		4
Cherry Point	Blaine		1
Edmonds	Seattle, maritime port		2
Ellensburg		1	
Everett	Seattle, maritime port		3
Ferndale	Blaine		2
Fort Lewis	Tacoma		2
Grays Harbor	Seattle, maritime port		6
Grays Harbor	Tacoma		4
Hood River	Ellensburg		6
Lynden	Blaine		2
McChord AFB	Seattle, maritime port		3
McChord AFB	Tacoma		2
Moses Lake	Ellensburg		3
Moses Lake	Wenatchee		3
Olympia	Seattle, maritime port		3
Olympia	Tacoma		2
Oroville		1	
Paine Field	Seattle, maritime port		3
Pasco	Ellensburg		5
Pasco	Spokane		6
Pasco	Wenatchee		6
Point Wells	Seattle, maritime port		2
Port Angeles	Seattle, maritime port		6
Port Angeles	Tacoma		6
Port Townsend	Seattle, maritime port		4
Sawyer	Ellensburg		3
Sawyer	Wenatchee		5
SEA TAC Airport		2	
Seattle, maritime port		2	
Sumas	Blaine		2
Tacoma		2	
Wenatchee		1	
Wenatchee	Ellensburg		4
Wenatchee	Spokane		6
Yakima		1	
Yakima	Ellensburg		3
Yakima	Wenatchee		6
West Virginia:			
Kanawha Airport	Charleston	1	
Kanawha Airport	Clarksburg		6
Wisconsin:			
Green Bay	Milwaukee		4
Kenosha	Milwaukee		2
Madison	Milwaukee		4
Milwaukee		1	
Milwaukee	Madison		4½
Milwaukee	O'Hare International Airport, Chicago, IL		5

COMMUTED TRAVELTIME ALLOWANCES—Continued

[In hours]

Location covered	Served from—	Metropolitan area	
		Within	Outside
Racine	Milwaukee	....	2
Superior	Duluth, MN	1	
Undesignated ports	Duluth, MN or Milwaukee	....	3
Wyoming:			
Cheyanne	....		1
Temporary detail:			
Any inspection point to which an employee may be temporarily detailed.	....	1	....

[49 FR 32332, Aug. 14, 1984]

EDITORIAL NOTE: For FEDERAL REGISTER citations affecting §354.2, see the List of CFR Sections Affected, which appears in the Finding Aids section of the printed volume and at *www.govinfo.gov*.

§354.3 User fees for certain international services.

(a) *Definitions.* Whenever in this section the following terms are used, unless the context otherwise requires, they shall be construed, respectively, to mean:

APHIS. The Animal and Plant Health Inspection Service of the United States Department of Agriculture.

Arrival. Arrival at a port of entry in the customs territory of the United States, or at any place served by a port of entry as specified in 19 CFR 101.3.

Barge. A non-self-propelled commercial vessel that transports cargo that is not contained in shipping containers. This does not include integrated tug barge combinations.

Calendar year. The period from January 1 to December 31, inclusive, of any particular year.

Certificate. Any certificate issued by or on behalf of APHIS describing the condition of a shipment of plants or plant products for export, including but not limited to Phytosanitary Certificate (PPQ Form 577), Export Certificate for Processed Plant Products (PPQ Form 578), and Phytosanitary Certificate for Reexport (PPQ Form 579).

Commercial aircraft. Any aircraft used to transport persons or property for compensation or hire.

Commercial purpose. The intention of receiving compensation, or making a gain or profit.

Commercial railroad car. A railroad car used or capable of being used for transporting property for compensation or hire.

Commercial shipment. A shipment for gain or profit.

Commercial truck. A self-propelled vehicle, designed and used for transporting property for compensation or hire. Empty trucks and truck cabs without trailers fitting this description are included.

Commercial vessel. Any watercraft or other contrivance used or capable of being used as a means of transportation on water to transport property for compensation or hire, with the exception of any aircraft or ferry.

Customs. The Bureau of Customs and Border Protection, U.S. Department of Homeland Security.

Customs territory of the United States. The 50 States, the District of Columbia, and Puerto Rico.

Designated State or county inspector. A State or county plant regulatory official designated by the Secretary of Agriculture to inspect and certify to shippers and other interested parties as to the phytosanitary condition of plant products inspected under the Plant Protection Act.

Person. An individual, corporation, partnership, trust, association, or any other public or private entity, or any officer, employee, or agent thereof.

(b) *Fee for inspection of commercial vessels of 100 net tons or more.* (1) Except as provided in paragraph (b)(2) of this section, the master, licensed deck officer, or purser of any commercial vessel which is subject to inspection under part 330 of this chapter or 9 CFR chapter I, subchapter D, and which is either required to make entry at the customs house under 19 CFR 4.3 or is a United States-flag vessel proceeding coastwise under 19 CFR 4.85, shall, upon arrival, proceed to Customs and pay an agricultural quarantine and inspection (AQI) user fee. The AQI user fee for each arrival is shown in the following table. The AQI user fee shall be collected at each port of arrival.

Effective date	Amount
Beginning December 28, 2015	$825

(2) The following categories of commercial vessels are exempt from paying an AQI user fee:

(i) Foreign passenger vessels making at least three trips a week from a port in the United States to the high seas (including "cruises to nowhere") and returning to the same port in the United States, not having touched any foreign port or place, or taken on any stores;

(ii) Any vessel which, at the time of arrival, is being used solely as a tugboat;

(iii) Vessels used exclusively in the governmental service of the United States or a foreign government, including any agency or political subdivision of the United States or a foreign government, so long as the vessel is not carrying persons or merchandise for commercial purposes;

(iv) Vessels arriving in distress or to take on fuel, sea stores, or ship's stores; and

(v) Tugboats towing vessels on the Great Lakes.

(vi) Barges traveling solely between the United States and Canada that do not carry cargo originating from countries other than the United States or Canada and do not carry plants or plant products, or animals or animal products, and that do not carry soil or quarry products from areas in Canada listed in § 319.77-3 of this chapter as being infested with gypsy moth.

(vii) Vessels returning to the United States after traveling to Canada solely to take on fuel.

(c) *Fee for inspection of commercial trucks.* (1) The driver or other person in charge of a commercial truck that is enteing the customs territory of the United States and that is subject to inspection under part 330 of this chapter

or under 9CFR, chapter I, subchapter D, must, upon arrival, proceed to Customs and pay and AQI user fee for each arrival, as shown in the following table:

Effective date	Amount
Beginning December 28, 2015	$7.55

(2) [Reserved]

(3) Prepayment.

(i) The owner or operator of a commercial truck, *if* entering the customs territory of the United States *and* applying for a prepaid Customs permit for a calendar year, must apply for a prepaid AQI permit for the same calendar year. Applicants must apply to Customs for prepaid AQI permits.[1] The following information must be provided, together with payment of an amount 40 times the AQI user fee for each arrival:

(A) Vehicle make, model, and model year.

(B) Vehicle Identification Number (VIN).

(C) License numbers issued by State, Province, or country.

(D) Owner's name and address.

(ii) No credit toward the prepaid AQI permit will be given for user fees paid for individual arrivals.

(d) *Fee for inspection of commercial railroad cars.* (1) Except as provided in paragraph (d)(2) of this section, an AQI user fee will be charged for each loaded commercial railroad car which is subject to inspection under part 330 of this chapter or under 9 CFR chapter I, subchapter D, upon each arrival. The railroad company receiving a commercial railroad car in interchange at a port of entry or, barring interchange, the railroad company moving a commercial railroad car in line haul service into the customs territory of the United States, is responsible for paying the AQI user fee. The AQI user fee for each arrival of a loaded railroad car is shown in the following table. If the AQI user fee is prepaid for all arrivals of a commercial railroad car during a calendar year, the AQI user fee is an amount 20 times the AQI user fee for each arrival.

---

[1] Applicants should refer to Customs and Border Protection regulations (19 CFR part 24) for specific instructions.

Effective date	Amount
Beginning December 28, 2015	$2

(2) The following categories of commercial railroad cars are exempt from paying an AQI user fee:

(i) Any commercial railroad car that is part of a train whose journey originates and terminates in Canada if—

(A) The commercial railroad car is part of the train when the train departs Canada; and

(B) No passengers board or disembark from the commercial railroad car, and no cargo is loaded or unloaded from the commercial railroad car, while the train is within the United States.

(ii) Any commercial railroad car that is part of a train whose journey originates and terminates in the United States, if—

(A) The commercial railroad car is part of the train when the train departs the United States; and

(B) No passengers board or disembark from the commercial railroad car, and no cargo is loaded or unloaded from the commercial railroad car, while the train is within any country other than the United States; and

(iii) Locomotives and cabooses.

(3) Prepayment.

(i) Railroad companies may, at their option, prepay the AQI user fee for each commercial railroad car for a calendar year. This payment must be remitted in accordance with paragraph (d)(5) of this section.

(ii) No credit toward the calendar year AQI user fee will be given for AQI user fees paid for individual arrivals.

(4) Statement procedures. The Association of American Railroads (AAR), and the National Railroad Passenger Corporation (AMTRAK), shall file monthly statements with the U.S. Bank, United States Department of Agriculture (USDA), APHIS, AQI, P.O. Box 979044, St. Louis, MO 63197–9000, within 60 days after the end of each calendar month. Each statement shall indicate:

(i) The number of loaded commercial railroad cars entering the customs territory of the United States during the relevant period;

(ii) The number of those commercial railroad cars pulled by each railroad company; and

(iii) The total monthly AQI user fee due from each railroad company.

(5) Remittance procedures. Individual railroad companies shall remit the AQI user fees calculated by AAR, and AMTRAK shall remit the AQI user fees it has calculated, within 60 days after the end of each calendar month in which commercial railroad cars entered the customs territory of the United States. AQI user fees, together with monthly statements, must be remitted to the U.S. Bank, United States Department of Agriculture (USDA), APHIS, AQI, P.O. Box 979044, St. Louis, MO 63197–9000.

(6) Compliance. AAR, AMTRAK, and each railroad company responsible for making AQI user fee payments must allow APHIS personnel to verify the accuracy of AQI user fees collected and remitted and otherwise determine compliance with 21 U.S.C. 136a and this paragraph. The AAR, AMTRAK, and each railroad company responsible for making AQI user fee payments must advise the U.S. Bank, United States Department of Agriculture (USDA), APHIS, AQI, P.O. Box 979044, St. Louis, MO 63197–9000, of the name, address, and telephone number of a responsible officer who is authorized to verify AQI user fee calculations, collections, and remittances, as well as any changes in the identifying information submitted.

(e) *Fee for inspection of commercial aircraft.* (1) Except as provided in paragraph (e)(2) of this section, an AQI user fee will be charged for each commercial aircraft which is arriving, or which has arrived and is proceeding from one United States airport to another under a Bureau of Customs and Border Protection "Permit to Proceed," as specified in 19 CFR 122.81 through 122.85, or an "Agricultural Clearance or Safeguard Order" (PPQ Form 250), used pursuant to §330.400 of this chapter and 9 CFR 94.5, and which is subject to inspection under part 330 of this chapter or 9 CFR chapter I, subchapter D. Each carrier is responsible for paying the AQI user fee. The AQI user fee for each arrival is shown in the following table:

Effective date	Amount
Beginning December 28, 2015	$225

(2) The following categories of commercial aircraft are exempt from paying an AQI user fee:

(i) [Reserved]

(ii) Any aircraft used exclusively in the governmental services of the United States or a foreign government, including any Agency or political subdivision of the United States or a foreign government, as long as the aircraft is not carrying persons or merchandise for commercial purposes;

(iii) Any aircraft making an emergency or forced landing when the original destination of the aircraft was a foreign port;

(iv) Any passenger aircraft with 64 or fewer seats, which is not carrying the following cargo: Fresh fruits, fresh vegetables, plants, unprocessed plant products, cotton or covers, sugarcane, or fresh or processed meats; and which does not offer meal service other than beverages and prepackaged snacks that do not contain meats derived from ruminants, swine, or poultry or fresh fruits and fresh vegetables. Aircraft exempt from the user fee under this paragraph would still be subject to the garbage handling requirements found in § 330.400 of this chapter and 9 CFR 94.5;

(v) Any aircraft moving from the United States Virgin Islands to Puerto Rico; and

(vi) Any aircraft making an intransit stop at a port of entry, during which the aircraft does not proceed through any portion of the Federal clearance process, such as inspection or clearance by APHIS or the Bureau of Customs and Border Protection, no cargo is removed from or placed on the aircraft, no passengers get on or off the aircraft, no crew members get on or off the aircraft, no food is placed on the aircraft, and no garbage is removed from the aircraft.

(3) Remittance and statement procedures. (i) Each carrier must remit the appropriate fees to the U.S. Bank, United States Department of Agriculture (USDA), APHIS, AQI, P.O. Box 979044, St. Louis, MO 63197–9000, for receipt no later than 31 days after the close of the calendar quarter in which the aircraft arrivals occurred. Late payments will be subject to interest, penalty, and handling charges as provided in the Debt Collection Act of 1982, as amended by the Debt Collection Improvement Act of 1996 (31 U.S.C. 3717).

(ii) The remitter must mail with the remittance a written statement to the U.S. Bank, United States Department of Agriculture (USDA), APHIS, AQI, P.O. Box 979044, St. Louis, MO 63197–9000. The statement must include the following information:

(A) Name and address of the person remitting payment;

(B) Taxpayer identification number of the person remitting payment;

(C) Calendar quarter covered by the payment;

(D) Ports of entry at which inspections occurred;

(E) Number of arrivals at each port; and

(F) Amount remitted.

(iii) Remittances must be made by check or money order, payable in United States dollars, through a United States bank, to "The Animal and Plant Health Inspection Service."

(4) Compliance. Each carrier subject to this section must allow APHIS personnel to verify the accuracy of the AQI user fees remitted and to otherwise determine compliance with 21 U.S.C. 136a and this paragraph. Each carrier must advise the U.S. Bank, United States Department of Agriculture (USDA), APHIS, AQI, P.O. Box 979044, St. Louis, MO 63197–9000, of the name, address, and telephone number of a responsible officer who is authorized to verify AQI user fee calculations and remittances, as well as any changes in the identifying information submitted.

(5) Limitations on charges. (i) Airlines will not be charged reimbursable overtime for inspection of aircraft if the aircraft is subject to the AQI user fee for arriving aircraft as prescribed by this section.

(ii) Airlines will not be charged reimbursable overtime for inspection of cargo from an aircraft if:

(A) The aircraft is subject to the AQI user fee for arriving aircraft as prescribed by this section; and

(B) The cargo is inspected between 8 a.m. and 4:30 p.m., Monday through Friday; or

(C) The cargo is inspected concurrently with the aircraft.

(f) *Fee for inspection of international passengers.* (1) Except as specified in paragraph (f)(2) of this section, each passenger aboard a commercial aircraft or cruise ship who is subject to inspection under part 330 of this chapter or 9 CFR, chapter I, subchapter D, upon arrival from a place outside of the customs territory of the United States, must pay an AQI user fee. The AQI user fee will apply to tickets purchased beginning December 28, 2015. The fees are shown in the following table:

Effective dates [1]	Passenger type	Amount
Beginning December 28, 2015.	Commercial aircraft	$3.96
Beginning December 28, 2015.	Cruise ship	1.75

[1] Persons who issue international airline and cruise line tickets or travel documents are responsible for collecting the AQI international airline passenger user fee and the international cruise ship passenger user fee from ticket purchasers. Issuers must collect the fee applicable at the time tickets are sold. In the event that ticket sellers do not collect the AQI user fee when tickets are sold, the air carrier or cruise line must collect the user fee that is applicable at the time of departure from the passenger upon departure.

(2) The following categories of passengers are exempt from paying an AQI user fee:

(i) Crew members onboard for purposes related to the operation of the vessel;

(ii) Crew members who are on duty on a commercial aircraft;

(iii) Airline employees, including "deadheading" crew members, who are traveling on official airline business;

(iv) Diplomats, except for United States diplomats, who can show that their names appear on the accreditation listing maintained by the United States Department of State. In lieu of the accreditation listing, an individual diplomat may present appropriate proof of diplomatic status to include possession of a diplomatic passport or visa, or diplomatic identification card issued by a foreign government;

(v) Passengers departing and returning to the United States without having touched a foreign port or place;

(vi) Passengers arriving on any commercial aircraft used exclusively in the governmental service of the United States or a foreign government, including any agency or political subdivision of the United States or a foreign government, so long as the aircraft is not carrying persons or merchandise for commercial purposes. Passengers on commercial aircraft under contract to the United States Department of Defense (DOD) are exempted if they have been precleared abroad under the joint DOD/APHIS Military Inspection Program;

(vii) Passengers arriving on an aircraft due to an emergency or forced landing when the original destination of the aircraft was a foreign port;

(viii) Passengers transiting the United States and not subject to inspection; and

(ix) Passengers moving from the United States Virgin Islands to Puerto Rico.

(3) AQI user fees shall be collected under the following circumstances:

(i) When through tickets or travel documents are issued indicating travel to the customs territory of the United States that originates in any foreign country; and

(ii) When passengers arrive in the customs territory of the United States in transit from a foreign country and are inspected by APHIS or Customs.

(4) *Collection of fees.* (i) Any person who issues tickets or travel documents on or after May 13, 1991, is responsible for collecting the AQI user fee from all passengers transported into the customs territory of the United States to whom the AQI user fee applies.

(A) Tickets or travel documents must be marked by the person who collects the AQI user fee to indicate that the required AQI user fee has been collected from the passenger.

(B) If the AQI user fee applies to a passenger departing from the United States and if the passenger's tickets or travel documents were issued on or after May 13, 1991, but do not reflect collection of the AQI user fee at the time of issuance, then the carrier transporting the passenger from the United States must collect the AQI user fee upon departure.

(C) AQI user fees collected from international passengers pursuant to paragraph (f) of this section shall be held in trust for the United States by the person collecting such fees, by any person holding such fees, or by the person who is ultimately responsible for remittance of such fees to APHIS. AQI user fees collected from international passengers shall be accounted for separately and shall be regarded as trust funds held by the person possessing such fees as agents, for the beneficial interest of the United States. All such user fees held by any person shall be property in which the person holds only a possessory interest and not an equitable interest. As compensation for collecting, handling, and remitting the AQI user fees for international passengers, the person holding such user fees shall be entitled to any interest or other investment return earned on the user fees between the time of collection and the time the user fees are due to be remitted to APHIS under this section. Nothing in this section shall affect APHIS' right to collect interest for late remittance.

(5) *Remittance and statement procedures.* (i) The carrier whose ticket stock or travel document reflects collection of the AQI user fee must remit the fee to the U.S. Bank, United States Department of Agriculture (USDA), APHIS, AQI, P.O. Box 979044, St. Louis, MO 63197–9000. The travel agent, United States-based tour wholesaler, or other entity, which issues its own non-carrier related ticket or travel document to a passenger who is subject to an AQI user fee under this part, must remit the fee to APHIS, unless by contract the carrier will remit the fee.

(ii) AQI user fees must be remitted to the U.S. Bank, United States Department of Agriculture (USDA), APHIS, AQI, P.O. Box 979044, St. Louis, MO 63197–9000, for receipt no later than 31 days after the close of the calendar quarter in which the AQI user fees were collected. Late payments will be subject to interest, penalty, and handling charges as provided in the Debt Collection Act of 1982, as amended by the Debt Collection Improvement Act of 1996 (31 U.S.C. 3717). Refunds by a remitter of AQI user fees collected in conjunction with unused tickets or travel documents shall be netted against the next subsequent remittance.

(iii) The remitter must mail with the remittance a written statement to the U.S. Bank, United States Department of Agriculture (USDA), APHIS, AQI, P.O. Box 979044, St. Louis, MO 63197–

9000. The statement must include the following information:

(A) Name and address of the person remitting payment;

(B) Taxpayer identification number of the person remitting payment;

(C) Calendar quarter covered by the payment; and

(D) Amount collected and remitted.

(iv) Remittances must be made by check or money order, payable in United States dollars, through a United States bank, to "The Animal and Plant Health Inspection Service."

(6) Carriers contracting with United States-based tour wholesalers are responsible for notifying the U.S. Bank, United States Department of Agriculture (USDA), APHIS, AQI, P.O. Box 979044, St. Louis, MO 63197–9000, of all flights contracted, the number of spaces contracted for, and the name, address, and taxpayer identification number of the United States-based tour wholesaler, within 31 days after the close of the calendar quarter in which such a flight occurred; *except that*, carriers are not required to make notification if tickets, marked to show collection of the AQI user fee, are issued for the individual contracted spaces.

(7) *Compliance.* Each carrier, travel agent, United States-based tour wholesaler, or other entity subject to this section must allow APHIS personnel to verify the accuracy of the AQI user fees collected and remitted and to otherwise determine compliance with 21 U.S.C. 136a and this paragraph. Each carrier, travel agent, United States-based tour wholesaler, or other entity must advise the U.S. Bank, United States Department of Agriculture (USDA), APHIS, AQI, P.O. Box 979044, St. Louis, MO 63197–9000, of the name, address, and telephone number of a responsible officer who is authorized to verify AQI user fee calculations, collections, and remittances, as well as any changes in the identifying information submitted.

(8) *Limitation on charges.* Airlines and cruise lines will not be charged reimbursable overtime for passenger inspection services required for any aircraft or cruise ship on which a passenger arrived who has paid the international passenger AQI user fee for that flight or cruise.

(g) *Fees for export certification of plants and plant products.* (1) For each certificate issued by APHIS personnel, the recipient must pay the applicable AQI user fee at the time and place the certificate is issued.

(2) When the work necessary for the issuance of a certificate is performed by APHIS personnel on a Sunday or holiday, or at any other time outside the regular tour of duty of the APHIS personnel issuing the certificate, in addition to the applicable user fee, the recipient must pay the applicable overtime rate in accordance with §354.1.

(3)(i) Each exporter who receives a certificate issued on behalf of APHIS by a designated State or county inspector must pay an administrative user fee, as shown in the following table. The administrative fee can be remitted by the exporter directly to APHIS through the Phytosanitary Certificate Issuance and Tracking System (PCIT), provided that the exporter has a PCIT account and submits the application for the export certificate through the PCIT. If the PCIT is not used, the State or county issuing the certificate is responsible for collecting the fee and remitting it monthly to the U.S. Bank, United States Department of Agriculture, APHIS, AQI, P.O. Box 979043, St. Louis, MO 63197–9000.

Effective dates	Amount per shipment	
	PCIT used	PCIT not used
October 1, 2009, through September 30, 2010	$3	$6
October 1, 2010, through September 30, 2011	6	12
Beginning October 1, 2011	6	12

(ii) The AQI user fees for an export or reexport certificate for a commercial shipment are shown in the following table.

Effective dates	Amount per shipment
October 1, 2009, through September 30, 2010	$77
October 1, 2010, through September 30, 2011	104
Beginning October 1, 2011	106

(iii) The AQI user fees for an export or reexport certificate for a low-value commercial shipment are shown in the following table. A commercial shipment is a low-value commercial shipment if the items being shipped are identical to those identified on the certificate; the shipment is accompanied by an invoice which states that the items being shipped are worth less than $1,250; and the shipper requests that the user fee charged be based on the low value of the shipment.

Effective dates	Amount per shipment
October 1, 2009, through September 30, 2010	$42
October 1, 2010, through September 30, 2011	60
Beginning October 1, 2011	61

(iv) The AQI user fees for an export or reexport certificate for a non-commercial shipment are shown in the following table.

Effective dates	Amount per shipment
October 1, 2009, through September 30, 2010	$42
October 1, 2010, through September 30, 2011	60
Beginning October 1, 2011	61

(v) The AQI user fees for replacing any certificate are shown in the following table.

Effective dates	Amount per certificate
October 1, 2009, through September 30, 2010	$11
October 1, 2010, through September 30, 2011	15
Beginning October 1, 2011	15

(4) If a designated State inspector issues a certificate, the State where the certificate is issued may charge for inspection services provided in that State.

(5) Any State which wishes to charge a fee for services it provides to issue certificates must establish fees in accordance with one of the following guidelines:

(i) *Calculation of a "cost-per-certificate" fee.* The State must:

(A) Estimate the annual number of certificates to be issued;

(B) Determine the total cost of issuing certificates by adding together delivery,[2] support,[3] and administrative[4] costs; and

(C) Divide the cost of issuing certificates by the estimated number of certificates to be issued to obtain a "raw" fee. The State may round the "raw" fee up to the nearest quarter, if necessary for ease of calculation, collection, or billing; or

(ii) *Calculation of a "cost-per-hour" fee.* The State must:

(A) Estimate the annual number of hours taken to issue certificates by adding together delivery,[2] support,[3] and administrative[4] hours;

(B) Determine the total cost of issuing certificates by adding together delivery, support, and administrative costs; and

(C) Divide the cost of issuing certificates by the estimated number of hours taken to issue certificates to obtain a "cost-per-hour" fee. The State may round the "cost-per-hour" fee up to the nearest quarter, if necessary for ease of calculation, collection, or billing.

(h) *Fee for conducting and monitoring treatments.* (1) Each importer of a consignment of articles that require treatment upon arrival from a place outside of the customs territory of the United States, either as a preassigned condition of entry or as a remedial measure ordered following the inspection of the consignment, must pay an AQI user fee. The AQI user fee is charged on a per-treatment basis, *i.e.,* if two or more consignments are treated together, only a single fee will be charged, and if a single consignment is split or must be retreated, a fee will be charged for each separate treatment conducted. The AQI user fee for each treatment is shown in the following table:

Effective dates	Amount
Beginning December 28, 2015	$47
Beginning December 28, 2016	95
Beginning December 28, 2017	142
Beginning December 28, 2018	190
Beginning December 28, 2019	237

(2) *Treatment provider.* (i) Private entities that provide AQI treatment services to importers are responsible for collecting the AQI treatment user fee from the importer for whom the service is provided. Treatment providers must collect the AQI treatment fee applicable at the time the treatment is applied.

(ii) When AQI treatment services are provided by APHIS, APHIS will collect the AQI treatment fee applicable at the time the treatment is applied from the person receiving the services. Remittances must be made by check or money order, payable in United States dollars, through a United States bank, to "The Animal and Plant Health Inspection Service."

(3) *Collection of fees.* (i) In cases where APHIS is not providing the AQI treatment and collecting the associated fee, AQI user fees collected from importers pursuant to this paragraph shall be held in trust for the United States by the person collecting such fees, by any person holding such fees, or by the person who is ultimately responsible for remittance of such fees to APHIS. AQI user fees collected from importers shall be accounted for separately and shall be regarded as trust funds held by the person possessing such fees as agents, for the beneficial interest of the United States. All such user fees held by any person shall be property in which the person holds only a possessory interest and not an equitable interest. As compensation for collecting, handling, and remitting the AQI treatment user fees, the person holding such user fees shall

---

[2] Delivery costs are costs such as employee salary and benefits, transportation, per diem, travel, purchase of specialized equipment, and user fee costs associated with maintaining field offices. Delivery hours are similar hours taken by inspectors, including travel time, inspection time, and time taken to complete paperwork.

[3] Support costs are costs at supervisory levels which are similar to delivery costs, and user fee costs such as training, automated data processing, public affairs, enforcement, legal services, communications, postage, budget and accounting services, and payroll, purchasing, billing, and collecting services. Support hours are similar hours taken at supervisory levels, as well as hours taken in training, automated data processing, enforcement, legal services, communication, budgeting and accounting, payroll purchasing, billing, and collecting.

[4] Administrative costs are costs incurred as a direct result of collecting and monitoring Federal phytosanitary certificates. Administrative hours are hours taken as a direct result of collecting and monitoring Federal phytosanitary certificates.

be entitled to any interest or other investment return earned on the user fees between the time of collection and the time the user fees are due to be remitted to APHIS under this section. Nothing in this section shall affect APHIS' right to collect interest from the person holding such user fees for late remittance.

(4) *Remittance and statement procedures.* (i) The treatment provider that collects the AQI treatment user fee must remit the fee to USDA, APHIS, AQI, PO Box 979044, St. Louis, MO 63197–9000.

(ii) AQI treatment user fees must be remitted to [address to be added in final rule] for receipt no later than 31 days after the close of the calendar quarter in which the AQI user fees were collected. Late payments will be subject to interest, penalty, and handling charges as provided in the Debt Collection Act of 1982, as amended by the Debt Collection Improvement Act of 1996 (31 U.S.C. 3717).

(iii) The remitter must mail with the remittance a written statement to USDA, APHIS, AQI, PO Box 979044, St. Louis, MO 63197–9000. The statement must include the following information:

(A) Name and address of the person remitting payment;

(B) Taxpayer identification number of the person remitting payment;

(C) Calendar quarter covered by the payment; and

(D) Amount collected and remitted.

(iv) Remittances must be made by check or money order, payable in United States dollars, through a United States bank, to "The Animal and Plant Health Inspection Service."

(i) *Payment methods.* For payment of any of the AQI user fees required in paragraph (g) of this section, we will accept personal checks for amounts less than $100, and checks drawn on commercial accounts, cashier's checks, certified checks, traveler's checks, and money orders for any amount. All payments must be for the exact amount due.

(j) The person for whom the service is provided and the person requesting the service are jointly and severally liable for payment of user fees for any import or entry services listed below, of $56 per hour, or $14 per quarter hour, with a minimum fee of $14 for each employee required to perform the following services. If the services must be conducted on a Sunday or holiday or at any other time outside the normal tour of duty of the employee, then the premium user fee rate as listed below applies, as well as the 2-hour minimum charge and a commuted traveltime period required by §354.1(a)(2). If the services requested are performed on a Sunday, the hourly user fee rate will be $74, or $18.50 per quarter hour, with a $18.50 minimum. If the services requested are performed on a day other than Sunday outside the normal tour of duty of the employee providing the service, the hourly user fee rate will be $65, or $16.25 per quarter hour, with a $16.25 minimum:

(1) Conducting inspections, on vessels or in storage areas, of solid wood packing material or cargo when a shipment arrives without a certificate or exporter statement required under §319.40–5(g) or §319.40–5(h) of this chapter, or with an incomplete certificate or exporter statement; and

(2) Supervising the separation of cargo from solid wood packing material denied entry under this subpart and the destruction or reexportation of the solid wood packing material.

(Approved by the Office of Management and Budget under control numbers 1651–0019, 0579–0094, or 0579–0052).

[69 FR 71679. Dec. 9, 2004, as amended at 71 FR 50328, Aug. 25, 2006; 72 FR 70765, Dec. 13, 2007; 74 FR 32399, July 8, 2009; 75 FR 10644, Mar. 9, 2010; 80 FR 66778, Oct. 29, 2015]

§354.4 **User fees for certain domestic services.**

(a) *Individual agreements for inspection services at ports of entry.* (1) Operators and owners of vessels or aircraft, or their agents, may enter into agreements with APHIS to receive, at points of entry in the United States inspection services in addition to the regular or on-call services available in connection with such vessels or aircraft.

(2) Agreements may be made to cover the following types of services;

(i) Opening and operating a new inspection station at a port of entry; and

(ii) Providing one-time or occasional inspection services at a location where

APHIS does not normally provide such services.

(3) Owners and operators of vessels or aircraft, or their agents, must contact the Regional Director, USDA, APHIS, Plant Protection and Quarantine,[5] for the State where they want APHIS to provide services, to make an agreement.

(4) All agreements must include the following:

(i) Name, mailing address, and telephone number of the operator or owner of the vessel or aircraft, or, if applicable, the operator's or owner's agent;

(ii) Explanation of inspection services to be provided;

(iii) Date(s) and time(s) inspection services will be provided;

(iv) Location (street address, port of entry, berth, dock, gate, etc.) and if applicable, identity (identification number, name, etc.) of vessel or aircraft or other thing to be inspected;

(v) An estimate of the actual cost, as calculated by APHIS, to provide the described inspection services for 6 months;

(vi) A statement that APHIS agrees to provide the described inspection services;

(vii) A statement that the owner or operator of the vessel or aircraft, or if appropriate, his or her agent, agrees to pay, at the time the agreement is entered into, a user fee equal to the estimated cost of providing the described inspection services for 6 months;

(viii) A statement that APHIS will credit an amount equal to all user fees received for services provided at the location to the owner or operator's account, until the total amount of user fees credited to the account is equal to the amount of money paid into the account by the owner or operator of the vessel or aircraft, or if appropriate, his or her agent, at the time the agreement was entered into; and

(ix) A statement that the owner or operator of the vessel or aircraft, or if appropriate, his or her agent, agrees to maintain a balance in the user fee payment account equal to the cost of providing the services described for 6 months, as calculated monthly by APHIS.

(5) APHIS will enter into an agreement only if qualified personnel can be made available to provide the services to be provided.

(6) An agreement can be terminated by either party on 30 days written notice.

(7) If, at the time an agreement is terminated, any unobligated funds remain in the user fee account, APHIS will return them to the owner or operator, or his or her agent.

[57 FR 770, Jan. 9, 1992, as amended at 57 FR 14475, Apr. 21, 1992; 58 FR 38269, July 16, 1993; 59 FR 67611, Dec. 30, 1994]

§ 354.5 Penalties for nonpayment or late payment of user fees.

(a) If a person requesting a service for which an APHIS user fee is payable, is delinquent in paying any APHIS user fee due under either title 7 or title 9, Code of Federal Regulations, or is delinquent in paying the interest on any delinquent APHIS user fee, then APHIS will not provide the service requested.

(b) If APHIS is in the process of providing a service for which an APHIS user fee is due, and the user has not paid the fee within the time required, or if the payment offered by the user is insufficient or not in compliance with the regulations in this part, then APHIS will take the following action:

(1) If an APHIS user fee is due for a certificate or a certificate for reexport, APHIS will not issue the certificate.

(2) If an APHIS user fee is past due by more than 30 days, APHIS will impose a late payment penalty and interest charges in accordance with 31 U.S.C. 3717.

[57 FR 771, Jan. 9, 1992]

---

[5] A list of the Regional Directors, USDA, APHIS, Plant Protection and Quarantine and the States for which they are responsible, may be obtained from the Animal and Plant Health Inspection Service, Plant Protection and Quarantine, Operational Support—Director's Office, 4700 River Road, Unit 131, Riverdale, Maryland 20737-1236.

PART 355—ENDANGERED SPECIES REGULATIONS CONCERNING TERRESTRIAL PLANTS

Subpart A—Purpose and Definitions

Sec.
355.1 Purpose.

355.2 Definitions.

Subpart B—Permission to Engage in Business

355.10 Permission to engage in business concerning nonlisted terrestrial plants.
355.11 General permits.

Subpart C—Inspections and Related Provisions

355.20 Marking and notification requirements for plants imported, exported, or reexported by means other than mail.
355.21 Marking and mailing requirements for plants imported, exported, or reexported by mail.
355.22 Validation of documentation.
355.23 Recordkeeping, access, and reports.

AUTHORITY: 16 U.S.C. 1532, 1538, and 1540; 7 CFR 2.22, 2.80, and 371.3.

SOURCE: 49 FR 42912, Oct. 25, 1984, unless otherwise noted.

Subpart A—Purpose and Definitions

SOURCE: Redesignated at 84 FR 2430, Feb. 7, 2019.

§ 355.1 Purpose.

Pursuant to the Endangered Species Act of 1973, as amended (16 U.S.C. 1531 et seq.), the Secretary is responsible for the enforcement of the provisions of the Act and Convention that pertain to the importation, exportation, or reexportation of terrestrial plants.[1] The regulations in this part are for the purpose of implementing this authority. Regulations of the U.S. Department of the Interior that correlate with the regulations in this part are contained in 50 CFR chapter I.[2]

[66 FR 21060, Apr. 27, 2001]

___

[1] Under section 11 of the Act (16 U.S.C. 1540), it is unlawful for any person to knowingly violate any provision of the Act, any permit or certificate issued under the Act, or any regulation promulgated under the Act. Section 11 of the Act also provides for criminal, civil, and administrative penalties for any such violation.

[2] Plant Protection and Quarantine also administers programs under the Lacey Act Amendments of 1981, as amended (16 U.S.C. 3371 through 3378), 7 U.S.C. 2814, and the Plant Protection Act (7 U.S.C. 7701–7772), which authorize additional prohibitions and restrictions on the importation of plants

§ 355.2 Definitions.

Terms used in the singular form in this part shall be construed as the plural, and vice versa, as the case may demand. The following terms, when used in this part, shall be construed, respectively, to mean:

Act. The Endangered Species Act of 1973, as amended (16 U.S.C. 1531 et seq.).

Convention. The Convention on International Trade in Endangered Species of Wild Fauna and Flora, TIAS 8249, 27 U.S.T. 1087, signed on March 3, 1973, and the Appendices thereto.

Deputy Administrator. The Deputy Administrator of the Animal and Plant Health Inspection Service for Plant Protection and Quarantine, U.S. Department of Agriculture, or any other officer or employee of the Department to whom authority to act in his or her stead has been or may hereafter be delegated.

Engage in business as an importer, exporter, or reexporter of terrestrial plants. To import, export, or reexport terrestrial plants for the purpose of selling, bartering, collecting, or otherwise exchanging or acquiring the plants as a livelihood or enterprise engaged in for gain or profit. This term shall not include persons engaged in business merely as carriers or customhouse brokers.

Export (exported, exporting, exportation). To carry, send, take, transport or otherwise remove, or to attempt to carry, send, take, transport or otherwise remove from any place subject to the jurisdiction of the United States.

Import (imported, importing, importation). To land on, bring into, or introduce into, or attempt to land on, bring into, or introduce into, any place subject to the jurisdiction of the United States, whether or not such landing, bringing, or introduction constitutes an importation within the meaning of the customs laws of the United States.

Inspector. Any employee of Plant Protection and Quarantine, Animal and Plant Health Inspection Service, U.S. Department of Agriculture, or other

___

subject to this part (see other parts of 7 CFR chapter III for regulations containing prohibitions and restrictions under these authorities).

person, authorized by the Deputy Administrator in accordance with law to enforce the provisions of the Act and Convention, and regulations promulgated thereunder.

Person. Any individual, corporation, partnership, trust, association, or any other private entity; or any officer, employee, agent, department, or instrumentality of the Federal Government, of any State or political subdivision thereof or of any foreign government.

Plant. Any member of the plant kingdom, including seeds, roots and other parts thereof.

Plant Protection and Quarantine. The organizational unit within the Animal and Plant Health Inspection Service, U.S. Department of Agriculture, delegated responsibility for enforcing provisions of the Act and Convention, and regulations promulgated thereunder.

Protected plant permit. PPQ Form 622, "Protected Plant Permit to Engage in the Business of Importing, Exporting, or Reexporting Terrestrial Plants Regulated by 50 CFR 17.12 and 23.23."

Reexport (reexported, reexportation). To export following importation.

Secretary. The Secretary of Agriculture, or any other officer or employee of the Department of Agriculture to whom authority to act in his or her stead has been or may hereafter be delegated.

Terrestrial plants. Any plants (including epiphytic plants), except marine plants.

Validation. An original stamp, signature, and date of inspection placed upon documentation required by 50 CFR part 17 or part 23 by an inspector at the port where the terrestrial plants are to be imported, exported or reexported.

United States. Any of the several States, the District of Columbia, the Commonwealth of Puerto Rico, American Samoa, the U.S. Virgin Islands, Guam, and the Trust Territory of the Pacific Islands.

[49 FR 42912, Oct. 25, 1984, as amended at 70 FR 57995, Oct. 5, 2005]

Subpart B—Permission to Engage in Business

SOURCE: Redesignated at 84 FR 2430, Feb. 7, 2019.

§ 355.10   Permission to engage in business concerning nonlisted terrestrial plants.

The Secretary hereby grants permission for any person engaged in business as an importer, exporter, or reexporter of terrestrial plants, other than terrestrial plants listed in 50 CFR 17.12 or 23.23, to engage in such business without a protected plant permit issued under § 355.11.

[49 FR 42912, Oct. 25, 1984, as amended at 70 FR 57995, Oct. 5, 2005]

§ 355.11   Protected plant permits.

(a) On or after March 26, 1985 no person shall engage in business as an importer, exporter, or reexporter of any terrestrial plants listed in 50 CFR 17.12 or 23.23 unless such person has obtained a protected plant permit for engaging in such business from Plant Protection and Quarantine.

(b) An application for a protected plant permit shall be submitted to the Animal and Plant Health Inspection Service, Plant Protection and Quarantine, Permit Services, 4700 River Road Unit 133, Riverdale, MD 20737–1236. The completed application shall include the following information: [3]

(1) Date of application;

(2) Applicant's name, mailing address, and telephone number;

(3) If the applicant is an individual, the business affiliation, if any, having to do with the importation, exportation, or reexportation of terrestrial plants listed in 50 CFR 17.12 or 23.23;

(4) If the applicant is in the name of a business or if the applicant is affiliated with a business which imports, exports, or reexports terrestrial plants

---

[3] Application forms are available on the Internet (*http://www.aphis.usda.gov/ppq/permits*), by calling (877) 770–5990, or by writing to the address in this paragraph. Application forms may also be obtained from local offices at any of the ports designated in 50 CFR part 24. Telephone numbers and addresses of local offices are listed in telephone directories.

listed in 50 CFR 17.12 or 23.23, the form of the business, e.g., corporation, firm, partnership; and the name and address of each partner, officer, director, holder, and owner of 10 percent or more of the voting stock, and employee in a managerial or executive capacity;

(5) The address of all applicants' business locations, including but not limited to locations of nurseries, growing fields, propagating beds, holding beds and similar facilities where activities relating to terrestrial plants listed in 50 CFR 17.12 or 23.23 would be conducted;

(6) A brief and complete description of the nature of the applicant's business as it relates to engaging in business as an importer, exporter, or re-exporter of terrestrial plants listed in 50 CFR 17.12 or 23.23;

(7) Any address where books or records concerning the importation, exportation, or reexportation of terrestrial plants listed in 50 CFR 17.12 or 23.23 would be kept;

(8) Name, address, and telephone number of the person authorized to make records or plant inventories available for examination by inspectors or other duly authorized representatives of the Secretary; and

(9) Certification by signature of the applicant (must be a partner or officer if the applicant is a business) after the following language: "I hereby certify that the information in this application is complete and accurate to the best of my knowledge and belief."

(c) Each application for a protected plant permit must be accompanied by a check or money order for $70 made payable to Plant Protection and Quarantine. The fee shall not be refunded if the application is denied or abandoned.

(d) After receipt and review of the application by Plant Protection and Quarantine, a protected plant permit for the importation, exportation, and reexportation of terrestrial plants listed in 50 CFR 17.12 or 23.23 shall be issued if the applicant has submitted an application containing all information requested in paragraph (b) of this section, if the applicant has paid the fee in accordance with paragraph (c) of this section, and if a protected plant permit of the applicant or anyone responsibly connected with the business

of the applicant has not been and is not denied, suspended or revoked pursuant to paragraph (i) of this section.

(e) The applicant shall be notified in writing by Plant Protection and Quarantine of the approval or denial of any request for a protected plant permit. If a protected plant permit is denied, the notification shall state the reasons therefor. If a protected plant permit is denied, the applicant may request a hearing pursuant to paragraph (i)(1) of this section and may submit to Plant Protection and Quarantine, in writing, reasons why the permit should not have been denied. Such submissions of the applicant shall not be considered a new application if submitted within 60 days following the receipt of notification of the denial by the applicant.

(f) Upon receipt of an incomplete or improperly executed application, the applicant shall be notified by Plant Protection and Quarantine of the deficiency of the application. If the applicant fails to supply the deficient information or otherwise fails to correct the deficiency within 60 days following the receipt of the notification by the applicant, the application shall be considered abandoned.

(g) Upon receipt of an application filed with an insufficient fee, or without a fee, the application and any fee submitted will be returned to the applicant.

(h) A protected plant permit shall be valid for 2 years from the date of issuance unless suspended or revoked pursuant to paragraph (i) of this section. A new application must be submitted for the renewal of the protected plant permit. A protected plant permit shall not be transferred, tampered with, amended or otherwise altered in any manner or form by any person.

(i)(1) Any application for a protected plant permit may be denied and any protected plant permit which has been issued may be suspended or revoked for a time specified by the Deputy Administrator for any of the reasons provided in paragraph (i)(2) of this section. Before such action is taken, the applicant or permittee will be informed of the reasons for the proposed action, and upon request, shall be afforded an opportunity for a hearing with respect to the merits or validity of such action, in accordance with rules of practice which shall be adopted for the proceeding. However, such denial, suspension or revocation may become effective pending final determination in the proceeding, if the permittee has been convicted or a criminal violation of the Act, or of any regulation, permit, or certificate issued under the Act. Such denial, suspension or revocation shall be effective upon oral or written notification, whichever is earlier, to the permittee. In the event of oral notification of the denial, suspension or revocation, written confirmation shall be given to the permittee as promptly as circumstances allow. This denial, suspension or revocation shall continue in effect pending the completion of the proceeding and any judicial review thereof, unless otherwise ordered by the Deputy Administrator.

(2) An application for a protected plant permit may be denied and any protected plant permit which has been issued may be suspended or revoked if:

(i) Any requirement of this subpart is not complied with, or

(ii) The applicant, permittee, or a person responsibly connected with the business of the applicant or permittee has been criminally convicted or had a civil penalty imposed for a violation of the Act or of any regulation, permit, or certificate issued under the Act, or

(iii) The applicant, permittee, or a person responsibly connected with the business of the applicant or permittee has been convicted of any crime involving fraud, bribery, extortion, or any other crime involving a' lack of integrity needed for the conduct of operations concerning the importation, exportation, or reexportation of terrestrial plants listed in 50 CFR 17.12 or 23.23.

(3) For the purposes of this section, a person shall be deemed to be responsibly connected with the business of the applicant or permittee if the person is a partner, officer, director, holder, or owner of 10 percent or more or its vot-

ing stock, or an employee in a managerial or executive capacity.

(Information collection requirements were approved by the Office of Management and Budget under control number 0579–0076)

[49 FR 42912, Oct. 25, 1984, as amended at 59 FR 67611, Dec. 30, 1994; 66 FR 21060, Apr. 27, 2001; 70 FR 57995, Oct. 5, 2005]

Subpart C—Inspections and Related Provisions

SOURCE: Redesignated at 84 FR 2430, Feb. 7, 2019.

§ 355.20 Marketing and notification requirements for plants imported, exported, or reexported by means other than mail. [4]

(a) Any terrestrial plant which is to be imported, exported, or reexported by means other than mail and which may be imported, exported, or reexported under 50 CFR part 17 or part 23 only if accompanied by documentation, shall at the time of importation, exportation, or reexportation plainly and correctly bear on the outer container or on a tag, invoice, packing list, or other document accompanying the plant, the following information:

(1) Genus and species, and quantity of each (if a hybrid, genus of each parent, and quantity of each hybrid),

(2) Country and locality where collected from the wild or where produced from cultivated stock,

(3) Name and address (in the United States if exported or reexported) of shipper, owner or person shipping or forwarding the plants,

-----

[4] Certain terrestrial plants listed in Appendices I, II, or III of the Convention or determined by the U.S. Department of the Interior to be endangered or threatened or similar in appearance to endangered or threatened species are required to be accompanied by documentation at the time of importation, exportation, or reexportation (see 50 CFR chapter I). Plants are allowed to be imported, exported or reexported only at ports authorized for such purposes by the U.S. Department of the Interior, or, under certain circumstances as determined by the U.S. Department of the Interior, at nondesignated ports, pursuant to section 9(f) of the Act (16 U.S.C. 1538(f)). (see 50 CFR part 24 for a list of designated ports.)

(4) Name and address (in the United States if imported) of consignee,

(5) Identifying shipper's mark and number, and

(6) Serial number and type (e.g., permit, certificate) of document issued for the importation, exportation, or reexportation of the plant.

(b) Promptly upon arrival at a port of import (listed in 50 CFR part 24, or, if allowed by the U.S. Department of the Interior, at a nondesignated port) of any terrestrial plant which is imported by means other than mail and which may be imported under 50 CFR part 17 or part 23 only if accompanied by documentation, the importer shall notify Plant Protection and Quarantine of the arrival and of the genus and species of the plant by such means as a manifest, Customs entry document, commercial invoice, waybill, broker's document, or notice form provided for that purpose.

(c) Prior to the exportation or reexportation of any terrestrial plant which is to be exported or reexported by other than mail and which may be exported or reexported under 50 CFR part 17 or part 23 only if accompanied by documentation, the exporter or reexporter shall notify Plant Protection and Quarantine of the intended exportation or reexportation and of the genus and species of the plant by such means as a manifest, commercial invoice, waybill, broker's document, or notice form provided for that purpose.

(Information collection requirements were approved by the Office of Budget and Management under control number 0579–0076)

[49 FR 42912, Oct. 25, 1984, as amended at 70 FR 57995, Oct. 5, 2005]

§355.21 Marking and mailing requirements for plants imported, exported, or reexported by mail.[5]

(a) Any terrestrial plant which is to be imported by mail and which may be imported under 50 CFR part 17 or part 23 only if accompanied by documentation, shall be mailed to Plant Protection and Quarantine (at a port authorized for such purpose by the U.S. Department of the Interior in 50 CFR part 24 pursuant to section 9(f) of the Act (16 U.S.C. 1538 (f))); and shall be accompanied by a separate sheet of paper within the package plainly and correctly bearing the name, address, and telephone number of the intended recipient in the United States; and shall plainly and correctly bear on the outer container the following information:

(1) Genus and species, and quantity of each (if a hybrid, genus of each parent, and quantity of each hybrid),

(2) Country and locality where collected from the wild or where produced from cultivated stock,

(3) Name and address of shipper, owner, or person shipping or forwarding the plants, and

(4) Serial number and type (e.g. permit, certificate) of document issued for the importation of the plant.

(b) Any terrestrial plant which is to be exported or reexported by mail and which may be exported or reexported under 50 CFR part 17 or part 23 only if accompanied by documentation, shall be mailed to Plant Protection and Quarantine (at a port authorized for such purpose by the U.S. Department of the Interior in 50 CFR part 24 pursuant to section 9(f) of the Act (16 U.S.C. 1538(f))); shall be wrapped in double wrapping, with an unsealed inner wrapping addressed to the foreign recipient and bearing sufficient postage for mailing to the foreign destination; shall be accompanied by a separate sheet of paper within the package plainly and correctly bearing the following information:

(1) Genus and species, and quantity of each (if a hybrid, genus of each parent, and quantity of each hybrid),

(2) Country and locality where collected from the wild or where produced from cultivated stock,

(3) Name and address in the United States of shipper, owner, or person shipping or forwarding the plants, and

(4) Serial number and type (e.g. permit, certificate) of document issued for the exportation or reexportation of the plant.

(Information collection requirements were approved by the Office of Budget and Management under control number 0579–0076)

§355.22 Validation of documentation.

(a) Documentation for any mailed or nonmailed terrestrial plant which is required to have documentation under 50 CFR part 17 or part 23 at the time of importation, must be validated by an inspector prior to movement of such plant from the Customs inspection area at the port of entry. The original documentation must be surrendered to the inspector at the time of validation.

(b) Documentation for any mailed or nonmailed terrestrial plant which is listed in 50 CFR 17.12 or 23.23 and which is required to have documentation under 50 CFR part 17 or part 23 at the time of exportation or reexportation, must be validated at the port of export or reexport by an inspector prior to the exportation or reexportation of such plant.[5] The original and one copy of the documentation must be submitted for validation, and the copy must be surrendered to the inspector at the time of validation.

(c) Documentation for a plant shall be validated under this section upon endorsement of the documentation by an inspector when he or she determines

---

[5] It is the policy of the Department of Agriculture to allow, if inspectors are available, terrestrial plants listed in 50 CFR 17.12 or 23.23 which are intended for export to be inspected at the premises where such plants are grown. However, the documentation required for the export of such plants by 50 CFR part 17 or part 23 shall only be validated at the port of export and only when such plants are presented at the port for export together with the documents required by 50 CFR part 17 or part 23 and a certified statement by the inspector who inspected the plants that the plants are apparently eligible for exportation in accordance with the provisions of this part and provisions of 50 CFR chapter I relating to the Act and Convention. Plants which have been previously inspected must be exported through a designated port (unless allowed by the United States Department of the Interior to be exported through a nondesignated port) in order to comply with section 9(f) of the Act [16 U.S.C. 1538(f)]. Plants which are inspected at the premises of origin must be available at the port of export for monitoring inspections and for other inspections deemed need for enforcement purposes, but, unless so inspected, will not need to be unpacked, inspected and repacked at the port. Information concerning the availability of inspectors to conduct inspections at the premise of origin may be obtained by calling local offices of Plant Protection and Quarantine, which are listed in telephone directories, or by writing the Animal and Plant Health Inspection Service, Plant Protection and Quarantine, Operational Support—Director's Office, 4700 River Road, Unit 131, Riverdale, Maryland 20737–1236.

that the plant was apparently eligible for importation, exportation, or reexportation in accordance with the provisions of this part and the provisions of 50 CFR chapter I relating to the Act and Convention.

(d) To obtain validation of documentation, the importer, exporter, or reexporter, or agent thereof, shall make available to an inspector:

(1) All shipping documents (including bills of lading, waybills, packing lists, and invoices):

(2) All documents required by the Act and Convention; and

(3) The plant being imported, exported, or reexported.

(Information collection requirements were approved by the Office of Budget and Management under control number 0579–0076)

[49 FR 42912, Oct. 25, 1984, as amended at 59 FR 67611, Dec. 30, 1994; 70 FR 57995, Oct. 5, 2005]

§ 355.23 Recordkeeping, access, and reports.

(a) Any person engaged in business as an importer, exporter, or reexporter of terrestrial plants listed in 50 CFR part 17 or part 23 shall keep such records as will fully and correctly disclose each importation, exportation, or reexportation of terrestrial plants made by such person and the subsequent disposition made by such person of the plants. Such records shall include shipping documents for each shipment of plants imported, exported, or reexported; a description of the form of the plants (such as whole live plants, cuttings, seeds, or other specific parts or derivatives of plants); the scientific and common names of the plants; the country or place of origin of the plants; the date and place of importation, exportation, or reexportation of the plants; the number (weight if the plants cannot be quantified by number) and specific location of plants; the date and means of subsequent disposition of the plants, whether by sale, barter, consignment, loan, delivery, destruction, or other means; and names and addresses of persons to whom the plants were disposed, if applicable.

(b) Every record required to be kept under this section shall be kept for a period of 5 years after the occurrence of the transactions to which the records relate, and for such further time as the Deputy Administrator may require by written notice to the person required to keep such records under this part for purposes of any investigation, litigation, or other proceeding under the Act or this part.

(c) Any person engaged in business as an importer, exporter, or reexporter of terrestrial plants listed in 50 CFR part 17 or part 23 shall, upon presentation of credentials by an inspector or duly authorized representatives of the Secretary; during ordinary business hours of the person given notice, afford such inspector access to the person's place of business, the opportunity to examine the person's inventory of plants and the records required to be kept under paragraph (a) of this section, and the opportunity to copy such records. The use of a room, table, or other facilities (other than reproduction equipment) necessary for examination and copying of records and for such examination of inventory shall be afforded such inspector.

(d) Any person engaged in business as an importer, exporter, or reexporter of terrestrial plants listed in 50 CFR part 17 or part 23, upon written request by the Deputy Administrator, shall submit within 60 days of such request, a report concerning any of the information required to be maintained under paragraphs (a) and (b) of this section.

(Information collection requirements were approved by the Office of Budget and Management under control number 0579–0076)

PART 356—FORFEITURE PROCEDURES

AUTHORITY: 16 U.S.C. 1540(f), 16 U.S.C. 3374); 7 CFR 2.22, 2.80, 371.3.

SOURCE: 49 FR 42916, Oct. 25, 1984, unless otherwise noted.

§356.1 Property subject to forfeiture procedures.

This part sets forth procedures relating to the forfeiture of any plant, equipment, means of conveyance or other property[1] seized under the Endangered Species Act of 1973, as amended, (16 U.S.C. 1531 *et seq.*) or the Lacey Act Amendments of 1981 (16 U.S.C. 3371 *et seq.*),[2] in possession (actual or constructive) of the United States Department of Agriculture, and subject to forfeiture under these Acts because of activities pertaining to the importation, exportation, or reexportation of terrestrial plants.

[49 FR 46336, Nov. 26, 1984]

§356.2 Appraisement.

Promptly following the seizure or other receipt of property specified in §356.1, the Deputy Administrator shall

---

[1]Under section 11(e)(4) of the Endangered Species Act ("Act"; 16 U.S.C. 1540(e)(4)) any such equipment and means of conveyance would be subject to forfeiture upon conviction of a criminal violation pursuant to section 11(b)(1) of the Act (16 U.S.C. 1540(b)(1)); however, such a plant may be subject to forfeiture regardless of whether a criminal conviction is obtained.

[2]Under section 5(a)(2) of the Lacey Act Amendments of 1981 (16 U.S.C. 3374(a)(2)) USDA has authority to initiate forfeiture proceedings against all vessels, vehicles, aircraft, and other equipment used to aid in the importation or exportation of plants in a criminal violation of the Lacey Act Amendments of 1981 for which a felony conviction has been obtained if (a) the owner of such vessel, vehicle, aircraft, or equipment was at the time of the alleged illegal act a consenting party or privy thereto or in the exercise of due care should have known that such vessel, vehicle, aircraft, or equipment would be used in a criminal violation of the Lacey Act Amendments of 1981, and (b) the violation involved the sale or purchase of, the offer of sale or purchase of, or the intent to sell or purchase plants. However, under section 5(a)(1) of the Lacey Act Amendments of 1981 (16 U.S.C. 3474(a)(1) plants seized for violations of the Amendments are subject to forfeiture regardless of whether a civil penalty assessment or criminal conviction is obtained.

determine the retail value of such property in the same quantity or quantities as seized. If the property may lawfully be sold in the United States, the value thereof shall be determined by ascertaining the price at which the property or similar property in the ordinary course of trade is freely offered for sale at the time of appraisement, and at a principal market as close as possible to the place of appraisement. If the property may not lawfully be sold in the United States, the value thereof shall be determined by other reasonable means.

§ 356.3 Property valued at greater than $10,000; notice of seizure and civil action to obtain forfeiture.

Promptly following the seizures or other receipt of any property specified in § 356.1 and determined under § 356.2 to have a value greater than $10,000, the Deputy Administrator shall mail a notice of seizure by registered or certified mail to the current or last known or reasonable ascertainable address, return receipt requested, to persons known or reasonably ascertained to be the owner or agent of the seized property and to any other person having an interest in the property. Such notice shall describe the seized property, shall state the time, date, place, and reason for the seizure, that there is a right to petition for remission or mitigation of forfeiture pursuant to § 356.7, and shall state that action shall be taken in accordance with this part. Promptly following the seizure of such property, the Secretary shall also submit a report concerning such property to the U.S. Attorney for the district in which the seizure was made for institution of forfeiture proceedings in the U.S. District Court. The report shall provide a statement of all the relevant facts and circumstances of the case, including the names of the witnesses, and a citation to the laws believed to have been violated and on which reliance may be had for forfeiture.

§ 356.4 Property valued at $10,000 or less; notice of seizure administrative action to obtain forfeiture.

(a) *When authorized.* The Secretary shall take measures to obtain forfeiture in accordance with this section

of any property specified in § 356.1 and determined under § 356.2 to have a value of $10,000 or less.

(b) *Waiver of forfeiture procedures by owner of seized property.* A person claiming to be an owner or to have an interest in any property specified in § 356.1 with a value of $10,000 or less may waive any rights to any procedures relating to forfeiture under this subpart by signing a statement providing for waiver of such rights.

(1) The Deputy Administrator shall publish a copy of the notice of seizure and proposed forfeiture as provided in paragraph (c)(1) of this section, by posting for 21 days in a conspicuous place accessible to the public at the Plant Protection and Quarantine Enforcement office nearest the place of seizure. The time and date of posting shall be indicated on the notice.

(2) Upon the execution of such statement and following publication of the notice for 21 days as provided in paragraph (c)(1) of this section, any interest in such property by such owner shall become forfeited under the Act without further action under this subpart, and the Deputy Administrator shall not be required to send such owner any notices or declarations otherwise required by this subpart.

(c) *Procedure absent waiver of forfeiture procedures by owner.* (1) Notice of seizure and proposed forfeiture. Promptly following seizure of property, the Deputy Administrator shall issue a notice of seizure and proposed forfeiture. The notice shall be in substantially the same form as a complaint for forfeiture filed in the U.S. District Court. The notice shall describe the seized property, including any identification numbers, such as the license, registration, motor, and serial numbers for a motor vehicle. The notice shall state the time, date, and place of seizure; the reason for seizure; and shall specify the value of the property as determined under § 356.2. The notice shall contain specific reference to the provisions of the Act, permit, certificate, or regulations allegedly violated and under which the property is subject to forfeiture. The notice shall state that any person desiring to claim the property must file a claim and a bond in accordance with paragraph (c)(2) of this section, and shall state that if a proper claim and bond are not received by the specified office within the time prescribed by such paragraph, the property will be declared forfeited to the United States and disposed of according to law. The notice shall also advise interested persons of their right to file a petition for remission or mitigation of forfeiture in accordance with § 356.7.

(i) Promptly following the seizure, The Deputy Administrator shall mail a copy of the notice by registered or certified mail, return receipt requested, to persons known or reasonably ascertained to be the owner or agent of the seized property, and to any other person having an interest in the property, if such owner or agent or other person and their address is known or reasonably ascertainable.

(ii) *Publication.* Promptly following the seizure, the Deputy Administrator shall publish a copy of the notice by posting for 21 days in a conspicuous place accessible to the public at the Plant Protection and Quarantine enforcement office nearest the place of seizure. The time and date of posting shall be indicated on the notice.

(2) *Filing a claim and bond.* Upon issuance of the notice of proposed forfeiture, any person claiming ownership of or other interest in the seized property may file with the office specified in the notice a claim to the property and a bond in the amount of $250, with sureties to be approved by the Deputy Administrator, conditioned that in case of condemnation of the articles so claimed, the obligor shall pay all the costs and expenses of the proceedings to obtain such condemnation. Any claim and bond must be received in such office within 20 days after posting of the notice of proposed forfeiture, and shall state claimant's interest in the property. The Deputy Administrator may extend the 20 day period with an appropriate statement on the posted notice of proposed forfeiture, if necessary, to allow a person deemed to have an interest in the property at least 10 days to file such a claim and bond after receipt of a notice of proposed forfeiture. The bond shall be on a U.S. Customs Form 4615 or on a similar form provided by Plant Protection and Quarantine. There shall be endorsed on

the bond a list or schedule in substantially the following form which shall be signed by the claimant in the presence of the witnesses to the bond, and attested by the witnesses:

List or schedule containing a description of seized articles, claim for which is covered by the bond:

_____

The foregoing list is correct.

_____

Claimant

Attest: _____

The claim and bond referred to in the paragraph shall not entitle the claimant or any other person to possession of the property.

(3) *Transmittal to U.S. Attorney.* As soon as practicable after timely receipt by the specified office of a proper claim and bond in accordance with paragraph (c)(2) of this section, the Secretary shall transmit such claim, bond (with a duplicate list and description of the articles seized), and a report as described in § 356.3 to the U.S. Attorney for the district in which seizure was made for forfeiture proceedings in the U.S. District Court.

(d) *Summary forfeiture.* If a proper claim and bond are not received by the specified office within the time periods as specified in paragraph (c)(2) of this section, the property shall be forfeited and the Deputy Administrator shall prepare a declaration of forfeiture. The declaration of forfeiture shall be in writing, and the Deputy Administrator shall send such declaration by registered or certified mail, return receipt requested, to each person whose whereabouts and prior interests in the seized property are known or reasonably ascertainable. The declaration shall be in substantially the same form as a default judgment of forfeiture entered in U.S. District Court. The declaration shall describe the property and state the time, date, place, and reason for its seizure. The declaration shall identify the notice of proposed forfeiture, describing the dates and manner of publication of the notice and any efforts made to serve the notice personally or by mail. The declaration shall state that in response to the notice a proper

399

claim and bond were not timely received by the proper office from any claimant, and that, therefore, all potential claimants are deemed to admit the truth of the allegations of the notice. The declaration shall conclude with an order of condemnation and forfeiture of the property to the United States for disposition according to law.

(Information collection requirements were approved by the Office of Budget and Management under control number 0579–0076)

§ 356.5 Bonded release.

(a) The Deputy Administrator may accept a bond or other security, in the amount of the value of the property as determined under § 356.3, in place of any property specified in § 356.1 and release the property to the owner or agent of the property, if such action would not frustrate the purposes of the Act and Convention. As an example, this section does not allow the release of terrestrial plants that are without documentation required under 50 CFR chapter I.

(b) Any request for the return of property based on the acceptance of a bond or other security shall be submitted in writing to the Deputy Administrator. The request shall include evidence to establish that the person making the request is the sole owner of the property referred to in the request or is the agent of the sole owner of such property. A response in writing, granting or denying the request, and the reasons therefor, shall be sent to the person making the request.

§ 356.6 Storage of property.

Following the seizure or other receipt of any property specified in § 356.1 and valued at $10,000 or less, the property shall remain in the custody of the Deputy Administrator pending disposition. Pending such disposition, the property shall be stored in such place, as, in the opinion of the Deputy Administrator, is most convenient and appropriate with due regard to the expense involved, whether or not the place of storage is within the judicial district in which the property was seized.

§ 356.7 Petition for remission or mitigation of forfeiture.

(a) Any person who has an interest in any property specified in § 356.1 and valued at $10,000 or less, or any person who has incurred or is alleged to have incurred a forfeiture of any such property, may file with the Deputy Administrator a petition for remission or mitigation of forfeiture while the property is in the custody of the Deputy Administrator.

(b) A petition filed with the Deputy Administrator need not be in any particular form, but must contain the following:

(1) A description of the property:

(2) The time, date, and place of seizure;

(3) Evidence of the petitioner's interest in the property such as contracts, bills of sale, invoices, security interests, certificates of title; and

(4) A statement of all facts and circumstances relied upon by the petitioners to justify remission or mitigation of the forfeiture.

(c) The petition shall be signed by the petitioner or the petitioner's attorney at law. If the petitioner is a business, the petition must be signed by a partner, officer, or petitioner's attorney at law.

(d) Upon receiving the petition, the Deputy Administrator shall decide whether or not to grant relief. In making a decision, the Deputy Administrator shall consider the information submitted by the petitioner, as well as any other available information relating to the matter, and may require that testimony be taken concerning the petition.

(e) If the Deputy Administrator finds that the forfeiture was incurred without willful negligence or without any intention on the part of the petitioner to violate the law or finds the existence of such mitigating circumstances as to justify remission or mitigation of the forfeiture or alleged forfeiture, the Deputy Administrator may remit or mitigate the same upon terms and conditions as he deems reasonable and just. However, remission or mitigation will not be made if such action would frustrate the purposes of the Act or

Convention. As an example, this section does not allow remission or mitigation with respect to terrestrial plants that are without documentation required under 50 CFR chapter I.

(f) The Deputy Administrator shall notify the petitioner in writing concerning whether the petition was granted or denied, and shall state the reasons therefor. If the petition is denied fully or in part, the petitioner may then file a supplemental petition, but no supplemental petition shall be considered unless it is received within 60 days from the date of the Deputy Administrator's notification concerning the original petition. The Deputy Administrator shall notify the petitioner in writing concerning the action taken in response to the supplemental petition, and shall state the reasons therefor.

(Information collection requirements were approved by the Office of Budget and Management under control number 0579–0076)

§356.8 Return procedure.

If, at the conclusion of proceedings, seized property is to be returned to the person determined to be the owner or agent thereof, the Deputy Administrator shall issue a letter or other document to the person determined to be owner or agent thereof authorizing its return. This letter shall be delivered personally or sent by registered or certified mail, return receipt requested, and shall identify the person determined to be the owner or agent, the seized property, and if appropriate, the bailee of the seized property. It shall also provide that upon presentation of the letter or other document and proper identification, and the signing of a receipt provided by Plant Protection and Quarantine, the seized property is authorized to be released.

§356.9 Filing of documents.

(a) Any document required by this subpart to be filed or served within a certain period of time, will be considered filed or served as of the time of receipt by the party with or upon whom filing or service is required.

(b) Saturdays, Sundays, and federal holidays shall be included in computing the time allowed for the filing or serving of any document or paper;

except that when such time expires on a Saturday, Sunday or federal holiday, such period shall be extended to include the next following business day.

PART 357—CONTROL OF ILLEGALLY TAKEN PLANTS

Sec.
357.1 Purpose and scope.
357.2 Definitions.
357.3 Declaration requirement.
357.4 Exceptions from the declaration requirement.

AUTHORITY: 16 U.S.C. 3371 *et seq.;* 7 CFR 2.22, 2.80, and 371.2(d).

SOURCE: 78 FR 40944, July 9, 2013, unless otherwise noted.

§ 357.1 Purpose and scope.

The Lacey Act, as amended (16 U.S.C. 3371 *et seq.*), makes it unlawful to, among other things, import, export, transport, sell, receive, acquire, or purchase in interstate or foreign commerce any plant, with some limited exceptions, taken, possessed, transported or sold in violation of any Federal or Tribal law, or in violation of a State or foreign law that protects plants or that regulates certain specified plant-related activities. The Lacey Act also makes it unlawful to make or submit any false record, account, or label for, or any false identification of, any plant covered by the Act. Common cultivars (except trees) and common food crops are among the categorical exclusions to the provisions of the Act. The Act does not define the terms "common cultivar" and "common food crop" but instead authorizes the U.S. Department of Agriculture and the U.S. Department of the Interior to define these terms by regulation. The regulations in this part provide the required definitions. Additionally, the regulations in this part address the declaration requirement of the Act.

[85 FR 12212, Mar. 2, 2020]

§ 357.2 Definitions.

Artificial selection. The process of selecting plants for particular traits, through such means as breeding, cloning, or genetic modification.

Commercial scale. Production, in individual products or markets, that is typical of commercial activity, regardless of the production methods or amount of production of a particular facility or the purpose of an individual shipment.

Common cultivar. A plant (except a tree) that:

(1) Has been developed through artificial selection for specific morphological or physiological characteristics; and

(2) Is a species or hybrid, or a selection thereof, that is produced on a commercial scale; and

(3) Is not listed:

(i) In an appendix to the Convention on International Trade in Endangered Species of Wild Fauna and Flora (27 UST 1087; TIAS 8249);

(ii) As an endangered or threatened species under the Endangered Species Act of 1973 (16 U.S.C. 1531 *et seq.*); or

(iii) Pursuant to any State law that provides for the conservation of species that are indigenous to the State and are threatened with extinction.

Common food crop. A plant that:

(1) Is raised, grown, or cultivated for human or animal consumption; and

(2) Is a species or hybrid, or a selection thereof, that is produced on a commercial scale; and

(3) Is not listed:

(i) In an appendix to the Convention on International Trade in Endangered Species of Wild Fauna and Flora (27 UST 1087; TIAS 8249);

(ii) As an endangered or threatened species under the Endangered Species Act of 1973 (16 U.S.C. 1531 *et seq.*); or

(iii) Pursuant to any State law that provides for the conservation of species that are indigenous to the State and are threatened with extinction.

Import. To land on, bring into, or introduce into, any place subject to the jurisdiction of the United States, whether or not such landing, bringing, or introduction constitutes an importation within the meaning of the customs laws of the United States.

Person. Any individual, partnership, association, corporation, trust, or any officer, employee, agent, department, or instrumentality of the Federal Government or of any State or political subdivision thereof, or any other entity subject to the jurisdiction of the United States.

Plant. Any wild member of the plant kingdom, including roots, seeds, parts or products thereof, and including trees from either natural or planted forest stands. The term plant excludes:

(1) Common cultivars, except trees, and common food crops (including roots, seeds, parts, or products thereof);

(2) A scientific specimen of plant genetic material (including roots, seeds, germplasm, parts, or products thereof) that is to be used only for laboratory or field research; and

(3) Any plant that is to remain planted or to be planted or replanted.

(4) A plant is not eligible for these exclusions if it is listed:

(i) In an appendix to the Convention on International Trade in Endangered Species of Wild Fauna and Flora (27 UST 1087; TIAS 8249);

(ii) As an endangered or threatened species under the Endangered Species Act of 1973 (16 U.S.C. 1531 *et seq.*); or

(iii) Pursuant to any State law that provides for the conservation of species that are indigenous to the State and are threatened with extinction.

Taken. Captured, killed, or collected, and with respect to a plant, also harvested, cut, logged, or removed.

Tree. A woody perennial plant that has a well-defined stem or stems and a continuous cambium, and that exhibits true secondary growth.

[78 FR 40944, July 9, 2013, as amended at 85 FR 12212, Mar. 2, 2020]

§ 357.3 Declaration requirement.

(a) Any person importing any plant shall file upon importation a declaration that contains:

(1) The scientific name of any plant (including the genus and species of the plant) contained in the importation;

(2) A description of the value of the importation and the quantity, including the unit of measure, of the plant; and

(3) The name of the country from which the plant was taken.

(b) The declaration relating to a plant product shall also contain:

(1) If the species of plant used to produce the plant product that is the subject of the importation varies, and the species used to produce the plant product is unknown, the name of each

species of plant that may have been used to produce the plant product;

(2) If the species of plant used to produce the plant product that is the subject of the importation is commonly taken from more than one country, and the country from which the plant was taken and used to produce the plant product is unknown, the name of each country from which the plant may have been taken; and

(3) If a paper or paperboard plant product includes recycled plant product, the average percent recycled content without regard for the species or country of origin of the recycled plant product, in addition to the information for the non-recycled plant content otherwise required by this section.

(c) Guidance on completion and submission of the declaration form can be found on the APHIS website at *http://www.aphis.usda.gov/plant__health/lacey__act.*

(Approved by the Office of Management and Budget under control number 0579–0349)

[85 FR 12212, Mar. 2, 2020]

§ 357.4 **Exceptions from the declaration requirement.**

Plants and products containing plant materials are excepted from the declaration requirement if:

(a) The plant is used exclusively as packaging material to support, protect, or carry another item, unless the packaging material itself is the item being imported; or

(b) The plant material in a product represents no more than 5 percent of the total weight of the individual product unit, provided that the total weight of the plant material in an entry of products in the same 10-digit provision of the Harmonized Tariff Schedule of the United States does not exceed 2.9 kilograms.

(c) A product will not be eligible for an exception under paragraph (b) of this section if it contains plant material listed:

(1) In an appendix to the Convention on International Trade in Endangered Species of Wild Fauna and Flora (27 UST 1087; TIAS 8249);

(2) As an endangered or threatened species under the Endangered Species Act of 1973 (16 U.S.C. 1531 *et seq.*); or

(3) Pursuant to any State law that provides for the conservation of species that are indigenous to the State and are threatened with extinction.

[85 FR 12212, Mar. 2, 2020]

PART 360—NOXIOUS WEED REGULATIONS

AUTHORITY: 7 U.S.C. 7701–7772 and 7781–7786; 7 CFR 2.22, 2.80, and 371.3.

§ 360.100 **Definitions.**

As used in this part, words in the singular form shall be deemed to import the plural and vice versa, as the case may require.

Administrator. The Administrator, Animal and Plant Health Inspection Service, or any individual authorized to act for the Administrator.

APHIS. The Animal and Plant Health Inspection Service, United States Department of Agriculture.

Department. The U.S. Department of Agriculture.

Interstate. From one State into or through any other State; or within the District of Columbia, Guam, the Virgin Islands of the United States, or any other territory or possession of the United States.

Move. To carry, enter, import, mail, ship, or transport; to aid, abet, cause, or induce the carrying, entering, importing, mailing, shipping, or transporting; to offer to carry, enter, import, mail, ship, or transport; to receive to carry, enter, import, mail, ship, or transport; to release into the environment; or to allow any of the activities described in this definition.

Noxious weed. Any plant or plant product that can directly or indirectly injure or cause damage to crops (including nursery stock or plant products), livestock, poultry, or other interests of agriculture, irrigation, navigation, the natural resources of the United States, the public health, or the environment.

Permit. A written authorization, including by electronic methods, by the Administrator to move plants, plant products, biological control organisms, plant pests, noxious weeds, or articles under conditions prescribed by the Administrator.

Person. Any individual, partnership, corporation, association, joint venture, or other legal entity.

Plant Protection and Quarantine Programs. The Plant Protection and Quarantine Programs, Animal and Plant Health Inspection Service of the Department.

Responsible person. The person who has control over and will maintain control over the movement of the noxious weed and assure that all conditions contained in the permit and requirements in this part are complied with. A responsible person must be at least 18 years of age and must be a legal resident of the United States or designate an agent who is at least 18 years of age and a legal resident of the United States.

State. Any of the several States of the United States, the Commonwealth of the Northern Mariana Islands, the Commonwealth of Puerto Rico, the District of Columbia, Guam, the Virgin Islands of the United States, or any other territory or possession of the United States.

Taxon (taxa). Any grouping within botanical nomenclature, such as family, genus, species, or cultivar.

Through the United States. From and to places outside the United States.

United States. All of the States.

[41 FR 49988, Nov. 12, 1976, as amended at 75 FR 68953, Nov. 102, 2010]

§ 360.200 Designation of noxious weeds.

The Administrator has determined that it is necessary to designate the following plants[1] as noxious weeds to prevent their introduction into the United States or their dissemination within the United States:

(a) *Aquatic and wetland weeds:*

Azolla pinnata R. Brown (mosquito fern, water velvet)
Caulerpa taxifolia (Vahl) C. Agardh, Mediterranean strain (killer algae)
Eichhornia azurea (Swartz) Kunth
Hydrilla verticillata (Linnaeus f.) Royle (hydrilla)
Hygrophila polysperma T. Anderson (Miramar weed)
Ipomoea aquatica Forsskal (water-spinach, swamp morning-glory)
Lagarosiphon major (Ridley) Moss
Limnophila sessiliflora (Vahl) Blume (ambulia)
Melaleuca quinquenervia (Cavanilles) S.T. Blake
Monochoria hastata (Linnaeus) Solms-Laubach
Monochoria vaginalis (Burman f.) C. Presl
Ottelia alismoides (L.) Pers.
Sagittaria sagittifolia Linnaeus (arrowhead)
Salvinia auriculata Aublet (giant salvinia)
Salvinia biloba Raddi (giant salvinia)
Salvinia herzogii de la Sota (giant salvinia)
Salvinia molesta D.S. Mitchell (giant salvinia)
Solanum tampicense Dunal (wetland nightshade)
Sparganium erectum Linnaeus (exotic bur-reed)

(b) *Parasitic weeds:*

Aeginetia spp.
Alectra spp.
Cuscuta spp. (dodders), other than following species:
    Cuscuta americana Linnaeus
    Cuscuta applanata Engelmann
    Cuscuta approximata Babington
    Cuscuta attenuata Waterfall

---

[1] One or more of the common names of weeds are given in parentheses after most scientific names to help identify the weeds represented by such scientific names; however, a scientific name is intended to include all subordinate taxa within the taxon. For example, taxa listed at the genus level include all species, subspecies, varieties, and forms within the genus; taxa listed at the species level include all subspecies, varieties, and forms within the species.

Cuscuta boldinghii Urban
Cuscuta brachycalyx (Yuncker) Yuncker
Cuscuta californica Hooker & Arnott
Cuscuta campestris Yuncker
Cuscuta cassytoides Nees ex Engelmann
Cuscuta ceanothi Behr
Cuscuta cephalanthi Engelmann
Cuscuta compacta Jussieu
Cuscuta coryli Engelmann
Cuscuta cuspidata Engelmann
Cuscuta decipiens Yuncker
Cuscuta dentatasquamata Yuncker
Cuscuta denticulata Engelmann
Cuscuta epilinum Weihe
Cuscuta epithymum (Linnaeus) Linnaeus
Cuscuta erosa Yuncker
Cuscuta europaea Linnaeus
Cuscuta exaltata Engelmann
Cuscuta fasciculata Yuncker
Cuscuta glabrior (Engelmann) Yuncker
Cuscuta globulosa Bentham
Cuscuta glomerata Choisy
Cuscuta gronovii Willdenow
Cuscuta harperi Small
Cuscuta howelliana Rubtzoff
Cuscuta indecora Choisy
Cuscuta leptantha Engelmann
Cuscuta mitriformis Engelmann
Cuscuta obtusiflora Kunth
Cuscuta odontolepis Engelmann
Cuscuta pentagona Engelmann
Cuscuta planiflora Tenore
Cuscuta plattensis A. Nelson
Cuscuta polygonorum Engelmann
Cuscuta rostrata Shuttleworth ex Engelmann & Gray
Cuscuta runyonii Yuncker
Cuscuta salina Engelmann
Cuscuta sandwichiana Choisy
Cuscuta squamata Engelmann
Cuscuta suaveolens Seringe
Cuscuta suksdorfii Yuncker
Cuscuta tuberculata Brandegee
Cuscuta umbellata Kunth
Cuscuta umbrosa Beyrich ex Hooker
Cuscuta veatchii Brandegee
Cuscuta warneri Yuncker
Orobanche spp. (broomrapes), other than the following species:
Orobanche bulbosa (Gray) G. Beck
Orobanche californica Schlechtendal & Chamisso
Orobanche cooperi (Gray) Heller
Orobanche corymbosa (Rydberg) Ferris
Orobanche dugesii (S. Watson) Munz
Orobanche fasciculata Nuttall
Orobanche ludoviciana Nuttall
Orobanche multicaulis Brandegee
Orobanche parishii (Jepson) Heckard
Orobanche pinorum Geyer ex Hooker
Orobanche uniflora Linnaeus
Orobanche valida Jepson
Orobanche vallicola (Jepson) Heckard
Striga spp. (witchweeds)

(c) *Terrestrial weeds:*

Acacia nilotica (Linnaeus) Wildenow ex Delile (gum arabic tree, thorny acacia)
Ageratina adenophora (Sprengel) King & Robinson (crofton weed)
Ageratina riparia (Regel) R.M. King and H. Robinson (creeping croftonweed, mistflower)
Alternanthera sessilis (Linnaeus) R. Brown ex de Candolle (sessile joyweed)
Arctotheca calendula (Linnaeus) Levyns (capeweed)
Asphodelus fistulosus Linnaeus (onionweed)
Avena sterilis Linnaeus (including *Avena ludoviciana* Durieu) (animated oat, wild oat)
Carthamus oxyacantha M. Bieberstein (wild safflower)
Chrysopogon aciculatus (Retzius) Trinius (pilipiliula)
Commelina benghalensis Linnaeus (Benghal dayflower)
Crupina vulgaris Cassini (common crupina)
Digitaria abyssinica (Hochstetter ex A. Richard) Stapf (African couchgrass, fingergrass)
Digitaria velutina (Forsskal) Palisot de Beauvois (velvet fingergrass, annual couchgrass)
Drymaria arenariodes Humboldt & Bonpland ex J.A. Schultes (lightning weed)
Emex australis Steinheil (three-cornered jack)
Emex spinosa (Linnaeus) Campdera (devil's thorn)
Euphorbia terracina Linnaeus (false caper, Geraldton carnation weed)
Galega officinalis Linnaeus (goatsrue)
Heracleum mantegazzianum Sommier & Levier (giant hogweed)
Imperata brasiliensis Trinius (Brazilian satintail)
Imperata cylindrica (Linnaeus) Palisot de Beauvois (cogongrass)
Inula britannica Linnaeus (British elecampane, British yellowhead)
Ischaemum rugosum Salisbury (murainograss)
Leptochloa chinensis (Linnaeus) Nees (Asian sprangletop)
Lycium ferocissimum Miers (African boxthorn)
Lygodium flexuosum (Linnaeus) Swartz (maidenhair creeper)
Lygodium microphyllum (Cavanilles) R. Brown (Old World climbing fern)
Melastoma malabathricum Linnaeus
Mikania cordata (Burman f.) B. L. Robinson (mile-a-minute)
Mikania micrantha Kunth
Mimosa diplotricha C. Wright (giant sensitiveplant)
Mimosa pigra Linneaus var. *pigra* (catclaw mimosa)
Moraea collina Thunberg (apricot Cape-tulip)
Moraea flaccida (Sweet) Steudel (one-leaf Cape-tulip)
Moraea miniata Andrews (two-leaf Cape-tulip)
Moraea ochroleuca (Salisbury) Drapiez (red Cape-tulip)

Moraea pallida (Baker) Goldblatt (yellow Cape-tulip)

Nassella trichotoma (Nees) Hackel ex Arechavaleta (serrated tussock)

Onopordum acaulon Linnaeus (stemless thistle)

Onopordum illyricum Linnaeus (Illyrian thistle)

Opuntia aurantiaca Lindley (jointed prickly pear)

Oryza longistaminata A. Chevalier & Roehrich (red rice)

Oryza punctata Kotschy ex Steudel (red rice)

Oryza rufipogon Griffith (red rice)

Paspalum scrobiculatum Linnaeus (Kodo-millet)

Pennisetum clandestinum Hochstetter ex Chiovenda (kikuyugrass)

Pennisetum macrourum Trinius (African feathergrass)

Pennisetum pedicellatum Trinius (kyasumagrass)

Pennisetum polystachion (Linnaeus) Schultes (missiongrass, thin napiergrass)

Prosopis alpataco R. A. Philippi

Prosopis argentina Burkart

Prosopis articulata S. Watson

Prosopis burkartii Munoz

Prosopis caldenia Burkart

Prosopis calingastana Burkart

Prosopis campestris Griseback

Prosopis castellanosii Burkart

Prosopis denudans Bentham

Prosopis elata (Burkart) Burkart

Prosopis farcta (Banks & Solander) J.F. Macbride

Prosopis ferox Grisebach

Prosopis fiebrigii Harms

Prosopis hassleri Harms

Prosopis humilis Gillies ex Hooker & Arnott

Prosopis kuntzei Harms

Prosopis pallida (Humboldt & Bonpland ex Willdenow) Kunth

Prosopis palmeri S. Watson

Prosopis reptans Bentham var. *reptans*

Prosopis rojasiana Burkart

Prosopis ruizlealii Burkart

Prosopis ruscifolia Grisebach

Prosopis sericantha Gillies ex Hooker & Arnott

Prosopis strombulifera (Lamarck) Bentham

Prosopis torquata (Cavanilles ex Lagasca y Segura) de Candolle

Rottboellia cochinchinensis (Lour.) W. Clayton

Rubus fruticosus Linnaeus (complex) (wild blackberry)

Rubus moluccanus Linnaeus (wild raspberry)

Saccharum spontaneum Linnaeus (wild sugarcane)

Salsola vermiculata Linnaeus (wormleaf salsola)

Senecio inaequidens DC. (South African ragwort)

Senecio madagascariensis Poir. (Madagascar ragwort)

Setaria pumila (Poir.) Roem. & Schult. subsp. *pallidefusca* (Schumach.) B.K. Simon (cattail grass)

Solanum torvum Swartz (turkeyberry)

Solanum viarum Dunal (tropical soda apple)

Spermacoce alata Aublet

Tridax procumbens Linnaeus (coat buttons)

Urochloa panicoides Beauvois (liverseed grass)

[48 FR 20039, May 4, 1983, as amended at 49 FR 25223, June 20, 1984; 57 FR 8838, Mar. 13, 1992; 60 FR 35832, July 12, 1995; 64 FR 12883, Mar. 16, 1999; 65 FR 33743, May 25, 2000; 66 FR 21060, Apr. 27, 2001; 71 FR 35381, June 20, 2006; 74 FR 53400, Oct. 19, 2009; 75 FR 68953, Nov. 10, 2010]

§ 360.300 Notice of restrictions on movement of noxious weeds.

No person may move a Federal noxious weed into or through the United States, or interstate, unless:

(a) He or she applies for a permit to move a noxious weed in accordance with § 360.301;

(b) The permit application is approved; and

(c) The movement is consistent with the specific conditions contained in the permit.

(Approved by the Office of Management and Budget under control number 0579–0054)

[75 FR 68954, Nov. 10, 2010]

§ 360.301 Information required for applications for permits to move noxious weeds.

(a) *Permit to import a noxious weed into the United States.* A responsible person must apply for a permit to import a noxious weed into the United States.[2] The application must include the following information:

(1) The responsible person's name, address, telephone number, and (if available) e-mail address;

(2) The taxon of the noxious weed;

(3) Plant parts to be moved;

(4) Quantity of noxious weeds to be moved per shipment;

(5) Proposed number of shipments per year;

(6) Origin of the noxious weeds;

(7) Destination of the noxious weeds;

---

[2] Information on applying for a permit to import a noxious weed into the United States is available at *http://www.aphis.usda.gov/plant_health/permits/plantproducts.shtml.*

(8) Whether the noxious weed is established in the State of destination;

(9) Proposed method of shipment;

(10) Proposed port of first arrival in the United States;

(11) Approximate date of arrival;

(12) Intended use of the noxious weeds;

(13) Measures to be employed to prevent danger of noxious weed dissemination; and

(14) Proposed method of final disposition of the noxious weeds.

(b) *Permit to move noxious weeds interstate.* A responsible person must apply for a permit to move a noxious weed interstate.[3] The application must include the following information:

(1) The responsible person's name, address, telephone number, and (if available) e-mail address;

(2) The taxon of the noxious weed;

(3) Plant parts to be moved;

(4) Quantity of noxious weeds to be moved per shipment;

(5) Proposed number of shipments per year,

(6) Origin of the noxious weeds;

(7) Destination of the noxious weeds;

(8) Whether the noxious weed is established in the State of destination;

(9) Proposed method of shipment,

(10) Approximate date of movement;

(11) Intended use of the noxious weeds;

(12) Measures to be employed to prevent danger of noxious weed dissemination; and

(13) Proposed method of final disposition of the noxious weeds.

(c) *Permits to move noxious weeds through the United States.* Permits to move noxious weeds through the United States must be obtained in accordance with part 352 of this chapter.

[75 FR 68954, Nov. 10, 2010]

§360.302 Consideration of applications for permits to move noxious weeds.

Upon the receipt of an application made in accordance with §360.301 for a permit for movement of a noxious weed into the United States or interstate,

the Administrator will consider the application on its merits.

(a) *Consultation.* The Administrator may consult with other Federal agencies or entities, States or political subdivisions of States, national governments, local governments in other nations, domestic or international organizations, domestic or international associations, and other persons for views on the danger of noxious weed dissemination into the United States, or interstate, in connection with the proposed movement.

(b) *Inspection of premises.* The Administrator may inspect the site where noxious weeds are proposed to be handled in connection with or after their movement under permit to determine whether existing or proposed facilities will be adequate to prevent noxious weed dissemination if a permit is issued.

[75 FR 68954, Nov. 10, 2010]

§360.303 Approval of an application for a permit to move a noxious weed; conditions specified in permit.

The Administrator will approve or deny an application for a permit to move a noxious weed. If the application is approved, the Administrator will issue the permit including any conditions that the Administrator has determined are necessary to prevent dissemination of noxious weeds into the United States or interstate. Such conditions may include requirements for inspection of the premises where the noxious weed is to be handled after its movement under the permit, to determine whether the facilities there are adequate to prevent noxious weed dissemination and whether the conditions of the permit are otherwise being observed. Before the permit is issued, the Administrator will require the responsible person to agree in writing to the conditions under which the noxious weed will be safeguarded.

[75 FR 68954, Nov. 10, 2010]

---

[3] Information on applying for a permit to move a noxious weed interstate is available at *http://www.aphis.usda.gov/plant_health/permits/plantproducts.shtml.*

§ 360.304 Denial of an application for a permit to move a noxious weed; revocation of a permit to move a noxious weed.

(a) The Administrator may deny an application for a permit to move a noxious weed when the Administrator determines that:

(1) No safeguards adequate or appropriate to prevent dissemination of the noxious weed can be implemented; or

(2) The destructive potential of the noxious weed, should it escape despite proposed safeguards, outweighs the probable benefits to be derived from the proposed movement and use of the noxious weed; or

(3) The responsible person, or the responsible person's agent, as a previous permittee, failed to maintain the safeguards or otherwise observe the conditions prescribed in a previous permit and failed to demonstrate the ability or intent to observe them in the future; or

(4) The movement could impede an APHIS eradication, suppression, control, or regulatory program; or

(5) A State plant regulatory official objects to the issuance of the permit on the grounds that granting the permit will pose a risk of dissemination of the noxious weed into the State; or

(6) The application for the permit contains information that is found to be materially false, fraudulent, or deceptive; or

(7) APHIS may deny a permit to a person who has previously failed to comply with any APHIS regulation.

(b) The Administrator may revoke any outstanding permit when:

(1) After the issuance of the permit, information is received that constitutes cause for the denial of an application for permit under paragraph (a) of this section; or

(2) The responsible person has not maintained the safeguards or otherwise observed the conditions specified in the permit.

(c) If a permit is orally revoked, APHIS will provide the reasons for the withdrawal of the permit in writing within 10 days. Any person whose permit has been revoked or any person who has been denied a permit may appeal the decision in writing to the Administrator within 10 days after receiving the written notification of the revocation or denial. The appeal must state all of the facts and reasons upon which the person relies to show that the permit was wrongfully revoked or denied. The Administrator will grant or deny the appeal, in writing, stating the reasons for the decision as promptly as circumstances allow. If there is a conflict as to any material fact, a hearing will be held to resolve the conflict. Rules of practice concerning such a hearing will be adopted by the Administrator.

[75 FR 68954, Nov. 10, 2010, as amended at 79 FR 19812, Apr. 10, 2014]

§ 360.305 Disposal of noxious weeds when permits are revoked.

When a permit for the movement of a noxious weed is revoked by the Administrator and not reinstated under § 360.304(c), further movement of the noxious weed covered by the permit into or through the United States, or interstate, is prohibited unless authorized by another permit. The responsible person must arrange for disposal of the noxious weed in question in a manner that the Administrator determines is adequate to prevent noxious weed dissemination. The Administrator may seize, quarantine, treat, apply other remedial measures to, destroy, or otherwise dispose of, in such manner as the Administrator deems appropriate, any noxious weed that is moved without compliance with any conditions in the permit or after the permit has been revoked whenever the Administrator deems it necessary in order to prevent the dissemination of any noxious weed into or within the United States.

[75 FR 68954, Nov. 10, 2010, as amended at 79 FR 19812, Apr. 10, 2014]

§ 360.400 Treatments.

(a) Seeds of *Guizotia abyssinica* (niger seed) are commonly contaminated with noxious weed seeds listed in § 360.200, including (but not limited to) *Cuscuta* spp. Therefore, *Guizotia abyssinica* seeds may be imported into the United States only if:

(1) They are treated in accordance with part 305 of this chapter at the time of arrival at the port of first arrival in the United States; or

(2) They are treated prior to shipment to the United States at a facility that is approved by APHIS[4] and that operates in compliance with a written agreement between the treatment facility owner and the plant protection service of the exporting country, in which the treatment facility owner agrees to comply with the provisions of §319.37–9(c) of this chapter and allow inspectors and representatives of the plant protection service of the exporting country access to the treatment facility as necessary to monitor compliance with the regulations. Treatments must be certified in accordance with the conditions described in §319.37–9(c) of this chapter.

(b) [Reserved]

[75 FR 68955, Nov. 10, 2010, as amended at 83 FR 11867, Mar. 19, 2018]

§360.500 Petitions to add a taxon to the noxious weed list.

A person may petition the Administrator to have a taxon added to the noxious weeds lists in §360.200. Details of the petitioning process for adding a taxon to the lists are available on the Internet at *http://www.aphis.usda.gov/plant_health/plant_pest_info/weeds/downloads/listingguide.pdf.* Persons who submit a petition to add a taxon to the noxious weed lists must provide their name, address, telephone number, and (if available) e-mail address. Persons who submit a petition to add a taxon to the noxious weed lists are encouraged to provide the following information, which can help speed up the review process and help APHIS determine whether the specified plant taxon should be listed as a noxious weed:

(a) *Identification of the taxon.* (1) The taxon's scientific name and author;

(2) Common synonyms;

(3) Botanical classification;

(4) Common names;

(5) Summary of life history;

(6) Native and world distribution;

(7) Distribution in the United States, if any (specific States, localities, or Global Positioning System coordinates);

(8) Description of control efforts, if established in the United States; and

(9) Whether the taxon is regulated at the State or local level.

(b) *Potential consequences of the taxon's introduction or spread.* (1) The taxon's habitat suitability in the United States (predicted ecological range);

(2) Dispersal potential (biological characteristics associated with invasiveness);

(3) Potential economic impacts (*e.g.,* potential to reduce crop yields, lower commodity values, or cause loss of markets for U.S. goods); and

(4) Potential environmental impacts (e.g., impacts on ecosystem processes, natural community composition or structure, human health, recreation patterns, property values, or use of chemicals to control the taxon).

(c) *Likelihood of the taxon's introduction or spread.* (1) Potential pathways for the taxon's movement into and within the United States; and

(2) The likelihood of survival and spread of the taxon within each pathway.

(d) List of references.

[75 FR 68955, Nov. 10, 2010]

§360.501 Petitions to remove a taxon from the noxious weed lists.

A person may petition the Administrator to remove a taxon from the noxious weeds lists in §360.200. Details of the petitioning process for removing a taxon from the lists are available at *http://www.aphis.usda.gov/plant_health/plant_pest_info/weeds/downloads/delistingguide.pdf.* Persons who submit a petition to remove a taxon from the noxious weed lists would be required to provide their name, address, telephone number, and (if available) e-mail address. Persons who submit a petition to remove a taxon from the noxious weed lists are encouraged to provide the following information, which can help speed up the review process and help APHIS determine whether the specified plant taxon should not be listed as a noxious weed:

(a) Evidence that the species is distributed throughout its potential range or has spread too far to implement effective control.

---

[4] Criteria for the approval of heat treatment facilities are contained in part 305 of this chapter.

(b) Evidence that control efforts have been unsuccessful and further efforts are unlikely to succeed.

(c) For cultivars of a listed noxious weed, scientific evidence that the cultivar has a combination of risk elements that result in a low pest risk. For example, the cultivar may have a narrow habitat suitability, low dispersal potential, evidence of sterility, inability to cross-pollinate with introduced wild types, or few if any potential negative impacts on the economy or environment of the United States.

(d) List of references.

[75 FR 68955, Nov. 10, 2010]

§ 360.600 Preemption of State and local laws.

(a) Under section 436 of the Plant Protection Act (7 U.S.C. 7756), a State or political subdivision of a State may not regulate in foreign commerce any noxious weed in order to control it, eradicate it, or prevent its dissemination. A State or political subdivision of a State also may not impose prohibitions or restrictions upon the movement in interstate commerce of noxious weeds if the Secretary has issued a regulation or order to prevent the dissemination of the noxious weed within the United States. The only exceptions to this are:

(1) If the prohibitions or restrictions issued by the State or political subdivision of a State are consistent with and do not exceed the regulations or orders issued by the Secretary; or

(2) If the State or political subdivision of a State demonstrates to the Secretary and the Secretary finds that there is a special need for additional prohibitions or restrictions based on sound scientific data or a thorough risk assessment.

(b) Therefore, in accordance with section 436 of the Plant Protection Act, the regulations in this part preempt all State and local laws and regulations that are inconsistent with or exceed the regulations in this part unless a special need request has been granted in accordance with the regulations in §§ 301.1 through 301.13 of this chapter.

[74 FR 53400, Oct. 19, 2009. Redesignated at 75 FR 68955, Nov. 10, 2010]

PART 361—IMPORTATION OF SEED AND SCREENINGS UNDER THE FEDERAL SEED ACT

Sec.
361.1 Definitions.
361.2 Preemption of State and local laws; general restrictions on the importation of seed and screenings.
361.3 Declarations and labeling.
361.4 Inspection at the port of first arrival.
361.5 Sampling of seeds.
361.6 Noxious weed seeds.
361.7 Special provisions for Canadian-origin seed and screenings.
361.8 Cleaning of imported seed and processing of certain Canadian-origin screenings.
361.9 Recordkeeping.
361.10 Costs and charges.

AUTHORITY: 7 U.S.C. 1581–1610; 7 CFR 2.22, 2.80, and 371.3.

SOURCE: 62 FR 48460, Sept. 16, 1997, unless otherwise noted.

§ 361.1 Definitions.

Terms used in the singular form in this part shall be construed as the plural, and vice versa, as the case may demand. The following terms, when used in this part, shall be construed, respectively, to mean:

Administrator. The Administrator of the Animal and Plant Health Inspection Service, U.S. Department of Agriculture, or any other individual to whom the Administrator delegates authority to act in his or her stead.

Agricultural seed. The following kinds and varieties of grass, forage, and field crop seed that are used for seeding purposes in the United States:

Agrotricum—x *Agrotriticum* Ciferri and Giacom.
Alfalfa—*Medicago sativa* L.
Alfilaria—*Erodium cicutarium* (L.) L'Her.
Alyceclover—*Alysicarpus vaginalis* (L.) DC.
Bahiagrass—*Paspalum notatum* Fluegge
Barley—*Hordeum vulgare* L.
Barrelclover—*Medicago truncatula* Gaertn.
Bean, adzuki—*Vigna angularis* (Willd.) Ohwi and Ohashi
Bean, field—*Phaseolus vulgaris* L.
Bean, mung—*Vigna radiata* (L.) Wilczek
Beet, field—*Beta vulgaris* L. subsp. *vulgaris*
Beet, sugar—*Beta vulgaris* L. subsp. *vulgaris*
Beggarweed, Florida—*Desmodium tortuosum* (Sw.) DC.
Bentgrass, colonial—*Agrostis capillaris* L.
Bentgrass, creeping—*Agrostis stolonifera* L. var. *palustris* (Huds.) Farw.
Bentgrass, velvet—*Agrostis canina* L.

Bermudagrass—*Cynodon dactylon* (L.) Pers. var. *dactylon*

Bermudagrass, giant—*Cynodon dactylon* (L.) Pers. var. *aridus* Harlan and de Wet

Bluegrass, annual—*Poa annua* L.

Bluegrass, bulbous—*Poa bulbosa* L.

Bluegrass, Canada—*Poa compressa* L.

Bluegrass, glaucantha—*Poa glauca* Vahl

Bluegrass, Kentucky—*Poa pratensis* L.

Bluegrass, Nevada—*Poa secunda* J.S. Presl

Bluegrass, rough—*Poa trivialis* L.

Bluegrass, Texas—*Poa arachnifera* Torr.

Bluegrass, wood—*Poa nemoralis* L.

Bluejoint—*Calamagrostis canadensis* (Michx.) P. Beauv.

Bluestem, big—*Andropogon gerardii* Vitm. var. *gerardii*

Bluestem, little—*Schizachyrium scoparium* (Michx.) Nash

Bluestem, sand—*Andropogon hallii* Hack.

Bluestem, yellow—*Bothriochloa ischaemum* (L.) Keng

Bottlebrush-squirreltail—*Elymus elymoides* (Raf.) Swezey

Brome, field—*Bromus arvensis* L.

Brome, meadow—*Bromus biebersteinii* Roem. and Schult.

Brome, mountain—*Bromus marginatus* Steud.

Brome, smooth—*Bromus inermis* Leyss.

Broomcorn—*Sorghum bicolor* (L.) Moench

Buckwheat—*Fagopyrum esculentum* Moench

Buffalograss—*Buchloe dactyloides* (Nutt.) Engelm.

Buffelgrass—*Cenchrus ciliaris* L.

Burclover, California—*Medicago polymorpha* L.

Burclover, spotted—*Medicago arabica* (L.) Huds.

Burnet, little—*Sanguisorba minor* Scop.

Buttonclover—*Medicago orbicularis* (L.) Bartal.

Canarygrass—*Phalaris canariensis* L.

Canarygrass, reed—*Phalaris arundinacea* L.

Carpetgrass—*Axonopus fissifolius* (Raddi) Kuhlm.

Castorbean—*Ricinus communis* L.

Chess, soft—*Bromus hordeaceus* L.

Chickpea—*Cicer arietinum* L.

Clover, alsike—*Trifolium hybridum* L.

Clover, arrowleaf—*Trifolium vesiculosum* Savi

Clover, berseem—*Trifolium alexandrinum* L.

Clover, cluster—*Trifolium glomeratum* L.

Clover, crimson—*Trifolium incarnatum* L.

Clover, Kenya—*Trifolium semipilosum* Fresen.

Clover, ladino—*Trifolium repens* L.

Clover, lappa—*Trifolium lappaceum* L.

Clover, large hop—*Trifolium campestre* Schreb.

Clover, Persian—*Trifolium resupinatum* L.

Clover, red or

Red clover, mammoth—*Trifolium pratense* L.

Red clover, medium—*Trifolium pratense* L.

Clover, rose—*Trifolium hirtum* All.

Clover, small hop or suckling—*Trifolium dubium* Sibth.

Clover, strawberry—*Trifolium fragiferum* L.

Clover, sub or subterranean—*Trifolium subterraneum* L.

Clover, white—*Trifolium repens* L. (also see Clover, ladino)

Clover—(also see Alyceclover, Burclover, Buttonclover, Sourclover, Sweetclover)

Corn, field—*Zea mays* L.

Corn, pop—*Zea mays* L.

Cotton—*Gossypium* spp.

Cowpea—*Vigna unguiculata* (L.) Walp. subsp. *unguiculata*

Crambe—*Crambe abyssinica* R.E. Fries

Crested dogtail—*Cynosurus cristatus* L.

Crotalaria, lance—*Crotalaria lanceolata* E. Mey.

Crotalaria, showy—*Crotalaria spectabilis* Roth

Crotalaria, slenderleaf—*Crotalaria brevidens* Benth. var. *intermedia* (Kotschy) Polh.

Crotalaria, striped or smooth—*Crotalaria pallida* Ait.

Crotalaria, sunn—*Crotalaria juncea* L.

Crownvetch—*Coronilla varia* L.

Dallisgrass—*Paspalum dilatatum* Poir.

Dichondra—*Dichondra repens* Forst. and Forst. f.

Dropseed, sand—*Sporobolus cryptandrus* (Torr.) A. Gray

Emmer—*Triticum dicoccon* Schrank

Fescue, chewings—*Festuca rubra* L. subsp. *commutata* Gaud.

Fescue, hair—*Festuca tenuifolia* Sibth.

Fescue, hard—*Festuca brevipila* Tracey

Fescue, meadow—*Festuca pratensis* Huds.

Fescue, red—*Festuca rubra* L. subsp. *rubra*

Fescue, sheep—*Festuca ovina* L. var. *ovina*

Fescue, tall—*Festuca arundinacea* Schreb.

Flax—*Linum usitatissimum* L.

Galletagrass—*Hilaria jamesii* (Torr.) Benth.

Grama, blue—*Bouteloua gracilis* (Kunth) Steud.

Grama, side-oats—*Bouteloua curtipendula* (Michx.) Torr.

Guar—*Cyamopsis tetragonoloba* (L.) Taub.

Guineagrass—*Panicum maximum* Jacq. var. *maximum*

Hardinggrass—*Phalaris stenoptera* Hack.

Hemp—*Cannabis sativa* L.

Indiangrass, yellow—*Sorghastrum nutans* (L.) Nash

Indigo, hairy—*Indigofera hirsuta* L.

Japanese lawngrass—*Zoysia japonica* Steud.

Johnsongrass—*Sorghum halepense* (L.) Pers.

Kenaf—*Hibiscus cannabinus* L.

Kochia, forage—*Kochia prostrata* (L.) Schrad.

Kudzu—*Pueraria montana* (Lour.) Merr. var. *lobata* (Willd.) Maesen and S. Almeida

Lentil—*Lens culinaris* Medik.

Lespedeza, Korean—*Kummerowia stipulacea* (Maxim.) Makino

Lespedeza, sericea or Chinese—*Lespedeza cuneata* (Dum.-Cours.) G. Don

Lespedeza, Siberian—*Lespedeza juncea* (L. f.) Pers.

Lespedeza, striate—*Kummerowia striata* (Thunb.) Schindler

Lovegrass, sand—*Eragrostis trichodes* (Nutt.) Wood

Lovegrass, weeping—*Eragrostis curvula* (Schrad.) Nees

Lupine, blue—*Lupinus angustifolius* L.

Lupine, white—*Lupinus albus* L.

Lupine, yellow—*Lupinus luteus* L.

Manilagrass—*Zoysia matrella* (L.) Merr.

Meadow foxtail—*Alopecurus pratensis* L.

Medic, black—*Medicago lupulina* L.

Milkvetch or cicer milkvetch—*Astragalus cicer* L.

Millet, browntop—*Brachiaria ramosa* (L.) Stapf

Millet, foxtail—*Setaria italica* (L.) Beauv.

Millet, Japanese—*Echinochloa frumentacea* Link

Millet, pearl—*Pennisetum glaucum* (L.) R. Br.

Millet, proso—*Panicum miliaceum* L.

Molassesgrass—*Melinis minutiflora* Beauv.

Mustard, black—*Brassica nigra* (L.) Koch

Mustard, India—*Brassica juncea* (L.) Czernj. and Coss.

Mustard, white—*Sinapis alba* L.

Napiergrass—*Pennisetum purpureum* Schumach.

Needlegrass, green—*Stipa viridula* Trin.

Oat—*Avena byzantina* C. Koch, *A. sativa* L., *A. nuda* L.

Oatgrass, tall—*Arrhenatherum elatius* (L.) J.S. Presl and K.B. Presl

Orchardgrass—*Dactylis glomerata* L.

Panicgrass, blue—*Panicum antidotale* Retz.

Panicgrass, green—*Panicum maximum* Jacq. var. *trichoglume* Robyns

Pea, field—*Pisum sativum* L.

Peanut—*Arachis hypogaea* L.

Poa trivialis—(see Bluegrass, rough)

Rape, annual—*Brassica napus* L. var. *annua* Koch

Rape, bird—*Brassica rapa* L. subsp. *rapa*

Rape, turnip—*Brassica rapa* L. subsp. *silvestris* (Lam.) Janchen

Rape, winter—*Brassica napus* L. var. *biennis* (Schubl. and Mart.) Reichb.

Redtop—*Agrostis gigantea* Roth

Rescuegrass—*Bromus catharticus* Vahl

Rhodesgrass—*Chloris gayana* Kunth

Rice—*Oryza sativa* L.

Ricegrass, Indian—*Oryzopsis hymenoides* (Roem. and Schult.) Ricker

Roughpea—*Lathyrus hirsutus* L.

Rye—*Secale cereale* L.

Rye, mountain—*Secale strictum* (K.B. Presl) K.B. Presl subsp. *strictum*

Ryegrass, annual or Italian—*Lolium multiflorum* Lam.

Ryegrass, intermediate—*Lolium* × *hybridum* Hausskn.

Ryegrass, perennial—*Lolium perenne* L.

Ryegrass, Wimmera—*Lolium rigidum* Gaud.

Safflower—*Carthamus tinctorius* L.

Sagewort, Louisiana—*Artemisia ludoviciana* Nutt.

Sainfoin—*Onobrychis viciifolia* Scop.

Saltbush, fourwing—*Atriplex canescens* (Pursh) Nutt.

Sesame—*Sesamum indicum* L.

Sesbania—*Sesbania exaltata* (Raf.) A.W. Hill

Smilo—*Piptatherum miliaceum* (L.) Coss.

Sorghum—*Sorghum bicolor* (L.) Moench

Sorghum almum—*Sorghum* × *almum* L. Parodi

Sorghum-sudangrass—*Sorghum* × *drummondii* (Steud.) Millsp. and Chase

Sorgrass—*Rhizomatous* derivatives of a johnsongrass × sorghum cross or a johnsongrass × sudangrass cross

Southernpea—(See Cowpea)

Sourclover—*Melilotus indicus* (L.) All.

Soybean—*Glycine max* (L.) Merr.

Spelt—*Triticum spelta* L.

Sudangrass—*Sorghum* × *drummondii* (Steud.) Millsp. and Chase

Sunflower—*Helianthus annuus* L.

Sweetclover, white—*Melilotus albus* Medik.

Sweetclover, yellow—*Melilotus officinalis* Lam.

Sweet vernalgrass—*Anthoxanthum odoratum* L.

Sweetvetch, northern—*Hedysarum boreale* Nutt.

Switchgrass—*Panicum virgatum* L.

Timothy—*Phleum pratense* L.

Timothy, turf—*Phleum bertolonii* DC.

Tobacco—*Nicotiana tabacum* L.

Trefoil, big—*Lotus uliginosus* Schk.

Trefoil, birdsfoot—*Lotus corniculatus* L.

Triticale—x *Triticosecale* Wittm. (Secale × Triticum)

Vaseygrass—*Paspalum urvillei* Steud.

Veldtgrass—*Ehrharta calycina* J.E. Smith

Velvetbean—*Mucuna pruriens* (L.) DC. var. *utilis* (Wight) Burck

Velvetgrass—*Holcus lanatus* L.

Vetch, common—*Vicia sativa* L. subsp. *sativa*

Vetch, hairy—*Vicia villosa* Roth subsp. *villosa*

Vetch, Hungarian—*Vicia pannonica* Crantz

Vetch, monantha—*Vicia articulata* Hornem.

Vetch, narrowleaf or blackpod—*Vicia sativa* L. subsp. *nigra* (L.) Ehrh.

Vetch, purple—*Vicia benghalensis* L.

Vetch, woollypod or winter—*Vicia villosa* Roth subsp. *varia* (Host) Corb.

Wheat, common—*Triticum aestivum* L.

Wheat, club—*Triticum compactum* Host

Wheat, durum—*Triticum durum* Desf.

Wheat, Polish—*Triticum polonicum* L.

Wheat, poulard—*Triticum turgidum* L.

Wheat × Agrotricum—*Triticum* × *Agrotriticum*

Wheatgrass, beardless—*Pseudoroegneria spicata* (Pursh) A. Love

Wheatgrass, crested or fairway crested—*Agropyron cristatum* (L.) Gaertn.

Wheatgrass, crested or standard crested—*Agropyron desertorum* (Link) Schult.

Wheatgrass, intermediate—*Elytrigia intermedia* (Host) Nevski subsp. *intermedia*

Wheatgrass, pubescent—*Elytrigia intermedia* (Host) Nevski subsp. *intermedia*

Wheatgrass, Siberian—*Agropyron fragile* (Roth) Candargy subsp. *sibiricum* (Willd.) Meld.

Wheatgrass, slender—*Elymus trachycaulus* (Link) Shinn.

Wheatgrass, streambank—*Elymus lanceolatus* (Scribn. and J.G. Smith) Gould subsp. *lanceolatus*

Wheatgrass, tall—*Elytrigia elongata* (Host) Nevski

Wheatgrass, western—*Pascopyrum smithii* (Rydb.) A. Love

Wildrye, basin—*Leymus cinereus* (Scribn. and Merr.) A. Love

Wildrye, Canada—*Elymus canadensis* L.

Wildrye, Russian—*Psathyrostachys juncea* (Fisch.) Nevski

Zoysia japonica—(see Japanese lawngrass)

Zoysia matrella—(see Manilagrass)

Animal and Plant Health Inspection Service (APHIS). The Animal and Plant Health Inspection Service of the U.S. Department of Agriculture.

APHIS inspector. Any employee of the Animal and Plant Health Inspection Service or any other individual authorized by the Administrator to enforce this part.

Coated Seed. Any seed unit covered with any substance that changes the size, shape, or weight of the original seed. Seeds coated with ingredients such as, but not limited to, rhizobia, dyes, and pesticides are excluded.

Declaration. A written statement of a grower, shipper, processor, dealer, or importer giving for any lot of seed the kind, variety, type, origin, or the use for which the seed is intended.

Hybrid. When applied to kinds or varieties of seed means the first generation seed of a cross produced by controlling the pollination and by combining two or more inbred lines; one inbred or a single cross with an open-pollinated variety; or two selected clones, seed lines, varieties, or species. "Controlling the pollination" means to use a method of hybridization that will produce pure seed that is at least 75 percent hybrid seed. Hybrid designations shall be treated as variety names.

Import/importation. To bring into the territorial limits of the United States.

Kind. One or more related species or subspecies that singly or collectively is known by one common name, e.g., soybean, flax, or carrot.

Lot of seed. A definite quantity of seed identified by a lot number, every portion or bag of which is uniform, within permitted tolerances, for the factors that appear in the labeling.

Mixture. Seeds consisting of more than one kind or variety, each present in excess of 5 percent of the whole.

Official seed laboratory. An official laboratory member of the Association of Official Seed Analysts.

Pelleted seed. Any seed unit covered with a substance that changes the size, shape, or weight of the original seed in order to improve the plantability or singulation of the seed.

Person. Any individual, partnership, corporation, company, society, association, receiver, trustee, or other legal entity or organized group.

Port of first arrival. The land area (such as a seaport, airport, or land border station) where a person, or a land, water, or air vehicle, first arrives after entering the territorial limits of the United States, and where inspection of articles is carried out by APHIS inspectors.

Registered seed technologist. A registered member of the Society of Commercial Seed Technologists.

Screenings. Chaff, sterile florets, immature seed, weed seed, inert matter, and any other materials removed in any way from any seeds in any kind of cleaning or processing and which contains less than 25 percent of live agricultural or vegetable seeds.

State. Any State, the District of Columbia, American Samoa, Guam, the Northern Mariana Islands, Puerto Rico, the Virgin Islands of the United States, and any other territory or possession of the United States.

United States. All of the States.

Variety. A subdivision of a kind which is characterized by growth, plant, fruit, seed, or other characteristics by which it can be differentiated from other sorts of the same kind.

Vegetable seed. The seed of the following kinds and varieties that are or may be grown in gardens or on truck farms and are or may be generally known and sold under the name of vegetable seed:

Artichoke—*Cynara cardunculus* L. subsp. *cardunculus*

Asparagus—*Asparagus officinalis* Baker

Asparagusbean or yard-long bean—*Vigna unguiculata* (L.) Walp. subsp. *sesquipedalis* (L.) Verdc.

Bean, garden—*Phaseolus vulgaris* L.

Bean, lima—*Phaseolus lunatus* L.

Bean, runner or scarlet runner—*Phaseolus coccineus* L.
Beet—*Beta vulgaris* L. subsp. *vulgaris*
Broadbean—*Vicia faba* L.
Broccoli—*Brassica oleracea* L. var. *botrytis* L.
Brussels sprouts—*Brassica oleracea* L. var. *gemmifera* DC.
Burdock, great—*Arctium lappa* L.
Cabbage—*Brassica oleracea* L. var. *capitata* L.
Cabbage, Chinese—*Brassica rapa* L. subsp. *pekinensis* (Lour.) Hanelt
Cabbage, tronchuda—*Brassica oleracea* L. var. *costata* DC.
Cantaloupe—(see Melon)
Cardoon—*Cynara cardunculus* L. subsp. *cardunculus*
Carrot—*Daucus carota* L. subsp. *sativus* (Hoffm.) Arcang.
Cauliflower—*Brassica oleracea* L. var. *botrytis* L.
Celeriac—*Apium graveolens* L. var. *rapaceum* (Mill.) Gaud.
Celery—*Apium graveolens* L. var. *dulce* (Mill.) Pers.
Chard, Swiss—*Beta vulgaris* L. subsp. *cicla* (L.) Koch
Chicory—*Cichorium intybus* L.
Chives—*Allium schoenoprasum* L.
Citron—*Citrullus lanatus* (Thunb.) Matsum. and Nakai var. *citroides* (Bailey) Mansf.
Collards—*Brassica oleracea* L. var. *acephala* DC.
Corn, sweet—*Zea mays* L.
Cornsalad—*Valerianella locusta* (L.) Laterrade
Cowpea—*Vigna unguiculata* (L.) Walp. subsp. *unguiculata*
Cress, garden—*Lepidium sativum* L.
Cress, upland—*Barbarea verna* (Mill.) Asch.
Cress, water—*Rorippa nasturtium-aquaticum* (L.) Hayek
Cucumber—*Cucumis sativus* L.
Dandelion—*Taraxacum officinale* Wigg.
Dill—*Anethum graveolens* L.
Eggplant—*Solanum melongena* L.
Endive—*Cichorium endivia* L.
Gherkin, West India—*Cucumis anguria* L.
Kale—*Brassica oleracea* L. var. *acephala* DC.
Kale, Chinese—*Brassica oleracea* L. var. *alboglabra* (Bailey) Musil
Kale, Siberian—*Brassica napus* L. var. *pabularia* (DC.) Reichb.
Kohlrabi—*Brassica oleracea* L. var. *gongylodes* L.
Leek—*Allium porrum* L.
Lettuce—*Lactuca sativa* L.
Melon—*Cucumis melo* L.
Muskmelon—(see Melon).
Mustard, India—*Brassica juncea* (L.) Czernj. and Coss.
Mustard, spinach—*Brassica perviridis* (Bailey) Bailey
Okra—*Abelmoschus esculentus* (L.) Moench
Onion—*Allium cepa* L.
Onion, Welsh—*Allium fistulosum* L.
Pak-choi—*Brassica rapa* L. subsp. *chinensis* (L.) Hanelt

Parsley—*Petroselinum crispum* (Mill.) A.W. Hill
Parsnip—*Pastinaca sativa* L.
Pea—*Pisum sativum* L.
Pepper—*Capsicum* spp.
Pe-tsai—(see Chinese cabbage).
Pumpkin—*Cucurbita pepo* L., *C. moschata* (Duchesne) Poiret, and *C. maxima* Duchesne
Radish—*Raphanus sativus* L.
Rhubarb—*Rheum rhabarbarum* L.
Rutabaga—*Brassica napus* L. var. *napobrassica* (L.) Reichb.
Sage—*Salvia officinalis* L.
Salsify—*Tragopogon porrifolius* L.
Savory, summer—*Satureja hortensis* L.
Sorrel—*Rumex acetosa* L.
Southernpea—(see Cowpea).
Soybean—*Glycine max* (L.) Merr.
Spinach—*Spinacia oleracea* L.
Spinach, New Zealand—*Tetragonia tetragonioides* (Pall.) Ktze.
Squash—*Cucurbita pepo* L., *C. moschata* (Duchesne) Poiret, and *C. maxima* Duchesne
Tomato—*Lycopersicon esculentum* Mill.
Tomato, husk—*Physalis pubescens* L.
Turnip—*Brassica rapa* L. subsp. *rapa*
Watermelon—*Citrullus lanatus* (Thunb.) Matsum. and Nakai var. *lanatus*

§ 361.2 Preemption of State and local laws; general restrictions on the importation of seed and screenings.

(a) The regulations in this part preempt State and local laws regarding seed and screenings imported into the United States while the seed and screenings are in foreign commerce. Seed and screenings imported for immediate distribution and sale to the consuming public remain in foreign commerce until sold to the ultimate consumer. The question of when foreign commerce ceases in other cases must be considered on a case-by-case basis.

(b) No person shall import any agricultural seed, vegetable seed, or screenings into the United States unless the importation is in compliance with this part.

(c) Any agricultural seed, vegetable seed, or screenings imported into the United States not in compliance with this part shall be subject to exportation, destruction, disposal, or any remedial measures that the Administrator determines are necessary to prevent the dissemination into the United States of noxious weeds.

(d) Except as provided in §361.7(b), and in addition to the permit requirements of §319.37–5 of this chapter, coated or pelleted seed, or seed that is embedded in a substrate that obscures visibility may enter the United States only if each lot of seed is accompanied by an officially drawn and sealed sample of seed drawn from the lot before the seed was coated or pelleted. The sample must be drawn in a manner consistent with that described in §361.5 of this part.

(e) Except as provided in §§361.4(a)(3) and 361.7(c), screenings of all agricultural seed and vegetable seed are prohibited entry into the United States.

[62 FR 48460, Sept. 16, 1997, as amended at 74 FR 53400, Oct. 19, 2009; 79 FR 74594, Dec. 16, 2014; 83 FR 11867, Mar. 19, 2018]

§361.3 Declarations and labeling.

(a) All lots of agricultural seed, vegetable seed, and screenings imported into the United States must be accompanied by a declaration from the importer of the seed or screenings. The declaration must state the kind, variety, and origin of each lot of seed or screenings and the use for which the seed or screenings are being imported.

(b) Each container of agricultural seed and vegetable seed imported into the United States for seeding (planting) purposes must be labeled to indicate the identification code or designation for the lot of seed; the name of each kind or kind and variety of agricultural seed or the name of each kind and variety of vegetable seed present in the lot in excess of 5 percent of the whole; and the designation "hybrid" when the lot contains hybrid seed. Kind and variety names used on the label shall conform to the kind and variety names used in the definitions of "agricultural seed" and "vegetable seed" in §361.1. If any seed in the lot has been treated, each container must be further labeled, in type no smaller than 8 point, as follows:

(1) The label must indicate that the seed has been treated and provide the name of the substance or process used to treat the seed. Substance names used on the label shall be the commonly accepted coined, chemical (generic), or abbreviated chemical name.

(i) Commonly accepted coined names are commonly recognized as names of particular substances, e.g., thiram, captan, lindane, and dichlone.

(ii) Examples of commonly accepted chemical (generic) names are bluestone, calcium carbonate, cuprous oxide, zinc hydroxide, hexachlorobenzene, and ethyl mercury acetate. The terms "mercury" or "mercurial" may be used in labeling all types of mercurials.

(iii) Examples of commonly accepted abbreviated chemical names are BHC (1,2,3,4,5,6-Hexachlorocyclohexane) and DDT (dichloro diphenyl trichloroethane).

(2) If the seed has been treated with a mercurial or similarly toxic substance harmful to humans and vertebrate animals, the label must include a representation of a skull and crossbones and a statement indicating that the seed has been treated with poison. The skull and crossbones must be at least twice the size of the type used for the information provided on the label, and the poison warning statement must be written in red letters on a background of distinctly contrasting color. Mercurials and similarly toxic substances include the following:

Aldrin, technical
Demeton
Dieldrin
p-Dimethylaminobenzenediazo   sodium
  sulfonate
Endrin
Ethion
Heptachlor
Mercurials, all types
Parathion
Phorate
Toxaphene
O-O-Diethyl-O-(isopropyl-4-methyl-6-
  pyrimidyl) thiophosphate
O,O-Diethyl-S-2-(ethylthio)   ethyl
  phosphorodithioate

(3) If the seed has been treated with a substance other than one classified as a mercurial or similarly toxic substance under paragraph (b)(2) of this section, and the amount remaining with the seed is harmful to humans or other vertebrate animals, the label must indicate that the seed is not to be used for food, feed, or oil purposes. Any amount of any substance used to treat the seed that remains with the seed

will be considered harmful when the seed is in containers of more than 4 ounces, except that the following substances will not be deemed harmful when present at a rate less than the number of parts per million (p/m) indicated:

Allethrin—2 p/m
Malathion—8 p/m
Methoxyclor—2 p/m
Piperonyl butoxide—20 p/m (8 p/m on oat and sorghum)
Pyrethrins—3 p/m (1 p/m on oat and sorghum)

(c) In the case of seed in bulk, the information required under paragraph (b) of this section shall appear in the invoice or other records accompanying and pertaining to such seed. If the seed is in containers and in quantities of 20,000 pounds or more, regardless of the number of lots included, the information required on each container under paragraph (b) of this section need not be shown on each container if each container has stenciled upon it or bears a label containing a lot designation and the invoice or other records accompanying and pertaining to such seed bear the various statements required for the respective seeds.

(d) Each container of agricultural seed and vegetable seed imported into the United States for cleaning need not be labeled to show the information required under paragraph (b) of this section if:

(1) The seed is in bulk;

(2) The seed is in containers and in quantities of 20,000 pounds or more, regardless of the number of lots involved, and the invoice or other records accompanying and pertaining to the seed show that the seed is for cleaning; or

(3) The seed is in containers and in quantities of less than 20,000 pounds, and each container carries a label that bears the words "Seed for cleaning."

§ 361.4 Inspection at the port of first arrival.

(a) All agricultural seed, vegetable seed, and screenings imported into the United States shall be made available for examination by an APHIS inspector at the port of first arrival and shall remain at the port of first arrival until released by an APHIS inspector. Lots of agricultural seed, vegetable seed, or screenings may enter the United States without meeting the sampling requirements of paragraph (b) of this section if the lot is:

(1) Seed that is not being imported for seeding (planting) purposes and the declaration required by § 361.3(a) states the purpose for which the seed is being imported;

(2) Seed that is being shipped in bond through the United States;

(3) Screenings from seeds of wheat, oats, barley, rye, buckwheat, field corn, sorghum, broomcorn, flax, millet, proso, soybeans, cowpeas, field peas, or field beans that are not being imported for seeding (planting) purposes and the declaration accompanying the screenings as required under § 361.2(a) indicates that the screenings are being imported for processing or manufacturing purposes;

(4) Seed that is being imported for sowing for experimental or breeding purposes, is not for sale, is limited in quantity to the amount indicated in column 3 of table 1 of § 361.5, and is accompanied by a declaration stating the purpose for which it is being imported (seed imported for increase purposes only will not be considered as being imported for experimental or breeding purposes); or

(5) Seed that was grown in the United States, exported, and is now returning to the United States, provided that the person importing the seed into the United States furnishes APHIS with the following documentation:

(i) Export documents indicating the quantity of seed and number of containers, the date of exportation from the United States, the distinguishing marks on the containers at the time of exportation, and the name and address of the United States exporter;

(ii) A document issued by a Customs or other government official of the country to which the seed was exported indicating that the seed was not admitted into the commerce of that country; and

(iii) A document issued by a Customs or other government official of the country to which the seed was exported indicating that the seed was not commingled with other seed after being exported to that country.

(b) Except as provided in §§361.5(a)(2) and 361.7, samples will be taken from all agricultural seed and vegetable seed imported into the United States for seeding (planting) purposes prior to being released into the commerce of the United States.

(1) Samples of seed will be taken from each lot of seed in accordance with §361.5 to determine whether any seeds of noxious weeds listed in §361.6(a) are present. If seeds of noxious weeds are present at a level higher than the tolerances set forth in §361.6(b), the lot of seed will be deemed to be adulterated and will be rejected for entry into the United States for seeding (planting) purposes. Once deemed adulterated, the lot of seed must be:

(i) Exported from the United States;

(ii) Destroyed under the monitoring of an APHIS inspector;

(iii) Cleaned under APHIS monitoring at a seed-cleaning facility that is operated in accordance with §361.8(a); or

(iv) If the lot of seed is adulterated with the seeds of a noxious weed listed in §361.6(a)(2), the seed may be allowed entry into the United States for feeding or manufacturing purposes, provided the importer withdraws the original declaration and files a new declaration stating that the seed is being imported for feeding or manufacturing purposes and that no part of the seed will be used for seeding (planting) purposes.

(2) Seed deemed adulterated may not be mixed with any other seed unless the Administrator determines that two or more lots of seed deemed adulterated are of substantially the same quality and origin. In such cases, the Administrator may allow the adulterated lots of seed to be mixed for cleaning as provided in paragraph (b)(1)(iii) of this section.

(3) If the labeling of a lot of seed is false or misleading in any respect, the seed will be rejected for entry into the United States. A falsely labeled lot of seed must be:

(i) Exported from the United States;

(ii) Destroyed under the monitoring of an APHIS inspector; or

(iii) The seed may be allowed entry into the United States if the labeling is corrected under the monitoring of an APHIS inspector to accurately reflect the character of the lot of seed.

§361.5 Sampling of seeds.

(a) *Sample sizes.* As provided in §361.4(b), samples of seed will be taken from each lot of seed being imported for seeding (planting) purposes to determine whether any seeds of noxious weeds listed in §361.6(a) are present. The samples shall be drawn in the manner described in paragraphs (b) and (c) of this section. Unused portions of samples of rare or expensive seeds will be returned by APHIS upon request of the importer.

(1) A minimum sample of not less than 1 quart shall be drawn from each lot of agricultural seed; a minimum sample of not less than 1 pint shall be drawn from each lot of vegetable seed, except that a sample of ¼ pint will be sufficient for a vegetable seed importation of 5 pounds or less. The minimum sample shall be divided repeatedly until a working sample of proper weight has been obtained. If a mechanical divider cannot be used or is not available, the sample shall be thoroughly mixed, then placed in a pile; the pile shall be divided repeatedly into halves until a working sample of the proper weight remains. The weights of the working samples for noxious weed examination for each lot of seed are shown in column 1 of table 1 of this section. If the lot of seed is a mixture, the following methods shall be used to determine the weight of the working sample:

(i) If the lot of seed is a mixture consisting of one predominant kind of seed or a group of kinds of similar size, the weight of the working sample shall be the weight shown in column 1 of table 1 of this section for the kind or group of kinds that comprises more than 50 percent of the sample.

(ii) If the lot of seed is a mixture consisting of two or more kinds or groups of kinds of different sizes, none of which comprises over 50 percent of the sample, the weight of the working sample shall be the weighted average (to the nearest half gram) of the weight shown in column 1 of table 1 of this section for each of the kinds that comprise the sample, as determined by the following method:

(A) Multiply the percentage of each component of the mixture (rounded off to the nearest whole number) by the sample sizes shown in column 1 of table 1 of this section;

(B) Add all these products;

(C) Total the percentages of all components of the mixtures; and

(D) Divide the sum in paragraph (a)(1)(ii)(B) of this section by the total in paragraph (a)(1)(ii)(C) of this section.

(2) It is not ordinarily practical to sample and test small lots of seed offered for entry. The maximum sizes of lots of each kind of seed not ordinarily sampled are shown in column 2 of table 1 of this section.

(3) The maximum sizes of lots of each kind of seed allowed entry without sampling for sowing for experimental or breeding purposes as provided in § 361.4(a)(4) are shown in column 3 of table 1 of this section.

TABLE 1

Name of seed	Working weight for noxious weed examination (grams) (1)	Maximum weight of seed lot not ordinarily sampled (pounds) (2)	Maximum weight of seed lot permitted entry for experimental or breeding purposes without sampling (pounds) (3)
VEGETABLE SEED:			
Artichoke	500	25	50
Asparagus	500	25	50
Asparagusbean	500	25	50
Bean		25	200
Garden	500	100	500
Lima	500	25	200
Runner	500	25	200
Beet	300	25	50
Broadbean	500	25	200
Broccoli	50	5	10
Brussels sprouts	50	5	10
Burdock, great	150	10	50
Cabbage	50	5	10
Cabbage, Chinese	50	5	10
Cabbage, tronchuda	100	5	10
Cantaloupe (see Melon).			
Cardoon	500	25	50
Carrot	50	5	10
Cauliflower	50	5	10
Celeriac	25	5	10
Celery	25	5	10
Chard, Swiss	300	25	50
Chicory	50	5	10
Chives	50	5	10
Citron	500	25	50
Collards	50	5	10
Corn, sweet	500	25	200
Cornsalad	50	5	10
Cowpea	500	25	200
Cress, garden	50	5	10
Cress, upland	35	5	10
Cress, water	25	5	10
Cucumber	500	25	50
Dandelion	35	5	10
Dill	50	5	10
Eggplant	50	5	10
Endive	50	5	10
Gherkin, West India	160	25	50
Kale	50	5	10
Kale, Chinese	50	5	10
Kale, Siberian	80	5	10
Kohlrabi	50	5	10
Leek	50	5	10
Lettuce	50	5	10
Melon	500	25	50
Mustard, India	50	25	100

TABLE 1—Continued

Name of seed	Working weight for noxious weed examination (grams) (1)	Maximum weight of seed lot not ordinarily sampled (pounds) (2)	Maximum weight of seed lot permitted entry for experimental or breeding purposes without sampling (pounds) (3)
Mustard, spinach	50	5	10
Okra	500	25	50
Onion	50	5	10
Onion, Welsh	50	5	10
Pak-choi	50	5	10
Parsley	50	5	10
Parsnip	50	5	10
Pea	500	25	200
Pepper	150	5	10
Pumpkin	500	25	50
Radish	300	25	50
Rhubarb	300	5	10
Rutabaga	50	5	10
Sage	150	25	50
Salsify	300	25	50
Savory, summer	35	5	10
Sorrel	35	5	10
Soybean	500	25	200
Spinach	150	25	50
Spinach, New Zealand	500	25	50
Squash	500	25	50
Tomato	50	5	10
Tomato, husk	35	5	10
Turnip	50	5	10
Watermelon	500	25	50
AGRICULTURAL SEED:			
Agrotricum	500	100	500
Alfalfa	50	25	100
Alfilaria	50	25	100
Alyceclover	50	25	100
Bahiagrass	50	25	100
Barrelclover	100	25	100
Barley	500	100	500
Bean, adzuki	500	100	500
Bean, field	500	100	500
Bean, mung	500	100	500
Bean (see Velvetbean).			
Beet, field	500	100	500
Beet, sugar	500	100	1,000
Beggarweed	50	25	100
Bentgrass, colonial	2.5	25	100
Bentgrass, creeping	2.5	25	100
Bentgrass, velvet	2.5	25	100
Bermudagrass	10	25	100
Bermudagrass, giant	10	25	100
Bluegrass, annual	10	25	100
Bluegrass, bulbous	40	25	100
Bluegrass, Canada	5	25	100
Bluegrass, glaucantha	10	25	100
Bluegrass, Kentucky	10	25	100
Bluegrass, Nevada	10	25	100
Bluegrass, rough	5	25	100
Bluegrass, Texas	10	25	100
Bluegrass, wood	5	25	100
Bluejoint	5	25	100
Bluestem, big	70	25	100
Bluestem, little	50	25	100
Bluestem, sand	100	25	100
Bluestem, yellow	10	25	100
Bottlebrush-squirreltail	90	25	100
Brome, field	50	25	100
Brome, meadow	130	25	100
Brome, mountain	200	25	100
Brome, smooth	70	25	100
Broomcorn	400	100	500

419

TABLE 1—Continued

Name of seed	Working weight for noxious weed examination (grams) (1)	Maximum weight of seed lot not ordinarily sampled (pounds) (2)	Maximum weight of seed lot permitted entry for experimental or breeding purposes without sampling (pounds) (3)
Buckwheat	500	100	500
Buffalograss:			
(Burs)	200	25	100
(Caryopses)	30	25	100
Buffelgrass:			
(Fascicles)	66	25	100
(Caryopses)	20	25	100
Burclover, California:			
(In bur)	500	100	500
(Out of bur)	70	25	100
Burclover, spotted:			
(In bur)	500	100	500
(Out of bur)	50	25	100
Burnet, little	250	25	100
Buttonclover	70	25	100
Canarygrass	200	25	100
Canarygrass, reed	20	25	100
Carpetgrass	10	25	100
Castorbean	500	100	500
Chess, soft	50	25	100
Chickpea	500	100	500
Clover, alsike	20	25	100
Clover, arrowleaf	40	25	100
Clover, berseem	50	25	100
Clover, cluster	10	25	100
Clover, crimson	100	25	100
Clover, Kenya	20	25	100
Clover, Ladino	20	25	100
Clover, Lappa	20	25	100
Clover, large hop	10	25	100
Clover, Persian	20	25	100
Clover, red	50	25	100
Clover, rose	70	25	100
Clover, small hop (suckling)	20	25	100
Clover, strawberry	50	25	100
Clover, sub (subterranean)	250	25	100
Clover, white	20	25	100
Corn, field	500	100	1,000
Corn, pop	500	100	1,000
Cotton	500	100	500
Cowpea	500	100	500
Crambe	250	25	100
Crested dogtail	20	25	100
Crotalaria, lance	70	25	100
Crotalaria, showy	250	25	100
Crotalaria, slenderleaf	100	25	100
Crotalaria, striped	100	25	100
Crotalaria, Sunn	500	25	100
Crownvetch	100	25	100
Dallisgrass	40	25	100
Dichondra	50	25	100
Dropseed, sand	2.5	25	100
Emmer	500	100	500
Fescue, Chewings	30	25	100
Fescue, hair	10	25	100
Fescue, hard	20	25	100
Fescue, meadow	50	25	100
Fescue, red	30	25	100
Fescue, sheep	20	25	100
Fescue, tall	50	25	100
Flax	150	25	100
Galletagrass:			
(Other than caryopses)	100	25	100
(Caryopses)	50	25	100
Grama, blue	20	25	100
Grama, side-oats:			
(Other than caryopses)	60	25	100

TABLE 1—Continued

Name of seed	Working weight for noxious weed examination (grams) (1)	Maximum weight of seed lot not ordinarily sampled (pounds) (2)	Maximum weight of seed lot permitted entry for experimental or breeding purposes without sampling (pounds) (3)
(Caryopses)	20	25	100
Guar	500	25	100
Guineagrass	20	25	100
Hardinggrass	30	25	100
Hemp	500	100	500
Indiangrass, yellow	70	25	100
Indigo, hairy	70	25	100
Japanese lawngrass	20	25	100
Johnsongrass	100	25	100
Kenaf	500	100	500
Kochia, forage	20	25	100
Kudzu	250	25	100
Lentil	500	25	100
Lespedeza, Korean	50	25	100
Lespedeza, sericea or Chinese	30	25	100
Lespedeza, Siberian	30	25	100
Lespedeza, striate	50	25	100
Lovegrass, sand	10	25	100
Lovegrass, weeping	10	25	100
Lupine, blue	500	100	500
Lupine, white	500	100	500
Lupine, yellow	500	100	500
Manilagrass	20	25	100
Meadow foxtail	30	25	100
Medick, black	50	25	100
Milkvetch	90	25	100
Millet, browntop	80	25	100
Millet, foxtail	50	25	100
Millet, Japanese	90	25	100
Millet, pearl	150	25	100
Millet, proso	150	25	100
Molassesgrass	5	25	100
Mustard, black	20	25	100
Mustard, India	50	25	100
Mustard, white	150	25	100
Napiergrass	50	25	100
Needlegrass, green	70	25	100
Oat	500	100	500
Oatgrass, tall	60	25	100
Orchardgrass	30	25	100
Panicgrass, blue	20	25	100
Panicgrass, green	20	25	100
Pea, field	500	100	500
Peanut	500	100	500
Poa trivialis (see bluegrass, rough)			
Rape, annual	70	25	100
Rape, bird	70	25	100
Rape, turnip	50	25	100
Rape, winter	100	25	100
Redtop	2.5	25	100
Rescuegrass	200	25	100
Rhodesgrass	10	25	100
Rice	500	100	500
Ricegrass, Indian	70	25	100
Roughpea	500	100	500
Rye	500	100	500
Rye, mountain	280	25	100
Ryegrass, annual	50	25	100
Ryegrass, intermediate	80	25	100
Ryegrass, perennial	50	25	100
Ryegrass, Wimmera	50	25	100
Safflower	500	100	500
Sagewort, Louisiana	5	25	100
Sainfoin	500	100	500
Saltbush, fourwing	150	25	100
Seasame	70	25	100

TABLE 1—Continued

Name of seed	Working weight for noxious weed examination (grams) (1)	Maximum weight of seed lot not ordinarily sampled (pounds) (2)	Maximum weight of seed lot permitted entry for experimental or breeding purposes without sampling (pounds) (3)
Sesbania	250	25	100
Smilo	20	25	100
Sorghum	500	100	1,000
Sorghum almum	150	25	100
Sorghum-sudangrass hybrid	500	100	1,000
Sorgrass	150	25	100
Sourclover	50	25	100
Soybean	500	100	500
Spelt	500	100	500
Sudangrass	250	25	100
Sunflower	500	100	500
Sweetclover, white	50	25	100
Sweetclover, yellow	50	25	100
Sweet vernalgrass	20	25	100
Sweetvetch, northern	190	25	100
Switchgrass	40	25	100
Timothy	10	25	100
Timothy, turf	10	25	100
Tobacco	5	1	1
Trefoil, big	20	25	100
Trefoil, birdsfoot	30	25	100
Triticale	500	100	500
Vaseygrass	30	25	100
Veldtgrass	40	25	100
Velvetbean	500	100	500
Velvetgrass	10	25	100
Vetch, common	500	100	500
Vetch, hairy	500	100	500
Vetch, Hungarian	500	100	500
Vetch, Monantha	500	100	500
Vetch, narrowleaf	500	100	500
Vetch, purple	500	100	500
Vetch, woolypod	500	100	500
Wheat, common	500	100	500
Wheat, club	500	100	500
Wheat, durum	500	100	500
Wheat, Polish	500	100	500
Wheat, poulard	500	100	500
Wheat × Agrotricum	500	100	500
Wheatgrass, beardless	80	25	100
Wheatgrass, fairway crested	40	25	100
Wheatgrass, standard crested	50	25	100
Wheatgrass, intermediate	150	25	100
Wheatgrass, pubescent	150	25	100
Wheatgrass, Siberian	50	25	100
Wheatgrass, slender	70	25	100
Wheatgrass, streambank	50	25	100
Wheatgrass, tall	150	25	100
Wheatgrass, western	100	25	100
Wildrye, basin	80	25	100
Wild-rye, Canada	110	25	100
Wild-rye, Russian	60	25	100
Zoysia Japonica (see Japanese lawngrass)			
Zoysia matrella (see Manilagrass)			

(b) *Method of sampling.* (1) When an importation consists of more than one lot, each lot shall be sampled separately.

(2) For lots of six or fewer bags, each bag shall be sampled. A total of at least five trierfuls shall be taken from the lot.

(3) For lots of more than six bags, five bags plus at least 10 percent of the number of bags in the lot shall be sampled. (Round off numbers with decimals

to the nearest whole number, raising 0.5 to the next whole number.) Regardless of the lot size, it is not necessary to sample more than 30 bags.

(4) When the lot of seed to be sampled is comprised of seed in small containers that cannot practically be sampled as described in paragraph (b)(2) or (b)(3) of this section, entire unopened containers may be taken in sufficient number to supply a sample that meets the minimum size requirements of paragraph (a)(1) of this section.

(c) *Drawing samples.* Samples will not be drawn unless each container is labeled to show the lot designation and the name of the kind and variety of each agricultural seed, or kind and variety of each vegetable seed, appearing on the invoice and other entry papers, and a declaration has been filed by the importer as required under §361.2(a). In order to secure a representative sample, an APHIS inspector will draw equal portions from evenly distributed parts of the quantity of seed to be sampled; the APHIS inspector, therefore, must be given access to all parts of that quantity.

(1) For free-flowing seed in bags or in bulk, a probe or trier shall be used. For small free-flowing seed in bags, a probe or trier long enough to sample all portions of the bag shall be used. When drawing more than one trierful of seed from a bag, a different path through the seed shall be used when drawing each sample.

(2) For non-free-flowing seed in bags or bulk that may be difficult to sample with a probe or trier, samples shall be obtained by thrusting one's hand into the seed and withdrawing representative portions. The hand shall be inserted in an open position with the fingers held closely together while the hand is being inserted and the portion withdrawn. When more than one handful is taken from a bag, the handfuls shall be taken from well-separated points.

(3) When more than one sample is drawn from a single lot, the samples may be combined into a composite sample unless it appears that the quantity of seed represented as a lot is not of uniform quality, in which case the separate samples shall be forwarded together, but without being combined into a composite sample.

(d) In most cases, samples will be drawn and examined by an APHIS inspector at the port of first arrival. The APHIS inspector may release a shipment if no contaminants are found and the labeling is sufficient. If contaminants are found or the labeling of the seed is insufficient, the APHIS inspector may forward the sample to the USDA Seed Examination Facility (SEF), Beltsville, MD, for analysis, testing, or examination. APHIS will notify the owner or consignee of the seed that samples have been drawn and forwarded to the SEF and that the shipment must be held intact pending a decision by APHIS as to whether the seed is within the noxious weed seed tolerances of §361.6 and is accurately labeled. If the decision pending is with regard to the noxious weed seed content of the seed and the seed has been determined to be accurately labeled, the seed may be released for delivery to the owner or consignee under the following conditions:

(1) The owner or consignee executes with Customs either a Customs single-entry bond or a Customs term bond, as appropriate, in such amount as is prescribed by applicable Customs regulations;

(2) The bond must contain a condition for the redelivery of the seed or any part thereof upon demand of the Port Director of Customs at any time;

(3) Until the seed is approved for entry upon completion of APHIS' examination, the seed must be kept intact and not tampered with in any way, or removed from the containers except under the monitoring of an APHIS inspector; and

(4) The owner or consignee must keep APHIS informed as to the location of the seed until it is finally entered into the commerce of the United States.

§361.6 Noxious weed seeds.

(a) Seeds of the plants listed in paragraphs (a)(1) and (a)(2) of this section shall be considered noxious weed seeds.

(1) Seeds with no tolerances applicable to their introduction:

Acacia nilotica (Linnaeus) Wildenow ex Delile
Aeginetia spp.

Ageratina adenophora (Sprengel) King & Robinson
Ageratina riparia (Regel) R.M. King and H. Robinson
Alectra spp.
Alternanthera sessilis (L.) R. Brown ex de Candolle
Arctotheca calendula (Linnaeus) Levyns
Asphodelus fistulosus L.
Avena sterilis L. (including *Avena ludoviciana* Durieu)
Azolla pinnata R. Brown
Carthamus oxyacantha M. Bieberstein
Chrysopogon aciculatus (Retzius) Trinius
Commelina benghalensis L.
Crupina vulgaris Cassini
Cuscuta spp.
Digitaria abyssinica (Hochstetter ex A. Richard) Stapf
Digitaria velutina (Forsskal) Palisot de Beauvois
Drymaria arenariodes Humboldt & Bonpland ex J.A. Schultes
Eichhornia azurea (Swartz) Kunth
Emex australis Steinheil
Emex spinosa (L.) Campdera
Euphorbia terracina Linnaeus
Galega officinalis L.
Heracleum mantegazzianum Sommier & Levier
Hydrilla verticillata (Linnaeus f.) Royle
Hygrophila polysperma T. Anderson
Imperata brasiliensis Trinius
Imperata cylindrica (Linnaeus) Palisot de Beauvois
Inula britannica Linnaeus
Ipomoea aquatica Forsskal
Ischaemum rugosum Salisbury
Lagarosiphon major (Ridley) Moss
Leptochloa chinensis (L.) Nees
Limnophila sessiliflora (Vahl) Blume
Lycium ferocissimum Miers
Lygodium flexuosum (Linnaeus) Swartz (maidenhair creeper)
Lygodium microphyllum (Cavanilles) R. Brown (Old World climbing fern)
Melaleuca quinquenervia (Cav.) Blake
Melastoma malabathricum L.
Mikania cordata (Burman f.) B. L. Robinson
Mikania micrantha Kunth
Mimosa diplotricha C. Wright
Mimosa pigra L. var. *pigra*
Monochoria hastata (L.) Solms-Laubach
Monochoria vaginalis (Burman f.) C. Presl
Moraea collina Thunberg
Moraea flaccida (Sweet) Steudel
Moraea miniata Andrews
Moraea ochroleuca (Salisbury) Drapiez
Moraea pallida (Baker) Goldblatt
Nassella trichotoma (Nees) Hackel ex Arechavaleta
Onopordum acaulon Linnaeus
Onopordum illyricum Linnaeus
Opuntia aurantiaca Lindley
Orobanche spp.
Oryza longistaminata A. Chevalier & Roehrich
Oryza punctata Kotschy ex Steudel
Oryza rufipogon Griffith

Ottelia alismoides (L.) Pers.
Paspalum scrobiculatum L.
Pennisetum clandestinum Hochstetter ex Chiovenda
Pennisetum macrourum Trinius
Pennisetum pedicellatum Trinius
Pennisetum polystachion (L.) Schultes
Prosopis alapataco R. A. Philippi
Prosopis argentina Burkart
Prosopis articulata S. Watson
Prosopis burkartii Munoz
Prosopis caldenia Burkart
Prosopis calingastana Burkart
Prosopis campestris Grisebach
Prosopis castellanosii Burkart
Prosopis denudans Bentham
Prosopis elata (Burkart) Burkart
Prosopis farcta (Banks & Solander) J.F. Macbride
Prosopis ferox Grisebach
Prosopis fiebrigii Harms
Prosopis hassleri Harms
Prosopis humilis Gillies ex Hooker & Arnott
Prosopis kuntzei Harms
Prosopis pallida (Humboldt & Bonpland ex Willdenow) Kunth
Prosopis palmeri S. Watson
Prosopis reptans Bentham var. *reptans*
Prosopis rojasiana Burkart
Prosopis ruizlealii Burkart
Prosopis ruscifolia Grisebach
Prosopis sericantha Gillies ex Hooker & Arnott
Prosopis strombulifera (Lamarck) Bentham
Prosopis torquata (Cavanilles ex Lagasca y Segura) de Candolle
Rottboellia cochinchinensis (Lour.) W. Clayon
Rubus fruticosus L. (complex)
Rubus moluccanus L.
Saccharum spontaneum L.
Sagittaria sagittifolia L.
Salsola vermiculata L.
Salvinia auriculata Aublet
Salvinia biloba Raddi
Salvinia herzogii de la Sota
Salvinia molesta D.S. Mitchell
Senecio inaequidens DC.
Senecio madagascariensis Poir.
Setaria pumila (Poir.) Roem. & Schult. subsp. *pallidefusca* (Schumach.) B.K. Simon
Solanum tampicense Dunal (wetland nightshade)
Solanum torvum Swartz
Solanum viarum Dunal
Sparganium erectum L.
Spermacoce alata Aublet
Striga spp.
Tridax procumbens L.
Urochloa panicoides Beauvois

(2) Seeds with tolerances applicable to their introduction:

Acroptilon repens (L.) DC. (=*Centaurea repens* L.) (=*Centaurea picris*)
Cardaria draba (L.) Desv.
Cardaria pubescens (C. A. Mey.) Jarmol.

Convolvulus arvensis L.
Cirsium arvense (L.) Scop.
Elytrigia repens (L.) Desv. (=*Agropyron repens* (L.) Beauv.)
Euphorbia esula L.
Sonchus arvensis L.
Sorghum halepense (L.) Pers.

(b) The tolerance applicable to the prohibition of the noxious weed seeds listed in paragraph (a)(2) of this section shall be two seeds in the minimum amount required to be examined as shown in column 1 of table 1 of §361.5. If fewer than two seeds are found in an initial examination, the shipment from which the sample was drawn may be entered. If two seeds are found in an initial examination, a second sample must be examined. If two or fewer seeds are found in the second examination, the shipment from which the samples were drawn may be entered. If three or more seeds are found in the second examination, the shipment from which the samples were drawn may not be entered. If three or more seeds are found in an initial examination, the shipment from which the sample was drawn may not be entered.

(c) Any seed of any noxious weed that can be determined by visual inspection (including the use of transmitted light or dissection) to be within one of the following categories shall be considered inert matter and not counted as a weed seed:

(1) Damaged seed (other than grasses) with over one half of the embryo missing;

(2) Grass florets and caryopses classed as inert:

(i) Glumes and empty florets of weedy grasses;

(ii) Damaged caryopses, with over one-half the root-shoot axis missing (the scutellum excluded);

(iii) Immature free caryopses devoid of embryo or endosperm;

(iv) Free caryopses of quackgrass (*Elytrigia repens*) that are 2 mm or less in length; or

(v) Immature florets of quackgrass (*Elytrigia repens*) in which the caryopses are less than one-third the length of the palea. The caryopsis is measured from the base of the rachilla.

(3) Seeds of legumes (*Fabaceae*) with the seed coats entirely removed.

(4) Immature seed units, devoid of both embryo and endosperm, such as occur in (but not limited to) the following plant families: buckwheat (*Polygonaceae*), morning glory (*Convolvulaceae*), nightshade (*Solanaceae*), and sunflower (*Asteraceae*).

(5) Dodder (*Cuscuta* spp.) seeds devoid of embryos and seeds that are ashy gray to creamy white in color are inert matter. Dodder seeds should be sectioned when necessary to determine if an embryo is present, as when the seeds have a normal color but are slightly swollen, dimpled, or have minute holes.

[62 FR 48460, Sept. 16, 1997, as amended at 64 FR 12884, Mar. 16, 1999; 65 FR 33743, May 25, 2000; 71 FR 35381, June 20, 2006; 74 FR 53400, Oct. 19, 2009; 75 FR 68956, Nov. 10, 2010]

§361.7 Special provisions for Canadian-origin seed and screenings.

(a) In addition to meeting the declaration and labeling requirements of §361.2 and all other applicable provisions of this part, all Canadian-origin agricultural seed and Canadian-origin vegetable seed imported into the United States from Canada for seeding (planting) purposes or cleaning must be accompanied by a certificate of analysis issued by the Canadian Food Inspection Agency or by a private seed laboratory accredited by the Canadian Food Inspection Agency. Samples of seed shall be drawn using sampling methods comparable to those detailed in §361.5 of this part. The seed analyst who examines the seed at the laboratory must be accredited to analyze the kind of seed covered by the certificate.

(1) If the seed is being imported for seeding (planting) purposes, the certificate of analysis must verify that the seed meets the noxious weed seed tolerances of §361.6. Such seed will not be subject to the sampling requirements of §361.3(b).

(2) If the seed is being imported for cleaning, the certificate of analysis must name the kinds of noxious weed seeds that are to be removed from the lot of seed. Seed being imported for cleaning must be consigned to a facility operated in accordance with §361.8(a).

(b) Coated or pelleted agricultural seed and coated or pelleted vegetable seed of Canadian origin may be imported into the United States if the seed was analyzed prior to being coated or pelleted and is accompanied by a certificate of analysis issued in accordance with paragraph (a) of this section.

(c) Screenings otherwise prohibited under this part may be imported from Canada if the screenings are imported for processing or manufacture and are consigned to a facility operating under a compliance agreement as provided by § 361.8(b).

(Approved by the Office of Management and Budget under control number 0579–0124)

§ 361.8 Cleaning of imported seed and processing of certain Canadian-origin screenings.

(a) Imported seed that is found to contain noxious weed seeds at a level higher than the tolerances set forth in § 361.6(b) may be cleaned under the monitoring of an APHIS inspector. The cleaning will be at the expense of the owner or consignee.

(1) At the location where the seed is being cleaned, the identity of the seed must be maintained at all times to the satisfaction of the Administrator. The refuse from the cleaning must be placed in containers and securely sealed and identified. Upon completion of the cleaning, a representative sample of the seed will be analyzed by a registered seed technologist, an official seed laboratory, or by APHIS; if the seed is found to be within the noxious weed tolerances set forth in § 361.6(b), the seed may be allowed entry into the United States;

(2) The refuse from the cleaning must be destroyed under the monitoring of an APHIS inspector at the expense of the owner or consignee of the seed.

(3) Any person engaged in the business of cleaning imported seed may enter into a compliance agreement under paragraph (c) of this section to facilitate the cleaning of seed imported into the United States under this part.

(b) Any person engaged in the business of processing screenings who wishes to process screenings imported from Canada under § 361.7(c) that are otherwise prohibited under this part must enter into a compliance agreement under paragraph (c) of this section.

(c) A compliance agreement for the cleaning of imported seed or processing of otherwise prohibited screenings from Canada shall be a written agreement[1] between a person engaged in such a business, the State in which the business operates, and APHIS, wherein the person agrees to comply with the provisions of this part and any conditions imposed pursuant thereto. Any compliance agreement may be canceled orally or in writing by the APHIS inspector who is monitoring its enforcement whenever the inspector finds that the person who entered into the compliance agreement has failed to comply with the provisions of this part or any conditions imposed pursuant thereto. If the cancellation is oral, the decision and the reasons for the decision shall be confirmed in writing, as promptly as circumstances permit. Any person whose compliance agreement has been canceled may appeal the decision to the Administrator, in writing, within 10 days after receiving written notification of the cancellation. The appeal shall state all of the facts and reasons upon which the person relies to show that the compliance agreement was wrongfully canceled. The Administrator shall grant or deny the appeal, in writing, stating the reasons for such decision, as promptly as circumstances permit. If there is a conflict as to any material fact, a hearing shall be held to resolve such conflict. Rules of practice concerning such a hearing will be adopted by the Administrator.

§ 361.9 Recordkeeping.

(a) Each person importing agricultural seed or vegetable seed under this part must maintain a complete record, including copies of the declaration and labeling required under this part and a sample of seed, for each lot of seed imported. Except for the seed sample, which may be discarded 1 year after the entire lot represented by the sample has been disposed of by the person

---

[1] Compliance Agreement forms are available without charge from Permit Unit, PPQ, APHIS, 4700 River Road Unit 136, Riverdale, MD 20737–1236, and from local offices of the Plant Protection and Quarantine. (Local offices are listed in telephone directories).

who imported the seed, the records must be maintained for 3 years following the importation.

(b) Each sample of vegetable seed and each sample of agricultural seed must be at least equal in weight to the sample size prescribed for noxious weed seed examination in table 1 of § 361.5.

(c) An APHIS inspector shall, during normal business hours, be allowed to inspect and copy the records.

(Approved by the Office of Management and Budget under control number 0579–0124)

§ 361.10 Costs and charges.

Unless a user fee is payable under § 354.3 of this chapter, the services of an APHIS inspector during regularly assigned hours of duty and at the usual places of duty will be furnished without cost. The U.S. Department of Agriculture's provisions relating to overtime charges for an APHIS inspector's services are set forth in part 354 of this chapter. The U.S. Department of Agriculture will not be responsible for any costs or charges incident to inspections or compliance with this part, other than for the services of the APHIS inspector during regularly assigned hours of duty and at the usual places of duty. All expenses incurred by the U.S. Department of Agriculture (including travel, per diem or subsistence, and salaries of officers or employees of the Department) in connection with the monitoring of cleaning, labeling, other reconditioning, or destruction of seed, screenings, or refuse under this part shall be reimbursed by the owner or consignee of the seed or screenings.

PART 370—FREEDOM OF INFORMATION

AUTHORITY: 5 U.S.C. 552.

SOURCE: 40 FR 43223, Sept. 19, 1975, unless otherwise noted.

§ 370.1 Scope and purpose.

These regulations are issued pursuant to the Freedom of Information Act, as amended (5 U.S.C. 552), and in accordance with the requirements of the Department of Agriculture regulations in part 1, subpart A of this title. The availability of records of the Animal and Plant Health Inspection Service (APHIS), and the procedures by which the public may obtain such information, shall be governed by the Department regulations as implemented by the regulations in this part. It is the policy of APHIS to be an open agency and to promptly make available for public inspection any records or information which are required to be released under the Act. Material which is exempt from disclosure will also be promptly made available when the Agency in its discretion determines that release of such material is in the public interest.

§ 370.2 Published materials.

Rules and regulations of APHIS relating to its regulatory responsibilities are continuously published in the FEDERAL REGISTER, and codified in this chapter III, title 7, and in 9 CFR chapter I. APHIS issues publications explaining animal and plant health programs and the laws and regulations, including quarantines, under which the programs are conducted. These publications are, for the most part available free from the Office of Governmental and Public Affairs, USDA, Washington, DC 20250; or, in some cases from the Superintendent of Documents, U.S. Government Printing Office, Washington, DC 20402, at established rates.

[44 FR 53490, Sept. 14, 1979]

§ 370.3 Index.

Pursuant to the regulations in § 1.4(b) of this title, APHIS will maintain and make available for public inspection and copying a current index providing identifying information regarding the materials required to be published or made available under the Freedom of Information Act (5 U.S.C. 552(a)(2)). Notice is hereby given that publication of this index is unnecessary and impracticable, since the material is voluminous and does not change often enough to justify the expense of publication.

§ 370.4 Facilities for inspection and copying.

Facilities for public inspection and copying of the index and materials required to be made available under 5 U.S.C. 552(c)(2) will be provided by APHIS, on business days between 8 a.m. and 4:30 p.m. Requests for this information should be made to the FOIA Coordinator at the following address:

Freedom of Information Act Coordinator, Animal and Plant Health Inspection Service, Legislative and Public Affairs, Freedom of Information, 4700 River Road, Unit 50, Riverdale, Maryland 20737–1231.

Copies of such material may be obtained in person or by mail. Applicable fees for copies will be charged in accordance with the regulations prescribed by the Office of Operations and Finance, USDA, pursuant to § 2.75 of this title. See § 1.10 and appendix A— Fee Schedule in part 1, subtitle A of this title.

[44 FR 53490, Sept. 14, 1979, as amended at 51 FR 30836, Aug. 29, 1986; 59 FR 67611, Dec. 30, 1994]

§ 370.5 Requests for records.

(a) Requests for APHIS records or information other than material published or made available under the preceding sections, shall be made in writing in accordance with 7 CFR 1.3(a) and submitted to the APHIS Freedom of Information Act Coordinator at the following address:

Freedom of Information Act Coordinator, (FOIA Request), Animal and Plant Health Inspection Service, Legislative and Public Affairs, Freedom of Information, 4700 River Road, Unit 50, Riverdale, Maryland 20737–1231.

The request shall identify each record with reasonable specificity as prescribed in § 1.3(b) of this title. The APHIS FOIA Coordinator is hereby delegated authority to make determinations with respect to such requests in accordance with 7 CFR.

(b) The FOIA Coordinator or his designee is authorized to receive requests and to exercise the authority under § 1.4(c) of this title to:

(1) Make determinations to grant or deny requests,

(2) Extend the administrative deadline,

(3) Make discretionary releases of exempt records, and

(4) Make determinations regarding charges pursuant to the fee schedule.

(c) In exercising his authority under § 1.4(c) of this title to grant and deny requests, the Coordinator will comply with subsection (b) of the Freedom of Information Act, as amended (5 U.S.C. 552(b)), which requires that any reasonably segregable portion of a document shall be provided to a person requesting such document after deletion of any portions which are exempt under the Act. Therefore, unless the disclosable and non-disclosable portions are so inextricably linked that it is not reasonably possible to separate them, the document will be released with the non-disclosable portions deleted, except that the Coordinator may exercise discretion as limited by § 1.11 of this title, to release the entire document, or to make only a minimum number of deletions, e.g., the names of individuals.

[40 FR 43223, Sept. 19, 1975, as amended at 44 FR 53490, Sept. 14, 1979; 51 FR 30837, Aug. 29, 1986; 59 FR 67611, Dec. 30, 1994]

§ 370.6 Appeals.

If the request for information made under § 370.5 is denied in whole or in part, the requester may file an appeal pursuant to § 1.3(e) of this title. The appeal should be in writing and should be addressed as follows:

Administrator, Animal and Plant Health Inspection Service (FOIA Appeal), Room 313–E, U.S. Department of Agriculture, Washington, DC 20250.

[44 FR 53490, Sept. 14, 1979]

§ 370.7 Agency response to requests.

(a) The response to requests for information and to appeals shall be made in accordance with the Department regulations in § 1.5 of this title and the regulations in this part.

(b) Requests for records and information which have customarily been directed to field stations and agency

428

headquarters may continue to be directed to those locations, notwithstanding the provisions of these regulations. If the information is not available at the location at which the request is made, or the official receiving the request is in doubt as to whether the information should be released, the official shall (1) promptly forward the request to the FOIA Coordinator, or (2) inform the requester of the procedures established in these regulations by which the request may be sent directly to the FOIA Coordinator. The date of receipt of the request by the Coordinator shall be the determining date for purposes of the time limitations under the Freedom of Information Act and the regulations.

PART 371—ORGANIZATION, FUNCTIONS, AND DELEGATIONS OF AUTHORITY

Sec.
371.1   General Statement.
371.2   The Office of the Administrator.
371.3   Plant Protection and Quarantine.
371.4   Veterinary Services.
371.5   Marketing and Regulatory Programs Business Services.
371.6   Wildlife Services.
371.7   Animal Care.
371.8   International Services.
371.9   Policy and Program Development.
371.10 Legislative and Public Affairs.
371.11 Delegations of authority.
371.12 Concurrent authority and responsibility to the Administrator.
371.13 Reservation of authority.
371.14 Availability of information and records.

AUTHORITY: 5 U.S.C. 301.

SOURCE: 65 FR 1299, Jan. 10, 2000, unless otherwise noted.

§371.1   General statement.

(a) *The creation of APHIS.* The Animal and Plant Health Inspection Service (APHIS) was created by the Secretary of Agriculture on April 2, 1972 (37 FR 6327, March 28, 1972).

(b) *Central offices.* APHIS is headquartered in Washington, DC, and Riverdale, MD. The APHIS Management Team at these locations consists of the following:

Administrator
Associate Administrator

Deputy Administrator, Plant Protection and Quarantine (PPQ)
Deputy Administrator, Veterinary Services (VS)
Deputy Administrator, Marketing and Regulatory Programs Business Services (MRPBS)
Deputy Administrator, Wildlife Services (WS)
Deputy Administrator, Animal Care (AC)
Deputy Administrator, International Services (IS)
Director, Policy and Program Development (PPD)
Director, Legislative and Public Affairs (LPA)

(c) *Field organization.* AC, MRPBS, PPQ, VS, and WS all have field offices located throughout the United States. IS has field offices located throughout the world. A list of APHIS' field offices with addresses and telephone numbers is in the blue pages of local telephone books.

§371.2   The Office of the Administrator.

(a) *The Administrator.* (1) The Administrator of APHIS formulates, directs, and supervises the execution of APHIS policies, programs, and activities.

(2) The Administrator is authorized to take any action authorized by law and deemed necessary to carry out APHIS functions. Delegations of authority by the Administrator and provisions for redelegations of authority are stated in §371.11.

(b) *The Associate Administrator.* The Associate Administrator of APHIS shares responsibility with the Administrator for general direction and supervision of APHIS programs and activities. The Associate Administrator may act for the Administrator.

§371.3   Plant protection and quarantine.

(a) *General statement.* Plant Protection and Quarantine (PPQ) protects and safeguards the Nation's plant resources through programs and activities to prevent the introduction and spread of plant pests and diseases.

(b) *Deputy Administrator of PPQ.* The Deputy Administrator of PPQ is responsible for:

(1) Participating with the Administrator of APHIS and other officials in

the planning and formulation of policies, programs, procedures, and activities of APHIS.

(2) Providing direction and coordination for PPQ programs and activities. The authorities for PPQ programs include:

(i) The Terminal Inspection Act, as amended (7 U.S.C. 166);

(ii) The Honeybee Act, as amended (7 U.S.C. 281 through 286);

(iii) Sections 1 and 15 of the Federal Noxious Weed Act of 1974, (7 U.S.C. 2801 note and 7 U.S.C. 2814);

(iv) The Endangered Species Act of 1973 (16 U.S.C. 1531–1544);

(v) Executive Order 13112;

(vi) The responsibilities of the United States under the International Plant Protection Convention;

(vii) Lacey Act Amendments of 1981, as amended (16 U.S.C. 3371 through 3378);

(viii) Title III (and Title IV to the extent that it relates to activities under Title III) of the Federal Seed Act, as amended (7 U.S.C. 1581 through 1610);

(ix) Authority to prescribe and collect fees under The Act of August 31, 1951, as amended (31 U.S.C. 9701), and sections 2508 and 2509 of the Food, Agriculture, Conservation, and Trade Act of 1990, as amended (21 U.S.C. 136 and 136a);

(x) Plant Protection Act, as amended (7 U.S.C. 7701–7786).

(xi) Authority to collect reimbursement for overtime paid to employees for inspection or quarantine services (7 U.S.C. 2260).

(xii) Title V of the Agricultural Risk Protection Act of 2000 (7 U.S.C. 2279e and 2279f).

(xiii) title II, Subtitle B, of the Public Health Security and Bioterrorism Preparedness and Response Act of 2002 (7 U.S.C. 8401 not and 8401).

(3) Developing of regulations (including quarantines) regarding noxious weeds and plant pests and diseases.

(4) Cooperating with and providing technical assistance to State and local governments, farmer's associations, and individuals with regard to plant pest control. Cooperating with and providing technical assistance to foreign governments with regard to plant pests and diseases.

(5) Assisting in the development of sanitary and phytosanitary measures.

(6) Regulating the field release into the environment, interstate movement, and importation of genetically modified organisms.

(7) Serving as a member of the North American Plant Protection Organization (NAPPO). NAPPO is composed of plant protection officials and industry cooperators from Canada, Mexico, and the United States.

(8) Administering plant and animal pest and disease exclusion policies, procedures, and regulations at international ports of entry (land, sea, and air) relative to all plants and plant and animal products and associated materials (excluding live animals).

(9) Providing laboratory support, diagnostic services, methods development, and research activities in support of PPQ programs.

[65 FR 1299, Jan. 10, 2000, as amended at 65 FR 49471, Aug. 14, 2000; 66 FR 21060, Apr. 27, 2001; 68 FR 27449, May 20, 2003; 70 FR 55706, Sept. 23, 2005]

§ 371.4 Veterinary Services.

(a) *General statement.* Veterinary Services (VS) protects and safeguards the Nation's livestock and poultry through programs and activities to prevent the introduction and spread of pests and disease of livestock and poultry. VS also provides leadership and coordinates activities pertaining to veterinary biologics.

(b) *Deputy Administrator of VS.* The Deputy Administrator of VS is responsible for:

(1) Participating with the Administrator of APHIS and other officials in the planning and formulation of policies, programs, procedures, and activities of APHIS.

(2) Providing direction and coordination for the activities of the Center for Veterinary Biologics.

(3) Providing direction and coordination for VS programs and activities.

The authorities for VS programs include:

(i) Section 18 of the Federal Meat Inspection Act, as amended, as it pertains to the issuance of certificates of condition of live animals intended and offered for export (21 U.S.C. 618).

(ii) 28 Hour Law, as amended (49 U.S.C. 80502);

(iii) Act of August 26, 1983, as amended (46 U.S.C. 3901 through 3902);

(iv) Harmonized Tariff Schedule of the United States;

(v) Virus-Serum-Toxin Act (21 U.S.C. 151 through 159);

(vi) Sections 203 and 205 of the Agricultural Marketing Act of 1946, as amended, with respect to voluntary inspection and certification of animal products; inspection, testing, treatment, and certification of animals; and a program to investigate and develop solutions to the problems resulting from the use of sulfonamides in swine (7 U.S.C. 1622 and 1624);

(vii) Section 101(d) of the Organic Act of September 21, 1944 (7 U.S.C. 430);

(viii) The Swine Health Protection Act (7 U.S.C. 3801 through 3813);

(ix) Conducting diagnostic and related activities necessary to prevent, detect, control, or eradicate foot-and-mouth disease and other animal diseases (21 U.S.C. 113a);

(x) Authority to prescribe and collect fees under the Act of August 31, 1951, as amended (31 U.S.C. 9701), and sections 2508 and 2509 of the Food, Agriculture, Conservation, and Trade Act of 1990, as amended (21 U.S.C. 136 and 136a); and

(xi) Transportation of horses to slaughter under sections 901–905 of the Federal Agriculture Improvement and Reform Act of 1996 (7 U.S.C. 1901 note).

(xii) Animal Health Protection Act (7 U.S.C. 8301–8317).

(xiii) Section 10504 of the Farm Security and Rural Investment Act of 2002 (7 U.S.C. 8318).

(xiv) The responsibilities of the United States related to activities of the Office International des Epizooties.

(xv) Title II, Subtitles B and C, of the Public Health Security and Bioterrorism Preparedness and Response Act of 2002 (7 U.S.C. 8401 note, 8401, 8411).

(4) Directing and coordinating animal health information systems and maintaining a Federal-State program operation capable of responding to exotic livestock and poultry disease outbreaks.

(5) Cooperating with and providing technical assistance to State and local governments, farmer's associations and similar organizations, and individuals with regard to VS programs and activities. Cooperating with and providing technical assistance to foreign governments with regard to pests and diseases of livestock and poultry.

(6) Providing laboratory support, diagnostic services, methods development, and research activities in support of VS programs.

[65 FR 1299, Jan. 10, 2000, as amended at 68 FR 27449, May 20, 2003; 70 FR 55706, Sept. 23, 2005]

§371.5 **Marketing and Regulatory Programs Business Services.**

(a) *General statement.* Marketing and Regulatory Programs Business Services (MRPBS) plans and provides for the agency's human, financial, and physical resources.

(b) *Deputy Administrator of MRPBS.* The Deputy Administrator of MRPBS is responsible for:

(1) Assisting the Under Secretary for Marketing and Regulatory Programs, and the Administrators of APHIS, the Agricultural Marketing Service (AMS), and the Grain Inspection and Packers and Stockyards Administration (GIPSA), and other APHIS, AMS, and GIPSA officials in the planning and formulation of MRP policies, programs, and activities. Providing human resource, certain financial, and management services for AMS, APHIS, and GIPSA.

(2) Planning, formulating and coordinating policies, and directing management support functions for APHIS and designated functions for other MRP agencies, including finance, personnel, and management services.

(3) Conducting administrative reviews and inspections in APHIS to assess the implementation of policies and procedures and to assess the accomplishments of program objectives.

(4) Evaluating and issuing administrative directives.

(5) Serving as APHIS' liaison official with the General Accounting Office and the Office of the Inspector General.

(6) Preparing cooperative agreements, memoranda of understanding, agreements between APHIS and other agencies, and agreements that require the signature of more than one Deputy Administrator or Director.

(7) Directing and coordinating investigations related to APHIS program laws and regulations and coordinating enforcement of program laws and regulations with the Office of the General Counsel.

(8) Supporting and enforcing APHIS program activities, which include:

(i) Title 7, Code of Federal Regulations, §§ 371.3(b)(2)(i) through (xiv);

(ii) Title 7, Code of Federal Regulations, §§ 371.4(b)(3)(i) through (xx);

(iii) The Animal Welfare Act, as amended (7 U.S.C. 2131 through 2159); and

(iv) The Virus-Serum Toxin Act, as amended (21 U.S.C. 159).

(9) Formulating and recommending employee development and training policies.

(10) Developing, delivering, and administering organizational development, training, recruitment, and employee development programs for MRP agencies.

(11) Providing computer support and related services for APHIS.

§ 371.6　Wildlife Services.

(a) *General statement.* Wildlife Services (WS) manages problems caused by wildlife.

(b) *Deputy Administrator of WS.* The Deputy Administrator of WS is responsible for:

(1) Participating with the Administrator of APHIS and other officials in the planning and formulation of policies, programs, procedures, and activities of APHIS.

(2) Providing direction and coordination for programs authorized by the Act of March 2, 1931 (7 U.S.C. 426 and 426b, as amended) and the Act of December 22, 1987 (7 U.S.C. 426c).

(3) Assisting Federal, State, local, and foreign agencies and individuals with regard to wildlife damage and control.

(4) Conducting research to develop wildlife damage management methods.

[65 FR 1299, Jan. 10, 2000, as amended at 69 FR 76379, Dec. 21, 2004]

§ 371.7　Animal Care.

(a) *General statement.* Animal Care (AC) establishes acceptable standards of humane care and treatment for regulated animals and monitors and

achieves compliance through inspections, enforcement, education, and cooperative efforts under the Animal Welfare and Horse Protection Acts.

(b) *Deputy Administrator of AC.* The Deputy Administrator of AC is responsible for:

(1) Participating with the Administrator of APHIS and other officials in the planning and formulation of policies, programs, and activities of APHIS.

(2) Directing activities to ensure compliance with and enforcement of animal welfare and horse protection laws and regulations. These laws are:

(i) The Animal Welfare Act, as amended (7 U.S.C. 2131 through 2159); and

(ii) The Horse Protection Act (15 U.S.C. 1821 through 1831).

(3) Providing recommendations for policy and program changes and promulgating requirements, procedures, and guidelines for the conduct of field activities relating to AC programs.

§ 371.8　International Services.

(a) *General statement.* International Services (IS) protects U.S. agriculture and enhances agricultural trade with foreign countries.

(b) *Deputy Administrator of IS.* The Deputy Administrator of IS is responsible for:

(1) Participating with the Administrator of APHIS and other officials in the planning and formulation of international policies, programs, and activities of APHIS.

(2) Maintaining and administering the foreign service personnel system for employees of APHIS in accordance with section 202(a)(2) of the Foreign Service Act of 1980 (22 U.S.C. 3922), E.O. 12363, dated May 21, 1982, and the provisions of § 2.51(a)(1) of this title.

(3) Developing and maintaining systems for monitoring and reporting the presence and movement of plant and animal diseases and pests in foreign countries.

(4) Developing and maintaining cooperative relationships and programs with other Federal agencies, foreign governments, industry, and international organizations, such as the Food and Agriculture Organization of

the United Nations, with regard to APHIS activities in foreign countries.

(5) Developing and maintaining systems for observing the effects of plant and animal diseases in foreign countries and evaluating their effect on the agriculture industry.

(6) Developing and directing programs to enhance the trade in U.S. plants, animals, and their products in compliance with established international sanitary and phytosanitary standards.

(7) Providing recommendations for policy and program changes, and promulgating requirements, procedures, and guidelines for the conduct of field activities relating to IS programs.

§371.9 Policy and Program Development.

(a) *General statement.* Policy and Program Development (PPD) provides analytical support for agency decisions and plans.

(b) *Director of PPD.* The Director of PPD is responsible for:

(1) Participating with the Administrator of APHIS and other officials in the planning and formulation of APHIS policies, programs, and activities.

(2) Providing planning and evaluations; regulations development; and policy, risk, and economic analysis for APHIS programs.

(3) Analyzing the environmental effects of APHIS programs to ensure their compliance with environmental laws and regulations and providing support for pesticide registration and drug approval.

(4) Coordinating registration of chemicals and other substances used in APHIS control and eradication programs.

§371.10 Legislative and Public Affairs.

(a) *General statement.* Legislative and Public Affairs (LPA) is the communications arm of APHIS.

(b) *Director of LPA.* The Director of LPA is responsible for:

(1) Advising and assisting the Administrator and other officials on matters relating to agency legislative and media affairs.

(2) Preparing legislative proposals for APHIS programs and responsibilities. Assisting in compiling support mate-rial for agency witnesses for congressional hearings. Preparing legislative reports.

(3) Establishing and maintaining liaison with Members of Congress, various congressional committees and subcommittees, and their staffs on matters pertaining to APHIS.

(4) Planning and conducting an information program to promote interest in and increase the public knowledge of APHIS programs and activities.

(5) Drafting and administering policy guidelines on press contacts, photography, audiovisual activities, graphic design, radio-TV, and policy/editorial/graphics clearances for publications. Planning and conducting a program to explain APHIS policies in written form to Members of Congress, State and industry leaders, officials of foreign governments, and private citizens.

(6) Preparing replies to written inquiries and establishing and maintaining a system for the control of written inquiries referred by the Office of the Secretary or sent directly to the agency.

(7) Assisting in the preparation of position papers regarding APHIS programs.

(8) Assisting in the preparation of directives, procedural manuals, articles for publication, and agency correspondence. Coordinating APHIS activities within the scope of the Freedom of Information Act and the Privacy Act.

§371.11 Delegations of authority.

(a) *Associate Administrator.* The Associate Administrator is delegated the authority to perform the duties and to exercise the functions and powers that are now, or that may become, vested in the Administrator, including the power of redelegation except where prohibited, and including authority reserved to the Administrator in §371.14 of this part. The Associate Administrator is also authorized to act for the Administrator in the absence of the Administrator.

(b) *Deputy Administrators and Directors.* The Deputy Administrators of Plant Protection and Quarantine (PPQ), Veterinary Services (VS), Wildlife Services (WS), Marketing and Regulatory Programs Business Services

433

(MRPBS), Animal Care (AC), and International Services (IS); the Directors of Policy and Program Development (PPD) and Legislative and Public Affairs (LPA); and the officers they designate to act for them, with prior specific approval of the Administrator, are delegated the authority, severally, to perform duties and to exercise the functions and powers that are now, or that may become vested in the Administrator (including the power of redelegation, except where prohibited) except authority that is reserved to the Administrator. Each Deputy Administrator or Director shall be responsible for the programs and activities in APHIS assigned to that Deputy Administrator or Director.

§371.12 Concurrent authority and responsibility to the Administrator.

(a) *Delegations that preclude the Administrator or each Deputy Administrator or Director from exercising powers or functions.* No delegation or authorization in this part shall preclude the Administrator or each Deputy Administrator or Director from exercising any of the powers or functions or from performing any of the duties conferred upon each, respectively. Any delegation or authorization is subject, at all times, to withdrawal or amendment by the Administrator, and in their respective fields, by each Deputy Administrator or Director. The officers to whom authority is delegated in this part shall:

(1) Maintain close working relationships with the officers to whom they report.

(2) Keep them advised with respect to major problems and developments.

(3) Discuss with them proposed actions involving major policy questions or other important considerations or questions, including matters involving relationships with other Federal agencies, other agencies of the Department, other divisions, staffs, or offices of the agency, or other governmental, private organizations, or groups.

(b) *Prior authorizations and delegations.* All prior delegations and redelegations of authority relating to any function, program, or activity covered by the statement of Organization, Functions, and Delegations of Authority, shall remain in effect except as they are inconsistent with this part or are amended or revoked. Nothing in this part shall affect the validity of any action taken previously under prior delegations or redelegations of authority or assignments of functions.

§371.13 Reservation of authority.

The following are reserved to the Administrator, or to the individual designated to act for the Administrator:

(a) The initiation, change, or discontinuance of major program activities.

(b) The issuance of regulations pursuant to law.

(c) The transfer of functions between Deputy Administrators and Directors.

(d) The transfer of funds between Deputy Administrators and Directors.

(e) The transfer of funds between work projects within each Deputy Administrator's or Director's area, except those not exceeding 10 percent of base funds or $50,000 in either work project, whichever is less.

(f) The approval of any change in the formal organization, including a section, its equivalent, or higher level.

(g) The making of recommendations to the Department concerning establishment, consolidation, change in location, or abolishment of any regional, State, area, and other field headquarters, and any region or other program area that involves two or more States, or that crosses State lines.

(h) Authority to establish, consolidate, change a location, abolish any field office, or change program area boundaries not included in paragraph (g) of this section.

(i) Approval of all appointments, promotions, and reassignments at the GS–14 level and above.

(j) Authorization for foreign travel and for attendance at foreign and international meetings, including those held in the United States.

(k) Approval of all appointments, promotions, and reassignments of employees to foreign countries.

(l) Approval of program budgets.

(m) Authority to determine the circumstances under which commuted traveltime allowances may be paid to employees performing inspections and necessary auxiliary services after normal working hours or on holidays, when these services come within the scope of the Act of August 28, 1950 (7 U.S.C. 2260).

§ 371.14 Availability of information and records.

Any person desiring information or to comment on the programs and functions of the agency should address correspondence to the appropriate Deputy Administrator or Director, APHIS, U.S. Department of Agriculture, Washington, DC 20250. The availability of information and records of the agency is governed by the rules and regulations in part 370 of this chapter.

PART 372—NATIONAL ENVIRONMENTAL POLICY ACT IMPLEMENTING PROCEDURES

Sec.
372.1   Purpose.
372.2   Designation of responsible APHIS official.
372.3   Information and assistance.
372.4   Definitions.
372.5   Classification of actions.
372.6   Early planning.
372.7   Planning and decision points and public involvement.
372.8   Processing and use of environmental documents.
372.9   Supplementing environmental impact statements.
372.10   Process for rapid response to emergencies.

AUTHORITY: 42 U.S.C. 4321 et seq.; 40 CFR 1500–1508; 7 CFR 1b, 2.22, 2.80, and 371.9.

SOURCE: 60 FR 6002, Feb. 1, 1995, unless otherwise noted.

§ 372.1   Purpose.

These procedures implement section 102(2) of the National Environmental Policy Act (NEPA) by assuring early and adequate consideration of environmental factors in Animal and Plant Health Inspection Service planning and decisionmaking and by promoting the effective, efficient integration of all relevant environmental requirements under the NEPA. The goal of timely, relevant environmental analysis will be secured principally by adhering to the NEPA implementing regulations (40 CFR parts 1500–1508), especially provisions pertaining to timing (§ 1502.5),

integration (§ 1502.25), and scope of analysis (§ 1508.25).

[60 FR 6002, Feb. 1, 1995, as amended at 83 FR 24009, May 24, 2018]

§ 372.2   Designation of responsible APHIS official.

The Administrator of APHIS, or an agency official to whom the Administrator may formally delegate the task, is responsible for overall review of APHIS' NEPA compliance.

§ 372.3   Information and assistance.

Information, including the status of studies, and the availability of reference materials, as well as the informal interpretations of APHIS' NEPA procedures and other forms of assistance, will be made available upon request to the APHIS NEPA contact at: Policy and Program Development, APHIS, USDA, Attention: NEPA Contact, 4700 River Road Unit 149, Riverdale, MD 20737–1238, (301) 851–3043.

[83 FR 24010, May 24, 2018]

§ 372.4   Definitions.

The terminology and definitions set forth in the Council on Environmental Quality's (CEQ) implementing regulations at 40 CFR part 1508 are incorporated herein. In addition, the following terms, as used in these procedures, are defined as follows:

APHIS. The Animal and Plant Health Inspection Service (APHIS).

Decisionmaker. The agency official responsible for signing the document based on a categorical exclusion or findings of no significant impact (FONSI) and environmental assessment or the record of decision following the environmental impact statement (EIS) process.

Department. The United States Department of Agriculture (USDA).

Environmental unit. The analytical unit in Policy and Program Development responsible for coordinating APHIS' compliance with NEPA and other environmental laws and regulations.

[60 FR 6002, Feb. 1, 1995, as amended at 83 FR 24010, May 24, 2018]

§ 372.5 Classification of actions.

(a) *Actions normally requiring environmental impact statements.* This class of policymakings and rulemakings seeks to establish programmatic approaches to animal and plant health issues. Actions in this class typically involve the agency, an entire program, or a substantial program component and are characterized by their broad scope (often global or nationwide) and potential effect (impacting a wide range of environmental quality values or indicators, whether or not affected individuals or systems may be completely identified at the time). Ordinarily, new or untried methodologies, strategies, or techniques to deal with pervasive threats to animal and plant health are the subjects of this class of actions. Alternative means of dealing with those threats usually have not been well developed. Actions in this class include:

(1) Formulation of contingent response strategies to combat future widespread outbreaks of animal and plant diseases; and

(2) Adoption of strategic or other long-range plans that purport to adopt for future program application a preferred course of action.

(b) *Actions normally requiring environmental assessments but not necessarily environmental impact statements.* This class of APHIS actions may involve the agency as a whole or an entire program, but generally is related to a more discrete program component and is characterized by its limited scope (particular sites, species, or activities) and potential effect (impacting relatively few environmental values or systems). Potential environmental impacts associated with the proposed action are not considered potentially significant at the outset of the planning process. Any effects of the action on environmental resources (such as air, water, soil, plant communities, animal populations, or others) or indicators (such as dissolved oxygen content of water) can be reasonably identified, and mitigation measures are generally available and have been successfully employed. Unless the actions are categorically excluded as provided in paragraph (c) of this section, actions in this class include:

(1) Policymakings and rulemakings that seek to remedy specific animal and plant health risks or that may affect opportunities on the part of the public to influence agency environmental planning and decisionmaking. Examples of this category of actions include:

(i) Development of program plans that seek to adopt strategies, methods, and techniques as the means of dealing with particular animal and plant health risks that may arise in the future; and

(ii) Implementation of program plans at the site-specific, action level.

(2) Planning, design, construction, or acquisition of new facilities, or proposals for modifications to existing facilities.

(3) Disposition of waste and other hazardous or toxic materials at laboratories and other APHIS facilities.

(4) Approvals and issuance of permits for proposals involving regulated genetically engineered organisms or products, or regulated nonindigenous species.

(5) Programs or statewide activities to reduce damage or harm by a specific wildlife species or group of species, such as deer or birds, or to reduce a specific type of damage or harm, such as protection of agriculture from wildlife depredation and disease; for the management of rabies in wildlife; or for the protection of threatened or endangered species.

(6) Research or testing that will be conducted outside of a laboratory or other containment area or reaches a stage of development (e.g., formulation of premarketing strategies) that forecasts an irretrievable commitment to the resulting products or technology.

(c) *Categorically excluded actions.* This class of APHIS actions shares many of the same characteristics—particularly in terms of the extent of program involvement, as well as the scope, effect of, and the availability of alternatives to proposed actions—as the class of actions that normally requires environmental assessments but not necessarily environmental impact statements. The major difference is that the means through which adverse environmental impacts may be avoided or minimized have actually been built right into the actions themselves. The efficacy of this approach generally has been established through testing and/or monitoring. The Department of Agriculture has also promulgated a listing of categorical exclusions that are applicable to all agencies within the department unless their procedures provide otherwise. Those categorical exclusions, codified at 7 CFR 1b.3(a), are entirely appropriate for APHIS. Other actions in this class include:

(1) *Routine measures.* (i) Routine measures, such as identifications, inspections, surveys, sampling that does not cause physical alteration of the environment, testing, seizures, quarantines, removals, sanitizing, inoculations, control, and monitoring employed by agency programs to pursue their missions and functions. Such measures may include the use—according to any label instructions or other lawful requirements and consistent with standard, published program practices and precautions—of chemicals, pesticides, or other potentially hazardous or harmful substances, materials, and target-specific devices or remedies, provided that such use meets all of the following criteria (insofar as they may pertain to a particular action):

(A) The use is localized or contained in areas where humans are not likely to be exposed, and is limited in terms of quantity, i.e., individualized dosages and remedies;

(B) The use will not cause contaminants to enter water bodies, including wetlands;

(C) The use does not adversely affect any federally protected species or critical habitat; and

(D) The use does not cause bioaccumulation.

(ii) Examples of routine measures include:

(A) Inoculation or treatment of discrete herds of livestock or wildlife undertaken in contained areas (such as a barn or corral, a zoo, an exhibition, or an aviary);

(B) Use of vaccinations or inoculations including new vaccines (e.g., genetically engineered vaccines) and applications of existing vaccines to new species provided that the project is conducted in a controlled and limited

manner, and the impacts of the vaccine can be predicted; and

(C) Isolated (for example, along a highway) weed control efforts.

(2) *Research and development activities.* (i) Activities limited in magnitude, frequency, and scope that occur in laboratories, facilities, pens, or field sites. Examples are:

(A) Vaccination trials that occur on groups of animals in areas designed to limit interaction with similar animals, or include other controls needed to mitigate potential risk.

(B) Laboratory research involving the evaluation and use of chemicals in a manner not specifically listed on the product label pursuant to applicable Federal authorizations.

(C) The development and/or production (including formulation, packaging or repackaging, movement, and distribution) of articles such as program materials, devices, reagents, and biologics that were approved and/or licensed in accordance with existing regulations, or that are for evaluation in confined animal, plant, or insect populations under conditions that prevent exposure to the general population.

(D) Research evaluating wildlife management products or tools, such as animal repellents, frightening devices, or fencing, that is carried out in a manner and area designed to eliminate the potential for harmful environmental effects and in accordance with applicable regulatory requirements.

(ii) Development, production, and release of sterile insects.

(3) *Licensing and permitting.* (i) Issuance of a license, permit, authorization, or approval to ship or field test previously unlicensed veterinary biologics, including veterinary biologics containing genetically engineered organisms (such as vector-based vaccines and nucleic acid-based vaccines);

(ii) Issuance of a license, permit, authorization, or approval for movement or uses of pure cultures of organisms (relatively free of extraneous micro-organisms and extraneous material) that are not strains of quarantine concern and occur, or are likely to occur, in a State's environment; or

437

(iii) Permitting for confined field releases of genetically engineered organisms and products; and

(iv) Permitting of:

(A) Importation of nonindigenous species into containment facilities,

(B) Interstate movement of nonindigenous species between containment facilities, or

(C) Releases into a State's environment of pure cultures of organisms that are either native or are established introductions.

(4) [Reserved]

(5) *Minor renovation, improvement, and maintenance of facilities.* Examples are:

(i) Renovation of existing laboratories and other facilities.

(ii) Functional replacement of parts and equipment.

(iii) Minor additions to existing facilities.

(iv) Minor excavations of land and repairs to properties.

(d) *Exceptions for categorically excluded actions.* Whenever the decisionmaker determines that a categorically excluded action may have the potential to affect "significantly" the quality of the "human environment," as those terms are defined at 40 CFR 1508.27 and 1508.14, respectively, an environmental assessment or an environmental impact statement will be prepared. For example:

(1) When any routine measure, the incremental impact of which, when added to other past, present, and reasonably foreseeable future actions (regardless of what agency or person undertakes such actions), has the potential for significant environmental impact;

(2) When a previously licensed or approved biologic has been subsequently shown to be unsafe, or will be used at substantially higher dosage levels or for substantially different applications or circumstances than in the use for which the product was previously approved; or

(3) When a confined field release of genetically engineered organisms or products involves new species or organisms or novel modifications that raise new issues.

[60 FR 6002, Feb. 1, 1995; 60 FR 13212, Mar. 10, 1995; 83 FR 24010, May 24, 2018; 85 FR 29838, May 18, 2020]

§ 372.6   Early planning.

Prospective applicants are encouraged to contact APHIS program officials to determine what types of environmental analyses or documentation, if any, need to be prepared.

[83 FR 24011, May 24, 2018]

§ 372.7   Planning and decision points and public involvement.

(a) *Major planning and decisions points.* The NEPA process will be fully coordinated with APHIS planning in cooperation with program personnel. Specific decision points or milestones will be identified and communicated to the public and others in a notice of intent and in the context of the public scoping process.

(b) *Public involvement.* There will be an early and open process for determining the scope of issues to be addressed in the environmental impact statement process.

(1) A notice of intent to prepare an environmental impact statement will be published in the FEDERAL REGISTER as soon as it is determined that a proposed major Federal action has the potential to affect significantly the quality of the human environment. The notice may include a preliminary scope of environmental study. All public and other involvement in APHIS' environmental impact statement process, including the scoping process, commenting on draft documents, and participation in the preparation of any supplemental documents, will be pursuant to CEQ's implementing regulations.

(2) Opportunities for public involvement in the environmental assessment process will be announced in the same fashion as the availability of environmental assessments and findings of no significant impact.

(3) Notification of the availability of environmental assessments and findings of no significant impact for proposed activities will be published in the FEDERAL REGISTER, unless it is determined that the effects of the action are primarily of regional or local concern. Where the effects of the action are primarily of regional or local concern, notice will normally be provided through

publication in a local or area newspaper of general circulation and/or the procedures implementing Executive Order 12372, "Intergovernmental Review of Federal Programs."

(4) All environmental documents and comments received will be made available to the public via *Regulations.gov*.

[60 FR 6002, Feb. 1, 1995, Redesignated and amended at 83 FR 24011, May 24, 2018]

§372.8 Processing and use of environmental documents.

(a) Environmental assessments will be forwarded immediately upon completion to the decisionmaker for a determination of whether the proposed action may have significant effects on the quality of the human environment, and for the execution, as appropriate, of a finding of no significant impact or a notice of intent to prepare an environmental impact statement. This determination is based on information provided in the NEPA document and available in the record.

(1) The availability of environmental assessments will be announced by publishing a notice consistent with the notification provisions of §372.7.

(2) Comments, if any, will be transmitted, together with any analyses and recommendations, to the APHIS decisionmaker who may then take appropriate action.

(3) Changes to environmental assessments and findings of no significant impact that are prompted by comments, new information, or any other source, will normally be announced in the same manner as the notice of availability prior to implementing the proposed action or any alternative. APHIS will mail notice upon request.

(b) Environmental impact statements will be processed from inception (publication of the notice of intent) to completion (publication of a final environmental impact statement or a supplement) according to the Council on Environmental Quality implementing regulations.

(c) For rulemaking or adjudicatory proceedings, relevant environmental documents, comments, and responses will be a part of the administrative record.

(d) For all APHIS activity that is subject to the NEPA process, relevant environmental documents, comments, and responses will accompany proposals through the review process.

(e) The APHIS decisionmaker will consider the alternatives discussed in environmental documents in reaching a determination on the merits of proposed actions.

(f) APHIS will implement mitigation and other conditions established in environmental documentation and committed to as part of the decisionmaking process.

[60 FR 6002, Feb. 1, 1995. Redesignated and amended at 83 FR 24011, May 24, 2018]

§372.9 Supplementing environmental impact statements.

Once a decision to supplement an environmental impact statement is made, a notice of intent will be published. The supplemental document will then be processed in the same fashion (exclusive of scoping) as a draft and a final statement (unless alternative procedures are approved by CEQ) and will become part of the record.

[60 FR 6002, Feb. 1, 1995. Redesignated and amended at 83 FR 24011, May 24, 2018]

§372.10 Process for rapid response to emergencies.

When it is determined (by the Administrator or the delegated Agency official responsible for environmental review) that an emergency exists that requires immediate action before preparing and completing the usual NEPA review, then the provisions of this section apply.

(a) The Administrator or the delegated Agency official responsible for environmental review may take actions that are necessary to control the immediate impacts of the emergency and that are urgently needed to prevent imminent damage to public health or safety, or prevent threats to valuable resources. When taking such actions, the Administrator or the delegated Agency official responsible for environmental review will consider the probable environmental consequences of the emergency action and mitigate foreseeable adverse environmental effects to the extent practicable.

(b) If a proposed emergency action is normally analyzed in an environmental assessment as described in §372.5 and the nature and scope of proposed emergency actions are such that there is insufficient time to prepare an EA and FONSI before commencing the proposed action, the Administrator shall consult with APHIS' Chief of Environmental and Risk Analysis Services about alternative arrangements for NEPA compliance. APHIS' Chief of Environmental and Risk Analysis Services may authorize emergency alternative arrangements for completing the required NEPA compliance documentation. Any alternative arrangements must be documented and notice of their use provided to CEQ.

(c) If a proposed emergency action is likely to result in significant environmental impacts, then APHIS will immediately consult with CEQ and request alternative arrangements in accordance with CEQ regulations at 40 CFR 1506.11. Such alternative arrangements will apply only to the proposed actions necessary to control the immediate impacts of the emergency. Other proposed actions remain subject to NEPA analysis and documentation in accordance with the CEQ regulations and these regulations.

[83 FR 24011, May 24, 2018]

PART 380—RULES OF PRACTICE GOVERNING PROCEEDINGS UNDER CERTAIN ACTS

Subpart A—General

Sec.
380.1 Scope and applicability of rules of practice.

Subpart B—Supplemental Rules of Practice

380.10 Stipulations.

AUTHORITY: 7 U.S.C. 7701–7772 and 7781–7786; 16 U.S.C. 1540(a), 3373(a) and (b); 7 CFR 2.22, 2.80, and 371.3.

Subpart A—General

§ 380.1 Scope and applicability of rules of practice.

(a) The Uniform Rules of Practice for the Department of Agriculture promulgated in subpart H of part 1, subtitle A, title 7 CFR are the Rules of Practice applicable to adjudicatory administrative proceedings under the following statutory provisions:

(1) The Plant Protection Act, section 424 (7 U.S.C. 7734),

(2) Endangered Species Act Amendments of 1973, as amended, section 11(a), 16 U.S.C. 1540(a), and

(3) Lacey Act Amendments of 1981, as amended, section 4(a) and (b), (16 U.S.C. 3373 (a) and (b)).

(b) In addition, the Supplemental Rules of Practice set forth in subpart B of this part are applicable to such proceedings.

[66 FR 21061, Apr. 27, 2001]

Subpart B—Supplemental Rules of Practice

§ 380.10 Stipulations.

(a) At any time prior to the issuance of a complaint seeking a civil penalty under any of the Acts listed in § 380.1, the Administrator, in his discretion, may enter into a stipulation with any person in which:

(1) The Administrator or the Administrator's delegate gives notice of an apparent violation of the applicable Act, or the regulations issued thereunder, by such person and affords such person an opportunity for a hearing regarding the matter as provided by such Act;

(2) Such person expressly waives hearing and agrees to pay a specified penalty within a designated time; and

(3) The Administrator agrees to accept the specified penalty in settlement of the particular matter involved if the penalty is paid within the designated time.

(b) If the specified penalty is not paid within the time designated in such a stipulation, the amount of the stipulated penalty shall not be relevant in any respect to the penalty which may be assessed after issuance of a complaint.

[48 FR 33468, July 22, 1983]

PARTS 381–399 [RESERVED]

FINDING AIDS

A list of CFR titles, subtitles, chapters, subchapters and parts and an alphabetical list of agencies publishing in the CFR are included in the CFR Index and Finding Aids volume to the Code of Federal Regulations which is published separately and revised annually.

Table of CFR Titles and Chapters
Alphabetical List of Agencies Appearing in the CFR
List of CFR Sections Affected

Table of CFR Titles and Chapters

(Revised as of January 1, 2022)

Title 1—General Provisions

Title 2—Grants and Agreements

Title 2—Grants and Agreements—Continued

Title 3—The President

Title 4—Accounts

Title 5—Administrative Personnel

446

Title 7—Agriculture—Continued

Title 8—Aliens and Nationality

Title 9—Animals and Animal Products

Title 10—Energy

Title 11—Federal Elections

Title 12—Banks and Banking

Title 12—Banks and Banking—Continued

Title 13—Business Credit and Assistance

Title 14—Aeronautics and Space

Title 15—Commerce and Foreign Trade

Title 15—Commerce and Foreign Trade—Continued

Title 16—Commercial Practices

Title 17—Commodity and Securities Exchanges

Title 18—Conservation of Power and Water Resources

Title 19—Customs Duties

Title 20—Employees' Benefits

Title 21—Food and Drugs

Title 22—Foreign Relations

Title 28—Judicial Administration—Continued

Title 29—Labor

Title 30—Mineral Resources

Title 31—Money and Finance: Treasury

Title 31—Money and Finance: Treasury—Continued

Title 32—National Defense

Title 33—Navigation and Navigable Waters

Title 34—Education

Title 35 [Reserved]

Title 36—Parks, Forests, and Public Property

Title 37—Patents, Trademarks, and Copyrights

Title 37—Patents, Trademarks, and Copyrights—Continued

Title 38—Pensions, Bonuses, and Veterans' Relief

Title 39—Postal Service

Title 40—Protection of Environment

Title 41—Public Contracts and Property Management

Title 41—Public Contracts and Property Management—Continued

Title 42—Public Health

Title 43—Public Lands: Interior

Title 44—Emergency Management and Assistance

Title 45—Public Welfare

Title 46—Shipping

Title 47—Telecommunication

Title 48—Federal Acquisition Regulations System

Title 48—Federal Acquisition Regulations System—Continued

Chap.

Title 49—Transportation

Title 50—Wildlife and Fisheries

Title 50—Wildlife and Fisheries—Continued

Alphabetical List of Agencies Appearing in the CFR
(Revised as of January 1, 2022)

Agency	CFR Title, Subtitle or Chapter
Administrative Conference of the United States	1, III
Advisory Council on Historic Preservation	36, VIII
Advocacy and Outreach, Office of	7, XXV
Afghanistan Reconstruction, Special Inspector General for	5, LXXXIII
African Development Foundation	22, XV
Federal Acquisition Regulation	48, 57
Agency for International Development	2, VII; 22, II
Federal Acquisition Regulation	48, 7
Agricultural Marketing Service	7, I, VIII, IX, X, XI; 9, II
Agricultural Research Service	7, V
Agriculture, Department of	2, IV; 5, LXXIII
Advocacy and Outreach, Office of	7, XXV
Agricultural Marketing Service	7, I, VIII, IX, X, XI; 9, II
Agricultural Research Service	7, V
Animal and Plant Health Inspection Service	7, III; 9, I
Chief Financial Officer, Office of	7, XXX
Commodity Credit Corporation	7, XIV
Economic Research Service	7, XXXVII
Energy Policy and New Uses, Office of	2, IX; 7, XXIX
Environmental Quality, Office of	7, XXXI
Farm Service Agency	7, VII, XVIII
Federal Acquisition Regulation	48, 4
Federal Crop Insurance Corporation	7, IV
Food and Nutrition Service	7, II
Food Safety and Inspection Service	9, III
Foreign Agricultural Service	7, XV
Forest Service	36, II
Information Resources Management, Office of	7, XXVII
Inspector General, Office of	7, XXVI
National Agricultural Library	7, XLI
National Agricultural Statistics Service	7, XXXVI
National Institute of Food and Agriculture	7, XXXIV
Natural Resources Conservation Service	7, VI
Operations, Office of	7, XXVIII
Procurement and Property Management, Office of	7, XXXII
Rural Business-Cooperative Service	7, XVIII, XLII
Rural Development Administration	7, XLII
Rural Housing Service	7, XVIII, XXXV
Rural Utilities Service	7, XVII, XVIII, XLII
Secretary of Agriculture, Office of	7, Subtitle A
Transportation, Office of	7, XXXIII
World Agricultural Outlook Board	7, XXXVIII
Air Force, Department of	32, VII
Federal Acquisition Regulation Supplement	48, 53
Air Transportation Stabilization Board	14, VI
Alcohol and Tobacco Tax and Trade Bureau	27, I
Alcohol, Tobacco, Firearms, and Explosives, Bureau of	27, II
AMTRAK	49, VII
American Battle Monuments Commission	36, IV
American Indians, Office of the Special Trustee	25, VII
Animal and Plant Health Inspection Service	7, III; 9, I
Appalachian Regional Commission	5, IX
Architectural and Transportation Barriers Compliance Board	36, XI

465

468

469

List of CFR Sections Affected

All changes in this volume of the Code of Federal Regulations (CFR) that were made by documents published in the FEDERAL REGISTER since January 1, 2017 are enumerated in the following list. Entries indicate the nature of the changes effected. Page numbers refer to FEDERAL REGISTER pages. The user should consult the entries for chapters, parts and subparts as well as sections for revisions.

For changes to this volume of the CFR prior to this listing, consult the annual edition of the monthly List of CFR Sections Affected (LSA). The LSA is available at *www.govinfo.gov*. For changes to this volume of the CFR prior to 2001, see the "List of CFR Sections Affected, 1949–1963, 1964–1972, 1973–1985, and 1986–2000" published in 11 separate volumes. The "List of CFR Sections Affected 1986–2000" is available at *www.govinfo.gov*.

7 CFR—Continued

7 CFR—Continued

2021

7 CFR

○

Lightning Source UK Ltd.
Milton Keynes UK
UKHW021300260123
416011UK00033B/392

9 781636 711249